Finite Element and Discontinuous Galerkin Methods for Transient Wave Equations

Scientific Computation

More information about this series at http://www.springer.com/series/718

Gary Cohen · Sébastien Pernet

Finite Element and Discontinuous Galerkin Methods for Transient Wave Equations

Springer

Gary Cohen
Inria
Saclay
France

Sébastien Pernet
The French Aerospace Lab
ONERA
Toulouse
France

ISSN 1434-8322 ISSN 2198-2589 (electronic)
Scientific Computation
ISBN 978-94-017-7759-9 ISBN 978-94-017-7761-2 (eBook)
DOI 10.1007/978-94-017-7761-2

Library of Congress Control Number: 2016942523

Printed on acid-free paper

This Springer imprint is published by Springer Nature
The registered company is Springer Science+Business Media B.V. Dordrecht

Foreword

This book is the natural continuation of a previous book by Gary Cohen on numerical methods for time domain wave propagation problems, a topic to which Gary has dedicated three decades of his research career at INRIA. He has written this new book with Sébastien Pernet, one of his brightest Ph.D. students. Sébastien, in particular, has made many contributions to the field of discontinuous Galerkin methods for wave equations, and he currently is Research Engineer at ONERA. The importance of the topic of the book is indisputable: in most domains of physics, one is faced with the tasks of understanding, describing and simulating numerically the propagation of waves of various natures.

The content of this book is a harmonious blend of classical and very recent material. In the first book, several chapters were dedicated to higher order finite difference methods, a subject which is not considered in this second book. The intersection with the first book is reduced to so-called spectral element methods (Chap. 3), in other words higher order continuous finite element methods allowing for mass lumping. This already emphasizes the importance of using ad hoc quadrature formulas for the approximation of integrals over finite elements, a feature that recurs throughout the book.

Concerning the most valuable additions to this new book, I would like to emphasize a few: the construction (in Chap. 2) of 3D finite elements on nonclassical elementary shapes such as prisms or pyramids (which allows one to work with hybrid meshes involving tetrahedra and quadrilaterals); an extensive analysis of discontinuous Galerkin methods (this is Chap. 4); and Chap. 5 specifically dedicated to Maxwell's equations, which contains an extensive treatment of the delicate question of spurious modes (specific to electromagnetic waves).

Throughout the book, the emphasis is on the accuracy and computational efficiency of the numerical methods they present, with a nonnegligible attention brought to their practical aspects, making this book particularly useful for practitioners. On the other hand, the theoretical foundations and numerical analysis of these methods are also detailed.

If most of the book is dedicated to space discretization methods, the important questions of bounding artificially the computational domain (with absorbing boundary conditions or perfectly matched layers) and time discretization via finite differences (including modified equation techniques for higher order time stepping and local time stepping) are treated in Chaps. 6 and 7, in less but sufficient detail.

Gary Cohen has been my friend and colleague for more than 30 years. I am happy to see that Sébastien Pernet and he have written a book that is a major achievement of their respective careers. I have no doubt that this book will become a reference in the future for any student, applied mathematician, or engineer involved in the numerical simulation of waves.

December 2015 Patrick Joly

Preface

For a long time, wave equations in general and Maxwell's equations in particular were solved in the time-harmonic domain[1] by finite element methods (FEM). Equations in time domain were solved by using second-order finite difference methods (FDM) whose outstanding representative is the Yee scheme (also called FDTD) used for electromagnetic waves since 1966 and still alive.

Unfortunately, second-order FDM found their limits in modeling long time propagation which often occur in physical problems, either because the wavelength is small compared to the size of the efficient domain, or because we have trapped waves in cavities. Actually, the number of points required to get an accurate solution grows with the interval of resolution in time. A palliative to this drawback seemed to be the use of higher order FDM which enable us to increase the size of the space-step while keeping a satisfactory accuracy. However, this technique is very troublesome to model complex geometries because of the large size of the grid cells.

People were nevertheless reluctant to use FEM in the time domain[2] (called FETD in the electromagnetic community), which could ensure a good approximation of these geometries. The reason came from the presence of a mass matrix which is naturally diagonal for FDM, but n-diagonal for FEM, with n increasing with the dimension of space and the order of the method. This matrix requires to be inverted at each time-step and substantially slows down the performance of the method, even when using iterative algorithms of inversion.

An answer to this difficult problem was given in two ways. A first way, introduced by Cohen et al. [1] for wave equations, was based on mass lumping of FEM on quadrilateral and hexahedral meshes with Gauss–Lobatto points. Actually, this idea was first used for reservoir simulation [2] and neutronics [3] and was later

---

[1]We recall that time-harmonic problems are derived from the transient wave equations by Fourier transform in time which replaces time by frequency. Their solutions then depend on the frequency.

[2]Although resolution in the time domain became attractive since the sources became polychromatic and even broadband.

extended to other problems under the name of "spectral elements" [4]. This method was then combined to mixed formulations first for electromagnetic (using $H(curl)$-L^2 elements) [5] and then for acoustic (using H^1-L^2 elements) [6, 7] wave equations and led to efficient and low-storage approximations. These approximations were called "mixed spectral element methods."

A second way to overcome the inversion of an n-diagonal mass matrix was proposed by Hesthaven et al. [8, 9]. It was based on the use of high-order discontinuous Galerkin methods (DGM) on triangular and tetrahedral meshes and led to block-diagonal mass matrices. The use of high-order is here essential for Maxwell's equations since these equations produce parasitic waves which must be suppressed by adding a penalty term. This term being dissipative, higher order methods substantially reduce the dissipation. On the other hand, Hesthaven dramatically decreased the storage of DGM, providing a reasonable increase of computation time, by reconstructing the stiffness matrix at each time-step. Moreover, he computed quasi-optimal interpolation points to avoid the Runge phenomenon for high-order polynomial approximations [10].

This new point of view suggested to construct DGM by using spectral elements and mixed formulation. This new approach was attractive for Maxwell's equations which presented parasitic modes even when using edge ($H(curl)$) elements [11]. However, Cohen et al. [12] noticed that the penalty term used for DGM could be also used for the mixed spectral element approach of Maxwell's equations which led to an even more efficient method.

The main advantage of mixed spectral element methods is to produce very sparse mass and stiffness matrices, which leads to very fast algorithm. Unfortunately, it has an important drawback: hexahedral meshes are difficult to construct for very complex geometries (which cannot be regarded as deformation of a cube).[3] For cons, tetrahedral meshes are much easier to produce, which gives an important advantage to Hesthaven's DGM. However, matrices involved in this method are not very sparse and substantially slow down its performance compared to hexahedral mixed spectral elements.

The above remarks naturally led to an interest in hybrid meshes mostly composed of hexahedra and using tetrahedra around the complex domains. The problem then was to stick the two types of elements. A natural way to do it is to use pyramidal and even prismatic elements. The second elements are deduced from triangular and spectral elements by the use of Cartesian product. The pyramidal elements, in turn, were much less obvious to construct since they used rational functions. This was achieved in a general form [13] and then extended to edge elements [14].

Another and important step was to get error estimates for hexahedral and quadrilateral elements which are not obvious because of the presence of a

[3] On the other hand, as we shall see in this book, mass-lumped triangular and tetrahedral elements are difficult to construct and not really efficient.

nonconstant Jacobian matrix. A new approach was developed, which enabled us to analyze discontinuous and continuous spectral element approaches [15–17].

The scope of this book—which is essentially the continuation of G. Cohen's previous book [18]—is to provide a survey of all these methods and to describe how to implement them in physical models. It is divided into eight chapters.

The first chapter recalls the continuous formulations of the main wave equations (acoustics, Maxwell, linear elastodynamics) and the mathematical concepts needed to understand their approximation. Moreover, the important Hille–Yosida theorem is given and applied to the different equations.

The second chapter provides an exhaustive presentation of different finite elements classically used for wave equations approximations. These elements are of different shapes and are generally defined at any order.

The third chapter covers an approximation of the acoustics equation and the linear elastodynamics system by mixed spectral element methods. After their description, a plane wave analysis is given and error estimates are developed for the acoustics equation.

The fourth chapter is devoted to DGM approximations using spectral elements or Hesthaven's method. The method is first defined in a general framework containing both approaches and the three wave equations. Then the construction of the different methods for each equation is given in detail. Finally, plane wave analysis is done.

The fifth chapter deals with different approximations of Maxwell's equations and the important and not obvious problem of parasitic modes. The last section contains error estimates of spectral DGM on hexahedra.

The sixth chapter introduces the treatment of unbounded domains by absorbing boundary conditions (ABC) and perfectly matched layers (PML) for the three wave equations.

The seventh chapter defines different time approximation algorithms, first with a constant time-step, then by local time stepping methods which lead to a dramatic gain of computational time.

The eighth, and the last, chapter presents three more complex equations with specific properties: The linearized Euler equations (LEE) which model acoustics in flow and contains a convective term, the Cauchy–Poisson problem which models gravity waves and whose evolution equation is on a boundary, and two models of wave propagation in thin plates which are dispersive equations.

This book is the result of a fruitful (and harmonious) collaboration between G. Cohen and S. Pernet.[4] Although G. Cohen's contribution was mainly focused on construction of the methods and S. Pernet's work was mainly focused on their analysis, the book was written in a very interactive way which enriched both contributions.

On the other hand, this book could not have been written without the contributions of people with whom the authors worked and to whom they are grateful.

---

[4]Who is a former Ph.D. student of G. Cohen.

First, Marc Duruflé who brought a substantial contribution in several topics. We also thank Peter Monk, Sandrine Fauqueux, Manfred Kaltenbacher, Alexandre Sinding, Sébastien Imperiale, Pascal Grob, and many others.

A special mention is given to our friend and colleague Xavier Ferrieres, from the French Aerospace Lab, with whom we had a fruitful collaboration on Maxwell's equations approximation—in the frame of a cooperation with INRIA—for more than ten years and who enabled us to apply these methods to industrial topics. These topics were mainly proposed by Bernard Pecqueux from Centre d'Etudes de Gramat (CEG) who constantly motivated our research. This cooperation would not have been possible without the management of Vincent Gobin from the French Aerospace Lab, whom we thank here for his involvement.

The authors want to thank the POEMS team of INRIA and ENSTA, headed by Patrick Joly (with special thanks for his foreword) and Anne-Sophie Bonnet-Ben Dhia and the M2SN team of the French Aerospace Lab, headed by François Rogier, who supported them for the writing of this book.

Our preface would not be complete if we did not thank our wives Yaffa Cohen and Delphine Pernet for their encouragement throughout the difficult task of writing this book.

Saclay, France — Gary Cohen
Toulouse, France — Sébastien Pernet
November 2015

References

1. Cohen, G., Joly, P., Tordjman, N.: Higher-order finite elements with mass lumping for the 1-D wave equation. Finite Elem. Anal. Des. **17**(3–4), 329–336 (1994)
2. Young, L.C.: An efficient finite element method for reservoir simulation. Proceeding of the 53rd Annual Fall Technical Conference and Exhibition of the Society of Petroleum Engineers of AIME, Houston, Texas, 1–3 Oct. (1978)
3. Hennart, J.-P., Sainz, E., Villegas, M.: On the efficient use of the finite element method in static neutron diffusion calculations. Computational Methods in Nuclear Engineering **1**, pp. 3–87 (1979)
4. Maday, Y., Patera, A.T.: Spectral element methods for the imcompressible Navier-Stokes equations. In: Noor, A.K. (ed.) State of the Art Survey in Computational Mechanics, pp. 71–143 (1989)
5. Cohen, G., Monk, P.: Mur-Nédélec finite element schemes for Maxwell's equations. Comput. Methods Appl. Mech. Engrg. **169**(3–4), 197–217 (1999)
6. Cohen, G., Fauqueux, S.: Mixed finite elements with mass-lumping for the transient wave equation. J. Comput. Acoust. **8**(1), 171–188 (2000)
7. Cohen, G., Fauqueux, S.: Mixed spectral finite elements for the linear elasticity system in unbounded domains. SIAM J. Sci. Comput. **26**(3), 864–884 (2005)
8. Hesthaven, J.S., Teng, C.H.: Stable spectral methods on tetrahedral elements. SIAM J. Sci. Comp. **21**(6), 2352–2380 (2000)
9. Hesthaven, J.S., Warburton, T.: Nodal discontinuous Galerkin methods. Texts in Applied Mathematics, **54**, Springer ed., (2008)

10. Hesthaven, J.S.: From electrostatics to almost optimal nodal sets for polynomial interpolation in a simplex. SIAM J. Numer. Anal. **35**(2), 655–676 (1998)
11. Cohen, G., Ferrieres, X., Pernet, S.: A spatial high-order hexahedral discontinuous Galerkin method to solve Maxwell's equations in time domain. J. Comp. Phys. **217**(2), 340–363 (2006)
12. Cohen, G., Duruflé, M.: Non spurious spectral-like element methods for Maxwell's equations. J. Comp. Math. **25**(3), 282–304 (2007)
13. Bergot, M., Cohen, G., Duruflé, M.: Higher-Order finite element for hybrid meshes Using New Nodal Pyramidal Elements. J. Sci. Comput. **42**(3), 345–381 (2010)
14. Bergot, M., Duruflé, M.: High-Order Optimal edge elements for pyramids, prisms and hexahedra. J. Comput. Phys. **232**(1), 189–213 (2013)
15. Pernet, S.: Etude de méthodes d'ordre élevé pour résoudre les équations de Maxwell dans le domaine temporel : Application à la détection et à la compatibilité électromagnétique, thèse de doctorat, U. de Paris-Dauphine, (2004)
16. Pernet, S., Ferrieres, X.: HP a-priori error estimates for a non-dissipative spectral discontinuous Galerkin method to solve the Maxwell equations in the time domain. Math. Comp. **76**(260), 1801–1832 (2007)
17. Duruflé, M., Grob, P.: Joly, Influence of Gauss and Gauss-Lobatto quadrature rules on the accuracy of a quadrilateral finite element method in the time domain, Numer. Methods Partial Differential Equations **25**(3), pp. 526–551 (2009)
18. Cohen, G.: High order numerical methods for transient wave equations. Scientific Computation, Springer-Verlag (2001)

Contents

Chapter 1
Classical Continuous Models and Their Analysis

Abstract This chapter presents some important background concerning the transient wave problems. In the first part, the classical models, i.e. acoustics, linear elastodynamics and electromagnetism, are recalled. Some elements about their analysis are next given with a particular focus on their variational formulations and their well-posedness based on the Hille-Yosida theorem. Finally, their plane wave solutions, which are an important tool for understanding and analyzing wave phenomena, are derived.

1.1 The Basic Equations

The equations that model wave propagation can be classified into three physical categories. The acoustics equation and the elastodynamics system model mechanical waves in fluids and solids respectively. On the other hand, Maxwell's equations describe the propagation of electromagnetic waves (radio waves or light).

Several models for these three types of waves, which have various degrees of complexity, are given in the scientific literature. The purpose of this book is to construct numerical models for linear wave propagation in heterogeneous and even anisotropic media.

The three main models we consider are described in the following sections. Throughout the book, we use the notation[1] $x \in \mathbb{R}$ or $\underline{x} \in \mathbb{R}^d$, $d = 2, 3$ to denote spatial position and $t \in \mathbb{R}^+$ to denote time. Moreover, in all the following equations, when used with vectors, the symbols $\cdot$ and $\times$ denote the scalar and cross products. In particular, these products are applied to vector-valued differential operators.

[1] Underlined characters indicate a vector of $\mathbb{R}^d$.

G. Cohen and S. Pernet, *Finite Element and Discontinuous Galerkin Methods for Transient Wave Equations*, Scientific Computation,
DOI 10.1007/978-94-017-7761-2_1

1.1.1 The Acoustics Equation

Let u denote the acoustic pressure field and ρ the density of the medium in which the acoustic wave travels. In its scalar form, the acoustics equation, which is a 0th approximation of the Euler equations, can be written as

$$\frac{1}{\kappa(\underline{x})}\frac{\partial^2 u}{\partial t^2}(\underline{x},t) - \nabla\cdot\left(\frac{1}{\rho(\underline{x})}\nabla u(\underline{x},t)\right) = f(\underline{x},t), \tag{1.1}$$

where[2] $\nabla = (\partial/\partial x_1, \ldots, \partial/\partial x_d)^T$ and ρ and κ are strictly positive functions of position $\underline{x}$. This equation can be written as a first-order system by introducing the velocity $\underline{v}$:

$$\frac{1}{\kappa(\underline{x})}\frac{\partial u}{\partial t}(\underline{x},t) = \nabla\cdot\underline{v}(\underline{x},t) + F(\underline{x},t), \tag{1.2a}$$

$$\rho(\underline{x})\frac{\partial \underline{v}}{\partial t}(\underline{x},t) = \nabla u(\underline{x},t), \tag{1.2b}$$

where $F(\underline{x},t) = \displaystyle\int_0^t f(\underline{x},\tau)\mathrm{d}\tau$.

The velocity[3] c of sound wave propagation is given by the relation

$$c(\underline{x}) = \sqrt{\frac{\kappa(\underline{x})}{\rho(\underline{x})}}. \tag{1.3}$$

In order to obtain well-posed equations, we must add the initial conditions

$$u(\underline{x},0) = u_0(\underline{x}), \quad \frac{\partial u}{\partial t}(\underline{x},0) = u_1(\underline{x}) \tag{1.4}$$

to (1.1) and

$$u(\underline{x},0) = u_0(\underline{x}), \quad \underline{v}(\underline{x},0) = \underline{v}_0(\underline{x}) \tag{1.5}$$

to (1.2a) and (1.2b) where u_0, u_1 and $\underline{v}_0$ are given functions.

1.1.2 Maxwell's Equations

1.1.2.1 The 3D Case

We denote by $\underline{E}$, $\underline{H}$, $\underline{D}$ and $\underline{B}$ the electric and magnetic fields and the electric and magnetic inductions respectively. The equations of electromagnetism considered in

[2] T in upperscript denotes a transposed matrix.

[3] A more appropriate term would be "celerity" but "velocity" is widely used in literature.

this book can be written as

$$\frac{\partial \underline{D}}{\partial t}(\underline{x}, t) - \nabla \times \underline{H}(\underline{x}, t) = -\underline{J}(\underline{x}, t), \tag{1.6a}$$

$$\frac{\partial \underline{B}}{\partial t}(\underline{x}, t) + \nabla \times \underline{E}(\underline{x}, t) = 0, \tag{1.6b}$$

$$\underline{D} = \underline{\underline{\varepsilon}}(\underline{x})\underline{E}, \tag{1.6c}$$

$$\underline{B} = \underline{\underline{\mu}}(\underline{x})\underline{H}, \tag{1.6d}$$

where[4] $\underline{\underline{\varepsilon}}$ and $\underline{\underline{\mu}}$ are symmetric, positive definite, dielectric permittivity and magnetic permeability matrices depending on space which take into account anisotropic media, and $\underline{J}$ is the current density.

Moreover, the fields $\underline{D}$ and $\underline{B}$ satisfy the two relations

$$\mathrm{div}\underline{D} = \rho, \tag{1.7a}$$

$$\mathrm{div}\underline{B} = 0, \tag{1.7b}$$

where ρ is the charge density.[5]

Of course, the isotropic case is obtained when $\underline{\underline{\varepsilon}} = \varepsilon I_3$ and $\underline{\underline{\mu}} = \mu I_3$, where ε and μ are strictly positive scalar functions and I_3 is the identity matrix of $\mathbb{R}^3$.

We use (1.6c) and (1.6d) to eliminate $\underline{B}$ and $\underline{D}$ from the Maxwell's system. Hence in the remainder of the book we generally deal with equations involving $\underline{E}$ and $\underline{H}$ which read

$$\underline{\underline{\varepsilon}}\frac{\partial \underline{E}}{\partial t}(\underline{x}, t) - \nabla \times \underline{H}(\underline{x}, t) = -\underline{J}(\underline{x}, t), \tag{1.8a}$$

$$\underline{\underline{\mu}}\frac{\partial \underline{H}}{\partial t}(\underline{x}, t) + \nabla \times \underline{E}(\underline{x}, t) = 0. \tag{1.8b}$$

The initial conditions for these Maxwell's equations are

$$\underline{E}(\underline{x}, 0) = \underline{E_0}(\underline{x}), \quad \underline{H}(\underline{x}, 0) = \underline{H_0}(\underline{x}). \tag{1.9}$$

By combining (1.8a) and (1.8b), we obtain the second-order forms of Maxwell's equations

$$\underline{\underline{\varepsilon}}(\underline{x})\,\frac{\partial^2 \underline{E}}{\partial t^2}(\underline{x}, t) + \nabla \times \left(\underline{\underline{\mu}}^{-1}(\underline{x})\nabla \times \underline{E}(\underline{x}, t)\right) = -\underline{j}(\underline{x}, t), \tag{1.10}$$

---

[4]Double underline indicates a matrix or a tensor.

[5]When $\underline{D}$ and $\underline{E}$ and $\underline{B}$ and $\underline{H}$ are not explicitly related one to the other as in (1.6c) and (1.6d), the Maxwell's system is composed of Eqs. (1.6a), (1.6b), (1.7a) and (1.7b).

$$\underline{\underline{\mu}}(\underline{x}) \frac{\partial^2 \underline{H}}{\partial t^2}(\underline{x}, t) + \nabla \times \left(\underline{\underline{\varepsilon}}^{-1}(\underline{x}) \nabla \times \underline{H}(\underline{x}, t)\right) = \underline{J}'(\underline{x}, t), \tag{1.11}$$

where $\underline{j} = \partial \underline{J}/\partial t$ and $\underline{J}' = \nabla \times (\underline{\underline{\varepsilon}}^{-1} \underline{J})$.

Now, if we assume that the waves propagate in a homogeneous isotropic medium (the vacuum for instance) far enough from its source, so that $\underline{J} = 0$ and $\rho = 0$, (1.10) and (1.11) can be rewritten as

$$\frac{\partial^2 \underline{E}}{\partial t^2}(\underline{x}, t) + \frac{1}{\varepsilon \mu} \nabla \times \nabla \times \underline{E}(\underline{x}, t) = 0, \tag{1.12}$$

$$\frac{\partial^2 \underline{H}}{\partial t^2}(\underline{x}, t) + \frac{1}{\varepsilon \mu} \nabla \times \nabla \times \underline{H}(\underline{x}, t) = 0, \tag{1.13}$$

by using the fact that $\nabla \times \nabla \times \underline{V} = \nabla(\nabla \cdot \underline{V}) - \Delta \underline{V}$ and taking into account relations (1.7a) and (1.7b), we obtain

$$\frac{\partial^2 \underline{E}}{\partial t^2}(\underline{x}, t) - \frac{1}{\varepsilon \mu} \Delta \underline{E}(\underline{x}, t) = 0, \tag{1.14}$$

$$\frac{\partial^2 \underline{H}}{\partial t^2}(\underline{x}, t) - \frac{1}{\varepsilon \mu} \Delta \underline{H}(\underline{x}, t) = 0. \tag{1.15}$$

In other words, each component of $\underline{E}$ and $\underline{H}$ satisfies the wave equation with a velocity equal to $1/\sqrt{\varepsilon \mu}$.

More generally, the velocity c in a non-homogeneous isotropic medium is defined by

$$c^2(\underline{x}) \varepsilon(\underline{x}) \mu(\underline{x}) = 1. \tag{1.16}$$

Remark: The right-hand side $\underline{J}$ can take the form $\underline{J} = \underline{J}_1 + \underline{\underline{\sigma}}\, \underline{E}$, where $\underline{\underline{\sigma}}$ is the conductivity. This additional term introduces physical damping. In this case, we can only obtain a second-order formulation in $\underline{E}$ in the heterogeneous case (i.e. when $\underline{\underline{\varepsilon}}$ and $\underline{\underline{\mu}}$ depend on $\underline{x}$).

1.1.2.2 The 2D Case

The Transverse-Magnetic (TM) Case By considering an electric field polarized in the plane (x_1, x_2) and a magnetic field polarized in the direction orthogonal to this plane and assuming that both fields are independent of x_3, we obtain

$$\underline{\underline{\varepsilon}}(\underline{x}) \frac{\partial \underline{E}}{\partial t}(\underline{x}, t) - \mathbf{curl} H(\underline{x}, t) = -\underline{J}(\underline{x}, t), \tag{1.17a}$$

$$\mu(\underline{x}) \frac{\partial H}{\partial t}(\underline{x}, t) + \mathrm{curl} \underline{E}(\underline{x}, t) = 0, \tag{1.17b}$$

where $\mathbf{curl}H = (\partial H/\partial x_2, -\partial H/\partial x_1)^T$ and $\mathrm{curl}\underline{E} = \partial E_y/\partial x_1 - \partial E_x/\partial x_2$ and its two second-order versions

$$\underline{\underline{\varepsilon}}(\underline{x})\frac{\partial^2 \underline{E}}{\partial t^2}(\underline{x}, t) + \mathbf{curl}\left(\frac{1}{\mu(\underline{x})}\mathrm{curl}\underline{E}(\underline{x}, t)\right) = -\underline{j}(\underline{x}, t), \tag{1.18}$$

$$\mu(\underline{x})\frac{\partial^2 H}{\partial t^2}(\underline{x}, t) + \mathrm{curl}\left(\underline{\underline{\varepsilon}}^{-1}(\underline{x})\mathbf{curl}H(\underline{x}, t)\right) = J'(\underline{x}, t), \tag{1.19}$$

where $\underline{j}$ and J' are defined as in (1.10) and (1.11).

The Transverse-Electric (TE) Case In this case, the polarizations of the electric and magnetic fields are switched. We obtain

$$\varepsilon(\underline{x})\frac{\partial E}{\partial t}(\underline{x}, t) - \mathrm{curl}\underline{H}(\underline{x}, t) = -J(\underline{x}, t), \tag{1.20a}$$

$$\underline{\underline{\mu}}(\underline{x})\frac{\partial \underline{H}}{\partial t}(\underline{x}, t) + \mathbf{curl}E(\underline{x}, t) = 0, \tag{1.20b}$$

where **curl** and curl are defined as for the TM case, and its two second-order versions

$$\varepsilon(\underline{x})\frac{\partial^2 E}{\partial t^2}(\underline{x}, t) + \mathrm{curl}\left(\underline{\underline{\mu}}^{-1}(\underline{x})\mathbf{curl}E(\underline{x}, t)\right) = -j(\underline{x}, t), \tag{1.21}$$

$$\underline{\underline{\mu}}(\underline{x})\frac{\partial^2 \underline{H}}{\partial t^2}(\underline{x}, t) + \mathbf{curl}\left(\frac{1}{\varepsilon(\underline{x})}\mathrm{curl}\underline{H}(\underline{x}, t)\right) = \underline{J}'(\underline{x}, t). \tag{1.22}$$

Remarks:

1. In the isotropic case ($\underline{\underline{\varepsilon}} = \varepsilon I_2$, $\underline{\underline{\mu}} = \mu I_2$, I_2 identity of $\mathbb{R}^2$), (1.19) and (1.22) can be written as the two following second-order wave equations[6]

$$\mu(\underline{x})\frac{\partial^2 H}{\partial t^2}(\underline{x}, t) - \nabla\cdot\left(\frac{1}{\varepsilon(\underline{x})}\nabla H(\underline{x}, t)\right) = J'(\underline{x}, t). \tag{1.23}$$

$$\varepsilon(\underline{x})\frac{\partial^2 E}{\partial t^2}(\underline{x}, t) - \nabla\cdot\left(\frac{1}{\mu(\underline{x})}\nabla E(\underline{x}, t)\right) = -j(\underline{x}, t), \tag{1.24}$$

2. Some authors call "transverse-magnetic" the formulation given in (1.20a) and (1.20b) and "transverse-electric" the formulation given in (1.17a) and (1.17b).

[6]Similar but very complex forms can also be obtained for the anisotropic case.

1.1.3 The Linear Elastodynamics System

1.1.3.1 General Formulation

Let $\underline{v} \in \mathbb{R}^d$ denote the displacement vector and $\underline{\underline{\tau}}$ the stress tensor for the elastic medium. Then, the general formulation of the linear elastodynamics system in a non-homogeneous, anisotropic medium reads

$$\rho(\underline{x})\frac{\partial^2 \underline{v}}{\partial t^2}(\underline{x}, t) - \mathbf{div}\underline{\underline{\tau}}(\underline{x}, t) = \underline{f}(\underline{x}, t), \tag{1.25a}$$

$$\underline{\underline{\tau}}(\underline{x}, t) = \underline{\underline{C}}(\underline{x})\underline{\underline{\varepsilon}}(\underline{v})(\underline{x}, t) \quad \text{(Hooke's law)}. \tag{1.25b}$$

Now, if $(i, j, k, \ell) \in \{1, \ldots, d\}^4$, we have

$$\underline{\underline{\tau}} = (\underline{\tau}_1, \ldots, \tau_d), \tag{1.26a}$$

$$\underline{\tau}_i = (\tau_{i1}, \ldots, \tau_{id})^T, \tag{1.26b}$$

$$\mathbf{div}\underline{\underline{\tau}} = (\nabla \cdot \underline{\tau}_1, \ldots, \nabla \cdot \underline{\tau}_d)^T, \tag{1.26c}$$

$$\varepsilon_{ij}\underline{v} = \frac{1}{2}\left(\frac{\partial v_i}{\partial x_j} + \frac{\partial v_j}{\partial x_i}\right) \text{ (strain tensor)}, \tag{1.26d}$$

$$\left(\underline{\underline{C\varepsilon}}\right)_{ij} = \sum_{k=1}^{d}\sum_{\ell=1}^{d} C_{ijk\ell}\varepsilon_{k\ell}. \tag{1.26e}$$

Moreover, $\underline{\underline{C}}$ is a cyclic symmetric tensor, i.e. $C_{ijk\ell} = C_{k\ell ij} = C_{jik\ell}$. For this reason, the number of independent coefficients of $\underline{\underline{C}}$ is equal to 6 (instead of 16) for $d = 2$ and 21 (instead of 81) for $d = 3$. This symmetry of $\underline{\underline{C}}$ implies the symmetry of the stress tensor $\underline{\underline{\tau}}$.

Here, the initial conditions can be written as

$$\underline{v}(\underline{x}, 0) = \underline{v}_0(\underline{x}), \quad \frac{\partial \underline{v}}{\partial t}(\underline{x}, 0) = \underline{v}_1(\underline{x}). \tag{1.27}$$

1.1.3.2 The Isotropic Case

When one deals with an isotropic medium, $\underline{\underline{C}}$ can be written as

$$C_{ijk\ell} = \lambda\delta_{ij}\delta_{k\ell} + \mu(\delta_{ik}\delta_{j\ell} + \delta_{i\ell}\delta_{jk}), \tag{1.28}$$

where $\delta_{j\ell}$ is the Kronecker's symbol.

So, Hooke's law assumes the well-known form

$$\tau_{ij} = \lambda\delta_{ij}\sum_{k=1}^{d}\varepsilon_{kk} + 2\mu\varepsilon_{ij}, \tag{1.29}$$

where λ and μ are Lamé's coefficients.

In this case, system (1.25a) and (1.25b) can be written as

$$\rho\frac{\partial^2 \underline{v}}{\partial t^2} = \mu\Delta\underline{v} + (\lambda+\mu)\nabla(\nabla\cdot\underline{v}), \tag{1.30}$$

where $\Delta\underline{v} = (\Delta v_j)_{j=1..d}$.

Now, let us consider a decomposition of the displacement vector $\underline{v}$ of the form

$$\underline{v} = \nabla\varphi + \nabla\times\underline{\psi}. \tag{1.31}$$

We obtain, after inserting this decomposition into (1.30):

$$\rho\frac{\partial^2}{\partial t^2}(\nabla\varphi+\nabla\times\underline{\psi}) = \mu\Delta(\nabla\varphi+\nabla\times\underline{\psi}) + (\lambda+\mu)\nabla(\nabla\cdot(\nabla\varphi+\nabla\times\underline{\psi})). \tag{1.32}$$

Since $\nabla\cdot\nabla\varphi = \Delta\varphi$ and $\nabla\cdot(\nabla\times\underline{\psi}) = 0$, we get, after rearranging the terms of (1.32)

$$\nabla\left[\rho\frac{\partial^2\varphi}{\partial t^2} - (\lambda+2\mu)\Delta\varphi\right] + \nabla\times\left[\rho\frac{\partial^2\underline{\psi}}{\partial t^2} - \mu\Delta\underline{\psi}\right] = 0. \tag{1.33}$$

So, φ and $\underline{\psi}$ each satisfy one wave equation with a different velocity. These potentials actually correspond to two physical waves:

- The P-wave (or *pressure*[7] wave) whose velocity is

$$V_P = \sqrt{\frac{\lambda+2\mu}{\rho}}. \tag{1.34}$$

- The S-wave (or *shear*[8] wave) whose velocity is

$$V_S = \sqrt{\frac{\mu}{\rho}}. \tag{1.35}$$

Obviously, we have

$$V_P^2 \geq 2V_S^2. \tag{1.36}$$

---

[7] Also called *primary* or *longitudinal* wave.

[8] Also called *secondary* or *transverse* wave.

From the physical point of view, the P-wave corresponds to the propagation of a displacement parallel to the direction of propagation and the S-wave to the propagation of a distortion (described by the curl operator) in a plane orthogonal to the direction of propagation.

1.1.4 Boundary Conditions

The wave equations given in this chapter were all written in $\mathbb{R}^d$. However, in practice, the models are defined in bounded domains at the boundary of which some physical conditions must be written. These boundary conditions can be of different sorts but we provide here the most classical ones. The modeling of unbounded domains by boundary conditions will be treated in Chap. 6.

1.1.4.1 The Wave Equation

For the wave equation, the most classical conditions are

- *The Dirichlet condition*:

$$u = g(\underline{x}, t), \tag{1.37}$$

  which models, when $g = 0$, a soft boundary which can be, for instance, the surface of contact of a liquid with the air.

- *The Neumann condition*:

$$\frac{\partial u}{\partial n} = g(\underline{x}, t), \tag{1.38}$$

  where $\underline{n}$ is the unit outward normal to the boundary, which models, when $g = 0$ a rigid boundary which can be, for instance, the wall of a container.

For both conditions, the function g can be a source located on the boundary. When $g = 0$, both conditions provide perfectly reflecting boundaries.

More seldom, one can use impedance boundary conditions

$$\frac{\partial u}{\partial t} + \alpha \frac{\partial u}{\partial n} = 0, \tag{1.39}$$

which model semi-reflecting boundaries. This kind of condition appears, in particular, in the treatment of unbounded domains by absorbing boundary conditions (cf. Chap. 6).

1.1.4.2 Maxwell's Equations

The simplest and most classical boundary condition for Maxwell's equation is the perfectly conducting boundary condition

$$\underline{E} \times \underline{n} = 0, \tag{1.40}$$

where $\underline{n}$ is the outward normal unit, which is also a reflecting boundary condition. Equation (1.40) means that the tangential component of $\underline{E}$ is equal to 0.

In isotropic media, this condition can also take the form

$$\underline{H} \cdot \underline{n} = 0. \tag{1.41}$$

For these equations, the impedance-like conditions are more frequent. In particular, the Silver-Müller condition [1]

$$\sqrt{\varepsilon\mu}\left(\frac{\partial \underline{E}}{\partial t} \times \underline{n}\right) \times \underline{n} - (\nabla \times \underline{E}) \times \underline{n} = 0 \tag{1.42}$$

is a basic condition of physics which models a partially absorbing boundary in the time domain.

1.1.4.3 The Linear Elastodynamics System

For the linear elastodynamics system, two kinds of boundary conditions are classically used [2].

- *Displacement boundary conditions*:

$$v_j = g_j(\underline{x}, t), \quad j = 1..d, \tag{1.43}$$

for which the components of the displacement are prescribed on the boundary.

- *Traction boundary conditions*:

$$\tau_{jk} n_j = g_k(\underline{x}, t), \quad j = 1..d, k = 1..d, \tag{1.44}$$

which provide, when $g_k = 0 \, \forall k = 1..d$, a free surface condition which models the interface of a solid with the vacuum (or the air). In particular, this is the condition which models the surface of the earth in geophysics. This condition generates a surface wave called the *Rayleigh wave* whose velocity V_{R} is given by the following equation:

$$\left(2 - \frac{V_{\mathrm{R}}^2}{V_S^2}\right)^2 - 4\left(1 - \frac{V_{\mathrm{R}}^2}{V_P^2}\right)^{\frac{1}{2}}\left(1 - \frac{V_{\mathrm{R}}^2}{V_S^2}\right)^{\frac{1}{2}} = 0, \tag{1.45}$$

where V_P and V_S are defined as in (1.34) and (1.35). One can show that we have $0 < V_R < V_S$.

A detailed study of this wave can be found in [2].

From the theoretical point of view, one can find a general proof of existence of this class of equations in, among others, [3–5] and a broad discussion of the homogeneous case in [6]. From the physical point of view, one can consult [2, 7–9], for instance, and [10] for a general description of elastic waves in anisotropic media.

1.2 Functional Issues

Of course, the purpose of this chapter is not to provide an exhaustive theory of functional analysis. We only give the basic notions which enable us to better understand the finite element approximations. Although these approximations can be introduced in an intuitive way, their definition based on functional spaces provides a wider and more rigorous view of these methods.

1.2.1 Some Functional Spaces

The definition of functional spaces is based on the concept of distributions, which extends the notion of derivative to a much larger class of functions than classical differential calculus. An important application of this theory for us is the computation of the derivative of any continuous function, even not derivable. The presentation, even when summarized, of this theory is beyond of the framework of this book and the reader who has no knowledge of it could find its description in [6, 11–13] for instance. So, in the following, all the derivatives that we use are taken in the distributions sense.

1.2.1.1 Sobolev Spaces

The first and basic functional spaces used in the finite element theory are the Sobolev spaces H^m. Let us give their definitions for an open set Ω of $\mathbb{R}^d$ (in our case, we take $1 \leq d \leq 3$) whose boundary is $\partial\Omega$. As a first step, we define the general partial differential operator for a scalar function u of $\mathbb{R}^d$ ($\underline{x} = (x_1, .., x_d) \in \mathbb{R}^d$)

$$D^{\alpha} = \frac{\partial^p}{\partial x_1^{\alpha_1} .. \partial x_d^{\alpha_d}}, \tag{1.46}$$

where $p \in \mathbb{N}^*$ and

$$\alpha = \left\{ (\alpha_1, .., \alpha_d) \in \mathbb{N}^d \text{ such that } |\alpha| = \sum_{j=1}^{d} \alpha_j = p. \right\}.$$

Then, the Sobolev space H^m is defined as follows:

$$H^m(\Omega) = \left\{u \in L^2(\Omega) \text{ such that } \forall\alpha \text{ such that } |\alpha| \leq m,\ D^\alpha u \in L^2(\Omega)\right\}, \quad (1.47)$$

where, of course, $L^2(\Omega)$ is the space of functions whose square is integrable over Ω.

The space H^m is a Hilbert space when equipped with the scalar product

$$(u\ ,\ v)_m = \sum_{|\alpha|\leq m} (D^\alpha u\ ,\ D^\alpha v)_0 \quad (1.48)$$

with $(\cdot\ ,\ \cdot)_0$ denoting the standard scalar product on $L^2(\Omega)$.

From this scalar product, we define the norm: $\forall u \in H^m(\Omega)$,

$$\|u\|_m^2 = \sum_{|\alpha|\leq m} \|D^\alpha u\|_0^2. \quad (1.49)$$

We will use at some places the fractional Sobolev spaces $H^{m+\theta}(\Omega)$ where $m \in \mathbb{N}$ and $\theta \in]0, 1[$. These spaces can be defined in several equivalent ways (Fourier transform, interpolation technique, etc ...) and here, we use the Sobolev–Slobodeckij norm

$$H^{m+\theta}(\Omega) = \left\{u \in L^2(\Omega) \text{ such that } \|u\|_{m+\theta} < +\infty\right\} \quad (1.50)$$

where

$$\|u\|_{m+\theta}^2 = \|u\|_m^2 + \sum_{|\alpha|=m} \int_\Omega \int_\Omega \frac{|D^\alpha u(\underline{x}) - D^\alpha u(\underline{y})|^2}{|\underline{x} - \underline{y}|^{2\theta+d}} d\underline{x} d\underline{y}. \quad (1.51)$$

where $|\cdot|$ denotes the Euclidean norm on $\mathbb{R}$ or $\mathbb{R}^d$.

From (1.50), we immediately have the continuous embedding of $H^{m+\theta}(\Omega)$ in $H^m(\Omega)$.

In our case, the most useful Sobolev space is

$$H^1(\Omega) = \left\{u \in L^2(\Omega) \text{ such that } \forall j = 1..d,\ \frac{\partial u}{\partial x_j} \in L^2(\Omega)\right\} \quad (1.52)$$

and, in this context, the norm (1.49) can be rewritten as

$$\|u\|_1^2 = \|u\|_0^2 + \|\nabla u\|_0^2. \quad (1.53)$$

One can define a value (trace) of a function of $H^1(\Omega)$ on $\Gamma = \partial\Omega$ but not a value of its derivative. More precisely, the trace operator $\gamma : u \in C^1(\overline{\Omega}) \mapsto \gamma u = u_{|\Gamma} \in C^0(\Gamma)$ can be continuously extended to $\gamma : H^1(\Omega) \to H^{1/2}(\Gamma)$ i.e. $\exists C > 0$ such that $\forall u \in H^1(\Omega)$,

$$\|\gamma u\|_{1/2,\Gamma} \leq C \|u\|_1 \tag{1.54}$$

where the norm $\|\cdot\|_{1/2,\Gamma}$ is deduced from (1.51) by taking $d = 2$ if $\Omega \subset \mathbb{R}^3$ or $d = 1$ if $\Omega \subset \mathbb{R}^2$:

$$\|u\|^2_{1/2,\Gamma} = \|u\|^2_{0,\Gamma} + \int_\Gamma \int_\Gamma \frac{|u(\underline{x}) - u(\underline{y})|^2}{|\underline{x} - \underline{y}|^{1+d}} d\underline{x} d\underline{y}. \tag{1.55}$$

Remark: The trace operator γ is surjective onto $H^{1/2}(\Gamma)$, and is in a subspace of $L^2(\Gamma)$.

This property enables us to define the following subspace of $H^1(\Omega)$ (which is the kernel of γ)

$$H^1_0(\Omega) = \left\{u \in H^1(\Omega) \text{ such that } u = 0 \text{ on } \Gamma\right\}. \tag{1.56}$$

The continuity of γ implies that the space $H^1_0(\Omega)$ is a closed subspace of $H^1(\Omega)$ and consequently, it is also a Hilbert space for the scalar product $(\cdot\,,\,\cdot)_1$. This subspace is the appropriate framework for solving problems with homogeneous Dirichlet boundary conditions.

The main property of $H^1(\Omega)$ for us is contained in the theorem below that we give without proof:

Theorem 1 *Let Ω_1 and Ω_2 be two subsets of Ω such that $\bar{\Omega} = \bar{\Omega}_1 \cup \bar{\Omega}_2$, $\Omega_1 \cap \Omega_2 = \emptyset$ and $\bar{\Omega}_1 \cap \bar{\Omega}_2 = \Gamma$ and u a function such that $u_1 = u_{|\Omega_1} \in H^1(\Omega_1)$, $u_2 = u_{|\Omega_2} \in H^1(\Omega_2)$. Then $u \in H^1(\Omega)$ if and only if $u_1 = u_2$ on Γ.*

An immediate consequence of this theorem is:

Corollary 1 *Let $\left\{\bar{\Omega}_j\right\}_{j=1}^N$ be a partition of $\bar{\Omega}$ and u a real function defined on Ω. If $\forall j = 1..N$, $u_j = u_{|\Omega_j} \in H^1(\Omega_j)$ and $\forall \ell = 1..N, \forall m = 1..N$ such that $\bar{\Omega}_\ell \cap \bar{\Omega}_m = \Gamma_{\ell,m}$, $u_\ell = u_m$ on $\Gamma_{\ell,m}$, then $u \in H^1(\Omega)$.*

This corollary, which is a more general form of Theorem 1, has an important feature in terms of finite element approximation. It justifies the approximation of $H^1(\Omega)$ by Lagrangian finite elements. These elements are such that, on a mesh $\mathcal{M} = \bigcup K_j$, the restriction of an approximated function to K_j is a polynomial[9] and only the continuity is required at the interfaces of the mesh (cf. Chaps. 2 and 3).

[9] Or derived from a polynomial.

Remarks:

1. The equality $u_1 = u_2$ holds almost everywhere on Γ. This implies that for $d = 1$, $H^1(\Omega) \subset C^0(\Omega)$ but, for $d > 1$, we have no relation between $H^1(\Omega)$ and $C^0(\Omega)$ since the solution can be discontinuous at some points when $d = 2$ and on some curves when $d = 3$. This shows that the notion of derivability is applied here to a much larger class of functions than $C^1(\Omega)$.[10]
2. The spaces $H^m(\Omega)$ are actually a small part of the general Sobolev spaces which can be defined by using L^p spaces and for $m \in \mathbb{R}$ but their definition is far beyond the needs of this book. A compact presentation of these spaces with more references can be found in [11].

1.2.1.2 Spaces $H(\text{Curl}, \Omega)$ and $H(\text{div}, \Omega)$

The Sobolev spaces described in the previous section required that all the derivatives of the functions should be in $L^2(\Omega)$. For some equations, such as Maxwell's equations, it is useful to define functional spaces with fewer requirements.

The basic space for Maxwell's equations is, of course, the space in which one can define the curl of a vector function. In 3D, this space reads

$$H(\mathbf{curl}, \Omega) = \left\{ \underline{u} \in [L^2(\Omega)]^3 \text{ such that } \nabla \times \underline{u} \in [L^2(\Omega)]^3 \right\}. \tag{1.57}$$

and is a Hilbert space for the norm

$$\|\underline{u}\|^2_{curl} = \|\underline{u}\|^2_0 + \|\nabla \times \underline{u}\|^2_0. \tag{1.58}$$

As for $H^1(\Omega)$, one can define a notion of trace on Γ but only the tangential trace has a sense in $H(\mathbf{curl}, \Omega)$. More precisely, define the trace operator: $\underline{u} \in C^1(\overline{\Omega})^3$,

$$\gamma_t\, \underline{u} = \underline{n} \times \underline{u}_\Gamma. \tag{1.59}$$

Now, by taking $\underline{v} \in H^1(\Omega)^3$ and by using an integration by part, we have: $\forall \underline{u} \in C^1(\overline{\Omega})^3$,

$$\int_\Gamma \gamma_t\, \underline{u} \cdot \underline{v}\, d\sigma = \int_\Omega \nabla \times \underline{u} \cdot \underline{v}\, d\underline{x} - \int_\Omega \underline{u} \cdot \nabla \times \underline{v}\, d\underline{x}. \tag{1.60}$$

By using identity (1.60), we can first prove that γ_t is a bounded linear operator from $H(\mathbf{curl}, \Omega)$ to $H^{-1/2}(\Gamma)^3$ where $H^{-1/2}(\Gamma)$ is the topological dual space of $H^{1/2}(\Gamma)$ and it holds that $\forall \underline{u} \in H(\mathbf{curl}, \Omega)$ and $\forall \underline{v} \in H^1(\Omega)^3$,

[10] We recall that $C^0(\Omega)$ is the space of continuous functions on Ω and $C^n(\Omega)$, the space of functions whose derivatives are continuous to the nth-order.

$$< \gamma_t \underline{u}\,,\ \underline{v} >_{-1/2,1/2,\Gamma} = \int_\Omega \nabla \times \underline{u} \cdot \underline{v}\, d\underline{x} - \int_\Omega \underline{u} \cdot \nabla \times \underline{v}\, d\underline{x}, \tag{1.61}$$

where $< \cdot, \cdot >_{-1/2,1/2,\Gamma}$ denotes the duality bracket i.e. the extension of the L^2 scalar product, between $H^{1/2}(\Gamma)$ and its own dual.

Operator γ_t is not surjective onto $H^{-1/2}(\Gamma)^3$. It is in fact surjective onto the space

$$Z = \left\{ \underline{v} \in [H^{-1/2}(\Gamma)]^3 \,:\, \exists \underline{u} \in H(\mathbf{curl}, \Omega) \text{ such that } \gamma_t\, \underline{u} = \underline{v} \right\} \tag{1.62}$$

which is an eigen-subspace of $H^{-1/2}(\Gamma)^3$ and it is equipped with the norm

$$\|\underline{v}\|_Z = \inf_{\substack{\underline{u} \in H(\mathbf{curl}, \Omega) \\ \gamma_t\, \underline{u} = \underline{v}}} \|\underline{u}\|_{curl} \tag{1.63}$$

We can give a more explicit definition of Z. Actually, if $\Omega \subset \mathbb{R}^3$ is a bounded domain with a C^1-regularity of the boundary for example, we have

$$Z = H^{-1/2}(div, \Gamma) = \left\{ \underline{v} \in H_t^{-1/2}(\Gamma) \,:\, div_\Gamma\, \underline{v} \in H^{-1/2}(\Gamma) \right\} \tag{1.64}$$

where $H_t^{-1/2}(\Gamma) = \left\{ \underline{v} \in [H^{-1/2}(\Gamma)]^3 \,:\, \underline{v} \cdot \underline{n} = 0 \text{ on } \Gamma \right\}$ and div_Γ is the surface divergence operator.

We can also define more explicitly the space Z in the context of Lipschitz domains [14].

Remark: The space Z is in fact a Hilbert space.

We can define another tangential trace operator (i.e. the tangential component) γ_T on $H(\mathbf{curl}, \Omega)$ as

$$\gamma_T\, \underline{u} = \underline{n} \times (\underline{u}_\Gamma \times \underline{n}). \tag{1.65}$$

This operator is linear and bounded from $H(\mathbf{curl}, \Omega)$ to the dual space of Z. For regular domains, we have

$$Z' = H^{-1/2}(\mathbf{curl}, \Gamma) = \left\{ \underline{v} \in H_t^{-1/2}(\Gamma) \,:\, \mathbf{curl}_\Gamma\, \underline{v} \in H_t^{-1/2}(\Gamma) \right\}, \tag{1.66}$$

where $\mathbf{curl}_\Gamma$ is the surface curl operator which acts on tangential vector fields to Γ.

Finally, formula (1.61) can be extended in this way: it holds that $\forall \underline{u}, \underline{v} \in H(\mathbf{curl}, \Omega)$,

$$< \gamma_t\, \underline{u}\,,\ \gamma_T \underline{v} >_{Z,Z'} = \int_\Omega \nabla \times \underline{u} \cdot \underline{v}\, d\underline{x} - \int_\Omega \underline{u} \cdot \nabla \times \underline{v}\, d\underline{x}. \tag{1.67}$$

where $< \cdot, \cdot >_{Z,Z'}$ denotes the duality bracket i.e. the extension of the L^2 scalar product, between Z and its own dual Z'.

Properties of γ_t enable us to define the following subspace of $H(\mathbf{curl}, \Omega)$:

$$H_0(\mathbf{curl}, \Omega) = \{\underline{u} \in H(\mathbf{curl}, \Omega) \text{ such that } \underline{u} \times \underline{n} = 0 \text{ on } \Gamma\}. \tag{1.68}$$

This space is a closed subspace of $H(\mathbf{curl}, \Omega)$ and so, it is a Hilbert space.

Remark: The space $H_0(\mathbf{curl}, \Omega)$ allows to take into account a medium with a perfectly conducting boundary condition.

In 2D, $H(\mathbf{curl}, \Omega)$ can be defined in two ways which depend on the character of the function. For a vector function, we have

$$H(\mathrm{curl}, \Omega) = \{\underline{u} \in [L^2(\Omega)]^2 \text{ such that } \mathrm{curl}\, \underline{u} \in L^2(\Omega)\}, \tag{1.69}$$

where curl is defined as in Sect. 1.1.2.2.

For a scalar function, the space is based on the definition of **curl** in Sect. 1.1.2.2

$$H(\mathbf{curl}, \Omega) = \{u \in L^2(\Omega) \text{ such that } \mathbf{curl}\, u \in [L^2(\Omega)]^2\}. \tag{1.70}$$

The definition of **curl** shows that, in this case, we actually have $H(\mathbf{curl}, \Omega) = H^1(\Omega)$.

Another useful Hilbert space for the Maxwell's equations is the following space in which are the electric and magnetic inductions $\underline{D}$ and $\underline{B}$:

$$H(\mathrm{div}, \Omega) = \{\underline{u} \in [L^2(\Omega)]^3 \text{ such that } \nabla \cdot \underline{u} \in L^2(\Omega)\}, \tag{1.71}$$

which is equipped with the norm

$$\|\underline{u}\|_{div}^2 = \|\underline{u}\|_0^2 + \|\nabla \cdot \underline{u}\|_0^2. \tag{1.72}$$

One can define a normal trace of a function of $H(\mathrm{div}, \Omega)$. More precisely, $\forall \underline{v} \in C^1(\overline{\Omega})^3$, this trace operator is defined by

$$\gamma_n \underline{v} = \underline{v}_{|\Gamma} \cdot \underline{n}. \tag{1.73}$$

Now, by taking $v \in H^1(\Omega)$ and by using an integration by part, we have: $\forall \underline{u} \in C^1(\overline{\Omega})^3$,

$$\int_\Gamma \gamma_n \underline{u}\, v\, d\sigma = \int_\Omega \nabla \cdot \underline{u}\, v\, d\underline{x} + \int_\Omega \underline{u} \cdot \nabla v\, d\underline{x}. \tag{1.74}$$

By using identity (1.74), we can prove that γ_n is a bounded surjective linear operator from $H(\mathbf{div}, \Omega)$ to $H^{-1/2}(\Gamma)$ and it holds that $\forall \underline{u} \in H(\mathbf{div}, \Omega)$ and $\forall v \in H^1(\Omega)$,

$$< \gamma_n \underline{u}, v >_{-1/2, 1/2, \Gamma} = \int_\Omega \nabla \cdot \underline{u}\, v\, d\underline{x} + \int_\Omega \underline{u} \cdot \nabla v\, d\underline{x}. \tag{1.75}$$

The properties of γ_n enables us to define the following subspace of $H(\mathbf{div}, \Omega)$:

$$H_0(\mathbf{div}, \Omega) = \{\underline{u} \in H(\mathbf{div}, \Omega) \text{ such that } \underline{u} \cdot \underline{n} = 0 \text{ on } \Gamma\}. \tag{1.76}$$

This space is a closed subspace of $H(\mathbf{div}, \Omega)$ and therefore a Hilbert space.

As for $H^1(\Omega)$, we have the following important property of the functions of all the above spaces:

Theorem 2 *Let Ω_1 and Ω_2 be two subsets of Ω such that $\bar{\Omega} = \bar{\Omega}_1 \cup \bar{\Omega}_2$, $\Omega_1 \cap \Omega_2 = \emptyset$ and $\bar{\Omega}_1 \cap \bar{\Omega}_2 = \Gamma$ and $\underline{u}$ a function such that $\underline{u}_1 = \underline{u}_{|\Omega_1} \in H(\mathbf{curl}, \Omega_1)$ (respectively $\in H(\text{div}, \Omega_1)$), $\underline{u}_2 = \underline{u}_{|\Omega_2} \in H(\mathbf{curl}, \Omega_2)$ (respectively $\in H(\text{div}, \Omega_2)$). Then $\underline{u} \in H(\mathbf{curl}, \Omega)$ (respectively $\in H(\text{div}, \Omega)$) if and only if $\underline{u}_1 \times \underline{n} = \underline{u}_2 \times \underline{n}$ (respectively $\underline{u}_1 \cdot \underline{n} = \underline{u}_2 \cdot \underline{n}$) in the sense of distributions on Γ, where $\underline{n}$ denotes a unit normal to Γ.*

On this theorem is based the approximation of $H(\mathbf{curl}, \Omega)$ by the so-called *edge elements* which are continuous in each K_j of a mesh $\mathscr{M}$ and whose tangential components only are continuous through the interfaces. In the same way, $H(\text{div}, \Omega)$ is approximated by elements with continuous normal components through the interfaces (cf. Chap. 2).

1.2.2 Variational Formulations

1.2.2.1 The Acoustics Equation

On the basis of the functional spaces defined in the previous section, one can define *variational formulations* of the wave equations which are the first step of their finite element approximations. Let us first define it for the acoustics equation.

If we multiply the acoustics equation

$$\eta \frac{\partial^2 u}{\partial t^2} - \nabla \cdot (\gamma \nabla u) = f \tag{1.77}$$

by a function $v \in H^1(\Omega)$ and we integrate by parts the *stiffness integral*, i.e. the integral corresponding to the *stiffness term* $\nabla \cdot (\gamma \nabla u)$, we get

$$\frac{\mathrm{d}^2}{\mathrm{d}t^2} \int_\Omega \eta\, u\, v \,\mathrm{d}\underline{x} + \int_\Omega \gamma\, \nabla u \cdot \nabla v \,\mathrm{d}\underline{x} - \int_{\partial\Omega} \gamma\, v \frac{\partial u}{\partial n} \,\mathrm{d}\sigma = \int_\Omega f\, v \,\mathrm{d}\underline{x}. \tag{1.78}$$

Let us now suppose that we have a Neumann condition on the boundary $\partial\Omega$, i.e. $\partial u / \partial n = g$. We derive from (1.77) the following *variational problem*:

Find u such that $u(., t) \in H^1(\Omega)$ and

$$\frac{\mathrm{d}^2}{\mathrm{d}t^2} \int_\Omega \eta\, u\, v \,\mathrm{d}\underline{x} + \int_\Omega \gamma\, \nabla u \cdot \nabla v \,\mathrm{d}\underline{x} = \int_{\partial\Omega} \gamma\, g\, v \,\mathrm{d}\sigma + \int_\Omega f\, v \,\mathrm{d}\underline{x}, \ \forall v \in H^1(\Omega). \tag{1.79}$$

For the homogeneous Dirichlet condition $u = 0$, the variational problem is: Find u such that $u(.,t) \in H_0^1(\Omega)$ and

$$\frac{\mathrm{d}^2}{\mathrm{d}t^2}\int_\Omega \eta\, u\, v\, \mathrm{d}\underline{x} + \int_\Omega \gamma\, \nabla u \cdot \nabla v\, \mathrm{d}\underline{x} = \int_\Omega f v\, \mathrm{d}\underline{x},\ \forall v \in H_0^1(\Omega) \tag{1.80}$$

and, for the impedance condition given in (1.39), we obtain:
Find u such that $u(.,t) \in H^1(\Omega)$ and

$$\begin{cases} \dfrac{\mathrm{d}^2}{\mathrm{d}t^2}\displaystyle\int_\Omega \eta u v\, \mathrm{d}\underline{x} + \int_\Omega \gamma \nabla u \cdot \nabla v\, \mathrm{d}\underline{x} = -\int_{\partial\Omega} \frac{1}{\alpha}\gamma v \frac{\partial u}{\partial t}\mathrm{d}\sigma + \int_\Omega f v\, \mathrm{d}\underline{x}, \\ \forall v \in H^1(\Omega). \end{cases} \tag{1.81}$$

Of course, one must add the initial conditions defined in (1.4) to (1.79)–(1.81).

By using density properties of functional spaces, one shows that (1.79)–(1.81) are equivalent to the acoustics Eq. (1.77) with different boundary conditions when the solution is sought as a distribution. These formulations are also called *weak formulations* of the acoustics equation because their solution is sought in a space whose functions are required to be only once derivable whereas the stiffness term uses second derivatives.

Remarks:

1. The use of g and of the time derivative of u in the boundary integrals in (1.79) and (1.81) avoids the presence of the normal derivative of u which cannot be defined for a function of $H^1(\Omega)$.
2. The functional frame for non-homogeneous Dirichlet conditions is more difficult to define and is widely treated in [5].

1.2.2.2 Maxwell's Equations

Let us consider the 3D Maxwell's equations

$$\underline{\underline{\varepsilon}}\frac{\partial \underline{E}}{\partial t} - \nabla \times \underline{H} = -\underline{J}, \tag{1.82a}$$

$$\underline{\underline{\mu}}\frac{\partial \underline{H}}{\partial t} + \nabla \times \underline{E} = 0. \tag{1.82b}$$

We first multiply the first equation by $\underline{\varphi} \in H(\mathbf{curl}, \Omega)$ and the second one by $\underline{\psi} \in [L^2(\Omega)]^3$. Then, after integrating by parts the stiffness integral of (1.82a) (which corresponds to $\nabla \times \underline{H}$), we obtain

$$\frac{\mathrm{d}}{\mathrm{d}t}\int_{\Omega}\underline{\underline{\varepsilon}}\,\underline{E}\cdot\underline{\varphi}\mathrm{d}\underline{x}-\int_{\Omega}\underline{H}\cdot(\nabla\times\underline{\varphi})\,\mathrm{d}\underline{x}$$
$$=\int_{\partial\Omega}[\underline{n}\times(\underline{H}\times\underline{n})]\cdot(\underline{\varphi}\times\underline{n})\,\mathrm{d}\sigma-\int_{\Omega}\underline{J}\underline{\varphi}\,\mathrm{d}\underline{x}, \tag{1.83a}$$

$$\frac{\mathrm{d}}{\mathrm{d}t}\int_{\Omega}\underline{\underline{\mu}}\,\underline{H}\cdot\underline{\psi}+\int_{\Omega}(\nabla\times\underline{E})\cdot\underline{\psi}\,\mathrm{d}\underline{x}=0. \tag{1.83b}$$

Since $[\underline{n}\times(\underline{H}\times\underline{n})]\cdot(\underline{\varphi}\times\underline{n})=\underline{H}\cdot(\underline{\varphi}\times\underline{n})$, for the perfectly conducting boundary condition (1.40), we obtain the following variational problem:

Find $\underline{E}$ and $\underline{H}$ such that $\underline{E}(.,t)\in H_0(\mathbf{curl},\Omega)$ and $\underline{H}(.,t)\in\left[L^2(\Omega)\right]^3$ and

$$\frac{\mathrm{d}}{\mathrm{d}t}\int_{\Omega}\underline{\underline{\varepsilon}}\,\underline{E}\cdot\underline{\varphi}\mathrm{d}\underline{x}-\int_{\Omega}\underline{H}\cdot(\nabla\times\underline{\varphi})\,\mathrm{d}\underline{x}=-\int_{\Omega}\underline{J}\underline{\varphi}\,\mathrm{d}\underline{x},\ \forall\underline{\varphi}\in H_0(\mathbf{curl},\Omega), \tag{1.84a}$$
$$\frac{\mathrm{d}}{\mathrm{d}t}\int_{\Omega}\underline{\underline{\mu}}\,\underline{H}\cdot\underline{\psi}+\int_{\Omega}(\nabla\times\underline{E})\cdot\underline{\psi}\,\mathrm{d}\underline{x}=0,\ \forall\underline{\psi}\in[L^2(\Omega)]^3. \tag{1.84b}$$

Of course, one must add to (1.84a) and (1.84b), the initial conditions defined in (1.9).

Here also, the magnetic field $\underline{H}$ is sought in $[L^2(\Omega)]^3$ which is larger than $H(\mathbf{curl},\Omega)$ to which belongs $\underline{H}$.

Another variational formulation can be derived from the second-order formulation of the Maxwell's equation given in (1.10) for instance. If we multiply this equation by $\underline{\varphi}\in H(\mathbf{curl},\Omega)$ and we integrate by parts the stiffness term, we get similarly

$$\frac{\mathrm{d}^2}{\mathrm{d}t^2}\int_{\Omega}\underline{\underline{\varepsilon}}\,\underline{E}\cdot\underline{\varphi}\mathrm{d}\underline{x}+\int_{\Omega}\underline{\underline{\mu}}^{-1}(\nabla\times\underline{E})\cdot(\nabla\times\underline{\varphi})\,\mathrm{d}\underline{x}$$
$$=-\int_{\partial\Omega}[\underline{n}\times(\underline{\underline{\mu}}^{-1}(\nabla\times\underline{E})\times\underline{n})]\cdot(\underline{\varphi}\times\underline{n})\,\mathrm{d}\sigma-\int_{\Omega}\underline{j\varphi}\,\mathrm{d}\underline{x}, \tag{1.85}$$

which provides, for perfectly conducting boundary condition:
Find $\underline{E}$ such that $\underline{E}(.,t)\in H_0(\mathbf{curl},\Omega)$ and

$$\begin{cases}\dfrac{\mathrm{d}^2}{\mathrm{d}t^2}\displaystyle\int_{\Omega}\underline{\underline{\varepsilon}}\,\underline{E}\cdot\underline{\varphi}\mathrm{d}\underline{x}+\int_{\Omega}\underline{\underline{\mu}}^{-1}(\nabla\times\underline{E})\cdot(\nabla\times\underline{\varphi})\,\mathrm{d}\underline{x}=-\int_{\Omega}\underline{j\varphi}\,\mathrm{d}\underline{x},\\ \forall\underline{\varphi}\in H_0(\mathbf{curl},\Omega).\end{cases} \tag{1.86}$$

In this form, one can take into account the Silver-Müller condition defined in (1.42), in the isotropic case, as follows:
Find $\underline{E}$ such that $\underline{E}(.,t)\in H(\mathbf{curl},\Omega)$ and

$$\begin{cases} \dfrac{\mathrm{d}^2}{\mathrm{d}t^2}\displaystyle\int_\Omega \varepsilon\,\underline{E}\cdot\underline{\varphi}\,\mathrm{d}\underline{x} + \int_\Omega \mu^{-1}(\nabla\times\underline{E})\cdot(\nabla\times\underline{\varphi})\,\mathrm{d}\underline{x} \\ +\dfrac{\mathrm{d}}{\mathrm{d}t}\displaystyle\int_{\partial\Omega}\sqrt{\frac{\varepsilon}{\mu}}\,(\underline{E}\times\underline{n})\cdot(\underline{\varphi}\times\underline{n})\,\mathrm{d}\sigma = -\int_\Omega \underline{j\varphi}\,\mathrm{d}\underline{x}, \\ \forall\underline{\varphi}\in H(\mathbf{curl},\Omega). \end{cases} \tag{1.87}$$

Remark: One can also take $\underline{E}(.,t)\in\left[L^2(\Omega)\right]^3$ and $\underline{H}(.,t)\in H(\mathbf{curl},\Omega)$ in order to obtain a variational formulation of (1.82a) and (1.82b). In this case, the integration by parts is made on (1.82b) and the boundary integral is

$$\int_{\partial\Omega}[\underline{n}\times(\underline{E}\times\underline{n})]\cdot(\underline{\psi}\times\underline{n})\,\mathrm{d}\sigma,\ \ \underline{\psi}\in H(\mathbf{curl},\Omega).$$

This boundary integral vanishes when $\underline{E}\times\underline{n}=0$ in the variational formulation and so, $\underline{H}$ must be sought in $H(\mathbf{curl},\Omega)$ instead of $H_0(\mathbf{curl},\Omega)$. Then, the resulting variational problem for the perfectly conducting boundary condition can be written as

Find $\underline{E}$ and $\underline{H}$ such that $\underline{E}(.,t)\in\left[L^2(\Omega)\right]^3$ and $\underline{H}(.,t)\in H(\mathbf{curl},\Omega)$ and

$$\begin{aligned} &\frac{\mathrm{d}}{\mathrm{d}t}\int_\Omega \underline{\underline{\varepsilon}}\,\underline{E}\cdot\underline{\varphi}\mathrm{d}\underline{x} - \int_\Omega(\nabla\times\underline{H})\cdot\underline{\varphi}\,\mathrm{d}\underline{x} \\ &\quad = -\int_\Omega \underline{J\varphi}\,\mathrm{d}\underline{x},\ \ \forall\underline{\varphi}\in[L^2(\Omega)]^3, \end{aligned} \tag{1.88a}$$

$$\frac{\mathrm{d}}{\mathrm{d}t}\int_\Omega \underline{\underline{\mu}}\,\underline{H}\cdot\underline{\psi} + \int_\Omega \underline{E}\cdot(\nabla\times\underline{\psi})\,\mathrm{d}\underline{x} = 0,\ \ \forall\underline{\psi}\in H(\mathbf{curl},\Omega). \tag{1.88b}$$

Actually, the perfectly conducting boundary condition behaves as a Dirichlet condition versus $\underline{E}$ and a Neumann condition versus $\underline{H}$.

A similar variational formulation can be obtained for (1.11).

1.2.2.3 The Linear Elastodynamics System

In order to construct the variational formulation of the linear elastodynamics system, let us first introduce the following notations:

If $\underline{\underline{u}} = (\underline{u}_1,..,\underline{u}_d)$, where $\underline{u}_j = (u_{j,1},..,u_{j,d})$ is a d-dimensional tensor and $\underline{\underline{v}} = (\underline{v}_1,..,\underline{v}_d)^T$, where $\underline{v}_j = (v_{j,1},..,v_{j,d})^T$, we write

$$\underline{\underline{u}} : \underline{\underline{v}} = \sum_{j=1}^{d}\underline{u}_j\cdot\underline{v}_j. \tag{1.89}$$

With these notations, if $\underline{\varphi}$ is a function of $\left[H^1(\Omega)\right]^d$, after multiplying the stiffness term of (1.25a) and integrating by parts and by taking into account the symmetric character of $\underline{\underline{\tau}}$, we get

$$\int_\Omega \mathbf{div}\underline{\underline{\tau}} \cdot \underline{\varphi} \, \mathrm{d}\underline{x} = -\int_\Omega \underline{\underline{\tau}} : \underline{\underline{\varepsilon}}(\underline{\varphi}) \, \mathrm{d}\underline{x} + \int_{\partial\Omega} \underline{\underline{\tau}}\,\underline{n} \cdot \underline{\varphi} \, \mathrm{d}\sigma. \tag{1.90}$$

So, by multiplying (1.25a) by $\underline{\varphi}$ and taking into account (1.90) and (1.25b), we obtain, for a free surface condition (i.e. $\underline{\underline{\tau}}\underline{n} = 0$), the following variational problem: Find $\underline{v}$ such that $\underline{v}(.,t) \in \left[H^1(\Omega)\right]^d$ and

$$\frac{\mathrm{d}^2}{\mathrm{d}t^2}\int_\Omega \rho \underline{v} \cdot \underline{\varphi} \, \mathrm{d}\underline{x} + \int_\Omega \underline{\underline{C}}\,\underline{\underline{\varepsilon}}(\underline{v}) : \underline{\underline{\varepsilon}}(\underline{\varphi}) \, \mathrm{d}\underline{x} = \int_\Omega \underline{f} \cdot \underline{\varphi} \, \mathrm{d}\underline{x}, \ \forall \underline{\varphi} \in \left[H^1(\Omega)\right]^d \tag{1.91}$$

to which one must add the initial conditions defined in (1.27).

1.2.3 *Energy Identities*

From the variational formulations, one can derive *energy identities*. These identities are the basic features of the wave equations which ensure their well-posedness and, more concretely, the stability of the solutions.

1.2.3.1 The Acoustics Equation

Let us consider the variational formulation of the acoustics equation obtained in (1.80) in which we set $f = 0$. If we set $v = \partial u/\partial t$ (which is legitimate since $\forall t \geq 0$, $(\partial u/\partial t)(.,t) \in H_0^1(\Omega)$), we get

$$\int_\Omega \eta \frac{\partial^2 u}{\partial t^2} \frac{\partial u}{\partial t} \, \mathrm{d}\underline{x} + \int_\Omega \gamma \, \nabla u \cdot \nabla \frac{\partial u}{\partial t} \, \mathrm{d}\underline{x} = 0, \tag{1.92}$$

which can be rewritten as

$$\frac{\mathrm{d}}{\mathrm{d}t}\mathscr{E}(u) = \frac{\mathrm{d}}{\mathrm{d}t}\left\{\frac{1}{2}\int_\Omega \eta \left|\frac{\partial u}{\partial t}\right|^2 \mathrm{d}\underline{x} + \frac{1}{2}\int_\Omega \gamma \, |\nabla u|^2 \, \mathrm{d}\underline{x}\right\} = 0. \tag{1.93}$$

Equation (1.93) means that, for a given initial solution u_0, we have $\forall t \geq 0$, $\mathscr{E}(u) = \mathscr{E}(u_0)$, which is the principle of *energy conservation* for the wave equation. The same identity can also be obtained for the homogeneous Neumann condition

($g = 0$ in (1.79)). Actually, both Dirichlet and Neumann homogeneous conditions, which are perfectly reflecting conditions, ensure that no dissipation of the waves occurs in the domain Ω. This feature is given by the energy conservation principle. Of course, the positivity of $\mathscr{E}$ is fundamental.

1.2.3.2 Maxwell's Equations

Let us set $\underline{\varphi} = \underline{E}$ and $\underline{\psi} = \underline{H}$ in (1.84a) and (1.84b), where we have set $\underline{J} = 0$. We obtain

$$\int_{\Omega} \underline{\underline{\varepsilon}} \frac{\partial \underline{E}}{\partial t} \cdot \underline{E} \mathrm{d}\underline{x} - \int_{\Omega} \underline{H} \cdot (\nabla \times \underline{E}) \, \mathrm{d}\underline{x} = 0, \tag{1.94a}$$

$$\int_{\Omega} \underline{\underline{\mu}} \frac{\partial \underline{H}}{\partial t} \cdot \underline{H} \mathrm{d}\underline{x} + \int_{\Omega} (\nabla \times \underline{E}) \cdot \underline{H} \, \mathrm{d}\underline{x} = 0. \tag{1.94b}$$

Since $\underline{\underline{\varepsilon}}$ and $\underline{\underline{\mu}}$ are both symmetric, definite, positive, there exists two matrices $\underline{\underline{\tilde{\varepsilon}}}$ and $\underline{\underline{\tilde{\mu}}}$ such that $\underline{\underline{\varepsilon}} = \underline{\underline{\tilde{\varepsilon}}}^T \underline{\underline{\tilde{\varepsilon}}}$ and $\underline{\underline{\mu}} = \underline{\underline{\tilde{\mu}}}^T \underline{\underline{\tilde{\mu}}}$. So, by combining (1.94a) and (1.94b), we obtain the following energy identity for the Maxwell's equations

$$\frac{\mathrm{d}}{\mathrm{d}t} \mathscr{E}(\underline{E}, \underline{H}) = \frac{\mathrm{d}}{\mathrm{d}t} \left\{ \int_{\Omega} |\underline{\underline{\tilde{\varepsilon}}} \, \underline{E}|^2 \, \mathrm{d}\underline{x} + \int_{\Omega} |\underline{\underline{\tilde{\mu}}} \, \underline{H}|^2 \, \mathrm{d}\underline{x} \right\} = 0. \tag{1.95}$$

In the same way, by setting $\underline{\varphi} = \partial \underline{E}/\partial t$ in (1.86) (with $\underline{j} = 0$), we obtain, as for the wave equation

$$\frac{\mathrm{d}}{\mathrm{d}t} \mathscr{E}(\underline{E}) = \frac{\mathrm{d}}{\mathrm{d}t} \left\{ \frac{1}{2} \int_{\Omega} \left| \underline{\underline{\tilde{\varepsilon}}} \frac{\partial \underline{E}}{\partial t} \right|^2 \mathrm{d}\underline{x} + \frac{1}{2} \int_{\Omega} |\underline{\underline{\tilde{\mu}}}^{*-1} \, \nabla \times \underline{E}|^2 \, \mathrm{d}\underline{x} \right\} = 0. \tag{1.96}$$

In the case of the Silver-Müller condition given in (1.87), we obtain

$$\begin{aligned} \frac{\mathrm{d}}{\mathrm{d}t} \mathscr{E}(\underline{E}) &= \frac{\mathrm{d}}{\mathrm{d}t} \left\{ \frac{1}{2} \int_{\Omega} \left| \varepsilon^{\frac{1}{2}} \frac{\partial \underline{E}}{\partial t} \right|^2 \mathrm{d}\underline{x} + \frac{1}{2} \int_{\Omega} |\mu^{-\frac{1}{2}} \, \nabla \times \underline{E}|^2 \, \mathrm{d}\underline{x} \right\} \\ &= - \int_{\partial\Omega} \left(\frac{\varepsilon}{\mu} \right)^{\frac{1}{2}} \left| \frac{\partial \underline{E}}{\partial t} \times \underline{n} \right|^2 \mathrm{d}\sigma. \end{aligned} \tag{1.97}$$

Relation (1.97) shows that, unlike perfectly conducting boundary condition, the energy is decreasing here. This means that the waves vanish from the domain. In fact, they are absorbed by the boundary condition.

1.2.3.3 The Linear Elastodynamics System

Although using vector-valued unknowns, the linear elastodynamics system can be treated as the acoustics equation in order to obtain energy identities. By setting $\underline{\varphi} = \partial\underline{v}/\partial t$ in (1.91), we get

$$\int_{\Omega} \rho \frac{\partial^2 \underline{v}}{\partial t^2} \cdot \frac{\partial \underline{v}}{\partial t} \, \mathrm{d}\underline{x} + \int_{\Omega} \underline{\underline{C}} \, \underline{\underline{\varepsilon}}(\underline{v}) : \underline{\underline{\varepsilon}} \left(\frac{\partial \underline{v}}{\partial t} \right) \mathrm{d}\underline{x} = 0, \; \forall \underline{\varphi} \in \left[H^1(\Omega) \right]^d. \tag{1.98}$$

By using the linearity of $\underline{\underline{\varepsilon}}$, (1.98) can be rewritten as

$$\frac{\mathrm{d}}{\mathrm{d}t} \mathscr{E}(\underline{v}) = \frac{\mathrm{d}}{\mathrm{d}t} \left\{ \frac{1}{2} \int_{\Omega} \rho \left| \frac{\partial \underline{v}}{\partial t} \right|^2 \mathrm{d}\underline{x} + \frac{1}{2} \int_{\Omega} \underline{\underline{C}} \, \underline{\underline{\varepsilon}}(\underline{v}) : \underline{\underline{\varepsilon}}(\underline{v}) \, \mathrm{d}\underline{x} \right\} = 0. \tag{1.99}$$

Positivity of $\mathscr{E}$ is derived from the positivity of $\underline{\underline{C}}$.

Remark: One can easily check that the energy identities can be written for subspaces of the functional spaces. In particular, they hold for finite element subspaces and then provide natural conditions of stability for these approximations.

1.2.4 Well-Posedness Results of Waves Equations

In this part, we give the classical well-posedness results of previous acoustic, electromagnetism and elastodynamic problems. These results are derived from the so-called Hille–Yosida theorem [15] which is a powerful tool to analyze the evolution equations.

Theorem 3 (Hille–Yosida) *Let* $(H\,,\,(\cdot,\cdot))$ *be an Hilbert space and* $A\,:\,D(A) \subset H \to H$ *be a linear maximal monotone operator i.e.:*

1. $\forall \underline{v} \in D(A)$, $(A\underline{v}\,,\,\underline{v}) \geq 0$,

2. $\forall f \in H$, $\exists \underline{u} \in D(A)$ *such that* $\underline{u} + A\underline{u} = f$.

   Then $\forall \underline{u}_0 \in D(A)$ *and* $\forall \underline{f} \in C^1(\mathbb{R}^+; H)$, *there exists one function*

$$\underline{u} \in C^1(\mathbb{R}^+; H) \cap C^0(\mathbb{R}^+; D(A)) \tag{1.100}$$

and only one such that

$$\frac{d\underline{u}}{dt} + A\underline{u} = \underline{f}, \; t > 0, \tag{1.101a}$$

$$\underline{u}(0) = \underline{u}_0. \tag{1.101b}$$

1.2.4.1 The Acoustics Equations

We want to apply the Theorem 3 to the acoustic problem: find $(u\,,\,\underline{v})\,:,\,\Omega\times\mathbb{R}^{+}\rightarrow\mathbb{R}\times\mathbb{R}^{d}$ such that

$$\frac{1}{\kappa(\underline{x})}\frac{\partial u}{\partial t}(\underline{x},t)-\nabla\cdot\underline{v}(\underline{x},t)=F(\underline{x},t),\;\;\underline{x}\in\Omega,\;t>0, \tag{1.102a}$$

$$\rho(\underline{x})\frac{\partial\underline{v}}{\partial t}(\underline{x},t)-\nabla u(\underline{x},t)=0,\;\;\underline{x}\in\Omega,\;t>0, \tag{1.102b}$$

$$u(\underline{x},t)=0,\;\;\underline{x}\in\Gamma_{D},\;t>0, \tag{1.102c}$$

$$(\underline{v}\cdot\underline{n})(\underline{x},t)=0,\;\;\underline{x}\in\Gamma_{N},\;t>0, \tag{1.102d}$$

$$u(\underline{x},t)+\alpha\rho(\underline{v}\cdot\underline{n})(\underline{x},t)=0,\;\;\underline{x}\in\Gamma_{ABC},\;t>0, \tag{1.102e}$$

$$u(\underline{x},0)=u_{0}(\underline{x}),\;\;\underline{x}\in\Omega, \tag{1.102f}$$

$$\underline{v}(\underline{x},0)=\underline{v}_{0}(\underline{x}),\;\;\underline{x}\in\Omega, \tag{1.102g}$$

where $\Gamma=\Gamma_{D}\cup\Gamma_{N}\cup\Gamma_{ABC},\kappa,\;\rho,\in L^{\infty}(\Omega)$ such $\kappa(\underline{x})\geq\kappa_{0}>0$ and $\rho(\underline{x})\geq\rho_{0}>0$ almost everywhere and $\alpha\in L^{\infty}(\Gamma_{ABC})$ such that $\alpha(\underline{x})\geq\alpha_{0}>0$ almost everywhere.

We rewrite (1.102a)–(1.102g) in the framework of the Hille–Yosida theorem in the following way:

1. $H=[L^{2}(\Omega)]^{d+1}$ and is equipped with the scalar product

$$((u,\underline{v}),(\tilde{u},\tilde{\underline{v}}))=\left(\frac{1}{\kappa}u,\tilde{u}\right)_{0}+\left(\rho\underline{v},\tilde{\underline{v}}\right)_{0} \tag{1.103}$$

2. Operator A is defined by

$$A=\begin{bmatrix}0 & -\kappa\nabla\cdot\\ -\frac{1}{\rho}\nabla & 0\end{bmatrix}. \tag{1.104}$$

3. The domain of A relative to problem (1.102a)–(1.102g) is

$$D(A)=\{(u\,,\,\underline{v})\in H^{1}(\Omega)\times H(\mathbf{div},\Omega)\;:\;u=0\text{ on }\Gamma_{D},\\ \underline{v}\cdot\underline{n}=0\text{ on }\Gamma_{N}\text{ and }u+\alpha\rho(\underline{v}\cdot\underline{n})=0\text{ on }\Gamma_{ABC}\}. \tag{1.105}$$

We now verify that operator A is maximal monotone.

First, $\forall(u, \underline{v}) \in D(A)$, we have

$$
\begin{aligned}
(A(u, \underline{v}), (u, \underline{v})) &= -\int_\Omega \nabla \cdot \underline{v}\, u\, dx - \int_\Omega \nabla u \cdot \underline{v}\, dx \\
&= -\int_\Gamma (\underline{v} \cdot \underline{n})\, u\, d\sigma \\
&= \int_{\Gamma_{ABC}} \frac{1}{\alpha\rho} u^2\, d\sigma \geq 0. \qquad (1.106)
\end{aligned}
$$

Then, we prove that operator $I + A$ is surjective. For that, we introduce the bilinear form

$$
\begin{aligned}
b : D(A) \times H \to \mathbb{R}, ((u, \underline{v}), (\tilde{u}, \tilde{\underline{v}})) \mapsto\ & b((u, \underline{v}), (\tilde{u}, \tilde{\underline{v}})) \\
&= ((I + A)(u, \underline{v}), (\tilde{u}, \tilde{\underline{v}})) \qquad (1.107)
\end{aligned}
$$

and we are going to prove that $\forall(f, \underline{g}) \in H, \exists!(u, \underline{v}) \in D(A)$ such that

$$
b((u, \underline{v}), (\tilde{u}, \tilde{\underline{v}})) = ((f, \underline{g}), (\tilde{u}, \tilde{\underline{v}})), \ \forall(\tilde{u}, \tilde{\underline{v}}) \in H. \qquad (1.108)
$$

The so-called Banach–Nečas–Babuška theorem [16] says that problem (1.108) is well-posed if and only if the two following conditions hold.

1. There exists a constant $C > 0$ such that $\forall(u, \underline{v}) \in D(A)$,

$$
C\|(u, \underline{v})\|_{D(A)} \leq \sup_{(\tilde{u}, \tilde{\underline{v}}) \in H\setminus\{0\}} \frac{b((u, \underline{v}), (\tilde{u}, \tilde{\underline{v}}))}{\|(\tilde{u}, \tilde{\underline{v}})\|_H} \qquad (1.109)
$$

   where $\|(u, \underline{v})\|^2_{D(A)} = \|u\|_0^2 + \|\underline{v}\|_0^2 + \|\nabla u\|_0^2 + \|\nabla \cdot \underline{v}\|_0^2$.

2. For all $(\tilde{u}, \tilde{\underline{v}}) \in H$,

$$
(\forall(u, \underline{v}) \in D(A),\ b((u, \underline{v}), (\tilde{u}, \tilde{\underline{v}})) = 0) \Rightarrow (\tilde{u}, \tilde{\underline{v}}) = 0. \qquad (1.110)
$$

We establish (1.109) in three steps.

1. By taking $(\tilde{u}, \tilde{\underline{v}}) = (u, \underline{v}) \in D(A) \subset H$, we have by using (1.106)

$$
\begin{aligned}
b((u, \underline{v}), (u, \underline{v})) &= \|\kappa^{-1/2} u\|_0^2 + \|\rho^{1/2} \underline{v}\|_0^2 + \int_{\Gamma_{ABC}} \frac{1}{\alpha\rho} u^2\, d\sigma \\
&\geq C_1(\|u\|_0^2 + \|\underline{v}\|_0^2). \qquad (1.111)
\end{aligned}
$$

   where $C_1 = \min(\|\kappa\|_\infty^{-1}, \rho_0)$.
2. By taking $(\tilde{u}, \tilde{\underline{v}}) = -(\kappa\nabla \cdot \underline{v}, \rho^{-1}\nabla u) \in H$, we immediately obtain

$$b((u,\underline{v}),(\tilde{u},\tilde{\underline{v}})) = \int_{\Gamma_{ABC}} \frac{1}{\alpha\rho} u^2\, d\sigma + \|\rho^{-1/2}\nabla u\|_0^2 + \|\kappa^{1/2}\nabla\cdot\underline{v}\|_0^2$$
$$\geq C_2(\|\nabla u\|_0^2 + \|\nabla\cdot\underline{v}\|_0^2), \tag{1.112}$$

where $C_2 = \min(\kappa_0, \|\rho\|_\infty^{-1})$.

3. Now, by taking $(\tilde{u},\tilde{\underline{v}}) = (u,\underline{v}) - (\kappa\nabla\cdot\underline{v}, \rho^{-1}\nabla u) \in H$, (1.111) and (1.112) lead to

$$b((u,\underline{v}),(\tilde{u},\tilde{\underline{v}})) \geq C\,\underbrace{(\|u\|_0^2 + \|\underline{v}\|_0^2 + \|\nabla u\|_0^2 + \|\nabla\cdot\underline{v}\|_0^2)}_{=:\|(u,\underline{v})\|^2_{D(A)}} \tag{1.113}$$

where $C = \min(C_1, C_2)$.

Finally, it is easy to see that $\|(\tilde{u},\tilde{\underline{v}})\|_H \leq \tilde{C}\|(u,\underline{v})\|_{D(A)}$ and (1.113) implies (1.109).

Now, we establish property (1.110). Since $D(A)$ contains the space of C^∞ compactly supported functions in Ω, we know that $D(A)$ is dense in H. Let us take $(\tilde{u},\tilde{\underline{v}})$ satisfying (1.110). There exists $(\tilde{u}_n,\tilde{\underline{v}}_n)_{n\in\mathbb{N}} \subset D(A)$ such that $\lim\limits_{n\to+\infty}(\tilde{u}_n,\tilde{\underline{v}}_n) = (\tilde{u},\tilde{\underline{v}})$ in H. By using (1.111), we can write

$$\|\kappa^{-1/2}\tilde{u}_n\|_0^2 + \|\rho^{1/2}\tilde{\underline{v}}_n\|_0^2 + \int_{\Gamma_{ABC}} \frac{1}{\alpha\rho}\tilde{u}_n^2\, d\sigma = b((\tilde{u}_n,\tilde{\underline{v}}_n),(\tilde{u}_n,\tilde{\underline{v}}_n))$$
$$\leq \sup_{(u,\underline{v})\in D(A)} b((u,\underline{v}),(\tilde{u}_n,\tilde{\underline{v}}_n)). \tag{1.114}$$

Finally, since $b((u,\underline{v}),(\tilde{u}_n,\tilde{\underline{v}}_n)) \underset{n\to+\infty}{\to} b((u,\underline{v}),(\tilde{u},\tilde{\underline{v}})) = 0$, $\forall (u,\underline{v}) \in D(A)$, then (1.114) leads to the result i.e. $(\tilde{u},\tilde{\underline{v}}) = (0\,,\,0)$.

Now, the Hille–Yosida theorem can be applied: if $F \in C^1(\mathbb{R}^+; L^2(\Omega))$ and $(u_0,\underline{v}_0) \in D(A)$, then the problem (1.102a)–(1.102g) has one and only one solution

$$(u,\underline{v}) \in C^1(\mathbb{R}^+; L^2(\Omega)^{d+1}) \cap C^0(\mathbb{R}^+; D(A)). \tag{1.115}$$

In particular,

1. $u \in C^1(\mathbb{R}^+; L^2(\Omega)) \cap C^0(\mathbb{R}^+; H^1(\Omega))$,

2. $\underline{v} \in C^1(\mathbb{R}^+; L^2(\Omega)^d) \cap C^0(\mathbb{R}^+; H(div,\Omega))$.

1.2.4.2 Maxwell's Equations

Let us consider the following 3D Maxwell's equations

$$\underline{\underline{\varepsilon}}\frac{\partial \underline{E}}{\partial t} - \nabla \times \underline{H} = -\underline{J},\ \underline{x} \in \Omega,\ t > 0, \tag{1.116a}$$

$$\underline{\underline{\mu}}\frac{\partial \underline{H}}{\partial t} + \nabla \times \underline{E} = 0,\ \underline{x} \in \Omega,\ t > 0, \tag{1.116b}$$

$$\underline{E} \times \underline{n} = 0,\ \underline{x} \in \Gamma_0,\ t > 0, \tag{1.116c}$$

$$\underline{n} \times (\underline{E} \times \underline{n}) + \mathscr{Z}\,(\underline{H} \times \underline{n}) = 0,\ \underline{x} \in \Gamma_{SM},\ t > 0, \tag{1.116d}$$

$$\underline{E}(\underline{x}, 0) = \underline{E}_0(\underline{x}),\ \underline{x} \in \Omega, \tag{1.116e}$$

$$\underline{H}(\underline{x}, 0) = \underline{H}_0(\underline{x}),\ \underline{x} \in \Omega, \tag{1.116f}$$

where $\Gamma = \Gamma_0 \cup \Gamma_{SM}$, $\underline{\underline{\varepsilon}}$, $\underline{\underline{\mu}} \in L^\infty(\Omega, \mathbb{R}^{3\times 3})$ such that it exists $\varepsilon^* > 0$ and $\mu^* > 0$, $\forall U \in \mathbb{R}^3$, $U^T \underline{\underline{\varepsilon}}\, U \geq \varepsilon^* \|U\|^2$ and $U^T \underline{\underline{\mu}}\, U \geq \mu^* \|U\|^2$ almost everywhere and $\mathscr{Z} \in L^\infty(\Gamma_{SM})$ such that $\mathscr{Z} \geq 0$ almost everywhere.

This problem is written under the Hille–Yosida theorem framework as follows:

1. $H = [L^2(\Omega)]^6$ and is equipped with the scalar product

$$((\underline{u}, \underline{v}), (\tilde{\underline{u}}, \tilde{\underline{v}})) = \left(\underline{\underline{\varepsilon}}\,\underline{u}, \tilde{\underline{u}}\right)_0 + \left(\underline{\underline{\mu}}\,\underline{v}, \tilde{\underline{v}}\right)_0. \tag{1.117}$$

2. Operator A is defined by

$$A = \begin{bmatrix} 0 & -\underline{\underline{\varepsilon}}^{-1}\nabla\times \\ \underline{\underline{\mu}}^{-1}\nabla\times & 0 \end{bmatrix}. \tag{1.118}$$

3. The domain of A relative to the problem (1.116) is

$$D(A) = \{(\underline{u}\,,\ \underline{v}) \in H(\mathbf{curl}, \Omega) \times H(\mathbf{curl}, \Omega)\ :\ \underline{u} \times \underline{n} = 0 \text{ on } \Gamma_0, \\ \text{and } \underline{n} \times (\underline{u} \times \underline{n}) + \mathscr{Z}\,(\underline{v} \times \underline{n}) = 0 \text{ on } \Gamma_{SM}\}. \tag{1.119}$$

We now verify that operator A is maximal monotone.

First, $\forall(\underline{u}, \underline{v}) \in D(A)$, we have

$$\begin{aligned}(A(\underline{u}, \underline{v})\,,\ (\underline{u}, \underline{v})) &= -\int_\Omega \nabla \times \underline{v} \cdot \underline{u}\, dx + \int_\Omega \nabla \times \underline{u} \cdot \underline{v}\, dx \\ &= -\int_\Gamma (\underline{v} \times \underline{n}) \cdot (\underline{n} \times (\underline{u} \times \underline{n}))\, d\sigma \\ &= \int_{\Gamma_{SM}} \mathscr{Z}\,(\underline{v} \times \underline{n}) \cdot (\underline{v} \times \underline{n})\, d\sigma \geq 0. \end{aligned} \tag{1.120}$$

Then, we prove that the operator $I + A$ is surjective. For that, we introduce the bilinear form

$$b : D(A) \times H \to \mathbb{R}, ((u, \underline{v}), (\tilde{u}, \tilde{\underline{v}})) \mapsto b((u, \underline{v}), (\tilde{u}, \tilde{\underline{v}})) = ((I + A)(u, \underline{v}) \,,\, (\tilde{u}, \tilde{\underline{v}})) \qquad (1.121)$$

and we are going to prove that $\forall (f, \underline{g}) \in H, \exists! (u, \underline{v}) \in D(A)$ such that

$$b((u, \underline{v}), (\tilde{u}, \tilde{\underline{v}})) = ((f, \underline{g}) \,,\, (\tilde{u}, \tilde{\underline{v}})), \; \forall (\tilde{u}, \tilde{\underline{v}}) \in H. \qquad (1.122)$$

We proceed as for the acoustic case and we establish the conditions (1.109) and (1.110).

We establish (1.109) in three steps

1. By taking $(\tilde{\underline{u}}, \tilde{\underline{v}}) = (\underline{u}, \underline{v}) \in D(A) \subset H$, we have by using (1.120)

$$b((\underline{u}, \underline{v}), (\underline{u}, \underline{v})) = \|\underline{\underline{\varepsilon}}^{1/2}\underline{u}\|_0^2 + \|\underline{\underline{\mu}}^{1/2}\underline{v}\|_0^2 + \int_{\Gamma_{SM}} \mathscr{Z}(\underline{v} \times \underline{n}) \cdot (\underline{v} \times \underline{n})\, d\sigma$$
$$\geq C_1(\|\underline{u}\|_0^2 + \|\underline{v}\|_0^2), \qquad (1.123)$$

where $C_1 = \min(\varepsilon^*, \mu^*)$.

2. By taking $(\tilde{\underline{u}}, \tilde{\underline{v}}) = (-\underline{\underline{\varepsilon}}^{-1}\nabla \times \underline{v}, \underline{\underline{\mu}}^{-1}\nabla \underline{u}) \in H$, we immediately obtain

$$b((\underline{u}, \underline{v}), (\tilde{\underline{u}}, \tilde{\underline{v}})) = \int_{\Gamma_{SM}} \mathscr{Z}(\underline{v} \times \underline{n}) \cdot (\underline{v} \times \underline{n})\, d\sigma + \|\underline{\underline{\mu}}^{-1/2}\nabla \times \underline{u}\|_0^2 + \|\underline{\underline{\varepsilon}}^{-1/2}\nabla \times \underline{v}\|_0^2$$
$$\geq C_2(\|\nabla \times \underline{u}\|_0^2 + \|\nabla \times \underline{v}\|_0^2), \qquad (1.124)$$

where $C_2 = \min(\|\underline{\underline{\varepsilon}}\|_\infty^{-1}, \|\underline{\underline{\mu}}\|_\infty^{-1})$.

3. Now, by taking $(\tilde{\underline{u}}, \tilde{\underline{v}}) = (\underline{u}, \underline{v}) + (-\underline{\underline{\varepsilon}}^{-1}\nabla \times \underline{v}, \underline{\underline{\mu}}^{-1}\nabla \underline{u}) \in H$, (1.123) and (1.124) lead to

$$b((\underline{u}, \underline{v}), (\tilde{\underline{u}}, \tilde{\underline{v}})) \geq C(\|\underline{u}\|_0^2 + \|\underline{v}\|_0^2 + \|\nabla \times \underline{u}\|_0^2 + \|\nabla \times \underline{v}\|_0^2)$$
$$\geq C\|(\underline{u}, \underline{v})\|_{D(A)}^2, \qquad (1.125)$$

where $C = \min(C_1, C_2)$.

Finally, it is easy to see that $\|(\tilde{\underline{u}}, \tilde{\underline{v}})\|_H \leq \tilde{C}\|(\underline{u}, \underline{v})\|_{D(A)}$ and (1.125) implies (1.109).

Property (1.110) is derived by using exactly the same arguments as those used for the acoustic case.

Now, the Hille–Yosida theorem can be applied: if $\underline{J} \in C^1(\mathbb{R}^+; L^2(\Omega)^3)$ and $(\underline{E}_0, \underline{H}_0) \in D(A)$, then problem (1.116) has one and only one solution

$$(\underline{E}, \underline{H}) \in C^1(\mathbb{R}^+; L^2(\Omega)^6) \cap C^0(\mathbb{R}^+; D(A)). \tag{1.126}$$

In particular,

$$\underline{E}, \underline{H} \in C^1(\mathbb{R}^+; L^2(\Omega)^3) \cap C^0(\mathbb{R}^+; H(\mathbf{curl}, \Omega)). \tag{1.127}$$

1.2.4.3 The Linear Elastodynamics System

In order to apply the Hille–Yosida theorem, we first rewrite the system (1.25a) and (1.25b)

$$\frac{\partial \underline{\underline{\sigma}}}{\partial t} - \underline{\underline{C}}\,\underline{\underline{\varepsilon}}(\underline{v}) = 0, \ \underline{x} \in \Omega, \ t > 0, \tag{1.128a}$$

$$\rho \frac{\partial \underline{v}}{\partial t} - \mathbf{div}\underline{\underline{\sigma}} = \underline{f}, \ \underline{x} \in \Omega, \ t > 0, \tag{1.128b}$$

$$\underline{v} = 0, \ \underline{x} \in \Gamma_D, \ t > 0, \tag{1.128c}$$

$$\underline{\underline{\sigma}}\,\underline{n} = 0, \ \underline{x} \in \Gamma_T, \ t > 0, \tag{1.128d}$$

$$\underline{\underline{\sigma}}(\underline{x}, 0) = \underline{\underline{\sigma}}_0(\underline{x}), \ \underline{x} \in \Omega, \tag{1.128e}$$

$$\underline{v}(\underline{x}, 0) = \underline{v}_0(\underline{x}), \ \underline{x} \in \Omega, \tag{1.128f}$$

where $\Gamma = \Gamma_D \cup \Gamma_T$, $(\underline{\underline{\sigma}}\,\underline{n})_j = \sigma_{jk}\, n_k$, $\underline{\underline{C}} \in L^\infty(\Omega)$ such that it exists $C_0 > 0$, $\forall \underline{\underline{\xi}}$ is a symmetric tensor of order 2, $\xi_{ij}\, C_{ijkl}\xi_{kl} \geq C_0 \xi_{kl}\xi_{kl}$, a.e. and $\rho \in L^\infty(\Omega)$ such that $\rho(\underline{x}) \geq \rho_0 > 0$, a.e.

In the following, we assume that the measure of boundary Γ_D is not equal to 0. An important consequence is that the norms $\|\cdot\|_1$ and $\underline{u} \to \|\underline{\underline{\varepsilon}}(\underline{u})\|_0$ are equivalent on space $\tilde{H}_0^1(\Omega)^3 = \{\underline{u} \in H(\Omega)^3 : \underline{u} = 0 \text{ on } \Gamma_D\}$. This result comes from Korn's inequality [17]: $\exists\, C_\Omega > 0$ such that $\forall \underline{u} \in \tilde{H}_0^1(\Omega)^3$,

$$\|\underline{u}\|_1 \leq C_\Omega \, \|\underline{\underline{\varepsilon}}(\underline{u})\|_0. \tag{1.129}$$

This problem is written under the Hille–Yosida theorem framework as follows:

1. $H = L^2(\Omega; \mathbb{M}) \times L^2(\Omega)^3$ where $\mathbb{M}$ is the space of all $d \times d$ symmetric matrices. This space is equipped with the scalar product

$$((\underline{\underline{u}}, \underline{v}), (\tilde{\underline{\underline{u}}}, \tilde{\underline{v}})) = \left(\underline{\underline{C}}^{-1}\, \underline{\underline{u}}, \tilde{\underline{\underline{u}}}\right)_0 + (\rho \underline{v}, \tilde{\underline{v}})_0 \tag{1.130}$$

with $\left(\underline{\underline{u}}, \tilde{\underline{\underline{u}}}\right)_0 = \sum_{i,j=1}^{d} (u_{ij}\,,\ \tilde{u}_{ij})_0$.

2. Operator A is defined by

$$A = \begin{bmatrix} 0 & -\underline{\underline{C}}\,\underline{\underline{\varepsilon}} \\ -\rho^{-1}\,\mathbf{div} & 0 \end{bmatrix}. \tag{1.131}$$

3. The domain of A relative to the problem (1.128) is

$$D(A) = \{(\underline{\underline{u}},\ \underline{v}) \in H(\mathbf{div}, \Omega; \mathbb{M}) \times H^1(\Omega)^3 : \underline{v} = 0 \text{ on } \Gamma_D, \text{ and } \underline{\underline{u}}\,\underline{n} = 0 \text{ on } \Gamma_T\}, \tag{1.132}$$

where $H(\mathbf{div}, \Omega; \mathbb{M}) = \{\underline{\underline{u}} \in L^2(\Omega; \mathbb{M})\ :\ \mathbf{div}\,\underline{\underline{u}} \in L^2(\Omega)^3\}$.

We now verify that the operator A is maximal monotone.
First, $\forall(\underline{\underline{u}}, \underline{v}) \in D(A)$, we have

$$\begin{aligned}(A(\underline{\underline{u}}, \underline{v})\,,\ (\underline{\underline{u}}, \underline{v})) &= -\int_\Omega \underline{\underline{\varepsilon}}(\underline{v})\ :\ \underline{\underline{u}}\,dx - \int_\Omega \mathbf{div}\underline{\underline{u}} \cdot \underline{v}\,dx \\ &= -\int_\Omega \nabla\underline{v}\ :\ \underline{\underline{u}}\,dx - \int_\Omega \mathbf{div}\underline{\underline{u}} \cdot \underline{v}\,dx \\ &= \int_\Gamma \underline{\underline{u}}\,\underline{n} \cdot \underline{v}\,d\sigma = 0, \end{aligned} \tag{1.133}$$

where the second equation of (1.133) comes from the symmetry of the tensor $\underline{\underline{u}}$.

Then, we prove that operator $I + A$ is surjective. For that, we introduce the bilinear form

$$b:\ D(A) \times H \to \mathbb{R}, ((u, \underline{v}), (\tilde{u}, \tilde{\underline{v}})) \mapsto b((u, \underline{v}), (\tilde{u}, \tilde{\underline{v}})) = ((I + A)(u, \underline{v})\,,\ (\tilde{u}, \tilde{\underline{v}})) \tag{1.134}$$

and we are going to prove that $\forall(f, \underline{g}) \in H$, $\exists!(u, \underline{v}) \in D(A)$ such that

$$b((u, \underline{v}), (\tilde{u}, \tilde{\underline{v}})) = ((f, \underline{g})\,,\ (\tilde{u}, \tilde{\underline{v}})),\ \forall(\tilde{u}, \tilde{\underline{v}}) \in H. \tag{1.135}$$

We now establish (1.109) in three steps:

1. By taking $(\tilde{\underline{\underline{u}}}, \tilde{\underline{v}}) = (\underline{\underline{u}}, \underline{v}) \in D(A) \subset H$, we have by using (1.133),

$$b((\underline{\underline{u}}, \underline{v}), (\underline{\underline{u}}, \underline{v})) = \|\underline{\underline{C}}^{-1/2}\underline{\underline{u}}\|_0^2 + \|\rho^{1/2}\underline{v}\|_0^2 \geq C_1(\|\underline{\underline{u}}\|_0^2 + \|\underline{v}\|_0^2), \tag{1.136}$$

where $C_1 = \min(\|\underline{\underline{C}}\|_\infty^{-1}, \rho_0)$.

2. By taking $(\tilde{\underline{\underline{u}}}, \tilde{\underline{v}}) = (-\underline{\underline{C}}\,\underline{\underline{\varepsilon}}(\underline{v}), -\rho^{-1}\mathbf{div}\underline{\underline{u}}) \in H$, we immediately obtain

$$b((\underline{u}, \underline{v}), (\tilde{\underline{u}}, \tilde{\underline{v}})) = \|\underline{\underline{C}}^{1/2}\underline{\underline{\varepsilon}}(\underline{v})\|_0^2 + \|\rho^{1/2}\mathbf{div}\underline{u}\|_0^2$$
$$\geq C_2(\|\underline{v}\|_1^2 + \|\mathbf{div}\underline{u}\|_0^2), \tag{1.137}$$

where $C_2 = \min(C_0/C_\Omega^2, \|\rho\|_\infty^{-1})$.

3. Now, by taking $(\tilde{\underline{u}}, \tilde{\underline{v}}) = (\underline{u}, \underline{v}) + (-\underline{\underline{C}}\underline{\underline{\varepsilon}}(\underline{v}), -\rho^{-1}\mathbf{div}\underline{u}) \in H$, (1.136) and (1.138) lead to

$$b((\underline{u}, \underline{v}), (\tilde{\underline{u}}, \tilde{\underline{v}})) \geq C(\|\underline{u}\|_0^2 + \|\underline{v}\|_0^2 + \|\mathbf{div}\underline{u}\|_0^2 + \|\underline{v}\|_1^2)$$
$$\geq C\|(u, \underline{v})\|_{D(A)}^2, \tag{1.138}$$

where $C = \min(C_1, C_2)$.

Finally, it is easy to see that $\|(\tilde{\underline{u}}, \tilde{\underline{v}})\|_H \leq \tilde{C}\|(\underline{u}, \underline{v})\|_{D(A)}$ and (1.138) implies (1.109).

Property (1.110) is derived by using exactly the same arguments as those used for the acoustic case.

Now, the Hille–Yosida theorem can be applied: if $\underline{f} \in C^1(\mathbb{R}^+; L^2(\Omega)^3)$ and $(\underline{\underline{\sigma}}_0, \underline{v}_0) \in D(A)$, then the problem (1.128) has one and only one solution.

$$(\underline{\underline{\sigma}}, \underline{u}) \in C^1(\mathbb{R}^+; L^2(\Omega; \mathbb{M}) \times L^2(\Omega)^3) \cap C^0(\mathbb{R}^+; D(A)). \tag{1.139}$$

In particular,

1. $\underline{u} \in C^1(\mathbb{R}^+; L^2(\Omega)^3) \cap C^0(\mathbb{R}^+; \tilde{H}_0^1(\Omega))$,
2. $\underline{\underline{\sigma}} \in C^1(\mathbb{R}^+; L^2(\Omega; \mathbb{M})) \cap C^0(\mathbb{R}^+; H(\mathbf{div}, \Omega; \mathbb{M}))$.

1.3 Plane Wave Solutions

1.3.1 A General Solution of the Homogeneous Wave Equation

Let us consider the homogeneous wave equation in $\mathbb{R}^d$[11]:

---

[11] With a right-hand side, the correct equation would actually be

$$\frac{1}{c^2}\frac{\partial^2 u}{\partial t^2} - \Delta u = f$$

but the formulation given below is equivalent and easier to manipulate when $f = 0$.

$$\frac{\partial^2 u}{\partial t^2} - c^2 \Delta u = 0 \tag{1.140}$$

and the direct and inverse *Fourier transforms* in space

$$\mathscr{F}_x u = \hat{u}(\underline{k}) = \frac{1}{(2\pi)^{d/2}} \int_{\mathbb{R}^d} u(\underline{x}) \mathrm{e}^{-\mathrm{i}\underline{k}\cdot\underline{x}} \mathrm{d}\underline{x}, \tag{1.141a}$$

$$\mathscr{F}_x^{-1} \hat{u} = u(\underline{x}) = \frac{1}{(2\pi)^{d/2}} \int_{\mathbb{R}^d} \hat{u}(\underline{k}) \mathrm{e}^{\mathrm{i}\underline{k}\cdot\underline{x}} \mathrm{d}\underline{k}, \tag{1.141b}$$

where $\mathrm{i}^2 = -1$, $\underline{x} \in \mathbb{R}^d$ and $\underline{k} \in \mathbb{R}^d$. By applying the Fourier transform in space to (1.140), we obtain the following ODE:

$$\frac{\mathrm{d}^2 \hat{u}}{\mathrm{d}t^2} + c^2 |\underline{k}|^2 \hat{u} = 0, \tag{1.142}$$

whose solution is of the form

$$\hat{u}(\underline{k}, t) = A(\underline{k}) \mathrm{e}^{\mathrm{i}c|\underline{k}|t} + B(\underline{k}) \mathrm{e}^{-\mathrm{i}c|\underline{k}|t}. \tag{1.143}$$

On the other hand, by applying to (1.142) the Fourier transform in time $\mathscr{F}_t : L^2(\mathbb{R}) \to L^2(\mathbb{R})$ where

$$\mathscr{F}_t u = \hat{u}(\omega) = \frac{1}{\sqrt{2\pi}} \int_{\mathbb{R}} u(t) \mathrm{e}^{-\mathrm{i}\omega t} \mathrm{d}t, \tag{1.144}$$

we obtain the *dispersion relation* of (1.140)

$$\omega^2 = c^2 |\underline{k}|^2. \tag{1.145}$$

Now, by assuming $\omega > 0$, the solution can be written as

$$\hat{u}(\underline{k}, t) = A(\underline{k}) \mathrm{e}^{\mathrm{i}\omega t} + B(\underline{k}) \mathrm{e}^{-\mathrm{i}\omega t}. \tag{1.146}$$

The inverse Fourier transform in space applied to $\hat{u}$ provides the following form of the solution of (1.142):

$$u(\underline{x}, t) = \frac{1}{(2\pi)^{d/2}} \int_{\mathbb{R}^d} \left[A(\underline{k}) \mathrm{e}^{\mathrm{i}(\omega t + \underline{k}\cdot\underline{x})} + B(\underline{k}) \mathrm{e}^{\mathrm{i}(-\omega t + \underline{k}\cdot\underline{x})} \right] \mathrm{d}\underline{k}. \tag{1.147}$$

Equation (1.147) shows that the solution of the homogeneous wave equation can be expressed as a continuous superposition of the plane waves

$$\mathrm{e}^{\mathrm{i}(\omega t + \underline{k}\cdot\underline{x})}, \tag{1.148}$$

whose amplitudes are $A(\underline{k})$ and $B(\underline{k})$. Hence the study of properties of the solutions of the wave equation can be carried out by considering the plane wave solution defined in (1.148).

Remarks:

1. ω is the *pulsation* and $\underline{k}$ the *wave vector* which indicates the direction of propagation of the plane wave. Obviously, $\omega/|\underline{k}|$ is the velocity of the propagated wave.

2. This result given in the case of the scalar wave equation, can be extended to the other equations.

1.3.2 Application to Maxwell's Equations

1.3.2.1 The 3D Case

Now, let us look for a plane wave solution of the homogeneous anisotropic Maxwell's equations

$$\frac{\partial \underline{D}}{\partial t} - \nabla \times \underline{H} = 0, \tag{1.149a}$$

$$\frac{\partial \underline{B}}{\partial t} + \nabla \times \underline{E} = 0, \tag{1.149b}$$

$$\underline{D} = \underline{\underline{\varepsilon_0}}\underline{E} \tag{1.149c}$$

$$\underline{B} = \underline{\underline{\mu_0}}\underline{H}, \tag{1.149d}$$

of the form

$$\underline{E} = \underline{E}_0 \mathrm{e}^{\mathrm{i}(\omega t + \underline{k}\cdot\underline{x})}, \tag{1.150a}$$

$$\underline{H} = \underline{H}_0 \mathrm{e}^{\mathrm{i}(\omega t + \underline{k}\cdot\underline{x})}, \tag{1.150b}$$

$$\underline{D} = \underline{D}_0 \mathrm{e}^{\mathrm{i}(\omega t + \underline{k}\cdot\underline{x})}, \tag{1.150c}$$

$$\underline{B} = \underline{B}_0 \mathrm{e}^{\mathrm{i}(\omega t + \underline{k}\cdot\underline{x})}. \tag{1.150d}$$

By inserting (1.150a)–(1.150d) into (1.149a)–(1.149d), we obtain the following relations:

$$\omega \underline{D}_0 - \underline{k} \times \underline{H}_0 = 0, \tag{1.151a}$$

$$\omega \underline{B}_0 + \underline{k} \times \underline{E}_0 = 0, \tag{1.151b}$$

$$\underline{D}_0 = \underline{\underline{\varepsilon_0}}\,\underline{E}_0, \tag{1.151c}$$

$$\underline{B}_0 = \underline{\underline{\mu_0}}\,\underline{H}_0. \tag{1.151d}$$

If $\omega \neq 0$, (1.151a) and (1.151b) show that $\underline{D}$ is orthogonal to $\underline{H}$ and $\underline{k}$ and $\underline{B}$ is orthogonal to $\underline{E}$ and $\underline{k}$. The case $\omega = 0$ will be discussed below.

On the other hand, since $\underline{\underline{\varepsilon_0}}$ and $\underline{\underline{\mu_0}}$ are symmetric positive definite, we can write $\underline{\underline{\varepsilon_0}} = \underline{\underline{\widetilde{\varepsilon}_0^*}}\,\underline{\underline{\widetilde{\varepsilon}_0}}$ and $\underline{\underline{\mu_0}} = \underline{\underline{\widetilde{\mu}_0^*}}\,\underline{\underline{\widetilde{\mu}_0}}$. By multiplying (1.151c) and (1.151d) by $\underline{H}_0$ and $\underline{E}_0$ respectively, we obtain, thanks to these decompositions

$$\underline{\underline{\widetilde{\varepsilon}_0}}\,\underline{E}_0 \cdot \underline{\underline{\widetilde{\varepsilon}_0}}\,\underline{H}_0 = 0, \tag{1.152a}$$
$$\underline{\underline{\widetilde{\mu}_0}}\,\underline{E}_0 \cdot \underline{\underline{\widetilde{\mu}_0}}\,\underline{H}_0 = 0. \tag{1.152b}$$

A similar process provides

$$\underline{\underline{\widetilde{\varepsilon}_0}}\,\underline{E}_0 \cdot \underline{\underline{\widetilde{\varepsilon}_0}}\,\underline{k} = 0, \tag{1.153a}$$
$$\underline{\underline{\widetilde{\mu}_0}}\,\underline{H}_0 \cdot \underline{\underline{\widetilde{\mu}_0}}\,\underline{k} = 0, \tag{1.153b}$$

which shows that the vector $\underline{\underline{\widetilde{\varepsilon}_0}}\,\underline{E}$ is orthogonal to $\underline{\underline{\widetilde{\varepsilon}_0}}\,\underline{H}$ and $\underline{\underline{\widetilde{\varepsilon}_0}}\,\underline{k}$ and the vector $\underline{\underline{\widetilde{\mu}_0}}\,\underline{H}$ is orthogonal to $\underline{\underline{\widetilde{\mu}_0}}\,\underline{k}$ and $\underline{\underline{\widetilde{\mu}_0}}\,\underline{E}$.

Now, by eliminating $\underline{E}_0$, $\underline{D}_0$ and $\underline{B}_0$ in (1.151a)–(1.151d), we obtain the following dispersion relation:

$$\omega^2 \underline{H}_0 = -\underline{\underline{\mu_0}}^{-1}\left(\underline{k} \times \left(\underline{\underline{\varepsilon_0}}^{-1}\left(\underline{k} \times \underline{H}_0\right)\right)\right). \tag{1.154}$$

In other words, there are three velocities which are the square roots of the eigenvalues of matrix M_0 defined by

$$M_0 \underline{H}_0 = -\frac{1}{|\underline{k}|^2}\underline{\underline{\mu_0}}^{-1}\left(\underline{k} \times \left(\underline{\underline{\varepsilon_0}}^{-1}\left(\underline{k} \times \underline{H}_0\right)\right)\right). \tag{1.155}$$

Obviously, 0 is an eigenvalue of M_0. The corresponding eigenvector is collinear to $\underline{k}$. So $\omega = 0$ provides a stationary mode for which the field $\underline{H}$ is parallel to $\underline{k}$. This stationary solution can be written as the gradient of a scalar potential $\underline{\psi}$. The other two modes lead to two dispersion relations which correspond to two waves with different polarizations and different velocities. As we shall see later, this zero eigenvalue will be troublesome for the approximation of Maxwell's equations.

In the isotropic case, one can easily check that the electric and magnetic fields are orthogonal and are both orthogonal to $\underline{k}$. The stationary mode still exists but M_0 has a double eigenvalue. Actually, when $\omega \neq 0$, the problem reads

$$\omega^2 \underline{H}_0 = \frac{1}{\varepsilon_0 \mu_0} |\underline{k}|^2 \underline{H}_0. \tag{1.156}$$

Equation (1.156) leads to the following dispersion relation

$$\varepsilon_0 \mu_0 \omega^2 = |\underline{k}|^2, \tag{1.157}$$

which shows that the velocity c of the waves is equal to $1/\sqrt{\varepsilon_0 \mu_0}$.

Remark: A similar development can be carried out for the electric field. In the isotropic case the same dispersion relation is obtained.

1.3.3 The 2D Case

In the 2D case, we present the TM polarization defined in (1.17a) and (1.17b) but an equivalent study can be carried out for the TE polarization.

$$\omega \underline{D}_0 - \underline{k} \times \underline{H}_0 = 0, \tag{1.158a}$$

$$\omega \underline{B}_0 + \underline{k} \times \underline{E}_0 = 0, \tag{1.158b}$$

$$\underline{D}_0 = \underline{\underline{\varepsilon_0}} \underline{E}_0, \tag{1.158c}$$

$$\underline{B}_0 = \underline{\underline{\mu_0}} \underline{H}_0. \tag{1.158d}$$

So, let us consider the 2D Maxwell's equations in a homogeneous anisotropic medium

$$\underline{\underline{\varepsilon_0}} \frac{\partial \underline{E}}{\partial t} - \mathbf{curl} H = 0, \tag{1.159a}$$

$$\mu_0 \frac{\partial H}{\partial t} + \mathrm{curl} \underline{E} = 0, \tag{1.159b}$$

for which we are looking for a plane wave solution derived from (1.150a)–(1.150d).

Let us set:

$$\underline{\underline{\varepsilon_0}}^{-1} = \begin{pmatrix} \alpha & \beta \\ \beta & \gamma \end{pmatrix}. \tag{1.160}$$

After inserting this solution into (1.159a) and (1.159b) and eliminating $\underline{E}_0$, we obtain the following dispersion relation

$$\mu_0 \omega^2 = \gamma k_1^2 + \alpha k_2^2 - 2\beta k_1 k_2, \tag{1.161}$$

Fig. 1.1 An example of an anisotropy curve of the velocity for the 2D Maxwell's system

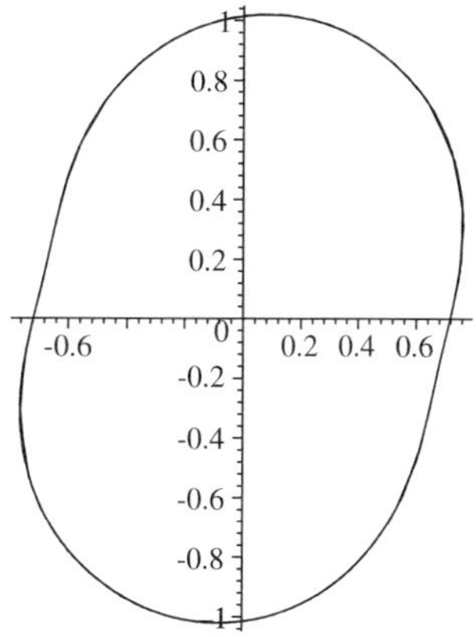

where k_1 and k_2 are the components of the vector $\underline{k}$.

Now, if we set $k_1 = |\underline{k}| \cos\theta$ and $k_2 = |\underline{k}| \sin\theta$, we obtain[12]

$$c = \frac{\omega}{|\underline{k}|} = \sqrt{\frac{1}{\mu_0}\left(\alpha \sin^2\theta + \gamma \cos^2\theta - 2\beta \sin\theta\cos\theta\right)}, \tag{1.162}$$

which provides an equation in polar coordinates of the velocity c. This equation indicates the anisotropy of c. For instance, when $\mu_0 = 1, \alpha = 32/31, \beta = -4/31$ and $\gamma = 16/31$ (which provides a matrix whose inverse is a positive definite matrix A), we obtain the curve drawn in Fig. 1.1.

1.3.4 Application to the Isotropic Linear Elastodynamics System

We are looking for a plane wave solution of (1.30) when Lamé's coefficients are constant. The plane wave solution is

$$\underline{v} = \underline{v}_0 \mathrm{e}^{\mathrm{i}(\omega t + \underline{k}\cdot\underline{x})}. \tag{1.163}$$

By inserting (1.163) into (1.30), we obtain the dispersion relation

$$\left(\mu|\underline{k}|^2 - s\right)\underline{v}_0 + (\lambda + \mu)(\underline{k}\cdot\underline{v}_0)\underline{k} = 0, \tag{1.164}$$

---

[12] The positivity of the expression under the square root comes from the positive character of the matrix $\underline{\underline{\varepsilon_0}}^{-1}$.

where $s = \rho\omega^2$.

Equation (1.164) has two classes of solutions:

1. $\underline{v}_0 \parallel \underline{k} \Rightarrow s = (\lambda + 2\mu)|\underline{k}|^2$.

2. $\underline{v}_0 \perp \underline{k} \Rightarrow s = \mu|\underline{k}|^2$.

These two classes provide two kinds of waves with two different dispersion relations:

1. Waves parallel to the direction of propagation whose dispersion relation is

$$\omega_P^2 = \frac{\lambda + 2\mu}{\rho}|\underline{k}|^2. \tag{1.165}$$

2. Waves perpendicular to the direction of propagation whose dispersion relation is

$$\omega_S^2 = \frac{\mu}{\rho}|\underline{k}|^2. \tag{1.166}$$

So, by setting $V_P = \sqrt{(\lambda + 2\mu)/\rho}$ and $V_S = \sqrt{\mu/\rho}$, we obtain the P-wave and S-wave defined in (1.34) and (1.35).

References

1. Bendali, A., Halpern, L.: Conditions aux limites absorbantes pour le système de Maxwell dans le vide en dimension trois d'espace. C. R. Acad. Sci. Paris Ser. I, Math. **307**(20), 1011–1013 (1988)
2. Achenbach, J.D.: Wave Propagation in Elastic Solids. North-Holland, Amsterdam (1984)
3. Dautray, R., Lions, J.-L.: Mathematical Analysis and Numerical Methods for Science and Technology, vol. 5. Springer, Berlin (1990)
4. Leis, R.: Initial Boundary Value Problems in Mathematical Physics. Wiley, New York (1988)
5. Lions, J.-L., Magenes, E.: Problèmes aux limites non homogènes et applications, vol. 1. Dunod, Paris (1968)
6. Taylor, M.E.: Partial Differential Equations, vol. 1–3. Springer, Berlin (1990)
7. Eringen, C.A., Suhubi, E.S.: Elastodynamics, vol. 1 and 2. Academic Press, New York (1975)
8. Feynman, R.P., Leighton, R.B., Sand, M.: The Feynman Lectures on Physics. Addison-Wesley, Reading (1963)
9. Lamb, H.: Hydrodynamics. Cambridge University Press, Cambridge (1974)
10. Auld, B.A.: Acoustic Fields and Waves in Solids, vol. 2. R. E. Krieger, Malabar (1990)
11. Dautray, R., Lions, J.-L.: Mathematical Analysis and Numerical Methods for Science and Technology, vol. 2. Springer, Berlin (1988)
12. Rudin, W.: Functional Analysis. McGraw-Hill, New York (1991)
13. Schwartz, L.: Théorie des distributions, Publications de l'Institut de Mathématique de l'Université de Strasbourg, no. IX–X, new edition. Hermann, Paris (1966)

14. Buffa, A., Costabel, M., Sheen, D.: On traces for $H(\mathbf{curl}, \boldsymbol{\Omega})$ in Lipschitz domains. J. Math. Anal. Appl. **276**, 845–867 (2002)
15. Brezis, H.: Functional Analysis, Sobolev Spaces and PDEs. Springer, New York (2010)
16. Ern, A., Guermond, J.-L.: Theory and Practice of Finite Elements. Applied Mathematical Series, vol. 159. Springer, New York (2004)
17. Ciarlet, P.G.: Mathematical Elasticity, I. Three-Dimensional Elasticity. North-Holland, Amsterdam (1988)

Chapter 2
Definition of Different Types of Finite Elements

Abstract This chapter presents a wide variety of finite elements of different shapes (quadrilaterals, hexahedra, triangles, tetrahedra, pyramids and wedges) useful for the numerical resolution of wave equations. More precisely, the H1, H(curl) and H(div) conforming finite elements are described in details by focusing on their spectral version which induces the important concept of mass-lumping for quadrilaterals and hexahedra. This concept enables us to construct performant algorithms.

2.1 1D Mass-Lumping and Spectral Elements

2.1.1 A Complex Solution for a Simple Problem

We can get the exact solution of the 1D wave equation:
Find $u : \mathbb{R}\times]0, T[\to \mathbb{R}$ such that:

$$\frac{\partial^2 u}{\partial t^2}(x,t) - c^2 \frac{\partial^2 u}{\partial x^2}(x,t) = 0 \quad \text{in } \mathbb{R}\times]0,T[, \tag{2.1a}$$

$$u(x,0) = u_0(x), \qquad \frac{\partial u}{\partial t}(x,0) = u_1(x) \quad \text{in } \mathbb{R}, \tag{2.1b}$$

by using a centered second-order finite difference method with a CFL equal to 1 (i.e. $c\Delta t/h = 1$, where Δt and h are the time-step and the space-step respectively).

Although simple, this solution does not hold in higher dimensions and finite difference methods are not satisfactory for complex domains. This is the reason why we are going to construct a finite element method to solve (2.1a) and (2.1b). For this purpose, we first write the variational formulation of this equation:
Find $u(.,t) \in H^1(\mathbb{R})$, $t \in]0,T[$ such that:

$$\frac{\mathrm{d}^2}{\mathrm{d}t^2}\int_{\mathbb{R}} u\, v\, \mathrm{d}x \; + \; c^2 \int_{\mathbb{R}} \frac{\partial u}{\partial x}\frac{\partial v}{\partial x}\mathrm{d}x = 0 \quad \forall v \in H^1(\mathbb{R}), \tag{2.2a}$$

G. Cohen and S. Pernet, *Finite Element and Discontinuous Galerkin Methods for Transient Wave Equations*, Scientific Computation,
DOI 10.1007/978-94-017-7761-2_2

$$u(x,0) = u_0(x), \qquad \frac{\partial u}{\partial t}(x,0) = u_1(x) \quad \text{in } \mathbb{R}. \tag{2.2b}$$

For the moment, we only develop the semi-discretization in space of this problem. Approximation in time will be introduced later.

Let

$$V_h^r(\mathbb{R}) = \{v \in C^0(\mathbb{R}) \mid \forall p \in \mathbb{Z},\ v_{|[x_p, x_{p+1}]} \in P_r\} \cap H^1(\mathbb{R}), \tag{2.3}$$

where P_r is the space of polynomials of degree r or less on $\mathbb{R}$, be the (continuous) finite element space of rth-order associated with a mesh $\{[x_p, x_{p+1}]\}_{p\in\mathbb{Z}}$ of $\mathbb{R}$. We have $V_h^r(\mathbb{R}) \subset H^1(\mathbb{R})$ [1]. By using these notations, the semi-discretized formulation in space of (2.2a) and (2.2b) reads

Find $u_h(.,t) \in V_h^r(\mathbb{R})$, $t \in]0,T[$ such that:

$$\frac{d^2}{dt^2}\int_{\mathbb{R}} u_h v_h dx + c^2 \int_{\mathbb{R}} \frac{\partial u_h}{\partial x}\frac{\partial v_h}{\partial x} dx = 0, \quad \forall v_h \in V_h^r(\mathbb{R}), \tag{2.4a}$$

$$u_h(x,0) = u_0(x), \qquad \frac{\partial u_h}{\partial t}(x,0) = u_1(x) \quad \text{in } \mathbb{R}. \tag{2.4b}$$

For each segment $[x_p, x_{p+1}]$ of $\mathbb{R}$, we have a set of $r+1$ interpolation points which are its two ends $x_p = x_{p,1}$ and $x_{p+1} = x_{p,r+1}$ and $r-1$ (a priori) regularly spaced interior points denoted $x_{p,j}$, such that $x_{p,j+1} = x_p + j(x_{p+1} - x_p)/r$, $j = 1\dots r-1$. The degrees of freedom of the finite element method are the values of the functions of $V_h^r(\mathbb{R})$ at these interpolation points. The restriction to an interval $[x_p, x_{p+1}]$ of a basis function of $V_h^r(\mathbb{R})$ corresponding to a degree of freedom located at the point $x_{p,j}$ is defined by the Lagrange polynomial $\varphi_{p,j}(x)$ defined as

$$\varphi_{p,j}(x) = \frac{\displaystyle\prod_{\ell=1,\ell\neq j}^{r+1} (x - x_{p,\ell})}{\displaystyle\prod_{\ell=1,\ell\neq j}^{r+1} (x_{p,j} - x_{p,\ell})}, \quad j = 1\dots r+1. \tag{2.5}$$

By taking basis functions λ_i based on (2.5) for v_h and computing all the integrals of (2.4a), we get the discrete system

$$M_{1,r}\frac{d^2\underline{U}}{dt^2}(t) + K_{1,r}\,\underline{U}(t) = 0, \tag{2.6}$$

with

$$(M_{1,r})_{\ell,m} = \int_{\mathbb{R}} \lambda_\ell(x)\,\lambda_m(x)\,dx, \tag{2.7a}$$

$$(K_{1,r})_{\ell,m} = \int_{\mathbb{R}} \frac{\partial\lambda_\ell}{\partial x}\frac{\partial\lambda_m}{\partial x} dx, \ (\ell,m) \in \mathbb{Z}^2, \tag{2.7b}$$

$$\underline{U}^T = (u_q, u_{q,1}, \dots, u_{q,r-1})_{q\in\mathbb{Z}}. \tag{2.7c}$$

$M_{1,r}$ is the *mass matrix* and $K_{1,r}$ is the *stiffness matrix* of the discrete problem. By using a centered second-order scheme in time, we get

$$M_{1,r}\,\frac{\underline{U}^{n+1} - 2\underline{U}^n + \underline{U}^{n-1}}{\Delta t^2} + K_{1,r}\,\underline{U}^n = 0. \tag{2.8}$$

From (2.8), we deduce

$$\underline{U}^{n+1} = 2\underline{U}^n - \underline{U}^{n-1} - \Delta t^2 M_{1,r}^{-1} K_{1,r}\,\underline{U}^n. \tag{2.9}$$

Now, $M_{1,r}$ is a symmetric $(2r+1)$-diagonal matrix that must be inverted at each time-step. Moreover, the number of diagonal rows grows for higher dimensions in space, which leads to a very costly algorithm.[1]

2.1.2 Mass-Lumping

A palliative to this inversion is to replace $M_{1,r}$ by a diagonal matrix. This can be done by using a technique of mass-lumping which consists of two steps:

1. Locate the interpolation points to the quadrature points,
2. Compute the mass integrals by using the corresponding quadrature rule.

Let us denote $\left(\xi_m^p\right)_{m=1}^{r+1}$ a set of quadrature points on $[x_p, x_{p+1}]$ and $\left(\omega_m^p\right)_{m=1}^{r+1}$ the corresponding weights. We have

$$\begin{aligned}\int_{x_p}^{x_{p+1}} u_h\varphi_{p,j}\mathrm{d}x &\simeq \sum_{\ell=1}^{r+1} u_{p,\ell}\int_{x_p}^{x_{p+1}} \varphi_{p,\ell}(x)\varphi_{p,j}(x)\mathrm{d}x\\ &\simeq \sum_{\ell=1}^{r+1} u_{p,\ell}\sum_{m=1}^{r+1}\omega_m^p\varphi_{p,\ell}(\xi_m^p)\varphi_{p,j}(\xi_m^p),\end{aligned} \tag{2.10}$$

where $u_{p,\ell}$ are the values of u_h at the quadrature points.

By taking $x_{p,m} = \xi_m^p$, we have $\varphi_{p,j}(\xi_m^p) = \delta_{j,m}$. So,

$$\int_{x_p}^{x_{p+1}} u_h\varphi_{p,j}\mathrm{d}x \simeq \sum_{\ell=1}^{r+1} u_{p,\ell}\sum_{m=1}^{r+1}\omega_m^p\delta_{\ell,m}\delta_{j,m} = \omega_j^p u_{p,j}, \tag{2.11}$$

which shows that conditions 1 and 2 ensure mass-lumping.

[1]The inverse of a n-diagonal matrix is not n-diagonal in general, which avoids the storage of $M_{1,r}^{-1}$. The correct approach would be to invert $M_{1,r}$ at each time-step by an iterative method. Since Δt is generally small, moving from t_n to t_{n+1} requires few iterations. However, each iteration is expensive compared to the product by a diagonal matrix.

An important point is to be sure that the error committed by using a quadrature rule is not greater than the error committed by the finite element method to keep the order of the approximation. The answer to this difficult question is given by Ciarlet in [1, 2] as follows: In order to keep the approximation of a finite element method of order r, one must use a quadrature rule at least exact for polynomials of P_{2r-1}.

A first candidate would be a Newton–Cotes quadrature rule, but such a rule is only exact for P_r with $r + 1$ quadrature points. Moreover, for $r > 6$, some of their weights are negative, which induces some stability problems, as we shall see later.

Gauss quadrature rules are exact for P_{2r+1} with $r + 1$ quadrature points, but they cannot include the ends of the intervals. However, they will be useful for discontinuous Galerkin methods, as shown in Chap. 4.

Actually, the good candidates are the Gauss-Lobatto rules which are exact for P_{2r-1} with $r + 1$ quadrature points, which are the $r - 1$ roots of the derivatives of the Legendre polynomial of order r (while Gauss points are the roots of the polynomials themselves) and the ends of the interval $[x_p, x_{p+1}]$. This technique of mass-lumping was first introduced independently in [3] for reservoir simulation and in [4] for neutronics and then renamed "spectral elements" [5]. It was then applied to the wave equation in [6, 7].

In Table 2.1, we give the exact values for $1 \leq r \leq 6$ and in Table 2.2, the approximate values for $7 \leq r \leq 10$ of the abscissae of the points and the corresponding weights of the Gauss-Lobatto rules when $x_p = 0$ and $x_{p+1} = 1$ on the semi-interval $[0, 1/2]$. The values on the interval $[1/2, 1]$ are deduced by symmetry.

In the same way, we give in Table 2.3 the exact values for $1 \leq r \leq 5$ and in Table 2.4, the approximate values for $6 \leq r \leq 9$ of the abscissae of the points and the corresponding weights of the Gauss rules when $x_p = 0$ and $x_{p+1} = 1$ on the semi-interval $[0, 1/2]$. Here also, the values on the interval $[1/2, 1]$ are deduced by symmetry.

The Gauss or Gauss-Lobatto points on an interval $[x_p, x_{p+1}]$ are deduced from those on the unit interval $[0, 1]$ by the following affine mapping:

$$F_p(\hat{x}) = (x_{p+1} - x_p)\hat{x} + x_p \tag{2.12}$$

and the weights are multiplied by $(x_{p+1} - x_p)$.

More details about mass-lumping can be found in [8].

Remark: We recall the definition of Legendre polynomials on the interval $[-1, 1]$

$$\begin{cases} L_0(x) = 1, \\ L_1(x) = x, \\ L_r(x) = \dfrac{(2r-1)}{r} x L_{r-1}(x) - \dfrac{(r-1)}{r} L_{r-2}(x), \quad \forall r > 1. \end{cases} \tag{2.13}$$

Table 2.1 Exact values of the abscissae of the points and of the weights of the Gauss-Lobatto quadrature rules for the semi-interval [0, 1/2] for $1 \le r \le 6$

	$j=1$	$j=2$	$j=3$	$j=4$
$r=1\ (\hat{\xi})$	0			
$r=1\ (\hat{\omega})$	$\frac{1}{2}$			
$r=2\ (\hat{\xi})$	0	$\frac{1}{2}$		
$r=2\ (\hat{\omega})$	$\frac{1}{6}$	$\frac{2}{3}$		
$r=3\ (\hat{\xi})$	0	$\frac{5-\sqrt{5}}{10}$		
$r=3\ (\hat{\omega})$	$\frac{1}{12}$	$\frac{5}{12}$		
$r=4\ (\hat{\xi})$	0	$\frac{7-\sqrt{21}}{14}$	$\frac{1}{2}$	
$r=4\ (\hat{\omega})$	$\frac{1}{20}$	$\frac{49}{180}$	$\frac{16}{45}$	
$r=5\ (\hat{\xi})$	0	$\frac{21-\sqrt{147+42\sqrt{7}}}{42}$	$\frac{21-\sqrt{147-42\sqrt{7}}}{42}$	
$r=5\ (\hat{\omega})$	$\frac{1}{30}$	$\frac{14-\sqrt{7}}{60}$	$\frac{14+\sqrt{7}}{60}$	
$r=6\ (\hat{\xi})$	0	$\frac{33-\sqrt{495+66\sqrt{15}}}{66}$	$\frac{33-\sqrt{495-66\sqrt{15}}}{66}$	$\frac{1}{2}$
$r=6\ (\hat{\omega})$	$\frac{1}{42}$	$\frac{31}{175}-\frac{\sqrt{15}}{100}$	$\frac{31}{175}+\frac{\sqrt{15}}{100}$	$\frac{128}{525}$

Table 2.2 Approximate values of the abscissae of the points and of the weights of the Gauss-Lobatto quadrature rules for the semi-interval [0, 1/2] for $7 \le r \le 10$

r	$\hat{\xi}$	$\hat{\omega}$	r	$\hat{\xi}$	$\hat{\omega}$
7	0	0.017 857 143	9	0	0.011 111 111
	0.064 129 926	0.105 352 114		0.040 233 046	0.066 652 995
	0.204 149 909	0.170 561 346		0.130 613 067	0.112 444 671
	0.395 350 391	0.206 229 397		0.261 037 525	0.146 021 342
				0.417 360 521	0.163 769 881
8	0	0.013 888 889	10	0	0.009 090 909
	0.055 012 100	0.082 747 681		0.032 999 285	0.054 806 137
	0.161 406 860	0.137 269 356		0.107 758 263	0.093 584 941
	0.318 441 268	0.173 214 255		0.217 382 337	0.124 024 052
	0.5	0.185 759 637		0.352 120 932	0.143 439 562
				0.5	0.150 108 798

Table 2.3 Exact values of the abscissae of the points and of the weights of the Gauss quadrature rules for the semi-interval [0, 1/2]

	$j=1$	$j=2$	$j=3$
$r=1\,(\hat{\xi})$	$\frac{1}{2}$		
$r=1\,(\hat{\omega})$	1		
$r=2\,(\hat{\xi})$	$\frac{3-\sqrt{3}}{6}$		
$r=2\,(\hat{\omega})$	$\frac{1}{2}$		
$r=3\,(\hat{\xi})$	$\frac{5-\sqrt{15}}{10}$	$\frac{1}{2}$	
$r=3\,(\hat{\omega})$	$\frac{5}{18}$	$\frac{4}{9}$	
$r=4\,(\hat{\xi})$	$\frac{35-\sqrt{525+70\sqrt{30}}}{70}$	$\frac{35-\sqrt{525-70\sqrt{30}}}{70}$	
$r=4\,(\hat{\omega})$	$\frac{18-\sqrt{30}}{72}$	$\frac{18+\sqrt{30}}{72}$	
$r=5\,(\hat{\xi})$	$\frac{21-\sqrt{245+14\sqrt{70}}}{42}$	$\frac{21-\sqrt{245-14\sqrt{70}}}{42}$	$\frac{1}{2}$
$r=5\,(\hat{\omega})$	$\frac{322-13\sqrt{70}}{1800}$	$\frac{322+13\sqrt{70}}{1800}$	$\frac{64}{225}$

Table 2.4 Approximate values of the abscissae of the points and of the weights of the Gauss quadrature rules for the semi-interval [0, 1/2] for $6 \le r \le 9$

r	$\hat{\xi}$	$\hat{\omega}$	r	$\hat{\xi}$	$\hat{\omega}$
6	0.033 765 243	0.085 662 246	8	0.019 855 072	0.050 614 268
	0.169 395 307	0.180 380 787		0.101 666 761	0.111 190 517
	0.380 690 407	0.233 956 967		0.237 233 795	0.156 853 323
				0.408 282 679	0.181 341 892
7	0.025 446 044	0.064 742 483	9	0.015 919 880	0.040 637 194
	0.129 234 407	0.139 852 696		0.081 984 446	0.090 324 080
	0.297 077 424	0.190 915 025		0.193 314 284	0.130 305 348
	0.5	0.208 979 592		0.337 873 288	0.156 173 539
				0.5	0.165 119 677

2.1.3 Spectral Elements

The above defined mass-lumping technique can be seen from a different point of view: In fact, the Gauss or Gauss-Lobatto points avoid the Runge phenomenon. This phenomenon is classically illustrated in the following way:

If you want to interpolate the function $f = 1/(x^2 + 1)$ on an interval containing $[-5, 5]$ by using Lagrange polynomials defined on regularly spaced points, the norm of the resulting polynomial grows with the order of interpolation. Actually, the Lagrange polynomial coincides with f at all the interpolation points but oscillates more and more between these points at the ends of the interval, as shown in Fig. 2.1.

From a theoretical point of view, avoiding Runge phenomenon depends on the minimization of the supremum norm $||.||_\infty$ of the difference of a function f and its rth-order interpolation denoted $\mathscr{I}_r^1 f$ over an interval $I_1 = [a, b]$. In practice, this minimum p^* is not obtained, but one can get an approximative knowledge by using the inequality

$$||f(x) - \mathscr{I}_r^1 f(x)||_\infty \leq [1 + \Lambda(\Pi_r^1)]\, ||f(x) - p^*(x)||_\infty, \tag{2.14}$$

where Π_r^1 is a given set of $r + 1$ points and

$$\Lambda(\Pi_r^1) = \max_{x \in I_1} \sum_{i=1}^{r+1} |L_i(\Pi_r^1, x)|, \tag{2.15}$$

L_i being a Lagrange polynomial on interval I_1.

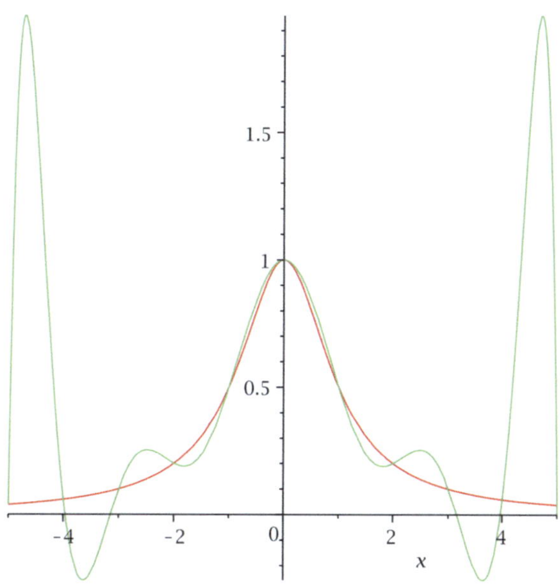

Fig. 2.1 Tenth order Lagrange polynomial interpolation on regularly spaced points of $f = 1/(x^2 + 1)$ on $[-5, 5]$ (in *green*) compared to the function (in *red*)

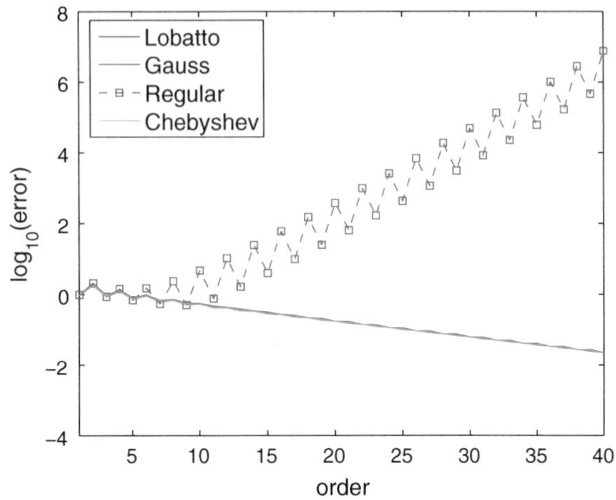

Fig. 2.2 L^2-norm logarithmic error of a Lagrange interpolation of f by polynomials of order 1 to 40 using different kinds of points. One can notice the exponential divergence of the regularly-spaced points

$\Lambda(\Pi_r^1)$ is called the Lebesgue constant. The approximation $\mathscr{I}_r^1 f$ is all the better that this constant is small. More details and references on this process can be found in [9, 10].

A classical answer to this problem was to use Chebyshev points to avoid these oscillations. For a long time, these points were regarded as the "optimal" ones and, since they enabled an exponential convergence in r of the methods, a Lagrange interpolation using Chebyshev points was defined as a "spectral element method" [11]. Unfortunately, Chebyshev points cannot support an efficient quadrature formula.

The question can be asked: is the use of Gauss or Gauss-Lobatto points for Lagrange interpolation able to get the so-called spectral convergence? A first answer is given by the good behaviour of their Lebesgue coefficients [9]. On the other hand, we interpolated f by using Chebyshev, Gauss and Gauss-Lobatto points on interval $[-5, 5]$ and we had very similar results for the three kinds of points as shown in Fig. 2.2.

In all the following, spectral elements will implicitly denote finite elements whose interpolation points coincide with quadrature points for quadrilaterals or hexahedra and elements with "optimal" nodes for other elements.

2.1.4 Nodal and Modal Elements

The construction of basis functions in this section corresponds to a "nodal" approach, i.e. the basis functions are defined by their values at the nodes of the elements. In this approach, the degrees of freedom are the values of the functions at these nodes. This enables us to get continuous finite element methods by enforcing the solution to have a unique value at the boundaries nodes (when they exist). However, as we shall see later, this type of elements can also be used for discontinuous formulations.

Another approach is to use basis functions corresponding to other criteria (for instance, values of linear forms [2] or orthogonal basis [12]). Then, the solution is given by a linear combination of the basis functions whose coefficients do not represent values of the functions. This approach, called the "modal" approach, leads to discontinuous formulations in general but can provide partially continuous elements in some cases, as in mixed formulations which will be introduced below.

The following three sections deal with nodal elements.

2.2 Quadrilaterals and Hexahedra

As we said in the previous section, our purpose is to solve wave equations on complex geometries in an efficient way. For this reason, we want to extend the concept of our spectral element approach to higher dimensions. This must be done in two steps:

1. Define a spectral unit element.
2. Extend this definition to any element.

As we shall see, the appropriate unit elements in 2D and 3D are the unit square $[0, 1]^2$ and unit cube $[0, 1]^3$ respectively. Before defining the approximation, let us first define the Gauss or Gauss-Lobatto quadrature rules for these unit elements, which is denoted $\widehat{K}$ in the following.

2.2.1 *Higher-Dimensional Tensor Quadrature Rules*

Let us first study the 2D case. We want to integrate an integrable function $f : \mathbb{R}^2 \to \mathbb{R}$ on the unit square by using a 1D quadrature rule on $[0, 1]$ whose quadrature points are $\left(\hat{\xi}_p\right)_{p=1}^{r+1}$ and whose weights are $\left(\hat{\omega}_p\right)_{p=1}^{r+1}$. By using Fubini's theorem, we have

$$
\begin{aligned}
\int_{\widehat{K}} f(\underline{\hat{x}})\,\mathrm{d}\underline{\hat{x}} &= \int_0^1\int_0^1 f(\underline{\hat{x}})\,\mathrm{d}\underline{\hat{x}} = \int_0^1\left(\int_0^1 f(\hat{x}_1,\hat{x}_2)\,\mathrm{d}\hat{x}_2\right)\mathrm{d}\hat{x}_1 \\
&\simeq \int_0^1\left(\sum_{q=1}^{r+1}\hat{\omega}_q f(\hat{x}_1,\hat{\xi}_q)\right)\mathrm{d}\hat{x}_1 \\
&\simeq \sum_{p=1}^{r+1}\hat{\omega}_p\left(\sum_{q=1}^{r+1}\hat{\omega}_q f(\hat{\xi}_p,\hat{\xi}_q)\right) \\
&= \sum_{p=1}^{r+1}\sum_{q=1}^{r+1}\hat{\omega}_p\hat{\omega}_q f(\hat{\xi}_p,\hat{\xi}_q) = \sum_{p=1}^{r+1}\sum_{q=1}^{r+1}\hat{\omega}_{p,q} f(\underline{\hat{\xi}}_{p,q}).
\end{aligned}
\tag{2.16}
$$

So, we see that we can define a 2D quadrature rule whose quadrature points are

– $\left(\hat{\underline{\xi}}_{p,q} = (\hat{\xi}_p, \hat{\xi}_q)\right)_{(p,q)=(1,1)}^{(r+1,r+1)}$

and whose weights are

– $\left(\hat{\omega}_{p,q} = \hat{\omega}_p \hat{\omega}_q\right)_{(p,q)=(1,1)}^{(r+1,r+1)}$.

Now, it is easy to see that, if this 1D quadrature rule is exact for polynomials of P_r, $r \in \mathbb{N}$, the corresponding 2D quadrature rule is exact for Q_r, where Q_r is the set of polynomials of degree r or less in each variable which reads, in dimension d:

$$Q_r = \left\{ v(\underline{\hat{x}}) = \sum_{\underline{j}=(j_1,\ldots j_d)\in\{0,\ldots r\}^d} a_{\underline{j}} \prod_{k=1}^{d} \hat{x}_k^{j_k},\ a_{\underline{j}} \in \mathbb{R} \right\}. \tag{2.17}$$

In fact, we can write for any monomial $\hat{x}_1^i \hat{x}_2^j$, so that $0 \leq i \leq r$ and $0 \leq j \leq r$,

$$\begin{aligned} \int_{\widehat{K}} \hat{x}_1^i \hat{x}_2^j \,\mathrm{d}\underline{\hat{x}} &= \int_0^1 \hat{x}_1^i \,\mathrm{d}\hat{x}_1 \times \int_0^1 \hat{x}_2^j \,\mathrm{d}\hat{x}_2 \\ &= \sum_{p=1}^{r+1} \hat{\omega}_p \hat{\xi}_p^i \sum_{q=1}^{r+1} \hat{\omega}_q \hat{\xi}_q^j = \sum_{p=1}^{r+1}\sum_{q=1}^{r+1} \hat{\omega}_{p,q} \hat{\xi}_p^i \hat{\xi}_q^j. \end{aligned} \tag{2.18}$$

In the same way, it is easy to show that the 3D quadrature rule whose quadrature points are

– $\left(\hat{\underline{\xi}}_{p,q,s} = (\hat{\xi}_p, \hat{\xi}_q, \hat{\xi}_s)\right)_{(p,q,s)=(1,1,1)}^{(r+1,r+1,r+1)}$

and whose weights are

– $\left(\hat{\omega}_{p,q,s} = \hat{\omega}_p \hat{\omega}_q \hat{\omega}_s\right)_{(p,q,s)=(1,1,1)}^{(r+1,r+1;r+1)}$,

is a 3D extension of our 1D quadrature rule, exact for polynomials of Q_r.

2.2.2 *Tensor Unit Spectral Elements*

Basis functions on the unit element are classically defined as the product of 1D basis functions. So, if we denote by $\hat{\varphi}_i(\hat{x})$ the Lagrange polynomial defined in (2.5) in which we set $x_p = 0$, $x_{p+1} = 1$ and $\hat{x} \in [0, 1]$ such that $\hat{\varphi}_i(\hat{\xi}_j) = \delta_{ij}$, we can define the basis functions $\hat{\varphi}_{\ell,m}(\hat{x}_1, \hat{x}_2)$ on the unit square as

$$\hat{\varphi}_{\ell,m}(\hat{x}_1, \hat{x}_2) = \hat{\varphi}_\ell(\hat{x}_1)\hat{\varphi}_m(\hat{x}_2), \tag{2.19}$$

and, in the same way, on the unit cube

$$\hat{\varphi}_{\ell,m,n}(\hat{x}_1, \hat{x}_2, \hat{x}_3) = \hat{\varphi}_\ell(\hat{x}_1)\hat{\varphi}_m(\hat{x}_2)\hat{\varphi}_n(\hat{x}_3), \tag{2.20}$$

where $1 \leq \ell,\ m,\ n \leq r + 1$.

Since $\hat{\varphi}_i \in P_r$, we have $\hat{\varphi}_{\ell,m} \in Q_r$ and $\hat{\varphi}_{\ell,m,n} \in Q_r$, with Q_r defined as in (2.17).

Obviously, we have

$$\hat{\varphi}_{\ell,m}(\hat{\underline{\xi}}_{i,j}) = \delta_{\ell,i}\, \delta_{m,j}, \tag{2.21a}$$

$$\hat{\varphi}_{\ell,m,n}(\hat{\underline{\xi}}_{i,j,k}) = \delta_{\ell,i}\, \delta_{m,j}\, \delta_{n,k}. \tag{2.21b}$$

Now, following Ciarlet's criteria given in [2], for a square or a cube, one must use a quadrature rule at least exact for polynomials of Q_{2r-1} in order to keep the approximation of finite element method of order r. So, using the results of the previous section, we see that the Gauss-Lobatto rule is still the good candidate for the unit elements in 2D and 3D. An example of Gauss-Lobatto points in 2D is given in Fig. 2.3.

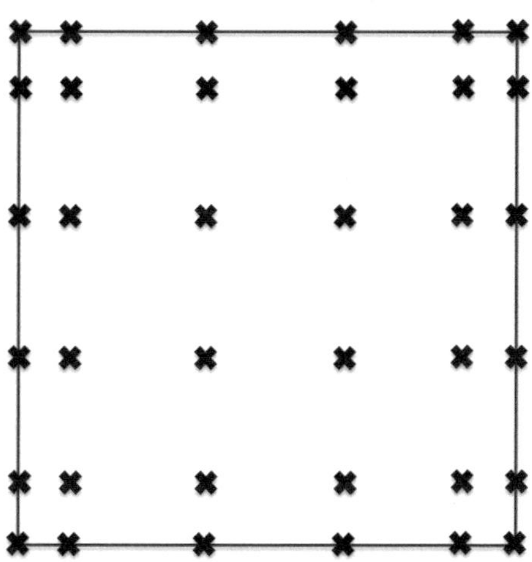

Fig. 2.3 Fifth-order Gauss-Lobatto points on the unit square

Remark: Even for non regularly-spaced, basis functions on the unit elements remain products of equations of lines or planes (parallel to the axes). This is not true for non tensor elements.

2.2.3 *Extension to Quadrilaterals and Hexahedra*

We extend our basis functions to a quadrilateral or an hexahedron K_i in a classical way (see for instance [2]) by using a mapping $\underline{F}_i = (F_i^{(1)}, \dots F_i^{(d)})^T$ such that

$$\underline{F}_i(\widehat{K}) = K_i. \tag{2.22}$$

The components $F_i^{(j)}$ of $\underline{F}_i$ are generally taken in Q_1, which provides quadrilaterals and hexahedra with straight edges.[2] However, one can also take these components in $Q_r, r > 1$. Such a choice leads to elements with curved edges (quadratic for $r = 2$, cubic for $r = 3$, etc.) which can be more accurate to approximate curved boundaries. In practice, we have, for $r = 1$ in 2D:

$$\underline{F}_i(\widehat{K})(\hat{x}_1, \hat{x}_2) = \sum_{\ell=1}^{2}\sum_{m=1}^{2} \underline{c}_{\ell,m}^{(i)} \hat{\varphi}_{\ell,m,}(\hat{x}_1, \hat{x}_2), \tag{2.23}$$

and, in 3D,

$$\underline{F}_i(\widehat{K})(\hat{x}_1, \hat{x}_2, \hat{x}_3) = \sum_{\ell=1}^{2}\sum_{m=1}^{2}\sum_{n=1}^{2} \underline{c}_{\ell,m,n}^{(i)} \hat{\varphi}_{\ell,m,n}(\hat{x}_1, \hat{x}_2, \hat{x}_3), \tag{2.24}$$

where $\underline{c}_{\ell,m}^{(i)}$ and $\underline{c}_{\ell,m,n}^{(i)}$ are the summits of K_i in 2D and 3D (Fig. 2.4).

As we said, $\underline{F}_i$ can be defined in the same way for $r > 1$ by using basis functions on $\widehat{K}$, provided the knowledge of the analytical definition of the edges of K_i and the use of the Gordon-Hall transform [13, 14].

One can easily see that, for $r = 1$, $\underline{F}_i$ is bilinear in 2D and trilinear in 3D[3] unlike mappings used for triangles and tetrahedra which are linear. This can be somehow troublesome since its Jacobian matrix and its Jacobian are not constant. This is a fortiori true for $r > 1$.

A basis functions φ on an element K_i is classically (for functions in H^1) defined as follows

$$\varphi = \hat{\varphi} \circ \underline{F}_i^{-1}, \tag{2.25}$$

where $\hat{\varphi}$ is a basis function on the unit element.

---

[2]But, of course, not plane faces for hexahedra, unless the summits of a face are coplanar.

[3]Unless all the faces are planes, which provides a linear mapping.

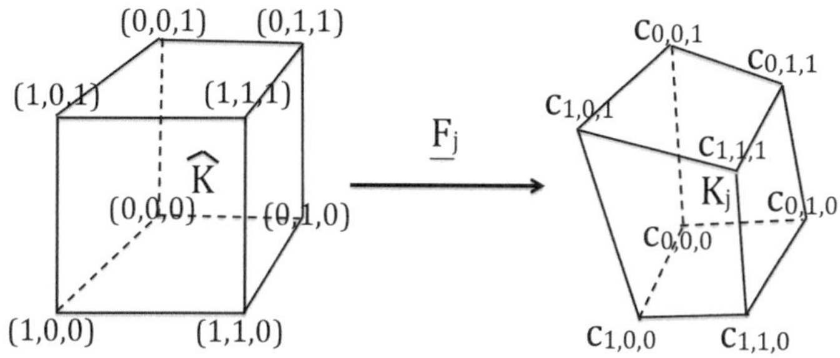

Fig. 2.4 The 3D mapping for $r = 1$

One can notice that such a definition provides basis functions which are in general no longer polynomial, since the components of $\underline{F}_i^{-1}$ are fractions involving polynomials and square roots of polynomials when $\underline{F}_i$ is not linear.

Now, the question can be asked: Does this definition of finite elements still provide mass-lumping?

To answer this question, we integrate the product of two basis functions $\varphi_{j,k}$ and $\varphi_{\ell,m}$ on K_i such that $\varphi_{j,k} = \hat{\varphi}_{j,k} \circ \underline{F}_i^{-1}$ and $\varphi_{\ell,m} = \hat{\varphi}_{\ell,m} \circ \underline{F}_i^{-1}$. By using (2.21a), we have

$$\begin{aligned}\int_{K_i} \varphi_{j,k}(\underline{x})\, \varphi_{\ell,m}(\underline{x}) \mathrm{d}\underline{x} &= \int_{\widehat{K}} |J_i(\hat{\underline{x}})|\, \hat{\varphi}_{j,k}(\hat{\underline{x}})\, \hat{\varphi}_{\ell,m}(\hat{\underline{x}})\, \mathrm{d}\hat{\underline{x}} \\ &\simeq \sum_{p=1}^{r+1} \sum_{q=1}^{r+1} \hat{\omega}_{p,q}\, |J_i(\hat{\underline{\xi}}_{p,q})|\, \hat{\varphi}_{j,k}(\hat{\underline{\xi}}_{p,q})\, \hat{\varphi}_{\ell,m}(\hat{\underline{\xi}}_{p,q}) \\ &= \sum_{p=1}^{r+1} \sum_{q=1}^{r+1} \hat{\omega}_{p,q}\, |J_i(\hat{\underline{\xi}}_{p,q})|\, \delta_{j,p}\, \delta_{k,q}\, \delta_{\ell,p}\, \delta_{m,q}.\end{aligned} \tag{2.26}$$

where $J_i(\hat{x}) := \det\, (DF_i(\hat{x}))$ with $DF_i(\hat{x})$ is the Jacobian matrix of $\underline{F}_i$.

From (2.26), we can easily deduce that

$$\int_{K_i} \varphi_{j,k}(\underline{x})\, \varphi_{\ell,m}(\underline{x}) \mathrm{d}\underline{x} \simeq \begin{cases} \hat{\omega}_{j,k}\, |J_i(\hat{\underline{\xi}}_{j,k})| \text{ if } \ell = j \text{ and } m = k, \\ \\ 0 \text{ otherwise,} \end{cases} \tag{2.27}$$

which obviously provides a diagonal mass-matrix.

A similar computation leads to the same conclusion in 3D.

Another and difficult question can be asked: Do we still keep the good order of approximation on distorted elements?

Actually, Ciarlet's criterion only applies to orthogonal elements (squares or cubes). So, the use of a mapping introduces a new difficulty to ensure the order of our approximation. This question will be addressed in the next chapter.

2.3 Triangles and Tetrahedra

Unlike tensor elements, the notions of spectral element and mass-lumping do not naturally coincide for other elements as triangles and tetrahedra. For such elements, mass-lumping cannot be obtained on the basis of classical interpolation points which lead to negative weights. Such weights lead, as we said above, to unconditionally unstable schemes. So, stable mass-lumping requires additional interpolation points for these elements, which induces some problems for their construction and their performance.

For this reason, we first present spectrally convergent elements, i.e. elements avoiding Runge phenomenon, based on Hesthaven's works [9, 15]. In a second part, we define mass-lumped elements introduced in [16, 17] for triangles up to third order and extended in [18] for higher-order tetrahedra and triangles.

2.3.1 Spectral Triangles and Tetrahedra

2.3.1.1 Finding Optimal Points

Natural Lagrange interpolation on triangles and tetrahedra is based on the P_r polynomial set defined, in dimension d, as

$$P_r = \left\{ v(\underline{\hat{x}}) = \sum_{\underline{j}=(j_1,\dots j_d),\, \sum_{\ell=1}^{d} j_\ell \le r} a_{\underline{j}} \prod_{k=1}^{d} \hat{x}_k^{j_k},\; a_{\underline{j}} \in \mathbb{R} \right\}. \tag{2.28}$$

Computation of quasi-optimal points (in terms of Runge phenomenon) for triangles and tetrahedra are based on an analogy with an optimization problem in electrostatics. This problems reads as follows:
Minimize the functional $W(\underline{x}_1, \dots, \underline{x}_{N_p})$ (which physically represents the electrostatic energy of N_p unit charges located at points $\{\underline{x}_1, \dots, \underline{x}_{N_p}\}$),

$$W(\underline{x}_1, \dots, \underline{x}_{N_p}) = \sum_{i=1}^{N_p} \left(\sum_{\ell=1}^{N_{\partial K}} \phi_\ell^L(\underline{x}_i) + \frac{1}{2} \sum_{j=1, j\neq i}^{N_p} \phi(\underline{x}_i, \underline{x}_j) \right), \tag{2.29}$$

where

$$\phi(\underline{x}_i, \underline{x}_j) = \frac{1}{|\underline{x}_i - \underline{x}_j|} \tag{2.30}$$

and

$$\phi_\ell^L(\underline{x}_i) = \int_{\gamma_\ell} \frac{\rho_\gamma}{|\underline{x} - \underline{x}_i|^2} \,\mathrm{d}\underline{x}. \tag{2.31}$$

N_p is the total number of points of interpolation, $N_{\partial K}$, the number of elementary components γ_ℓ of the boundary of the element (edges in 2D and faces in 3D), ρ_γ is a parameter of optimization on γ_ℓ.

Optimization is then made by solving numerically a N_p-body problem by integrating Newton's second law

$$\ddot{\underline{x}}_i + \sum_{\ell=1}^{N_{\partial K}} \nabla \phi_\ell^L(\underline{x}_i) + \sum_{j=1, j\neq i}^{N_p} \nabla \phi(\underline{x}_i, \underline{x}_j) + \varepsilon \dot{\underline{x}}_i = 0, \tag{2.32}$$

where ε is a positive constant which controls a dissipative term introduced for a computational purpose.

More details about this technique can be found in [9, 10, 15].

2.3.1.2 Optimal Points for the Unit Triangle and Tetrahedron

In Tables 2.5 and 2.6, we give the locations of the interior points for triangles and tetrahedra in barycentric coordinates. In these tables, r_m indicates the number of classes of points having an m-multiplicity (due to symmetry) in the element. For both elements, the Gauss-Lobatto points are used for the edges and do not appear in the tables. For this reason, we do not provide the points for the two first orders whose points are located on the edges. The number of interpolation points for P_r is equal to

- $\dfrac{(r+1)(r+2)}{2} = 3r + r_1 + 3r_3 + 6r_6$ for triangles,
- $\dfrac{(r+1)(r+2)(r+3)}{6} = 4 + 6(r-1) + r_1 + 4r_4 + 6r_6 + 12r_{12} + 24r_{24}$ for tetrahedra.

2.3.1.3 Computation of Basis Functions

The definition of the basis functions for these elements is slightly more complex than for tensor elements, for which basis functions can be constructed from products of 1D Lagrange polynomials, as mentioned in (2.19) and (2.20), or even triangle and tetrahedra with regularly spaced interpolation points. Actually, for these last

Table 2.5 Quasi-optimal interior points for triangles (to which summits and Gauss-Lobatto points on the edges (given in Tables 2.1 and 2.2) must be added)

r	r_1	r_3	r_6	λ_1	λ_2	λ_3
3	1			0.3333333333	0.3333333333	0.3333333333
4		1		0.2410021998	0.2410021998	0.5179956004
5		2		0.1591570023	0.1591570023	0.6816859954
				0.4099016620	0.4099016620	0.1801966760
6	1			0.3333333333	0.3333333333	0.3333333333
		1		0.1048904342	0.1048904342	0.7902191316
			1	0.3095036860	0.5582114022	0.1322849118
7		3		0.0666479037	0.0666479037	0.8667041924
				0.4474963910	0.4474963910	0.1050072180
				0.2606379453	0.2606379453	0.4787241094
			1	0.2328951264	0.6750593997	0.0920454739
8		3		0.0467325482	0.0467325482	0.9065349036
				0.2031909379	0.2031909379	0.5936181242
				0.3906323571	0.3906323571	0.2187352858
			2	0.3618069634	0.5544121321	0.0837809045
				0.1799524415	0.7523326932	0.0677148653
9	1			0.3333333333	0.3333333333	0.3333333333
		3		0.0354284515	0.0354284515	0.9291430970
				0.4641239296	0.4641239296	0.0717521408
				0.1632684184	0.1632684184	0.6734631632
			3	0.2965866112	0.6351919517	0.0682214371
				0.1437685751	0.8034472485	0.0527841764
				0.3225938344	0.4969468296	0.1804593360

elements, the basis functions are naturally defined by products of affine functions which are equations of straight lines or planes containing the interpolation points. In our case, we do not have such properties since the points are randomly spaced.

So, in order to get the basis functions, one must look for the coefficients of a complete polynomials of P_r, which implies to solve a linear system of dimension $(r+1)(r+2)/2$ for triangles and $(r+1)(r+2)(r+3)/6$ for tetrahedra. For high-order polynomials, this system is not well-conditioned and can be difficult to solve accurately. A palliative to this difficulty is the use of orthogonal basis functions of P_r, defined in [12] and used in [10]. Let us set $P_i^n = \widetilde{P}_i^{n,0}$, where $\widetilde{P}_i^{n,0}$ is the Jacobi polynomial defined by

$$\widetilde{P}_i^{(\ell,m)}(x) = \frac{(-1)^i}{2^i i!(1-x)^\ell (1+x)^m} \frac{\mathrm{d}^i}{\mathrm{d}x^i}((1-x)^\ell (1+x)^m (1-x^2)^i) \qquad (2.33)$$

Table 2.6 Quasi-optimal interior points for tetrahedra (to which summits and Gauss-Lobatto points on the edges (given in Tables 2.1 and 2.2) must be added)

r	r_1	r_4	r_6	r_{12}	r_{24}	λ_1	λ_2	λ_3	λ_4
3		1				0.3333333333	0.3333333333	0.3333333333	0.0000000000
4	1					0.2500000000	0.2500000000	0.2500000000	0.2500000000
				1		0.2371200168	0.2371200168	0.5257599664	0.0000000000
5		1				0.1834903473	0.1834903473	0.1834903473	0.4495289581
				2		0.1575181512	0.1575181512	0.6849636976	0.0000000000
						0.4105151510	0.4105151510	0.1789696980	0.0000000000
6		2				0.3333333333	0.3333333333	0.3333333333	0.0000000000
						0.1402705801	0.1402705801	0.1402705801	0.5791882597
			1			0.3542052583	0.3542052583	0.1457947417	0.1457947417
				1		0.1061169285	0.1061169285	0.7877661430	0.0000000000
					1	0.3097982151	0.5569099204	0.1332918645	0.0000000000
7		2				0.1144606542	0.1144606542	0.1144606542	0.6566180374
						0.2917002822	0.2917002822	0.2917002822	0.1248991534
				4		0.0660520784	0.0660520784	0.8678958432	0.0000000000
						0.4477725053	0.4477725053	0.1044549894	0.0000000000
						0.2604038024	0.2604038024	0.4791923952	0.0000000000
						0.1208429970	0.1208429970	0.4770203357	0.0000000000
					1	0.2325524777	0.6759625951	0.0914849272	0.0000000000
8	1					0.2500000000	0.2500000000	0.2500000000	0.2500000000
		1				0.0991203900	0.0991203900	0.0991203900	0.7026388300
			1			0.3920531037	0.3920531037	0.1079468963	0.1079468963
				5		0.0660520784	0.0660520784	0.8678958432	0.0000000000
						0.2033467796	0.2033467796	0.5933064408	0.0000000000
						0.3905496216	0.3905496216	0.2189007568	0.0000000000
						0.1047451941	0.1047451941	0.5581946462	0.2323149656
						0.2419418605	0.2419418605	0.4062097450	0.1099065340
					2	0.3617970895	0.5541643672	0.0840385433	0.0000000000
						0.1801396087	0.7519065566	0.0679538347	0.0000000000
9		3				0.3333333333	0.3333333333	0.3333333333	0.0000000000
						0.0823287303	0.0823287303	0.0823287303	0.7530138091
						0.2123055477	0.2123055477	0.2123055477	0.3630833569
				7		0.0355775717	0.0355775717	0.9288448566	0.0000000000
						0.4640303025	0.4640303025	0.0719393950	0.0000000000
						0.1633923069	0.1633923069	0.6732153862	0.0000000000
						0.0873980781	0.0873980781	0.6297057875	0.1954980564
						0.0916714679	0.0916714679	0.4819523024	0.3347047619
						0.2040338880	0.2040338880	0.4996292993	0.0923029247
						0.3483881173	0.3483881173	0.2075502723	0.0956734931
					3	0.2966333890	0.6349633653	0.0684032457	0.0000000000
						0.1439089974	0.8031490682	0.0529419344	0.0000000000
						0.3225890045	0.4968009397	0.1806100558	0.0000000000

These polynomials are orthogonal for the scalar product

$$(P, Q) = \int_{-1}^{1} (1-x)^{\ell}(1+x)^m P(x)Q(x)\mathrm{d}x. \tag{2.34}$$

We can now construct our orthogonal basis as follows

- For the unit triangle

$$\hat{T}_2 = \{(x_1, x_2) | \, x_1, \, x_2 \geq -1, \, x_1 + x_2 \leq -1\}, \tag{2.35}$$

  we have

$$\varphi_{i,j} = P_i^0(a) P_j^{2i+1}(b)(1-b)^i, \quad i + j \leq r, \tag{2.36}$$

  where $a = (2x_1 + x_2 + 1)/(1 - x_2)$ and $b = x_2$.
- For the unit tetrahedron

$$\hat{T}_3 = \{(x_1, x_2, x_3) | \, x_1, \, x_2, \, x_3 \geq -1, \, x_1 + x_2 + x_3 \leq -1\}, \tag{2.37}$$

  we have

$$\begin{aligned} &\varphi_{i,j,k} = P_i^0(a) P_j^{2i+1}(b) P_k^{2i+2j+2}(c)(1-b)^i(1-c)^{i+j}, \\ &i + j + k \leq r, \end{aligned} \tag{2.38}$$

  where $a = -(2x_1 + x_2 + x_3 + 2)/(x_2 + x_3)$, $b = (2x_2 + x_3 + 1)/(1 - x_3)$ and $c = x_3$.

Remarks:

1. a, b, c are related to x_1, x_2, x_3 by a mapping $\underline{F}_c$ from the unit cube $[-1, 1]^3$ (square $[-1, 1]^2$ in 2D) to our unit element (also called the Duffy's transform [19]). Actually, the functions $\varphi_{i,j,k} \circ \underline{F}_c^{-1}$ provide an orthogonal polynomial basis, as shown in [20].
2. Any other triangle or tetrahedron can be chosen for $\hat{T}_2$ and $\hat{T}_3$.
3. An easy way to get polynomials P_j^k is given by the following algorithm [21]

$$\begin{aligned} &P_0^k(x) = 1; \\ &P_1^k(x) = \frac{1}{2}[(k+2)x + k]; \\ &\forall j \geq 2, a_{j-2} = 2(j+k-1)(j-1)(2j+k), \\ &a_{j-1} = (2j+k-1)[k^2 + (2j+k)(2j+k-2)x], \\ &a_j = 2j(j+k)(2j+k-2), \\ &P_j^k(x) = \frac{1}{a_j}(a_{j-1}P_{j-1}^k(x) - a_{j-2}P_{j-2}^k(x)). \end{aligned} \tag{2.39}$$

4. A Matlab procedure to compute the points is given in [10].

2.3.1.4 Extension to Triangles and Tetrahedra of Any Shape

In order to get an (H^1) extension of the basis functions to triangles and tetrahedra of any shape, we must first construct a mapping as follows for tetrahedra.

Let T be a tetrahedron of any shape and $\underline{F}_T$ the mapping such that $\underline{F}_T(\hat{T}_3) = T$. $\{a_1, a_2, a_3, a_4\}$ being the four vertices of T, we get

$$\underline{F}_T = \sum_{i=1}^{4} a_i \hat{\theta}_i, \tag{2.40}$$

where $\hat{\theta}_i$ are the basis functions of P_1 associated to $\hat{T}_3$, i.e.

$$\hat{\theta}_1 = \hat{x}_1 + \hat{x}_2 + \hat{x}_3 + 1, \quad \hat{\theta}_2 = \hat{x}_1 + 1, \quad \hat{\theta}_3 = \hat{x}_2 + 1, \quad \hat{\theta}_4 = \hat{x}_3 + 1$$

and as follows, for triangles:

Let T be a triangle of any shape and $\underline{F}_T$ the mapping such that $\underline{F}_T(\hat{T}_2) = T$. $\{a_1, a_2, a_3\}$ being the three vertices of T, we get

$$\underline{F}_T = \sum_{i=1}^{3} a_i \hat{\theta}_i, \tag{2.41}$$

where $\hat{\theta}_i$ are the basis functions of P_1 associated to $\hat{T}_2$, i.e.

$$\hat{\theta}_1 = \hat{x}_1 + \hat{x}_2 + 1, \quad \hat{\theta}_2 = \hat{x}_1 + 1, \quad \hat{\theta}_3 = \hat{x}_2 + 1.$$

2.3.2 *Mass-Lumped Triangles and Tetrahedra*

As we said, for triangles and tetrahedra, mass-lumping does not coincide with the spectral character of the elements. Actually, this difficulty arises from the fact that Gauss-Lobatto-like rules based on P_r interpolation points do exist, but they have some negative weights. The presence of such weights prohibits the use of these quadratures for transient wave equations since they lead to non positive approximations of the operator in space, which provides unconditionally unstable approximations.[4]

For instance, as shown in [7, 8, 16, 17], we get the following quadratures for P_2 and P_3 equilateral triangles[5]:

- For P_2, the points are the summits S_ℓ and the midpoints of the edges $M_\ell, \ell = 1 \ldots 3$ affected of weights $\omega_S = 0$ and $\omega_M = 1/6$, which is of course not acceptable.

[4]This difficulty is also troublesome for time-harmonic problems.

[5] P_1 is naturally mass-lumped by using the Trapezoidal rule for triangles and tetrahedra.

- For P_3, the points are the summits S_ℓ, $\ell = 1\dots 3$, the center of the triangle G and the points of the edges $M_{12}(\theta)$, $M_{21}(\theta)$, $M_{13}(\theta)$, $M_{31}(\theta)$, $M_{23}(\theta)$, $M_{32}(\theta)$ such that $M_{\ell m}(\theta)$ is the barycenter of S_ℓ and S_m with the weights θ and $1-\theta$ (classically, $\theta = 1/3$). After some computations, we get $\theta = (3-\sqrt{3})/6$. The corresponding weights are then $\omega_S = -1/120$, $\omega_G = -9/40$, $\omega_M = 1/20$, which are also not acceptable.

In the same way, we have, as shown in [8], for the P_2 tetrahedron whose quadrature points are the summits S_ℓ, $\ell = 1\dots 4$ and the midpoints of the edges M_ℓ, $\ell = 1\dots 6$ affected of weights $\omega_S = -1/20$ and $\omega_M = 1/5$.

The palliative to this problem is to add degrees of freedom which implies to enrich the corresponding polynomial spaces P_r up to $P_{r'}, r' > r$. The new polynomial space $\check{P}_r$ is then such that $P_r \subset \check{P}_r \subset P_{r'}$. As shown in [17, 18], up to $r = 4$ for triangles, one must take $r' = r+1$ to get the appropriate quadrature rules with positive weights. In this case, the additional points represent about 50% of the initial points.

For $r > 4$ for triangles and for tetrahedra at any order, additional polynomials from P_{r+1} are not able to provide positive weights. One must take some other ones in $P_{r'}$, where $r' \geq r+2$. In this case, we multiply by about 2.5 points the initial number of points. At this stage, the increase of degrees of freedom seriously handicaps the method. Actually, besides the additional computational time induced by the important number of points, this number dramatically reduces the stability condition of the methods using these elements. Moreover, some experiments realized with continuous elements show that mass-lumped tetrahedra are less performant than classical ones.

For triangles, the polynomial space of interpolation is

$$\check{P}_r = P_r \oplus [b]P_{r'-3}, \tag{2.42}$$

where b is the so-called bubble function associated to the triangle and $[b] = \text{span}(b)$.

Remark: The bubble function of a triangle is defined as the product of its barycentric coordinates. For example, $b = x_1x_2(1-x_1-x_2)$ if we consider the triangle $\{(x_1,x_2)\,|\,x_1,\ x_2 \geq 0,\ x_1+x_2 \leq 1\}$.

This definition shows that we have $3r + (r'-2)(r'-1)/2$ nodes which can be decomposed into

- $3r$ nodes on the edges,
- $(r'-2)(r'-1)/2$ interior nodes.

For tetrahedra, one must add polynomials related to the four faces as well as polynomials related to the interior of the tetrahedra. We get

$$\check{P}_r = P_r \oplus \sum_{i=1}^{4} [b_j]P_{r'-3} \oplus [\widetilde{b}]P_{r''-4}, \tag{2.43}$$

where $r' > r''$, $\widetilde{b}$ is the 3D bubble function and b_j is the bubble function related to the face opposite to the jth summit of the tetrahedron.

Remark: $\widetilde{b} = x_1 x_2 x_3 (1 - x_1 - x_2 - x_3)$ for the tetrahedron $\{(x_1, x_2, x_3) \,|\, x_1, x_2, x_3 \geq 0,\ x_1 + x_2 + x_3 \leq 1\}$.

As for triangles, we have $6r - 2 + (r'' - 3)(r'' - 2)(r'' - 1)/6 + 2(r' - 2)(r' - 1)$ nodes which can be decomposed into

- $6r - 2$ nodes on the edges,
- $2(r' - 2)(r' - 1)$ on the faces,
- $(r'' - 3)(r'' - 2)(r'' - 1)/6$ interior nodes.

In order to construct the basis functions, one must determine the coefficients of a complete polynomial of $P_{r'}$. No orthogonal polynomial basis based on Jacobi polynomials was constructed for mass-lumped elements.

For such elements, Ciarlet's criterion reads as follows: In order to keep the approximation of finite element method of order r, one must use a quadrature rule at least exact for polynomials of P_N, where $N = r + r' - 2$. This criterion is fulfilled by the elements presented below.

In Figs. 2.5 and 2.6, we represent some mass-lumped triangles and tetrahedra. In Tables 2.7 and 2.8, we give the quadrature points and weights with their multiplicity for triangles and tetrahedra. The coordinates are given for the reference elements (i.e. with summits (0,0), (0,1) and (1,0) in 2D and (0,0,0), (1,0,0), (0,1,0) and (0,0,1) in 3D) for the closest points to the origin. Other points are obtained by central symmetry. Mass-lumped triangles for $r = 6$ and $r = 7$ can be found in [22].

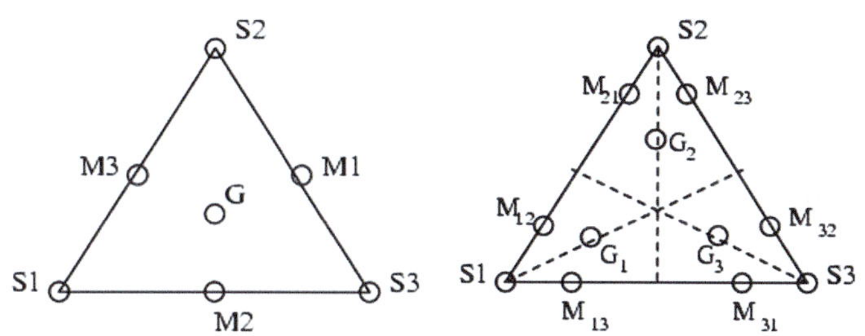

Fig. 2.5 The $\check{P}_2$ (*left*) and $\check{P}_3$ (*right*) triangular finite elements

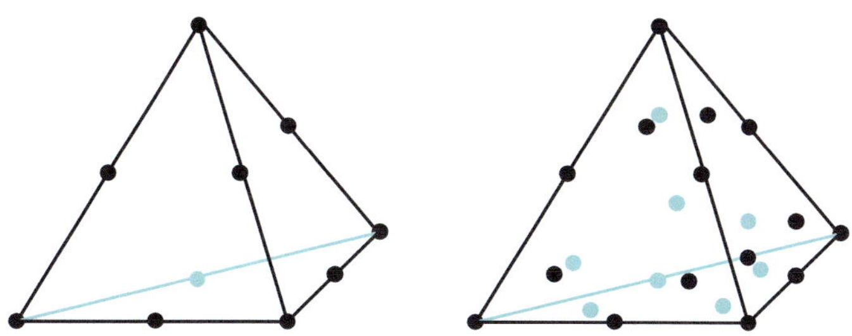

Fig. 2.6 The P_2 (*left*) and $\check{P}_2$ (*right*) tetrahedral elements

Table 2.7 Quadrature points and weights with their multiplicity for mass-lumped triangles

Element	Nodes	Mult.	Weights	Parameters
$r = r' = N = 1$	(0,0)	3	$\frac{1}{6}$	–
$r = 2, r' = 3, N = 3$	(0,0)	3	$\frac{1}{40}$	–
	$\left(\frac{1}{2}, 0\right)$	3	$\frac{1}{15}$	–
	$\left(\frac{1}{3}, \frac{1}{3}\right)$	1	$\frac{9}{40}$	–
$r = 3, r' = 4, N = 5$	(0,0)	3	$\frac{1}{90} - \frac{\sqrt{7}}{720}$	–
	$(\alpha,0)$	6	$\frac{7}{720} - \frac{\sqrt{7}}{180}$	$\frac{1}{2} - \frac{\sqrt{441 - 84(7 - \sqrt{7})}}{42}$
	(β,β)	3	$\frac{49}{360} - \frac{7\sqrt{7}}{720}$	$\frac{1}{3}\left(1 - \frac{\sqrt{7}}{7}\right)$
$r = 4, r' = 5, N = 7$	(0,0)	3	$\frac{1}{315}$	–
	$\left(\frac{1}{2}, 0\right)$	3	$\frac{4}{315}$	–
	$(\alpha,0)$	6	$\frac{3}{280}$	$\frac{1}{2}\left(1 - \frac{\sqrt{3}}{3}\right)$
	(β_1,β_1)	3	$\frac{163}{2520} + \frac{47\sqrt{7}}{8820}$	$\frac{5 + \sqrt{7}}{18}$
	(β_2,β_2)	3	$\frac{163}{2520} - \frac{47\sqrt{7}}{8820}$	$\frac{5 - \sqrt{7}}{18}$
$r = 5, r' = 7, N = 10$	(0,0)	3	0.709423970679E-03	–
	$(\alpha_1,0)$	6	0.619056500367E-02	0.363298074154
	$(\alpha_2,0)$	6	0.348057864050E-02	0.132264581633
	(β_1,β_1)	3	0.345043037728E-01	0.457836838079
	(β_2,β_2)	3	0.459012376308E-01	0.256859107620
	(β_2,β_2)	3	0.116261354596E-01	0.575276844114
	(γ,δ)	6	0.272785759700E-01	γ =0.781925836255
				δ =0.221001218760

Table 2.8 Quadrature points and weights with their multiplicity for mass-lumped tetrahedra

Element	Nodes	Mult.	Weights	Parameters
$r = r' = 1$ $r'' = N = 1$	(0,0,0)	4	$\frac{1}{24}$	–
$r = 2$, $r' = 4$, $r'' = N = 4$	(0,0,0)	4	$\frac{13 - 3\sqrt{13}}{10080}$	–
	$\left(\frac{1}{2}, 0, 0\right)$	6	$\frac{4 - \sqrt{13}}{315}$	–
	$(\alpha, \alpha, 0)$	12	$\frac{29 + 17\sqrt{13}}{10080}$	$\frac{7 - \sqrt{13}}{18}$
	$\left(\frac{1}{4}, \frac{1}{4}, \frac{1}{4}\right)$	1	$\frac{16}{315}$	–
$r = 3$, $r' = 6$, $r'' = 5$, $N = 7$	(0,0,0)	4	0.214360866805E-03	–
	$(\alpha, 0, 0)$	12	0.826817951780E-03	0.292829404767
	$(\beta_1, \beta_1, 0)$	12	0.184017790419E-02	0.197286228026
	$(\beta_2, \beta_2, 0)$	12	0.183132432925E-02	0.425646124314
	(γ, γ, γ)	4	0.754246890465E-02	0.950377585839E-01
	$\left(\delta, \delta, \frac{1}{2} - \delta\right)$	4	0.136099175597E-01	0.125246236258

2.4 Purely 3D Elements

It seems obvious that hexahedra hold interesting properties which enable us to get both spectral and mass-lumped elements. Unfortunately, it is very difficult to produce a purely hexahedral mesh for complex geometries.[6] A palliative to this difficulty is to construct hybrid meshes, containing tetrahedra and (mainly) hexahedra. In order to hybridize these two kinds of elements, it is useful to use pyramids and even wedges to stick together the different shapes.[7]

[6]Unless by splitting tetrahedra into four hexahedra, which provides a very distorted mesh. This distortion leads to very bad performance in terms of accuracy and stability.

[7]One must keep in mind that, besides the fact that the faces of tetrahedra lean on three edges and those of hexahedra on four, the first faces are plane while the second ones are not always plane. Thereby, even by sticking the edges of two triangular faces of two tetrahedra to the four edges of the quadrangular face of a an hexahedron, we get overlapping or underlapping of the elements. For this reason, pyramids and wedge are useful interfaces between tetrahedra and hexahedra.

2.4.1 *Wedges*

H^1-wedges are easy to construct. We first define a unit prism $\hat{W}$ by a tensor product of an unit triangle and a 1D spectral element of the same order (one can also mix the orders) which provides a regular element with plane faces. The 1D element is in the $\hat{x}_3$-direction. In other words, the polynomial space of order r associated to the unit prism is

$$P_r(\hat{x}_1, \hat{x}_2) \otimes P_{r'}(\hat{x}_3), \tag{2.44}$$

where $P_r(\hat{x}_1, \hat{x}_2)$ is the polynomial space on the unit triangle and $P_{r'}(\hat{x}_3)$ the 1D polynomial space in variable $\hat{x}_3$. Its dimension is $(r+1)(r+2)(r'+1)/2$.

The basis functions are the Cartesian product of the set of the basis functions on the unit triangle and the basis functions on [0, 1]. In general, one takes $r' = r$.

In a second step, in order to construct a wedge W of any shape (with straight edges) we define a transform $\underline{F}^w$ based on the six Q_1 basis functions of the unit prism, i.e.,

$$\hat{p}_1(\hat{x}_1, \hat{x}_2, \hat{x}_3) = (1 - \hat{x}_1 + \hat{x}_2)(1 - \hat{x}_3),\ \hat{p}_2(\hat{x}_1, \hat{x}_2, \hat{x}_3) = \hat{x}_1(1 - \hat{x}_3),$$

$$\hat{p}_3(\hat{x}_1, \hat{x}_2, \hat{x}_3) = \hat{x}_2(1 - \hat{x}_3),\ \hat{p}_4(\hat{x}_1, \hat{x}_2, \hat{x}_3) = (1 - \hat{x}_1 + \hat{x}_2)\hat{x}_3,$$

$$\hat{p}_5(\hat{x}_1, \hat{x}_2, \hat{x}_3) = \hat{x}_1\hat{x}_3,\ \hat{p}_6(\hat{x}_1, \hat{x}_2, \hat{x}_3) = \hat{x}_2\hat{x}_3,$$

which reads

$$\underline{F}_i^w(\hat{x}_1, \hat{x}_2, \hat{x}_3) = \sum_{j=1}^{6} \underline{a}_j \hat{p}_j(\hat{x}_1, \hat{x}_2, \hat{x}_3), \tag{2.45}$$

where $\underline{a}_j$ are the summits of a given wedge ($(\underline{a}_1, \underline{a}_2, \underline{a}_3)$ and $(\underline{a}_4, \underline{a}_5, \underline{a}_6)$ respectively being the summits of the two triangular faces of the wedge).

As for other elements, basis functions p_i on W are defined by

$$p_i \circ \underline{F}_i^w = \hat{p}_i.$$

Remark: One can use higher-order basis functions to get curved edges.

2.4.2 *Pyramids*

Constructing nodal pyramidal finite elements is not an obvious problem since its direct definition cannot be based on a polynomial space if we want to preserve the conformity of pyramids with other elements[8] [23]. For this reason, Bedrosian

[8] More precisely, the use of polynomial functions does not keep the plane character of the triangular faces when one transforms a reference pyramid into a pyramid of any shape.

introduced a space of rational functions and constructed rational basis functions of order 1 and 2 [23]. However, the quadrilateral face had no node at its center, which avoids the conformity with hexahedra. Zgainski et al. [24] added a center node which was not optimal. Graglia et al. [25] introduced a new basis function at the center of the face which provides an optimal element. Chatzi and Preparata [26] were the first to introduce a basis of any order. Unfortunately, their basis are consistent up to third order.

An important step, based on the *hp* approach [27, 28], was realized by Karniadakis, Sherwin et al. and Warburton [20, 29–31] who constructed consistent pyramidal finite elements of any order with an optimal error for first-order mapping.

In the following, we provide a way to construct pyramidal finite elements of any order following [32] (and close to that introduced by Karniadakis et al. and Zaglmayr [33]) which seems to have optimal error estimates for high-order mappings.

2.4.2.1 The Space of Rational Functions on the Unit Pyramid

The main difficulty to construct a pyramidal finite element is the definition of the space to which belong the functions. Let us first define the space $\mathscr{P}_r$ corresponding to the unit pyramid $\hat{\Pi}$, i.e. the (symmetric) pyramid whose summits are $\underline{\hat{a}}_1 = (-1, -1, 0)$, $\underline{\hat{a}}_2 = (1, -1, 0)$, $\underline{\hat{a}}_3 = (1, 1, 0)$, $\underline{\hat{a}}_4 = (-1, 1, 0)$, $\underline{\hat{a}}_5 = (0, 0, 1)$ (Fig. 2.7)

$$\mathscr{P}_r = P_r(\hat{x}_1, \hat{x}_2, \hat{x}_3) \oplus \sum_{k=0}^{r-1} \left(\frac{\hat{x}_1 \hat{x}_2}{1 - \hat{x}_3} \right)^{r-k} P_k(\hat{x}_1, \hat{x}_2), \tag{2.46}$$

where $P_k(\hat{x}_1, \hat{x}_2)$ and $P_r(\hat{x}_1, \hat{x}_2, \hat{x}_3)$ are the two polynomial spaces introduced in (2.28) for $d = 2$ and $d = 3$. After some computations (using the formula $\sum_{k=1}^{n} k^2 =$

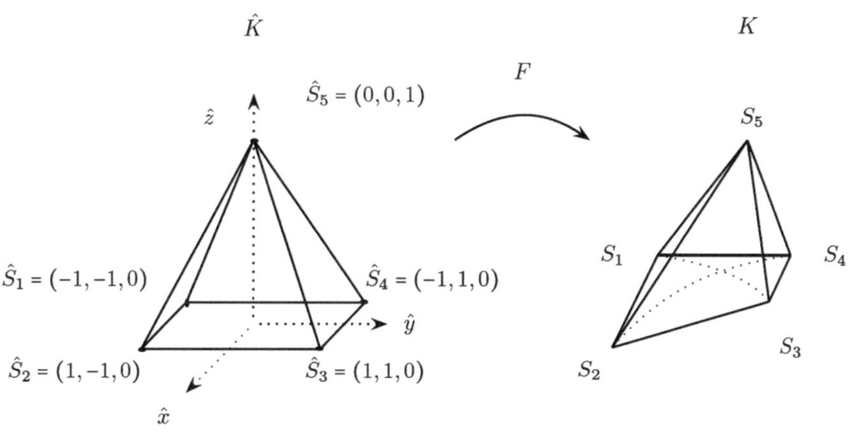

Fig. 2.7 First-order mapping for a pyramid

$n(n+1)(n+2)/6$), one can check that

$$\dim \mathscr{P}_r = \frac{1}{6}(r+1)(r+2)(2r+3).$$

Remark: We have $\dim P_r < \dim \mathscr{P}_r < \dim Q_r$.

2.4.2.2 Construction of Pyramids of Any Shape

In a second step, we construct a pyramid Π of any shape (with straight edges) by using the mapping based on the five shape functions $\left(\hat{p}_i\right)_{i=1}^{5}$ of $\mathscr{P}_1$ such that $\hat{p}_i(\hat{\underline{a}}_j) = \delta_{ij}$, δ_{ij} being the Kronecker symbol (Fig. 2.7). This mapping reads

$$\underline{F}^{\pi}(\hat{x}_1, \hat{x}_2, \hat{x}_3) = \sum_{j=1}^{5} \underline{a}_j \hat{p}_j(\hat{x}_1, \hat{x}_2, \hat{x}_3), \tag{2.47}$$

where $\left(\underline{a}_i\right)_{i=1}^{5}$ are the summits of Π and

$$\begin{cases} \hat{p}_1(\hat{x}_1, \hat{x}_2, \hat{x}_3) = \dfrac{1}{4}\left(1 - \hat{x}_1 - \hat{x}_2 - \hat{x}_3 + \dfrac{\hat{x}_1\hat{x}_2}{1-\hat{x}_3}\right), \\ \hat{p}_2(\hat{x}_1, \hat{x}_2, \hat{x}_3) = \dfrac{1}{4}\left(1 + \hat{x}_1 - \hat{x}_2 - \hat{x}_3 - \dfrac{\hat{x}_1\hat{x}_2}{1-\hat{x}_3}\right), \\ \hat{p}_3(\hat{x}_1, \hat{x}_2, \hat{x}_3) = \dfrac{1}{4}\left(1 + \hat{x}_1 + \hat{x}_2 - \hat{x}_3 + \dfrac{\hat{x}_1\hat{x}_2}{1-\hat{x}_3}\right), \\ \hat{p}_4(\hat{x}_1, \hat{x}_2, \hat{x}_3) = \dfrac{1}{4}\left(1 - \hat{x}_1 + \hat{x}_2 - \hat{x}_3 - \dfrac{\hat{x}_1\hat{x}_2}{1-\hat{x}_3}\right), \\ \hat{p}_5(\hat{x}_1, \hat{x}_2, \hat{x}_3) = \hat{x}_3. \end{cases} \tag{2.48}$$

$\underline{F}^{\pi}$ can also be written on the canonical basis of $\mathscr{P}_1$ as follows:

$$\begin{aligned} \underline{F}^{\pi}(\hat{x}_1, \hat{x}_2, \hat{x}_3) = \frac{1}{4}\Big[&(\underline{a}_1 + \underline{a}_2 + \underline{a}_3 + \underline{a}_4) \\ &+\hat{x}_1(-\underline{a}_1 + \underline{a}_2 + \underline{a}_3 - \underline{a}_4) \\ &+\hat{x}_2(-\underline{a}_1 - \underline{a}_2 + \underline{a}_3 + \underline{a}_4) \\ &+\hat{x}_3(4\underline{a}_5 - \underline{a}_1 - \underline{a}_2 - \underline{a}_3 - \underline{a}_4) \\ &+\frac{\hat{x}_1\hat{x}_2}{1-\hat{x}_3}(\underline{a}_1 - \underline{a}_2 + \underline{a}_3 - \underline{a}_4)\Big]. \end{aligned} \tag{2.49}$$

Here also, we have, for any basis function p_i on Π,

$$p_i \circ \underline{F}_i^{\pi} = \hat{p}_i.$$

Remarks:

1. Pyramids are not convenient for they have the drawback of hexahedra since $\underline{F}^{\pi}$ is not linear, which implies a storage of the Jacobian and the Jacobian matrix at each point of interpolation and the drawback of tetrahedra since they are not based on tensor products, which makes it difficult, even impossible, to construct mass-lumped pyramids. A palliative to this problem is the use of the Duffy's transform [19] which enables us to map a cube to a pyramid. The way to get (discontinuous) mass-lumped pyramids is to map the unit cube with Gauss points defined in Sect. 2.2.2 to a pyramid. The corresponding method of quadrature introduced by Radau [34] is quoted in [35] for triangles and was applied to pyramids [36].
2. Another feature of Duffy's transform is to provide a simple definition of $\mathscr{P}_r$. This space is merely the transform of the following polynomial space

$$C_r = \text{span}\{x^i y^j (1-z)^k, \quad 0 \le i, j \le k \le r\} \tag{2.50}$$

from the unit cube $[0, 1]^3$ to $\hat{\Pi}$.
3. One can use higher-order basis functions to get pyramids with curved edges.

2.4.2.3 Construction of Higher Order Basis Functions

In order to ensure the continuity between pyramids and the other types of elements (and also to simplify the computation for discontinuous methods), the nodes on the four triangular faces of $\hat{\Pi}$ coincide with Hesthaven's points on the faces of a tetrahedron and the nodes of the base of $\hat{\Pi}$ are located at the Gauss-Lobatto coordinates. Interior nodes (if any) are added by taken into account symmetry (Fig. 2.8). Let us call $\left(\hat{\underline{c}}_i\right)_{i=1}^{\dim \mathscr{P}_r}$ these nodes.

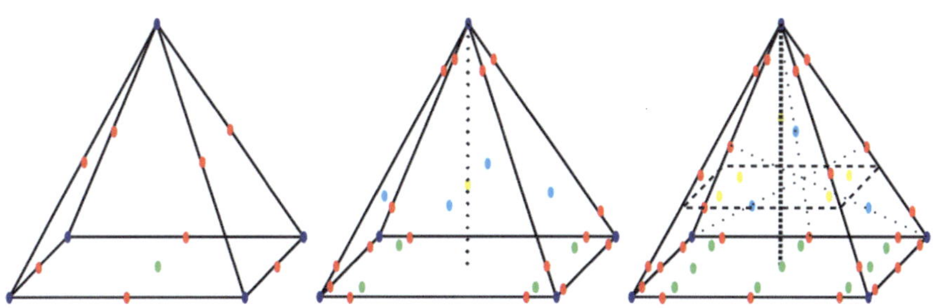

Fig. 2.8 Nodes for pyramids of order 2, 3, 4

For $\mathscr{P}_2$, we have $\dim \mathscr{P}_2 = 14$ and its canonical basis $\left(p_i^c\right)_{i=1}^{14}$ is

$$\{1, \hat{x}_1, \hat{x}_2, \hat{x}_3, \hat{x}_1\hat{x}_2, \hat{x}_1\hat{x}_3, \hat{x}_2\hat{x}_3, \hat{x}_1^2, \hat{x}_2^2, \hat{x}_3^2, \hat{r}, \hat{x}_1\hat{r}, \hat{x}_2\hat{r}, \hat{r}^2\},$$

where $\hat{r} = \hat{x}_1\hat{x}_2/(1 - \hat{x}_3)$.

So, a basis functions of $\hat{p}_i \in \mathscr{P}_2$ such that $\hat{p}_i(\hat{\underline{c}}_j) = \delta_{ij}$ can be decomposed as

$$\hat{p}_i = \sum_{j=1}^{14} \lambda_j^i p_j^c, \quad \lambda_j^i \in \mathbb{R}. \tag{2.51}$$

A natural way to compute these basis functions is to solve, for each $i = 1 \dots 14$ the following (Vandermonde-like) linear systems in λ_j^i:

$$\sum_{j=1}^{14} \lambda_j^i p_j^c(\hat{\underline{c}}_k) = \delta_{ik}, \quad \forall k = 1 \dots 14. \tag{2.52}$$

Unfortunately, as for tetrahedra, this system is ill-conditioned for high orders (Fig. 2.9). A palliative to this problem is to use an extension of the orthogonal basis introduced in (2.38). This basis is given by

$$\begin{aligned} \varphi_{i,j,k} &= P_i^0(2a-1)P_j^0(2b-1)(1-c)^{\max(i,j)} P_k^{2\max(i,j)+2}(2c-1), \\ 0 &\le i, j \le r, 0 \le k \le r - \max(i,j), \end{aligned} \tag{2.53}$$

where $a = (\hat{x}_1 - \hat{x}_3 + 1)/2(1 - \hat{x}_3)$, $b = (\hat{x}_2 - \hat{x}_3 + 1)/2(1 - \hat{x}_3)$, $c = \hat{x}_3$ and P_k^2 is defined as in (2.39).

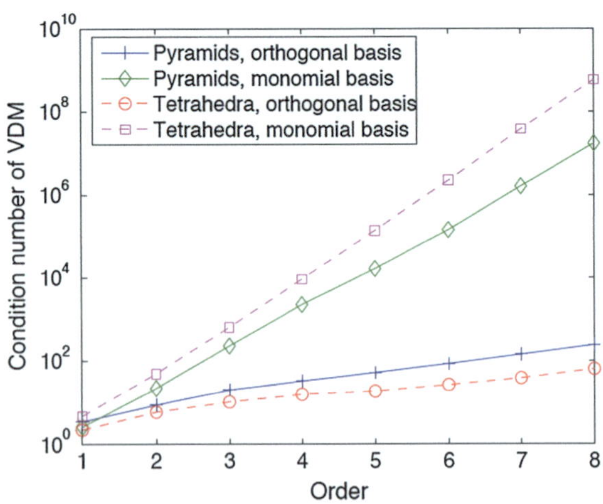

Fig. 2.9 Condition number of the linear systems providing nodal basis functions for tetrahedra and pyramids

Remark: An orthogonal basis can be used to define modal finite elements, which provide mass-lumping in discontinuous Galerkin methods [37].

2.5 Tetrahedral and Triangular Edge Elements

2.5.1 Mixed Formulation

Edge elements belong to the family of mixed finite elements which seem to have been derived from the Hellinger-Reissner principle for elasticity by Fraeijs de Veubeke [38]. This principle is to split a second order PDE system into two first order equations and to use different finite element approximations for each equation. First used for elastics, this approach was studied for the 2D harmonic problem [39] and eventual other problems and extended for 3D elements in [40]. For instance, the harmonic problem

$$-\Delta u = f \tag{2.54}$$

is replaced by

$$\begin{aligned} -\nabla \cdot \underline{q} &= f, \\ \underline{q} &= \nabla u. \end{aligned} \tag{2.55}$$

In a second step, u is sought in L^2 and $\underline{q}$ in $H(div)$. One can notice that, in such formulation, only one variable $\underline{q}$ keeps its physical character (normal continuity), the (continuous) character of the second one being weakened. Although classical, we shall see later that this choice is not unique.

An important extension of this approach for Maxwell's equations was given by Nédélec [40, 41] who constructed two families of edge elements, i.e. finite elements which belong to $H(curl)$ and then keep the important tangential continuity of the electric field.

Despite the anteriority of $H(div)$-conform elements, we first describe the $H(curl)$-conform elements which are more suited to wave phenomena. In the following, we just give a description of these elements. A exhaustive study of their properties can be found in [42].

2.5.2 A First Family

We describe in this section the first family of edge elements defined in [40], which is the most popular and which is also the most suited to Maxwell's equations.

2.5.2.1 The Polynomial Space

Let us first define the polynomial space $\mathscr{R}_r$ involved for the unit tetrahedron $\hat{T}_3$ defined at (2.37). We have

$$\mathscr{R}_r = (P_{r-1})^3 \oplus \underline{S}_r, \tag{2.56}$$

where

$$\underline{S}_r = \left\{ \underline{p} \in (\widetilde{P}_r)^3 \text{ such that } \underline{\hat{x}} \cdot \underline{p} = 0,\ \underline{x} \in \mathbb{R}^3 \right\}, \tag{2.57}$$

with

$$\widetilde{P}_r = \left\{ \sum_{i+j+k=r} a_{i,j,k}\, \hat{x}_1^i\, \hat{x}_2^j\, \hat{x}_3^k, \quad a_{i,j,k} \in \mathbb{R} \right\} \tag{2.58}$$

is the space of homogeneous polynomials of order r.

Obviously, $\dim (P_{r-1})^3 = r(r+1)(r+2)/2$. On the other hand, it is easy to see that $\widetilde{P}_{r+1} = \operatorname{span}\{\underline{\hat{x}} \cdot \underline{p},\ \underline{p} \in (\widetilde{P}_r)^3\}$ and $\dim \widetilde{P}_r = (r+1)(r+2)/2$. So, $\dim \underline{S}_r = \dim (\widetilde{P}_r)^3 - \dim \widetilde{P}_{r+1} = 3(r+1)(r+2)/2 - (r+2)(r+3)/2 = r(r+2)$. Finally

$$\dim \mathscr{R}_r = \frac{1}{2} r(r+2)(r+3). \tag{2.59}$$

An important issue is the construction of $\underline{S}_r$ which is not natural because of the constraint $\underline{\hat{x}} \cdot \underline{p} = 0$. In [43], $\underline{S}_r$ is characterized as follows

$$\underline{S}_r = \operatorname{span}\{\underline{V}_1, \underline{V}_2, \underline{V}_3\}, \tag{2.60}$$

with

$$\underline{V}_1 = \begin{pmatrix} \hat{x}_1^{m-1}\, \hat{x}_2^n\, \hat{x}_3^{r-m-n+1} \\ 0 \\ -\hat{x}_1^m\, \hat{x}_2^n\, \hat{x}_3^{r-m-n} \end{pmatrix}, \quad 0 \le m+n \le r, m \ne 0, n \ne r, \tag{2.61}$$

$$\underline{V}_2 = \begin{pmatrix} 0 \\ \hat{x}_1^m\, \hat{x}_2^{n-1}\, \hat{x}_3^{r-m-n+1} \\ -\hat{x}_1^m\, \hat{x}_2^n\, \hat{x}_3^{r-m-n} \end{pmatrix}, \quad 0 \le m+n \le r, m \ne r, n \ne 0, \tag{2.62}$$

$$\underline{V}_3 = \begin{pmatrix} \hat{x}_1^{m-1}\, \hat{x}_2^n \\ -\hat{x}_1^m\, \hat{x}_2^{n-1} \\ 0 \end{pmatrix}, \quad m+n = r+1, m \ne 0, n \ne 0. \tag{2.63}$$

Thanks to (2.60)–(2.63), one can construct the basis functions of the method. In order to get a better conditioned system, one can use an orthogonal basis of $(P_{r-1})^3$ and even express (2.61)–(2.63) in terms of orthogonal polynomials.

2.5.2.2 The Degrees of Freedom

In order to construct the basis functions, one must define the degrees of freedom for our method. In our case, these degrees of freedom are not values of the functions at some interpolation points, but tangent and volume momenta which a priori easier define the tangential continuity of the solution.

Following [40], we have three types of degrees of freedom:

1. $6r$ integrals on the edges defined by

$$\int_e \underline{u} \cdot \underline{t} q \, \mathrm{d}\hat{\sigma}, \quad \forall q \in P_{r-1},$$

where $\underline{t}$ is an unit vector along an edge $e \in \hat{T}_3$.

2. $4r(r-1)$ integrals on the faces defined by

$$\int_f (\underline{u} \times \underline{n}) \cdot \underline{q} \, \mathrm{d}\hat{\underline{\gamma}}, \quad \forall \underline{q} \in (P_{r-2})^2,$$

where $\underline{n}$ is the unit vector normal to a face $f \in \hat{T}_3$.

3. $r(r-1)(r-2)/2$ integrals in $\hat{T}_3$ defined by

$$\int_{\hat{T}_3} \underline{u} \cdot \underline{q} \, \mathrm{d}\hat{\underline{x}}, \quad \forall \underline{q} \in (P_{r-3})^3.$$

One can easily check that we have $r(r+2)(r+3)/2$ degrees of freedom. One can find the two first order basis functions in [40] and the third order in [44].

2.5.2.3 Extension to Any Tetrahedron

The extension to any tetrahedron of these functions must take into account the tangential continuity of the solution all over a mesh.

If $\hat{B}$ and B are the sets of basis functions on $\hat{T}_3$ and T respectively, for $\hat{\underline{\varphi}} \in \hat{B}$ and $\underline{\varphi} \in B$, the classical transform $\underline{\varphi} \circ \underline{F}_T = \hat{\underline{\varphi}}$ (used for H^1 functions which are globally continuous) would not ensure the tangential continuity. For this reason, this transform is replaced by the $H(curl)$-conform transform which ensures the tangential continuity

$$\underline{\varphi} \circ \underline{F}_T = DF_T^{-T} \hat{\underline{\varphi}}, \tag{2.64}$$

where DF_T^{-T} is the inverse of the transposed Jacobian matrix of $\underline{F}_T$ defined in (2.40).

Remarks:

1. DF_T is a constant matrix. One can use higher-order basis functions to define $\underline{F}_T$ in order to construct elements with curved faces by using the Gordon-Hall transform [13, 14]. In this case, DF_T is no longer constant.
2. Combined with the first family of $H(div)$-conform finite elements, this family provides a spurious-free approximation of Maxwell's equations.
3. The definition of the degrees of freedom seems to indicate that we have a family of modal elements. Actually, one can construct basis functions by using r vector-valued functions tangent to each edge, $r(r-1)$ functions tangent to each face and $r(r-1)(r-2)/2$ functions inside the tetrahedron. Then the $H(curl)$-conform transform would ensure the tangential continuity. This second definition is very useful for mass-lumping, as described below.

2.5.2.4 Triangles

As for tetrahedra, the edge element can be defined in a modal way on the unit triangle $\hat{T}_2$ defined at (2.35) by using the following degrees of freedom:

1. $3r$ integrals on the edges defined by

$$\int_e \underline{u} \cdot \underline{t} q \, d\hat{\sigma}, \quad \forall q \in P_{r-1},$$

   where $\underline{t}$ is the unit vector along an edge $e \in \hat{T}_2$.
2. $r(r-1)$ integrals in $\hat{T}_2$ defined by

$$\int_{\hat{T}_2} \underline{u} \cdot \underline{q} \, d\hat{\underline{x}}, \quad \forall \underline{q} \in (P_{r-2})^2,$$

or in a nodal way, by using pointwise values of r vector-valued functions tangent to each edge and $r(r-1)$ interior vector-valued functions collinear to the canonical basis of $\mathbb{R}^2$.

The corresponding polynomial space $\mathscr{R}_r^T$ is defined, as for tetrahedra, by

$$\mathscr{R}_r^T = (P_{r-1})^2 \oplus \underline{S}_r^T, \tag{2.65}$$

where

$$\underline{S}_r^T = \left\{ \underline{p} \in (\widetilde{P}_r)^2 \text{ such that } \hat{\underline{x}} \cdot \underline{p} = 0, \ \underline{x} \in \mathbb{R}^2 \right\}. \tag{2.66}$$

As for tetrahedra, one can show that $\dim \underline{S}_r^T = r$ and $\dim \mathscr{R}_r^T = r(r+2) = 3r + r(r-1)$.

More explicitly, we can write

$$\underline{S}_r^T = \mathrm{span}\{\underline{V}\}, \tag{2.67}$$

with

$$\underline{V} = \begin{pmatrix} \hat{x}_1^{m-1}\,\hat{x}_2^{n} \\ -\hat{x}_1^{m}\,\hat{x}_2^{n-1} \end{pmatrix}, \quad m+n=r+1, m \neq 0, n \neq 0. \tag{2.68}$$

Extension to any triangle is made by using (2.64), where $\underline{F}_T$ is then defined as in (2.41).

2.5.3 A Second Family

The second family of edge elements introduced by Nédélec [41] is composed of polynomials of $(P_r)^3$. In order to define the corresponding degrees of freedom, let us first introduce the polynomial spaces

$$\mathscr{D}_r^d = (P_{r-1})^d \oplus \underline{x}\,\widetilde{P}_{r-1}, \tag{2.69}$$

where d is the dimension of space and $\widetilde{P}_r$ is defined as in Sect. 2.5.2. We have $\dim \mathscr{D}_r^2 = r(r+2)$ and $\dim \mathscr{D}_r^3 = r(r+1)(r+3)/2$.

We can now define the degrees of freedom for this family:

1. $6(r+1)$ integrals on the edges defined by

$$\int_e \underline{u} \cdot \underline{t} q \,\mathrm{d}\hat{\sigma}, \quad \forall q \in P_r(e),$$

where $\underline{t}$ is the unit vector along an edge $e \in \hat{T}_3$.

2. $4(r^2-1)$ integrals on the faces defined by

$$\int_f \underline{u} \cdot \underline{q} \,\mathrm{d}\hat{\underline{\gamma}}, \quad \forall \underline{q} \in \mathscr{D}_{r-1}^2(f), \text{ for any face } f \text{ of } \hat{T}_3.$$

2. $(r^2-1)(r-2)/2$ integrals in $\hat{T}_3$ defined by

$$\int_{\hat{T}_3} \underline{u} \cdot \underline{q} \,\mathrm{d}\hat{\underline{x}}, \quad \forall \underline{q} \in \mathscr{D}_{r-2}^3.$$

We get $6(r+1) + 4(r^2-1) + (r^2-1)(r-2)/2 = \dim(P_r)^3$.

Extension to any element T is made in the same way as the first family.

It seems that this family does not provide a spurious-free approximation of Maxwell's equations.

Remarks:

1. The above degrees of freedom can be replaced by values at Hesthaven's points of three vector-valued functions. These functions are tangent to the edges at the summits, tangent to the edge and to the two faces (and orthogonal to the edge) for points located on an edge, tangent to a face and normal to this face (and orthogonal to each other) for points located on a face and collinear to the canonical basis vectors for an interior point (cf. Fig. 2.10).
2. This element was also suggested by Mur et al. [45].

2.5.3.1 Triangles

For the unit triangle $\hat{T}_2$, the polynomial space is $(P_r)^2$ and degrees of freedom are:

1. $3(r+1)$ integrals on the edges defined by

$$\int_e \underline{u} \cdot \underline{t} q \, \mathrm{d}\hat{\sigma}, \quad \forall q \in P_r(e),$$

   where $\underline{t}$ is the unit vector along an edge $e \in \hat{T}_2$.
2. (r^2-1) integrals inside the triangle defined by

$$\int_{\hat{T}_2} \underline{u} \cdot \underline{q} \, \mathrm{d}\hat{\underline{x}}, \quad \forall \underline{q} \in \mathscr{D}^2_{r-1}.$$

In a nodal way, the degrees of freedom are the values of $6r$ vector-valued functions tangent or orthogonal to each edge and $(r-1)(r-2)$ interior vector-valued functions collinear to the canonical basis of $\mathbb{R}^2$ at Hesthaven's points.

Obviously, we get $3(r+1)+(r^2-1) = 6r+(r-1)(r-2) = \dim(P_r)^2$.

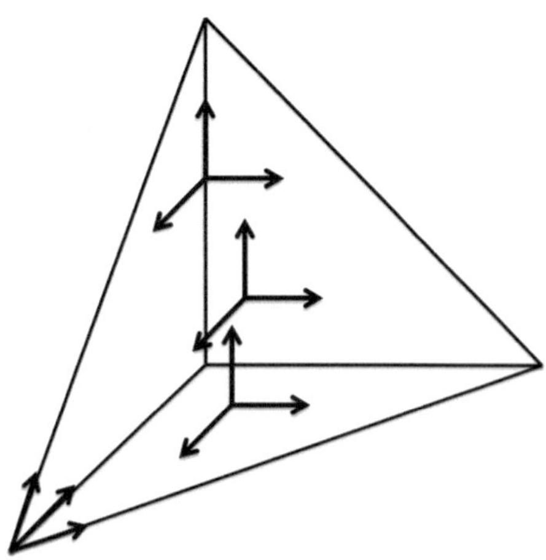

Fig. 2.10 The locations of the degrees of freedom for the second family of edge elements

2.5.4 *Tetrahedral Mass-Lumped Edge Elements*

2.5.4.1 The First-Order Element

Let (S_1, S_2, S_3, S_4) be the vertices of $\hat{T}_3$ and $(\lambda_1, \lambda_2, \lambda_3, \lambda_4)$ the corresponding barycentric coordinates. An edge is defined by its ends and each face is assigned to the index of its opposite vertex. So, a face Φ_j is such that $S_j \notin \Phi_j$. At each midpoint M_{pq} of an edge $S_p S_q$, we define the three unit vectors $(\underline{e}_{pq,1}, \underline{e}_{pq,2}, \underline{e}_{pq,3})$ such that $\underline{e}_{pq,1}$ is collinear to $S_p S_q$, and $\underline{e}_{pq,2}, \underline{e}_{pq,3}$ are the two vectors orthogonal to $\underline{e}_{pq,1}$ and parallel to the two faces containing $S_p S_q$. The degrees of freedom are the three components of a function at the point M_{pq} in this local basis, for each edge of the tetrahedron T (Fig. 2.11). So, we have 18 degrees of freedom (instead of 6 for the regular first family).

Now, let us set $I = \{(\{\ell, m\}, n)$ such that $(\ell, m, n) \in \{1, 2, 3, 4\}^3, \ell \neq m, n \neq \ell, n \neq m\}$. We can define the following spaces of polynomial vector-valued functions

$$\mathscr{S}_1 = \{\underline{u}(\hat{\underline{x}}) = \underline{a} \times \hat{\underline{x}} + \underline{b},\ (\underline{a}, \underline{b}) \in \mathbb{R}^3 \times \mathbb{R}^3\}, \tag{2.70}$$

$$\widetilde{\mathscr{S}}_1 = \mathscr{S}_1 \oplus \{\underline{w}^n_{\ell m}\}_{(\{\ell,m\},n)\in I}, \tag{2.71}$$

where

$$\underline{w}^n_{\ell m} = \lambda_\ell \lambda_m \nabla \lambda_n, \tag{2.72}$$

and ∇ is the gradient versus $\hat{\underline{x}}$.

Let $\underline{\nu}_j$ be the unit outward normal to a face $\Phi_j = S_\ell S_m S_n$. If we set

$$\nabla_j \underline{w} = \underline{\nu}_j \times ((\nabla \underline{w})_{|\Phi_j} \times \underline{\nu}_j), \tag{2.73}$$

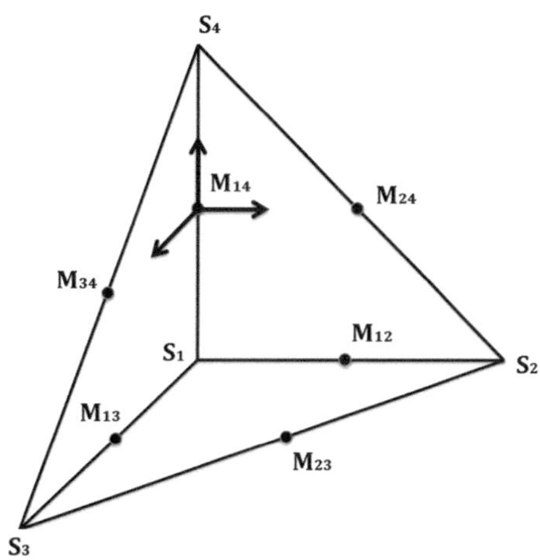

Fig. 2.11 The locations of the degrees of freedom and the three degrees of freedom at the point M_{14} for $\widetilde{\mathscr{S}}_1$ (see (2.71))

which is actually the surface gradient of $\underline{w}$ on Φ_j, we have

$$\underline{\nu}_j \times ((\underline{w}^n_{\ell m})_{|\Phi_j} \times \underline{\nu}_j) = \underline{t}^n_{\ell m} = \lambda_\ell \lambda_m \nabla_j \lambda_j. \tag{2.74}$$

Let

$$\mathscr{R}_1(\Phi_j) = \{\underline{\nu}_j \times (\underline{u}_{|\Phi_j} \times \underline{\nu}_j),\ \underline{u} \in \mathscr{S}_1\} \tag{2.75}$$

be the space of the tangential traces of functions of $\mathscr{S}_1$ and

$$\widetilde{\mathscr{R}}_1(\Phi_j) = \mathscr{R}_1(\Phi_j) \oplus \{\underline{t}^n_{\ell m}, \underline{t}^m_{n\ell}, \underline{t}^\ell_{mn}\}. \tag{2.76}$$

Equation (2.74) implies that $\widetilde{\mathscr{R}}_1(\Phi_j)$ is the space of the tangential traces of functions of $\widetilde{\mathscr{S}}_1$. One can easily see that $\mathscr{R}_1(\Phi_j)$ is isomorphic to the space $\widetilde{\mathscr{R}}_1$ defined in (2.84), which shows that the restriction to a face of the space of approximation is isomorphic to a subspace of the 2D $H(curl, \Omega)$ space.

The corresponding quadrature formula is

$$\int_T f(\underline{x})\,\mathrm{d}\underline{x} \simeq \frac{\mathrm{mes}(\hat{T}_3)}{6} \sum_{(\ell,m)\in\mathscr{I}} f(M_{\ell m}), \tag{2.77}$$

where $\mathscr{I} = \{(p,q) \in \{1,2,3,4\}^2 \text{ such that } p < q\}$.

2.5.4.2 The Second-Order Element

Paradoxically, the second-order element is easier to define than the first-order one. Its degrees of freedom are the three components of a function in the basis composed of the three unit vectors at each vertex and, on the other hand, its three components on an orthonormal basis at the center G_j of a face Φ_j, composed of the unit outward normal at this point and two arbitrary orthogonal unit vectors parallel to the face (Fig. 2.12). The space of polynomials for which this set of degrees of freedom is unisolvent can be constructed as follows:

Let $\mathscr{V}_2$ be the eight-dimensional space generated by the vectors

$$\begin{pmatrix} \hat{x}_2^2 \\ -\hat{x}_1\hat{x}_2 \\ 0 \end{pmatrix}, \begin{pmatrix} 0 \\ -\hat{x}_2\hat{x}_3 \\ \hat{x}_2^2 \end{pmatrix}, \begin{pmatrix} -\hat{x}_1\hat{x}_2 \\ \hat{x}_1^2 \\ 0 \end{pmatrix}, \begin{pmatrix} -\hat{x}_2\hat{x}_3 \\ 0 \\ \hat{x}_1^2 \end{pmatrix},$$

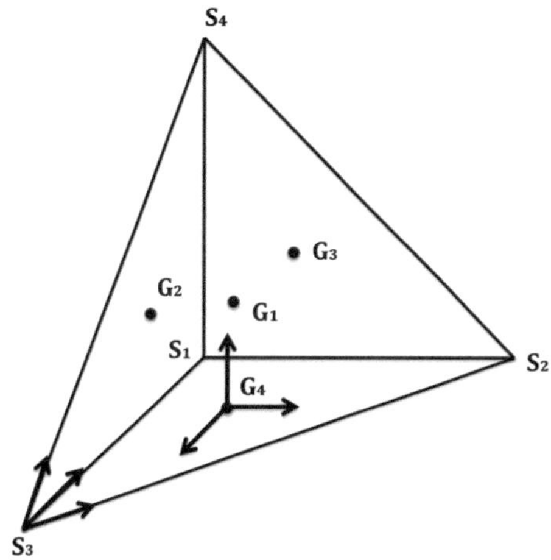

Fig. 2.12 The locations of the degrees of freedom and the three degrees of freedom at the point G_4 for $\widetilde{\mathscr{S}}_2$

$$\begin{pmatrix} \hat{x}_3^2 \\ 0 \\ -\hat{x}_1\hat{x}_3 \end{pmatrix}, \begin{pmatrix} 0 \\ \hat{x}_3^2 \\ -\hat{x}_2\hat{x}_3 \end{pmatrix}, \begin{pmatrix} \hat{x}_2\hat{x}_3 \\ -\hat{x}_1\hat{x}_3 \\ 0 \end{pmatrix}, \begin{pmatrix} 0 \\ \hat{x}_1\hat{x}_3 \\ -\hat{x}_1\hat{x}_2 \end{pmatrix}$$

and

$$\mathscr{S}_2 = [P_3(T)]^3 \oplus \mathscr{V}_2. \tag{2.78}$$

The polynomial space $\widetilde{\mathscr{S}}_2$ of dimension 24 (instead of 20 for the first family and 30 for the second family of second-order Nédélec's elements), for which the set of degrees of freedom is unisolvent, is

$$\widetilde{\mathscr{S}}_2 = \mathscr{S}_2 \oplus \{\underline{w}_1, \underline{w}_2, \underline{w}_3, \underline{w}_4\}, \tag{2.79}$$

where

$$\underline{w}_p = \lambda_\ell \lambda_m \lambda_n \nabla \lambda_p$$

and $\{\ell, m, n, p\}$ is a circular permutation of $\{1, 2, 3, 4\}$.

The corresponding quadrature formula is

$$\int_T f(\underline{x})\, \mathrm{d}\underline{x} \simeq \mathrm{mes}(\hat{T}_3) \left[\frac{3}{80} \sum_{p=1}^{4} f(G_p) + \frac{1}{720} \sum_{q=1}^{4} f(S_q) \right]. \tag{2.80}$$

One could a priori expect to obtain two points per edge with three degrees of freedom at each point. The quadrature formula was actually sought for this arrangement of points but the unique computed result was that the two points on the edges were moved to its ends. The dimension of $\widetilde{\mathscr{P}}_2$ shows that this element (also called a *vertex element*) seems to be between the first and the second families of Nédélec's elements. A similar element can be constructed in 2D.

The stability conditions for a regular mesh in a homogeneous isotropic medium such that $c = \sqrt{\varepsilon\mu}$ are, for a leapfrog scheme on a regular mesh composed of cubic cells divided into six tetrahedra (which provides 43 classes of degrees of freedom for $\widetilde{\mathscr{P}}_1$ and 62 for $\widetilde{\mathscr{P}}_2$!) [46]:

- $c\Delta t/h \leq 0.18$ for the first-order element,
- $c\Delta t/h \leq 0.21$ for the second-order element.

Remarks:

1. As for classical edge elements, one can deduce basis functions on an element of any shape by using (2.64).
2. This element has a surprisingly large stability condition and low number of degrees of freedom.
3. The third-order element contains too many degrees of freedom to be efficient.

2.5.5 *Triangular Mass-Lumped Edge Elements*

For a triangle, it is possible to define the basis functions on a triangle T of any shape. Therefore, we assume that the two basis functions at a point (which is not a vertex) are tangential and normal to the edges. At a vertex, they are tangential to the two edges. The degrees of freedom are defined in the same way.

2.5.5.1 The First-Order Element

The lower-order triangular edge element with mass-lumping is derived from the lower-order edge element given in [40]. Its six degrees of freedom are the values of the tangential and normal components at the midpoint of each edge of the triangle (Fig. 2.13). The number of its degrees of freedom would naturally suggest taking its basis functions in $(P_1)^2$. However, this set of degrees of freedom is not $(P_1)^2$-unisolvent. In order to construct the good space of approximation, we introduce the barycentric coordinates $(\lambda_1, \lambda_2, \lambda_3)$ of a point M of T such that, for any point O of $\mathbb{R}^2$, we have

$$\underline{OM} = \lambda_1 \underline{OS_1} + \lambda_2 \underline{OS_2} + \lambda_3 \underline{OS_3}, \tag{2.81}$$

where (S_1, S_2, S_3) are the vertices of T.

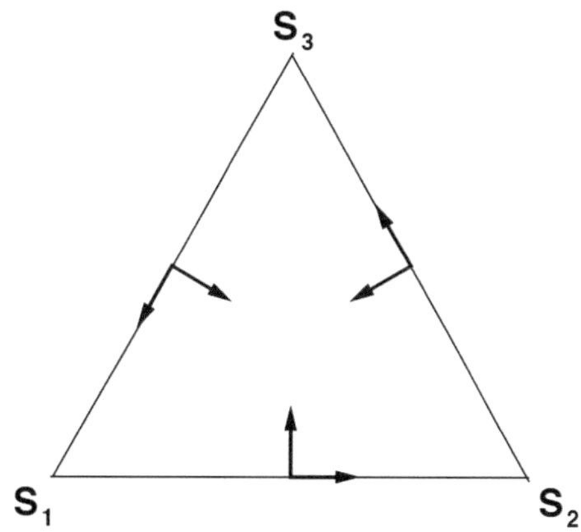

Fig. 2.13 The first-order triangular edge element with mass-lumping

We also introduce the set of functions

$$\mathscr{R}_1 = \{\underline{u} = \underline{a} + b(x_2, -x_1)^T,\ \underline{a} \in \mathbb{R}^2, b \in \mathbb{R}\} \tag{2.82}$$

and the three basis functions

$$\underline{w}_1 = \nabla\lambda_1\lambda_2\lambda_3, \tag{2.83a}$$

$$\underline{w}_2 = \lambda_1\nabla\lambda_2\lambda_3, \tag{2.83b}$$

$$\underline{w}_3 = \lambda_1\lambda_2\nabla\lambda_3. \tag{2.83c}$$

So, we can define

$$\widetilde{\mathscr{R}}_1 = \mathscr{R}_1 \oplus \{\underline{w}_1, \underline{w}_2, \underline{w}_3\}, \tag{2.84}$$

for which the set of degrees of freedom is unisolvent.

Of course, we have here two kinds of basis functions:

- 3 basis functions corresponding to the normal components:

$$\underline{w}_j^{\nu} = \frac{4}{||\nabla\lambda_j||}\underline{w}_j,\ \ j = 1\ldots3, \tag{2.85}$$

- 3 basis functions corresponding to the tangential components (which are actually the basis functions of $\mathscr{R}_1$):

$$\underline{w}_j^{\tau} = \underline{u}_j + \sum_{\ell=1}^{3}\alpha_{j\ell}\underline{w}_{\ell}^{\nu}, \tag{2.86}$$

where $\underline{u}_j = \nabla\lambda_m\lambda_n - \nabla\lambda_n\lambda_m$, $(m,n) \in \{1,2,3\}^2$, $m \neq j$, $n \neq j$, $\alpha_{j\ell} = -(\underline{u}_j \cdot \underline{\nu})(M_\ell)$. $\underline{\nu}$ is the unit outward normal to ∂T and M_ℓ is the midpoint of the edge opposite to the vertex S_ℓ.

The appropriate quadrature formula is, of course, given by

$$\int_T f(\underline{x})\,\mathrm{d}\underline{x} \simeq \frac{\mathrm{mes}(T)}{3}\sum_{\ell=1}^{3} f(M_\ell), \tag{2.87}$$

where $\mathrm{mes}(T)$ is the measure of the triangle.

2.5.5.2 The Second-Order Element

In order to define the second-order element, we first introduce some additional notations. For each edge of T, we denote by M_{pq} the point of the edge S_pS_q such that

$$\underline{S_pM_{pq}} = \alpha\underline{S_pS_q}, \tag{2.88}$$

where α is a given real constant. At each point M_{pq}, we define two degrees of freedom which are the values of the normal and tangential components of a function and at G, the two components of the functions in the two directions of space are the degrees of freedom (Fig. 2.14).

Let $\mathscr{V}_2$ be the space of polynomials generated by the vectors $(x_2^2, -x_1x_2)^T$, $(-x_1x_2, x_1^2)^T$,

$$\mathscr{R}_2 = (P_1(T))^2 \oplus \mathscr{V}_2, \tag{2.89}$$

and

$$\underline{w}_{mn} = \nabla\lambda_\ell\lambda_m\lambda_n\phi_{nm}, \tag{2.90a}$$

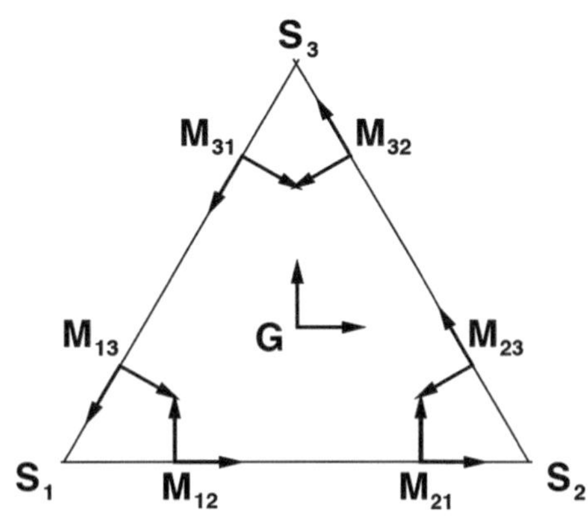

Fig. 2.14 The second-order triangular edge element with mass-lumping

$$\underline{w}_{nm} = \nabla \lambda_\ell \lambda_m \lambda_n \phi_{mn}, \tag{2.90b}$$

where $(\ell, m, n) \in \{1, 2, 3\}^3$ and are all different, and ϕ_{nm} is the polynomial of P_1 defined by $\phi_{nm}(G) = \phi_{nm}(M_{nm}) = 0$ and $\phi_{nm}(M_{mn}) = 1$.

With the above notations, we can define the space of polynomials $\widetilde{\mathscr{R}}_2$ for which the set of degrees of freedom is unisolvent

$$\widetilde{\mathscr{R}}_2 = \mathscr{R}_2 \oplus \{\underline{w}_{12}, \underline{w}_{21}, \underline{w}_{13}, \underline{w}_{31}, \underline{w}_{23}, \underline{w}_{32}\}. \tag{2.91}$$

The corresponding quadrature rule to obtain mass-lumping is given by

$$\int_T f(\underline{x})\, \mathrm{d}\underline{x} \simeq \mathrm{mes}(T) \left[\frac{9}{40} f(G) + \frac{11}{240} \sum_{(p,q)} f(M_{pq}) \right], \tag{2.92}$$

where $(p, q) \in \{1, 2, 3\}^2$ and $p \neq q$.

Stability conditions for a regular mesh in a homogeneous isotropic medium such that $c = \sqrt{\varepsilon\mu}$ are, for a leapfrog scheme on a regular mesh composed of rectangle triangles [46]:

- $c\Delta t / h \leq 0.2654$ for the first-order element,
- $c\Delta t / h \leq 0.1167$ for the second-order element.

Remark: A third-order element was studied in [46] but was less efficient than the second-order one because of its large number of degrees of freedom. For this reason, we do not describe it here.

2.6 Hexahedral and Quadrilateral Edge Elements

Both families of Nédélec's edge elements can be defined on hexahedra and quadrilaterals. However, the polynomial spaces involved in these elements are much easier to define. They actually are tensor products of 1D polynomial spaces.

2.6.1 First Family

2.6.1.1 Modal Definition

This family will be first defined in a modal way [40] on the unit cube $\widehat{K} = [0, 1]^3$. The polynomial space corresponding to this family reads

$$\mathscr{Q}_r = \left\{ \underline{u} \text{ such that } u_1 \in Q_{r-1,r,r},\ u_2 \in Q_{r,r-1,r},\ u_3 \in Q_{r,r,r-1} \right\}, \tag{2.93}$$

where

$$Q_{pqr}(\underline{\hat{x}}) = P_p(\hat{x}_1) \times P_q(\hat{x}_2) \times P_r(\hat{x}_3)$$
$$= \left\{ v(\underline{\hat{x}}) = \sum_{i=0}^{p}\sum_{j=0}^{q}\sum_{k=0}^{r} a_{i,j,k}\, \hat{x}_1^i \hat{x}_2^j \hat{x}_3^k, \quad a_{i,j,k} \in \mathbb{R} \right\}. \tag{2.94}$$

Obviously,

$$\dim \mathscr{Q}_r = 3r(r+1)^2. \tag{2.95}$$

The modal degrees of freedom related to this family are

1. $12r$ integrals on the edges defined by
$$\int_e \underline{u} \cdot \underline{t} q \,\mathrm{d}\hat{\sigma}, \quad \forall q \in P_{r-1},$$
where $\underline{t}$ is the unit vector along an edge $e \in \widehat{K}$.
2. $6 \times 2r(r-1)$ integrals on the faces defined by
$$\int_f (\underline{u} \times \underline{n}) \cdot \underline{q}\,\mathrm{d}\hat{\gamma}, \quad \forall \underline{q} = (q_1, q_2) \text{ such that } q_1 \in Q_{r-2,r-1},\ q_2 \in Q_{r-1,r-2},$$
where $\underline{n}$ is the unit vector normal to a face $f \in \widehat{K}$.
3. $3r(r-1)^2$ integrals in $\widehat{K}$ defined by
$$\int_{\widehat{K}} \underline{u} \cdot \underline{q}\,\mathrm{d}\underline{\hat{x}}, \quad \forall \underline{q} = (q_1, q_2, q_3)$$
such that $q_1 \in Q_{r-1,r-2,r-2}$, $q_2 \in Q_{r-2,r-1,r-2}$, $q_3 \in Q_{r-2,r-2,r-1}$.

Of course, $12r + 6 \times 2r(r-1) + 3r(r-1)^2 = 3r(r+1)^2$.

Extension to any hexahedron K_i is made by the $H(curl)$-conform transform defined in (2.64) in which DF_T is replaced by DF_i, where DF_i is the Jacobian matrix of the mapping $\underline{F}_i$ defined in (2.24).[9] This extension holds for any hexahedral or quadrilateral edge element.

2.6.1.2 Nodal Definition

The tensor character of $\mathscr{Q}_r$ enables us to define the degrees of freedom in a nodal way by using vector values of the basis functions in the three canonical directions.

---

[9]The (not obvious) $H(curl)$-conform character of this mapping for hexahedra is proven in the Annex of [8].

However, in order to agree with the definition of the spaces, we must use different sets of points for each direction. Actually, we need $r \times (r+1) \times (r+1)$ points in the $\hat{x}_1$ direction, $(r+1) \times r \times (r+1)$ in the $\hat{x}_2$ direction and $(r+1) \times (r+1) \times r$ in the $\hat{x}_3$ direction. This structure allows us to choose r Gauss points and $r+1$ Gauss-Lobatto points for the coordinates [47] (Fig. 2.15).

Let $\left(\hat{\xi}_\ell^G\right)_{\ell=1}^{r}$ and $\left(\hat{\xi}_\ell^{GL}\right)_{\ell=1}^{r+1}$ be respectively the sets of $(r-1)$-order Gauss and r-order Gauss-Lobatto points. We can define the $(n-1)$th-order 1D Lagrange polynomials

$$\varphi_j^A(x) = \frac{\prod\limits_{\ell=1,\ell\neq j}^{n} \left(x - \hat{\xi}_\ell^A\right)}{\prod\limits_{\ell=1,\ell\neq j}^{n} \left(\hat{\xi}_j^A - \hat{\xi}_\ell^A\right)}, \quad j = 1 \ldots n, \tag{2.96}$$

where $A = G$ and $n = r$ for Gauss points and $A = GL$ and $n = r+1$ for Gauss-Lobatto points.

With these notations, the basis functions read

$$\hat{\varphi}_{i,j,k}(\hat{x}_1, \hat{x}_2, \hat{x}_3)\, \underline{e}_1 = \hat{\varphi}_i^G(\hat{x}_1)\, \hat{\varphi}_j^{GL}(\hat{x}_2)\, \hat{\varphi}_k^{GL}(\hat{x}_3)\, \underline{e}_1, \tag{2.97a}$$

$$\hat{\varphi}_{i,j,k}(\hat{x}_1, \hat{x}_2, \hat{x}_3)\, \underline{e}_2 = \hat{\varphi}_i^{GL}(\hat{x}_1)\, \hat{\varphi}_j^{G}(\hat{x}_2)\, \hat{\varphi}_k^{GL}(\hat{x}_3)\, \underline{e}_2, \tag{2.97b}$$

$$\hat{\varphi}_{i,j,k}(\hat{x}_1, \hat{x}_2, \hat{x}_3)\, \underline{e}_3 = \hat{\varphi}_i^{GL}(\hat{x}_1)\, \hat{\varphi}_j^{GL}(\hat{x}_2)\, \hat{\varphi}_k^{G}(\hat{x}_3)\, \underline{e}_3. \tag{2.97c}$$

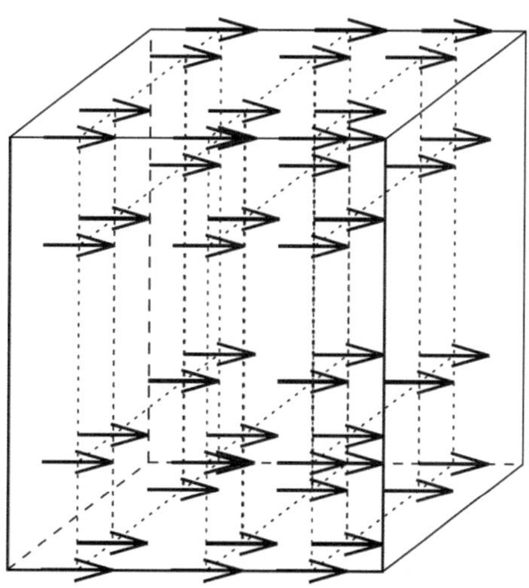

Fig. 2.15 The locations of the degrees of freedom of the first family of edge elements for $r = 3$ in the $\hat{x}_2$ direction on the unit cube

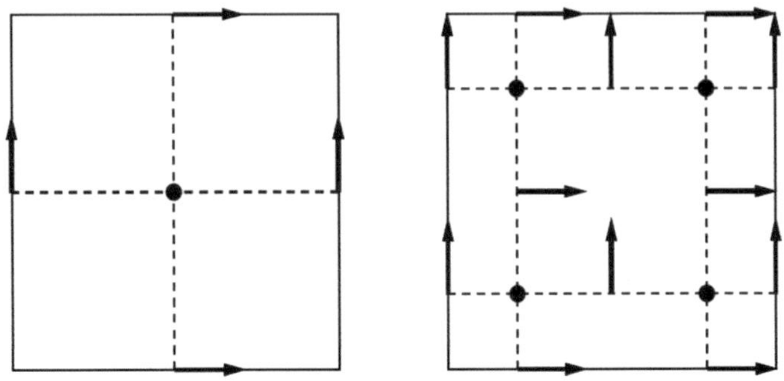

Fig. 2.16 The locations of the degrees of freedom of the first family of edge elements on the unit square for $r = 1$ (*left*) and $r = 2$ (*right*) in both directions. The *dashed lines* indicate the Gauss points

2.6.1.3 Quadrilateral Elements

A modal definition of quadrilateral edge elements is useless. So, we define them in a nodal way, which is very similar to hexahedra. The polynomial space on the unit square $\widehat{K} = [0, 1]^2$ is

$$\mathcal{Q}_r = \left\{ \underline{u} \text{ such that } u_1 \in Q_{r-1,r},\ u_2 \in Q_{r,r-1} \right\}. \tag{2.98}$$

As for the cube, we have $r \times (r+1)$ points in the $\hat{x}_1$ direction, $(r+1) \times r$ in the $\hat{x}_2$ direction, which enables us to mix Gauss and Gauss-Lobatto points (Fig. 2.16).

Basis functions obviously read

$$\hat{\varphi}_{i,j}(\hat{x}_1, \hat{x}_2)\, \underline{e}_1 = \hat{\varphi}_i^G(\hat{x}_1)\, \hat{\varphi}_j^{GL}(\hat{x}_2)\, \underline{e}_1, \tag{2.99a}$$

$$\hat{\varphi}_{i,j}(\hat{x}_1, \hat{x}_2)\, \underline{e}_2 = \hat{\varphi}_i^{GL}(\hat{x}_1)\, \hat{\varphi}_j^G(\hat{x}_2)\, \underline{e}_2. \tag{2.99b}$$

As we shall see later, this choice leads to mass-lumping on an orthogonal mesh, but not on elements of any shape (even with parallel faces). It provides a variational generalization of the Yee scheme [8, 48].

2.6.2 *Second Family*

2.6.2.1 Modal Definition

The polynomial space associated to the second family of hexahedra on the unit cube is simply $(Q_r)^3$, where Q_r is defined as in (2.17) with $d = 3$.

The degrees of freedom of this family are defined in [41] in a modal way as follows

1. $12(r+1)$ integrals on the edges defined by
$$\int_e \underline{u} \cdot \underline{t} q \, \mathrm{d}\hat{\sigma}, \quad \forall q \in P_r,$$
where $\underline{t}$ is the unit vector along an edge $e \in \widehat{K}$.
2. $6 \times 2r(r^2-1)$ integrals on the faces defined by
$$\int_f \underline{u} \cdot \underline{q} \, \mathrm{d}\hat{\gamma}, \quad \forall \underline{q} = (q_1, q_2) \text{ such that } q_1 \in Q_{r,r-2},\ q_2 \in Q_{r-2,r},$$
where f is a face of $\widehat{K}$.
3. $3(r^2-1)(r-1)$ integrals in $\widehat{K}$ defined by
$$\int_{\widehat{K}} \underline{u} \cdot \underline{q} \, \mathrm{d}\hat{\underline{x}}, \quad \forall \underline{q} = (q_1, q_2, q_3)$$
such that $q_1 \in Q_{r,r-2,r-2},\ q_2 \in Q_{r-2,r,r-2},\ q_3 \in Q_{r-2,r-2,r}$.

Obviously, $12(r+1)+6\times 2r(r^2-1)+3(r^2-1)(r-1) = 3(r+1)^3 = \dim (Q_r)^3$.

2.6.2.2 Nodal Definition

The nodal definition of the degrees of freedom is quite trivial. Following the polynomial space definition, it is sufficient to define the vector-valued basis functions by their values at $(r+1)^3$ Gauss-Lobatto points. So, the basis functions $\hat{\varphi}_{i,j,k}\, \underline{e}_p$, $\underline{e}_p$ being a vector of the canonical basis of $\mathbb{R}^3$, $p = 1 \ldots 3$ read

$$\hat{\varphi}_{i,j,k}(\hat{x}_1, \hat{x}_2, \hat{x}_3)\, \underline{e}_p = \hat{\varphi}_i^{GL}(\hat{x}_1)\, \hat{\varphi}_j^{GL}(\hat{x}_2)\, \hat{\varphi}_k^{GL}(\hat{x}_3)\, \underline{e}_p, \quad p = 1 \ldots 3. \tag{2.100}$$

Although trivial, by using the $H(curl)$-conform transform, this definition keeps the tangential continuity over the mesh since the tangents to the edges and the faces of the cube are actually in the directions of the vectors of the canonical basis of $\mathbb{R}^3$.

2.6.2.3 Quadrilateral Elements

The nodal definition of quadrilateral elements is quite obvious for this family. The polynomial space on unit square is $(Q_r)^2$ and the vector-valued basis functions (defined by their values at $(r+1)^2$ Gauss-Lobatto points) read (Fig. 2.17).

$$\hat{\varphi}_{i,j,k}(\hat{x}_1, \hat{x}_2)\, \underline{e}_p = \hat{\varphi}_i^{GL}(\hat{x}_1)\, \hat{\varphi}_j^{GL}(\hat{x}_2)\, \underline{e}_p, \quad p = 1, 2.$$

Fig. 2.17 The locations of the degrees of freedom on the unit square for $r = 5$ in both directions. The *black* points indicate the Gauss-Lobatto points

As we shall see, the nodal definition of the second family enables mass-lumping on any type of hexahedral or quadrilateral mesh [49] but does not provide spurious free approximations [14, 50].

2.7 $H(div)$ Finite Elements

$H(div)$ finite elements were constructed in order to keep the normal continuity of vector-valued functions. Their presence in this book is justified because they are the natural complementary elements to edge elements for Maxwell's equations [42, 51]. So, we just give in this section the guidelines to construct these elements.

2.7.1 Tetrahedral and Triangular Elements

2.7.1.1 First Family

The polynomial space associated to $\hat{T}_3$ is

$$D_r = (P_{r-1})^3 \oplus \underline{\hat{x}}\,\widetilde{P}_r \tag{2.101}$$

with $\widetilde{P}_r$ defined as in (2.58).

Obviously, $\dim D_r = \dfrac{1}{2} r(r+1)(r+3)$.

Following [40], we have two types of degrees of freedom:

1. $2r(r+1)$ integrals on the faces defined by

$$\int_f \underline{u} \cdot \underline{n}\, q \, \mathrm{d}\hat{\underline{\gamma}}, \quad \forall q \in P_{r-1},$$

where f is a face of $\widehat{K}$ and $\underline{n}$ the exterior unit normal to f.

2. $r(r^2-1)/2$ integrals in $\hat{T}_3$ defined by

$$\int_{\hat{T}_3} \underline{u} \cdot \underline{q} \, \mathrm{d}\hat{\underline{x}}, \quad \forall \underline{q} \in (P_{r-2})^3.$$

One can easily check that $2r(r+1) + r(r^2-1)/2 = r(r+1)(r+3)/2$.

Now, in order to keep the normal continuity of a function $\underline{\varphi}$ over a mesh, one must deduce $\underline{\varphi}$ on any tetrahedron T of the mesh from a function $\hat{\underline{\varphi}}$ on the unit element $\hat{T}_3$ by using the $H(div)$-conform transform (also called Piola's transform) which reads

$$\underline{\varphi} \circ \underline{F}_T = \frac{1}{J_T} DF_T \, \hat{\underline{\varphi}}, \tag{2.102}$$

where $J_T = \det DF_T$, $\underline{F}_T$ and DF_T being defined as in Sect. 2.5.2.

Here also, this element can be defined in a nodal way by using $2r(r+1)$ pointwise values of vector-valued functions orthogonal the faces and $r(r^2-1)/2$ values of vector-valued interior functions collinear to the canonical basis of $\mathbb{R}^3$.

The polynomial space corresponding to the unit triangular element $\hat{T}_2$ is the space $\mathscr{R}_r^T$ defined in (2.65). The degrees of freedom are

1. $3r$ integrals on the edges defined by

$$\int_e \underline{u} \cdot \underline{n}\, q \, \mathrm{d}\hat{\sigma}, \quad \forall q \in P_{r-1},$$

where is the exterior unit vector normal to an edge $e \in \hat{T}_2$.

2. $r(r-1)$ integrals in $\hat{T}_2$ defined by

$$\int_{\hat{T}_2} \underline{u} \cdot \underline{q} \, \mathrm{d}\hat{\underline{x}}, \quad \forall \underline{q} \in (P_{r-2})^2,$$

or in a nodal way, by using pointwise values of r vector-valued functions normal to each edge and $r(r-1)$ interior vector-valued functions collinear to the canonical basis of $\mathbb{R}^2$.

Extension to any triangle T is made by using (2.102).

2.7.1.2 Second Family

A modal definition of the second family would be tedious and useless in practice. Since the polynomial space for this family is $(P_r)^3$ on $\hat{T}_3$ ($(P_r)^2$ on $\hat{T}_2$), the degrees of freedom can be defined by pointwise values of three (two in 2D) vector-valued function at each Hesthaven's point as follows:

- In 3D

  1. Three functions normal to the faces at a summit,
  2. One function normal to the face and two functions tangent to the face (and orthogonal to each other) at a point on a face,
  3. Three functions collinear to the canonical basis of $\mathbb{R}^3$ at an interior point.

- In 2D

  1. Two functions normal to the edges at a summit,
  2. Two functions collinear to the canonical basis of $\mathbb{R}^2$ at an interior point.

  Extension to any element T is made by using (2.102).

2.7.2 *Hexahedral and Quadrilateral Elements*

2.7.2.1 First Family

Hexahedral and quadrilateral elements are constructed in a way very similar to edge elements. For hexahedra, the polynomial space is

$$\mathscr{Q}'_r = \left\{ \underline{u} \text{ s.t. } u_1 \in Q_{r,r-1,r-1},\ u_2 \in Q_{r-1,r,r-1},\ u_3 \in Q_{r-1,r-1,r} \right\}, \tag{2.103}$$

With the notations used in (2.96), the basis functions read

$$\hat{\varphi}_{i,j,k}(\hat{x}_1, \hat{x}_2, \hat{x}_3)\, \underline{e}_1 = \hat{\varphi}_i^{GL}(\hat{x}_1)\, \hat{\varphi}_j^{G}(\hat{x}_2)\, \hat{\varphi}_k^{G}(\hat{x}_3)\, \underline{e}_1, \tag{2.104a}$$

$$\hat{\varphi}_{i,j,k}(\hat{x}_1, \hat{x}_2, \hat{x}_3)\, \underline{e}_2 = \hat{\varphi}_i^{G}(\hat{x}_1)\, \hat{\varphi}_j^{GL}(\hat{x}_2)\, \hat{\varphi}_k^{G}(\hat{x}_3)\, \underline{e}_2, \tag{2.104b}$$

$$\hat{\varphi}_{i,j,k}(\hat{x}_1, \hat{x}_2, \hat{x}_3)\, \underline{e}_3 = \hat{\varphi}_i^{G}(\hat{x}_1)\, \hat{\varphi}_j^{G}(\hat{x}_2)\, \hat{\varphi}_k^{GL}(\hat{x}_3)\, \underline{e}_3. \tag{2.104c}$$

The degrees of freedom are defined in the same way as edge elements (Fig. 2.18). For quadrilaterals, the polynomial space is

$$\mathscr{Q}'_r = \left\{ \underline{u} \text{ such that } u_1 \in Q_{r,r-1},\ u_2 \in Q_{r-1,r} \right\}, \tag{2.105}$$

and basis functions read

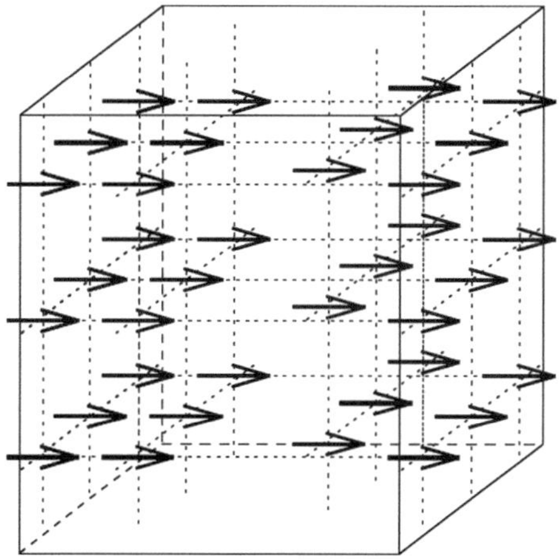

Fig. 2.18 The locations of the degrees of freedom of the first family of $H(div)$ elements for $r = 3$ in the $\hat{x}_2$ direction on the unit cube

$$\hat{\varphi}_{i,j,k}(\hat{x}_1, \hat{x}_2)\,\underline{e}_1 = \hat{\varphi}_i^{GL}(\hat{x}_1)\,\hat{\varphi}_j^{G}(\hat{x}_2)\,\underline{e}_1, \tag{2.106a}$$

$$\hat{\varphi}_{i,j,k}(\hat{x}_1, \hat{x}_2)\,\underline{e}_2 = \hat{\varphi}_i^{G}(\hat{x}_1)\,\hat{\varphi}_j^{GL}(\hat{x}_2)\,\underline{e}_2. \tag{2.106b}$$

The degrees of freedom are obvious (Fig. 2.19).

Extension to elements of any shape K_i in 2D or 3D is made by using the $H(div)$-conform transform

$$\underline{\varphi} \circ \underline{F}_i = \frac{1}{J_i} DF_i\,\hat{\underline{\varphi}}, \tag{2.107}$$

with $J_i = \det DF_i$ and $\underline{F}_i$ defined in (2.23) and (2.24).

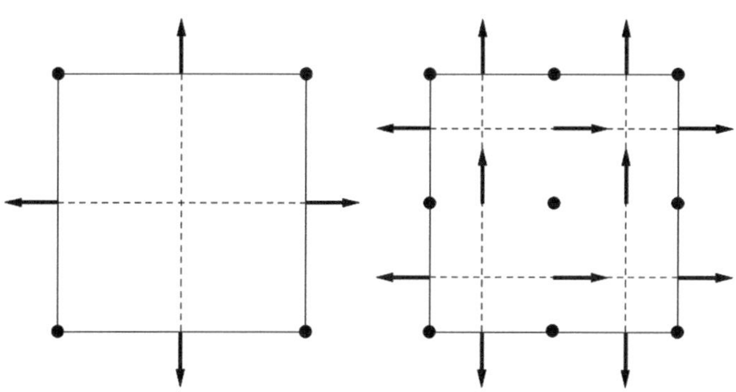

Fig. 2.19 The locations of the degrees of freedom of the first family of $H(div)$ elements on the unit square for $r = 1$ (*left*) and $r = 2$ (*right*) in both directions. The *dashed lines* indicate the Gauss points

2.7.2.2 Second Family

The second family of $H(div)$ elements is defined on the unit cube or square exactly as edge elements. The only difference lies in the the extension to elements of any shape which is made by using (2.107). This definition is justified by the both tangent and normal character of each nodal degree of freedom at any point for the second family.

Remark: Unlike 3D elements, 2D first families of edge and $H(div)$ elements have exactly the same number of degrees of freedom. This comes from the fact that 3D edge elements have degrees of freedom located on edges, faces and inside the elements whereas the degrees of freedom of $H(div)$ elements are only on the faces and inside the element. However, in 2D both types of element have their degrees of freedom on edges and inside. In 2D, we can say that $H(div)$ degrees of freedom are somehow orthogonal to these of edge elements (cf. Figs. 2.16 and 2.19).

2.8 Other Mixed Elements

In this section, we provide the guidelines to construct edge and $H(div)$ elements on pyramids and wedges, which can be useful for solving Maxwell's equations in frequency domain. A wider description of such elements can be found in [52, 53]

2.8.1 *Pyramidal and Prismatic Edge Elements*

2.8.1.1 Pyramids

The approximate space of edge elements on the unit pyramid corresponding to the first family of Nédélec's elements reads

$$\mathscr{P}_r^e = (\mathscr{P}_{r-1})^3 \oplus \left\{ \frac{\hat{x}_1^p \hat{x}_2^p}{(1-\hat{x}_3)^{p+2}} \, \underline{w}_1, \;\; 0 \le p \le r-1 \right\}$$

$$\oplus \left\{ \frac{\hat{x}_1^m \hat{x}_2^{n+2}}{(1-\hat{x}_3)^{m+2}} \, \underline{w}_2, \; \frac{\hat{x}_1^{n+2} \hat{x}_2^m}{(1-\hat{x}_3)^{m+2}} \, \underline{w}_3, \;\; 0 \le m \le n \le r-2 \right\} \tag{2.108}$$

$$\oplus \left\{ \frac{\hat{x}_1^p \hat{x}_2^q}{(1-\hat{x}_3)^{p+q-r+1}} \, \underline{w}_1, \; \frac{\hat{x}_1^q \hat{x}_2^p}{(1-\hat{x}_3)^{p+q-r+1}} \, \underline{w}_2, \;\; \begin{array}{l} 0 \le p \le r-1, \\ 0 \le q \le r \end{array} \right\},$$

where

$$\underline{w}_1 = \begin{pmatrix} \hat{x}_2(1-\hat{x}_3) \\ \hat{x}_1(1-\hat{x}_3) \\ \hat{x}_1\hat{x}_2 \end{pmatrix}, \quad \underline{w}_2 = \begin{pmatrix} 1-\hat{x}_3 \\ 0 \\ \hat{x}_1 \end{pmatrix}, \quad \underline{w}_3 = \begin{pmatrix} 0 \\ 1-\hat{x}_3 \\ \hat{x}_2 \end{pmatrix}$$

and $\mathscr{P}_r$ defined in (2.46). We have

$$\dim \mathscr{P}_r^e = \frac{r(2r^2+9r+5)}{2}.$$

The degrees of freedom on the faces of this element coincide with these of the first family tetrahedral and hexahedral edge elements. So, $\dim \mathscr{P}_r^e$ can be decomposed into

- r degrees of freedom on each edge ($8r$ for all the edges),
- $2r(r-1)$ degrees of freedom on the quadrilateral face,
- $r(r-1)$ degrees of freedom on a triangular face ($4r(r-1)$ for the four faces),
- $r(r-1)(2r-1)/2$ interior degrees of freedom.

As for other elements, a second family can be constructed by using $(\mathscr{P}_r)^3$.

For both families, a function $\underline{\varphi}$ on a pyramid of any shape can be deduced from a function $\hat{\underline{\varphi}} \in \mathscr{P}_r^e$ as follows

$$\underline{\varphi} \circ \underline{F}^\pi = (DF^\pi)^{*-1}\hat{\underline{\varphi}}, \tag{2.109}$$

where $\underline{F}^\pi$ is defined by (2.47).

2.8.1.2 Wedges

The polynomial space for edge elements on the unit prism $\hat{W}$ reads

$$\mathscr{P}_{e,r}^w = (\mathscr{R}_r^T(\hat{x}_1,\hat{x}_2) \otimes P_r(\hat{x}_3)) \times (P_r(\hat{x}_1,\hat{x}_2) \otimes P_{r-1}(\hat{x}_3)), \tag{2.110}$$

where $\mathscr{R}_r^T$ defined as in (2.65).

A simple computation shows that

$$\dim \mathscr{P}_{e,r}^w = 3r(r+1)(r+2)/2.$$

Since the degrees of freedom on the faces must fit with those of tetrahedra and hexahedra, they are defined in exactly the same way as on pyramids, the number of interior degrees of freedom being $r(r-1)(3r-4)/2$. As for pyramids also, the second family can be deduced from Sect. 2.4.1. The mapping to a wedge of any shape is given by

$$\underline{\varphi} \circ \underline{F}^w = (DF^w)^{*-1}\hat{\underline{\varphi}}, \tag{2.111}$$

where $\underline{F}^w$ is defined by (2.45).

2.8.2 *Pyramidal and Prismatic $H(div)$ Elements*

2.8.2.1 Pyramids

In this part, we give the minimal spaces which ensure the optimal convergence in $O(h^r)$ for $H(div)$ estimates (see [53] for more details about the construction). The approximate space on the unit pyramid corresponding to the first family of $H(div)$ elements reads

$$\begin{aligned}
\mathscr{P}_r^{div} &= (\mathscr{P}_{r-1})^3 \\
&\oplus \left\{ \frac{\hat{x}_1^{n+1}\hat{x}_2^m}{(1-\hat{x}_3)^{m+1}}\,\underline{e}_1 \oplus \frac{\hat{x}_1^m\hat{x}_2^{n+1}}{(1-\hat{x}_3)^{m+1}}\,\underline{e}_2,\ 0 \le m \le n \le r-1 \right\} \\
&\oplus \left\{ \frac{\hat{x}_1^m\hat{x}_2^{n+1}}{(1-\hat{x}_3)^{m+1}}\,\underline{u}_1 \oplus \frac{\hat{x}_1^{n+1}\hat{x}_2^m}{(1-\hat{x}_3)^{m+1}}\,\underline{u}_2,\ 0 \le m \le n \le r-1 \right\} \\
&\oplus \left\{ \frac{\hat{x}_1^p\hat{x}_2^q}{(1-\hat{x}_3)^{p+q-r}}\,\underline{u}_3,\ 0 \le p, q \le r \right\},
\end{aligned} \tag{2.112}$$

where $\underline{e}_1$, $\underline{e}_2$ are the two first vectors of the canonical basis of $\mathbb{R}^3$,

$$\underline{u}_1 = \begin{pmatrix} \dfrac{\hat{x}_1}{1-\hat{x}_3} \\ 0 \\ -1 \end{pmatrix}, \quad \underline{u}_2 = \begin{pmatrix} 0 \\ \dfrac{\hat{x}_2}{1-\hat{x}_3} \\ -1 \end{pmatrix}, \quad \underline{u}_3 = \begin{pmatrix} \dfrac{\hat{x}_1}{1-\hat{x}_3} \\ \dfrac{\hat{x}_2}{1-\hat{x}_3} \\ -1 \end{pmatrix}$$

and $\mathscr{P}_r$ defined in (2.46). We have

$$\dim \mathscr{P}_r^{div} = \frac{(r+1)\,(2r^2+7r+2)}{2}.$$

As for edge elements, the degrees of freedom on the faces must fit to these of the first family tetrahedral and hexahedral $H(div)$ elements. So, $\dim \mathscr{P}_r^{div}$ is decomposed into

- r^2 degrees of freedom on the quadrilateral face,
- $r(r+1)/2$ degrees of freedom on a triangular face ($2r(r+1)$ for the four faces),
- $(2r^3+3r^2+5r+2)/2$ interior degrees of freedom.

As for edge elements, a second family can be constructed by using $(\mathscr{P}_r)^3$.

2.8.2.2 Wedges

The polynomial space for $H(div)$ elements on the unit prism $\hat{W}$ reads

$$\mathscr{P}^{w}_{div,r} = (\mathscr{R}^{T}_{r}(\hat{x}_1, \hat{x}_2) \otimes P_{r-1}(\hat{x}_3)) \times (P_{r-1}(\hat{x}_1, \hat{x}_2) \otimes P_r(\hat{x}_3)), \tag{2.113}$$

where $\mathscr{R}^{T}_{r}$ defined as in (2.65). Obviously, we have

$$\dim \mathscr{P}^{w}_{div,r} = \frac{r\,(3r^2 + 6r + 1)}{2}.$$

The degrees of freedom are the same as those of pyramids on the faces and we have $r(r-1)(3r+1)/2$ interior degrees of freedom.

Mapping to an element of any shape is made by Piola's transform defined in (2.102), in which we use $\underline{F}^{\pi}$ for pyramids and $\underline{F}^{w}$ for wedges.

The second families of edge and $H(div)$ elements can immediately be deduced from Sects. 2.4.1 and 2.4.2.

Remark: References [52, 53] introduce "optimal" families by adding some degrees of freedom.

References

1. Ciarlet, P.G.: The Finite Element Method for Elliptic Problems. North-Holland (2002)
2. Ciarlet, P.G., Lions, J.-L.: Handbook of Numerical Analysis, vol. 2. North-Holland (1991)
3. Young, L.C.: An efficient finite element method for reservoir simulation. In: Proceedings of the 53rd Annual Fall Technical Conference and Exhibition of the Society of Petroleum Engineers of AIME. Houston, Texas Oct. 1–3 (1978)
4. Hennart, J.-P., Sainz, E., Villegas, M.: On the efficient use of the finite element method in static neutron diffusion calculations. Comput. Methods Nucl. Eng. **1**, 3–87 (1979)
5. Maday, Y., Patera, A.T.: Spectral element methods for the imcompressible Navier-Stokes equations. In: Noor, A.K. (ed.) State of the Art Survey in Computational Mechanics, pp. 71–143 (1989)
6. Cohen, G., Joly, P., Tordjman, N.: Higher-order finite elements with mass lumping for the 1-D wave equation. Finite Elem. Anal. Des. **17**(3–4), 329–336 (1994)
7. Tordjman, N.: Eléments finis d'ordre élevé avec condensation de masse pour l'équation des ondes, thèse de doctorat, U. Paris IX-Dauphine (1995)
8. Cohen, G.: High Order Numerical Methods for Transient Wave Equations, Scientific Computation. Springer, Heidelberg (2001)
9. Hesthaven, J.S.: From electrostatics to almost optimal nodal sets for polynomial interpolation in a simplex. SIAM J. Numer. Anal. **35**(2), 655–676 (1998)
10. Hesthaven, J.S., Warburton, T.: Nodal discontinuous Galerkin methods. In: Texts in Applied Mathematics, vol. 54. Springer, Heidelberg (2008)
11. Seriani, G., Priolo, E.: Spectral element method for acoustic wave simulation in heterogeneous media. Finite Elem. Anal. Des. **16**(3–4), 337–348 (1994)
12. Dubiner, M.: Spectral methods on triangles and other domains. J. Sci. Comput. **6**(4), 345–390 (1991)
13. Gordon, W., Hall, C.: Transfinite element methods: blending functions interpolation over arbitrary element domains. Numer. Math. **21**, 109–129 (1973)

14. Duruflé, M.: Intégration numérique et éléments finis d'ordre élevé appliqués aux équations de Maxwell en régime harmonique, thèse de doctorat, U. de Paris-Dauphine (2006)
15. Hesthaven, J.S., Teng, C.H.: Stable spectral methods on tetrahedral elements. SIAM J. Sci. Comput. **21**(6), 2352–2380 (2000)
16. Cohen, G., Elmkies, A.: Eléments finis triangulaires P_2 avec condensation de masse pour l'équation des ondes, INRIA Report RR-2418 (1994)
17. Cohen, G., Joly, P., Roberts, J.E., Tordjman, N.: Higher order triangular finite elements with mass lumping for the wave equation. SIAM J. Numer. Anal. **38**(6), 12047–2078 (2001)
18. Chin-Joe-Kong, M.J.S., Mulder, W.A., van Veldhuizen, M.: Higher-order triangular and tetrahedral finite elements with mass lumping for solving the wave equation. J. Eng. Math. **35**, 405–426 (1999)
19. Duffy, M.G.: Quadrature over a pyramid or cube of integrands with a singularity at a vertex. SIAM J. Numer. Anal. **19**(6), 1260–1262 (1982)
20. Karniadakis, G., Sherwin, S.J.: Spectral/hp Element Methods for CFD, 2nd edn. Oxford University Press, Oxford (2005)
21. Abramowitz, M., Stegun, I.A. (eds.): Handbook of Mathematical Functions. Dover Publications, New York (1972)
22. Giraldo, F.X., Taylor, M.A.: A diagonal mass matrix triangular spectral element method based on cubature points. J. Eng. Math. **56**(3), 307–322 (2006)
23. Bedrosian, G.: Shape functions and integration formulas for three-dimensional finite element analysis. Int. J. Numer. Methods Eng. **35**, 95–108 (1992)
24. Zgainski, F.X., Coulomb, J.C., Marchal, Y., Claeyssen, F., Brunotte, X.: A new family of finite elements: the pyramidal elements. IEEE Trans. Magn. **32**(3), 1393–1396 (1996)
25. Graglia, R.D., Wilton, D.R., Peterson, A.F., Gheorma, I.-L.: Higher order interpolatory vector bases on pyramidal elements. IEEE Trans. Antennas Propag. **47**(5), 775–782 (1999)
26. Chatzi, V., Preparata, F.P.: Using Pyramids in Mixed Meshes—Point Placement and Basis Functions. Brown University, Providence (2000)
27. Solín, P., Segeth, K., Dolezel, I.: Higher-order finite elements methods. In: Studies in Advanced Mathematics. Chapman and Hall, London (2003)
28. Szabó, B.A., Babuška, I.: Finite Element Analysis. Wiley, New York (1991)
29. Sherwin, S.J.: Hierarchical HP finite element in hybrid domains. Finite Elem. Anal. Des. **27**(1), 109–119 (1997)
30. Sherwin, S.J., Warburton, T., Karniadakis, G.E.: Spectral/HP methods for elliptic problems on hybrid grids. Contemp. Math. **218**, 191–216 (1998)
31. Warburton, T.: Spectral/HP Methods on Polymorphic Multi-Domains: Algorithms and Applications. Brown University, Providence (1999)
32. Bergot, M., Cohen, G., Duruflé, M.: Higher-order finite element for hybrid meshes using new nodal pyramidal elements. J. Sci. Comput. **42**(3), 345–381 (2010)
33. Zaglmayr, S.: High order finite element methods for electromagnetic field computation, Ph.D. thesis, Johannes Kepler University, Linz Austria (2006)
34. Radau, R.: Etudes sur les formules d'approximation qui servent à calculer la valeur d'une intégrale définie. J. Math. Pures Appl. **6**, 283–336 (1880)
35. Zienkiewicz, O.C., Taylor, R.L., Zhu, J.Z.: The Finite Element Method: Its Basis and Fundamentals. Elsevier, Amsterdam (2005)
36. Bergot, M.: Eléments finis d'ordre élevé pour maillages hybrides. Application à la résolution de systèmes hyperboliques linéaires en régimes harmonique et temporel, thèse de doctorat, U. de Paris-Dauphine (Paris IX) (2010)
37. Bergot, M., Duruflé, M.: Higher-order discontinuous Galerkin method for pyramidal elements using orthogonal basis. Numer. Methods Part. Differ. Eq. **29**(1), 144–169 (2013)
38. Fraeijs de Veubeke, B.M.: Displacement and Equilibrium Models in Stress Analysis. In: Zienkiewicz, O.C., Hollister, G. (eds.) pp. 145-197. Wiley, London (1965), (reprinted in Int. J. Numer. Methods Eng. **52**, 287–342, 2001)
39. Raviart, P.A., Thomas, J.M.: A mixed finite element method for 2nd order elliptic problems. In: Dold, A., Eckmann, B. (eds.) Mathematical Aspects of Finite Element Methods, Lecture Notes in Mathematics vol. 606. Springer, Heidelberg (1977)

40. Nédélec, J.-C.: Mixed finite elements in $I\!R^3$. Numer. Math. **35**(3), 315–341 (1980)
41. Nédélec, J.-C.: A new family of mixed finite elements in $I\!R^3$. Numer. Math. **50**(1), 57–81 (1986)
42. Monk, P.: Finite element methods for Maxwell's equations. In: Numerical Mathematics and Scientific Computation. Oxford University Press, Oxford (2003)
43. Bergot, M., Lacoste, P.: Generation of higher-order polynomial bases of Nédélec H(curl) finite elements for Maxwell's equations. J. Comput. Appl. Math. **234**(6), 1937–1944 (2010)
44. Savage, J.S., Peterson, A.F.: Higher-order vector finite element for tetrahedral cells. IEEE Trans. Microw. Theory Tech. **44**(6) (1996)
45. Mur, G., de Hoop, A.T.: A finite-element method for computing three-dimensional electromagnetic fields in inhomogeneous media. IEEE Trans. Magn. **21**(6), 2188–2191 (1985)
46. Elmkies, A.: Sur les éléments finis d'arête pour la résolution des équations de Maxwell en milieu anisotrope et pour des maillages quelconques, thèse de doctorat, U. de Paris Sud-Orsay (Paris XI) (1998)
47. Cohen, G., Monk, P.: Gauss point mass lumping schemes for Maxwell's equations. Numer. Methods Part. Differ. Eq. **14**(1), 63–88 (1998)
48. Yee, K.: Numerical solutions of initial boundary value problems involving Maxwell's equations in isotropic media. IEEE Trans. Antennas Propag. **14**(3), 302–307 (1966)
49. Cohen, G., Monk, P.: Mur-Nédélec finite element schemes for Maxwell's equations. Comput. Methods Appl. Mech. Eng. **169**(3–4), 197–217 (1999)
50. Cohen, G., Duruflé, M.: Non spurious spectral-like element methods for Maxwell's equations. J. Comput. Math. **25**(3), 282–304 (2007)
51. Demkowicz, L., Kurtz, J., Pardo, D., Paszynski, M., Rachowicz, W., Zdunek, A.: Computing with HP Finite Elements. II. Frontiers, Three-Dimensional Elliptic and Maxwell Problems with Applications. Chapman and Hall/CRC, London (2007)
52. Bergot, M., Duruflé, M.: High-order optimal edge elements for pyramids, prisms and hexahedra. J. Comput. Phys. **232**(1), 189–213 (2013)
53. Bergot, M., Duruflé, M.: Approximation of H(div) with high-order optimal finite elements for pyramids, prisms and hexahedra. Commun. Comput. Phys. **14**(5), 1372–1414 (2013)

Chapter 3
Hexahedral and Quadrilateral Spectral Elements for Acoustic Waves

Abstract This chapter covers an approximation of the acoustics equation and the linear elastodynamics system by (mixed) spectral element methods, i.e. finite element methods on quadrilaterals and hexahedra with mass-lumping. We first provide a detailed description of the method for second-order and first-order formulations of the acoustics equation and compare their performance. Then, a plane wave analysis in homogeneous and discontinuous media is given and error estimates are developed for the acoustics equation. We finally define a new first-order formulation of the linear elastodynamics system and we provide guidelines to its approximation.

The acoustics equation is the 0th-order approximation of Euler's equations and models a wave propagation in a fluid at rest. However, due to its simple formulation, it is often used as an approximate model in linear elastodynamics or for Maxwell's equations. So, in this section, we use the acoustics equation in its most general form, which is beyond its physical meaning. In a second step, we address the linear elastodynamics system and its approximations.

3.1 Second-Order Formulation of the Acoustics Equation

3.1.1 The Continuous and Approximate Problem

Let us first recall the second-order formulation of this problem in an open subset Ω of $\mathbb{R}^d$, $d = 2, 3$ (with a solid boundary):
Find $u : \Omega \times [0, T] \to \mathbb{R}$ such that

$$\begin{aligned} &\lambda(\underline{x}) \frac{\partial^2 u}{\partial t^2}(\underline{x}, t) - \nabla \cdot \left(\underline{\underline{\mu}} \nabla u(\underline{x}, t)\right) = f(\underline{x}, t), \\ &\frac{\partial u}{\partial n}(\underline{x}, t) = 0 \text{ on } \partial\Omega, \\ &u(\underline{x}, 0) = u_0(\underline{x}), \ \frac{\partial u}{\partial t}(\underline{x}, 0) = u_1(\underline{x}), \end{aligned} \tag{3.1}$$

G. Cohen and S. Pernet, *Finite Element and Discontinuous Galerkin Methods for Transient Wave Equations*, Scientific Computation,
DOI 10.1007/978-94-017-7761-2_3

where $\lambda \in \mathbb{R}$ and $\underline{\underline{\mu}}$ is a $d \times d$ real symmetric, definite, positive matrix (which can be the identity).[1]

The variational formulation of this problem in H^1 reads:
Find $u(.,t) \in H^1(\Omega)$ such that

$$\begin{aligned} &\frac{\mathrm{d}^2}{\mathrm{d}t^2}\int_\Omega \lambda\, u\, v\, \mathrm{d}\underline{x} + \int_\Omega \underline{\underline{\mu}} \nabla u \cdot \nabla v\, \mathrm{d}\underline{x} = \int_\Omega f v\, \mathrm{d}\underline{x},\ \forall v \in H^1(\Omega), \\ &u(\underline{x},0) = u_0(\underline{x}),\ \frac{\partial u}{\partial t}(\underline{x},0) = u_1(\underline{x}). \end{aligned} \tag{3.2}$$

Now, let $U_h^r \subset H^1(\Omega)$ be a r-order space of finite elements. The approximate problem reads:
Find $u_h(.,t) \in U_h^r$ such that

$$\begin{aligned} &\frac{\mathrm{d}^2}{\mathrm{d}t^2}\int_\Omega \lambda\, u_h\, v_h\, \mathrm{d}\underline{x} + \int_\Omega \underline{\underline{\mu}} \nabla u_h \cdot \nabla v_h\, \mathrm{d}\underline{x} = \int_\Omega f v_h\, \mathrm{d}\underline{x}, \\ &\forall v_h \in U_h^r, \\ &u_h(\underline{x},0) = u_0(\underline{x}),\ \frac{\partial u_h}{\partial t}(\underline{x},0) = u_1(\underline{x}). \end{aligned} \tag{3.3}$$

3.1.2 Discretization of the Integrals

As we saw in the previous chapter, the only worth continuous finite element approximation of this problem is quadrilateral and hexahedral elements, other elements providing an important mass matrix or need a too large additional number of degrees of freedom to get mass-lumping. However, other elements will be addressed in discontinuous Galerkin methods. So, if

$$\mathcal{T}_h = \bigcup_{\ell=1}^{N_e} K_\ell \tag{3.4}$$

denotes a quadrilateral or hexahedral mesh of Ω, $\widehat{K} = [0,1]^d$ and $\underline{F}_\ell$ the transform (defined in (2.23) and (2.24)) such that $\underline{F}_\ell(\widehat{K}) = K_\ell$, we set:

$$U_h^r = \left\{ v_h \in H^1(\Omega) \text{ such that } v_{h|K_\ell} \circ \underline{F}_\ell \in Q_r \right\}, \tag{3.5}$$

Q_r being defined in (2.17).

[1] The boundary condition, which can also be written $\nabla u \cdot \underline{n}$, $\underline{n}$ being the exterior normal on $\partial\Omega$ is set here because of its simplicity for the formulation of the variational problem.

With these notations, (3.3) can be rewritten as

$$\sum_{\ell=1}^{N_e}\left\{\frac{\mathrm{d}^2}{\mathrm{d}t^2}\int_{K_\ell}\lambda\, u_h\, v_h\,\mathrm{d}\underline{x}+\int_{K_\ell}\underline{\underline{\mu}}\nabla u_h\cdot\nabla v_h\,\mathrm{d}\underline{x}-\int_{K_\ell} f v_h\,\mathrm{d}\underline{x}\right\}=0,$$

$$\forall v_h\in U_h^r,\tag{3.6}$$

$$u_h(\underline{x},0)=u_0(\underline{x}),\ \frac{\partial u_h}{\partial t}(\underline{x},0)=u_1(\underline{x}).$$

Now, let n_d be the number of degrees of freedom in $\mathscr{T}_h$, we have:

$$u_h=\sum_{i=1}^{n_d}u_i v_i=\sum_{\ell=1}^{N_e}\sum_{j=1}^{(r+1)^d}u_{m(\ell,j)}v_{m(\ell,j)},\ v_i\in U_h^r,\ m(\ell,j)=i.\tag{3.7}$$

Let us set $\varphi_j^\ell=v_{m(\ell,j)|K_\ell}$. We get, $\forall K_\ell\in\mathscr{T}_h$,

$$u_{h|K_\ell}=\sum_{j=1}^{(r+1)^d}u_{m(\ell,j)}\varphi_j^\ell.\tag{3.8}$$

By replacing v_h by a test function φ_k^ℓ in (3.6) and taking into account (3.8), we obtain:

$$\int_{K_\ell}\lambda\, u_h\, v_h\,\mathrm{d}\underline{x}=\sum_{j=1}^{(r+1)^d}u_{m(\ell,j)}\int_{K_\ell}\lambda\,\varphi_j^\ell\,\varphi_k^\ell\,\mathrm{d}\underline{x}\tag{3.9}$$

and

$$\int_{K_\ell}\underline{\underline{\mu}}\nabla u_h\cdot\nabla v_h\,\mathrm{d}\underline{x}=\sum_{j=1}^{(r+1)^d}u_{m(\ell,j)}\int_{K_\ell}\underline{\underline{\mu}}\nabla\varphi_j^\ell\cdot\nabla\varphi_k^\ell\,\mathrm{d}\underline{x}.\tag{3.10}$$

Let us now set $\varphi_n^\ell\circ\underline{F}_\ell=\hat{\varphi}_n$ ($n=j$ or k). By assuming that

$$n=(r+1)(n_2-1)+n_1\tag{3.11}$$

when $d=2$ and

$$n=(r+1)[(r+1)(n_3-1)+n_2-1]+n_1\tag{3.12}$$

when $d=3$, we can write

$$\hat{\varphi}_n(\hat{x}_1,\hat{x}_2)=\hat{\varphi}_{n_1}(\hat{x}_1)\,\hat{\varphi}_{n_2}(\hat{x}_2)\text{ for }d=2,\tag{3.13a}$$

$$\hat{\varphi}_n(\hat{x}_1,\hat{x}_2,\hat{x}_3)=\hat{\varphi}_{n_1}(\hat{x}_1)\,\hat{\varphi}_{n_2}(\hat{x}_2)\,\hat{\varphi}_{n_3}(\hat{x}_3)\text{ for }d=3,\tag{3.13b}$$

where $\hat{\varphi}_{n_s}(\hat{x}_s)$ is defined as in (2.96) with $A = GL$.

Taking into account all the above definitions, (3.9) can be computed as in (2.26) and (2.27), which provides mass-lumping for the mass matrix.

Let us now compute the stiffness integral (3.10). By using the the change of variable induced by (2.22) and the relation

$$(\nabla \varphi_n^\ell) \circ \underline{F}_\ell = DF_\ell^{-T} \hat{\nabla} \hat{\varphi}_n, \tag{3.14}$$

where $\hat{\nabla} = (\partial/\partial \hat{x}_1, \partial/\partial \hat{x}_2, \partial/\partial \hat{x}_3)^T$ and DF_ℓ is the Jacobian matrix of $\underline{F}_\ell$, we get

$$\int_{K_\ell} \underline{\underline{\mu}} \nabla \varphi_j^\ell \cdot \nabla \varphi_k^\ell \, \mathrm{d}\underline{x} = \int_{\widehat{K}} DF_\ell^{-1} \underline{\underline{\mu}} \circ \underline{F}_\ell \, DF_\ell^{-T} \, \hat{\nabla} \hat{\varphi}_j \cdot \hat{\nabla} \hat{\varphi}_k \, \mathrm{d}\underline{\hat{x}}. \tag{3.15}$$

By setting $B = DF_\ell^{-1} \underline{\underline{\mu}} \circ \underline{F}_\ell \, DF_\ell^{-T}$ and using the Gauss–Lobatto rule to compute (3.15), we have, when $d = 2$, B being symmetric,

$$\begin{aligned} I_S &= \int_{\widehat{K}} B \, \hat{\nabla} \hat{\varphi}_j \cdot \hat{\nabla} \hat{\varphi}_k \, \mathrm{d}\underline{\hat{x}} \\ &\simeq \sum_{p=1}^{r+1} \sum_{q=1}^{r+1} \hat{\omega}_p \hat{\omega}_q B_{11}(\hat{\xi}_p, \hat{\xi}_q) \frac{\partial \hat{\varphi}_j}{\partial \hat{x}_1}(\hat{\xi}_p, \hat{\xi}_q) \frac{\partial \hat{\varphi}_k}{\partial \hat{x}_1}(\hat{\xi}_p, \hat{\xi}_q) \\ &\quad + \sum_{p=1}^{r+1} \sum_{q=1}^{r+1} \hat{\omega}_p \hat{\omega}_q B_{22}(\hat{\xi}_p, \hat{\xi}_q) \frac{\partial \hat{\varphi}_j}{\partial \hat{x}_2}(\hat{\xi}_p, \hat{\xi}_q) \frac{\partial \hat{\varphi}_k}{\partial \hat{x}_2}(\hat{\xi}_p, \hat{\xi}_q) \\ &\quad + \sum_{p=1}^{r+1} \sum_{q=1}^{r+1} \hat{\omega}_p \hat{\omega}_q B_{12}(\hat{\xi}_p, \hat{\xi}_q) \frac{\partial \hat{\varphi}_j}{\partial \hat{x}_2}(\hat{\xi}_p, \hat{\xi}_q) \frac{\partial \hat{\varphi}_k}{\partial \hat{x}_1}(\hat{\xi}_p, \hat{\xi}_q) \\ &\quad + \sum_{p=1}^{r+1} \sum_{q=1}^{r+1} \hat{\omega}_p \hat{\omega}_q B_{12}(\hat{\xi}_p, \hat{\xi}_q) \frac{\partial \hat{\varphi}_j}{\partial \hat{x}_1}(\hat{\xi}_p, \hat{\xi}_q) \frac{\partial \hat{\varphi}_k}{\partial \hat{x}_2}(\hat{\xi}_p, \hat{\xi}_q), \end{aligned} \tag{3.16}$$

where $\hat{\omega}_p$, $\hat{\omega}_q$ are the Gauss–Lobatto weights and $\hat{\xi}_p$, $\hat{\xi}_q$ the corresponding points

Now, by taking into account (3.13a), (3.16) can be written as

$$\begin{aligned} I_S \simeq &\sum_{p=1}^{r+1} \sum_{q=1}^{r+1} \hat{\omega}_p \hat{\omega}_q B_{11}(\hat{\xi}_p, \hat{\xi}_q) \hat{\varphi}'_{j_1}(\hat{\xi}_p) \hat{\varphi}_{j_2}(\hat{\xi}_q) \hat{\varphi}'_{k_1}(\hat{\xi}_p) \hat{\varphi}_{k_2}(\hat{\xi}_q) \\ &+ \sum_{p=1}^{r+1} \sum_{q=1}^{r+1} \hat{\omega}_p \hat{\omega}_q B_{22}(\hat{\xi}_p, \hat{\xi}_q) \hat{\varphi}_{j_1}(\hat{\xi}_p) \hat{\varphi}'_{j_2}(\hat{\xi}_q) \hat{\varphi}_{k_1}(\hat{\xi}_p) \hat{\varphi}'_{k_2}(\hat{\xi}_q) \\ &+ \sum_{p=1}^{r+1} \sum_{q=1}^{r+1} \hat{\omega}_p \hat{\omega}_q B_{12}(\hat{\xi}_p, \hat{\xi}_q) \hat{\varphi}_{j_1}(\hat{\xi}_p) \hat{\varphi}'_{j_2}(\hat{\xi}_q) \hat{\varphi}'_{k_1}(\hat{\xi}_p) \hat{\varphi}_{k_2}(\hat{\xi}_q) \end{aligned}$$

$$+\sum_{p=1}^{r+1}\sum_{q=1}^{r+1}\hat{\omega}_p\hat{\omega}_q B_{12}(\hat{\xi}_p,\hat{\xi}_q)\hat{\varphi}'_{j_1}(\hat{\xi}_p)\hat{\varphi}_{j_2}(\hat{\xi}_q)\hat{\varphi}_{k_1}(\hat{\xi}_p)\hat{\varphi}'_{k_2}(\hat{\xi}_q). \tag{3.17}$$

The Gauss–Lobatto points being the interpolation points, we get the following simplification of (3.17):

$$\begin{aligned} I_S \simeq & \sum_{p=1}^{r+1}\hat{\omega}_p\hat{\omega}_{k_2} B_{11}(\hat{\xi}_p,\hat{\xi}_{k_2})\hat{\varphi}'_{j_1}(\hat{\xi}_p)\hat{\varphi}'_{k_1}(\hat{\xi}_p)\delta_{j_2,k_2} \\ & + \sum_{q=1}^{r+1}\hat{\omega}_{k_1}\hat{\omega}_q B_{22}(\hat{\xi}_{k_1},\hat{\xi}_q)\hat{\varphi}'_{j_2}(\hat{\xi}_q)\hat{\varphi}'_{k_2}(\hat{\xi}_q)\delta_{j_1,k_1} \\ & + \hat{\omega}_{j_1}\hat{\omega}_{k_2} B_{12}(\hat{\xi}_{j_1},\hat{\xi}_{k_2})\hat{\varphi}'_{j_2}(\hat{\xi}_{k_2})\hat{\varphi}'_{k_1}(\hat{\xi}_{j_1}) \\ & + \hat{\omega}_{k_1}\hat{\omega}_{j_2} B_{12}(\hat{\xi}_{k_1},\hat{\xi}_{j_2})\hat{\varphi}'_{j_1}(\hat{\xi}_{k_1})\hat{\varphi}'_{k_2}(\hat{\xi}_{j_2}). \end{aligned} \tag{3.18}$$

A similar computation provides, when $d = 3$,

$$I_S \simeq \sum_{n=1}^{9} I_n, \tag{3.19}$$

with

$$\begin{aligned} I_1 &= \sum_{p=1}^{r+1}\sum_{q=1}^{r+1}\sum_{s=1}^{r+1}\hat{\omega}_p\hat{\omega}_q\hat{\omega}_s B_{11}(\hat{\xi}_p,\hat{\xi}_q,\hat{\xi}_s) \\ &\quad \times \hat{\varphi}'_{j_1}(\hat{\xi}_p)\hat{\varphi}_{j_2}(\hat{\xi}_q)\hat{\varphi}_{j_3}(\hat{\xi}_s)\hat{\varphi}'_{k_1}(\hat{\xi}_p)\hat{\varphi}_{k_2}(\hat{\xi}_q)\hat{\varphi}_{k_3}(\hat{\xi}_s) \\ &= \sum_{p=1}^{r+1}\hat{\omega}_p\hat{\omega}_{k_2}\hat{\omega}_{k_3} B_{11}(\hat{\xi}_p,\hat{\xi}_{k_2},\hat{\xi}_{k_3})\hat{\varphi}'_{j_1}(\hat{\xi}_p)\hat{\varphi}'_{k_1}(\hat{\xi}_p)\delta_{j_2,k_2}\delta_{j_3,k_3}. \end{aligned} \tag{3.20}$$

I_2 and I_3 can be computed in the same way:

$$I_2 = \sum_{q=1}^{r+1}\hat{\omega}_{k_1}\hat{\omega}_q\,\hat{\omega}_{k_3} B_{22}(\hat{\xi}_{k_1},\hat{\xi}_q,\hat{\xi}_{k_3})\hat{\varphi}'_{j_2}(\hat{\xi}_q)\hat{\varphi}'_{k_2}(\hat{\xi}_q)\delta_{j_1,k_1}\delta_{j_3,k_3}. \tag{3.21}$$

$$I_3 = \sum_{s=1}^{r+1}\hat{\omega}_{k_1}\hat{\omega}_{k_2}\hat{\omega}_s B_{33}(\hat{\xi}_{k_1},\hat{\xi}_{k_2},\hat{\xi}_s)\hat{\varphi}'_{j_3}(\hat{\xi}_s)\hat{\varphi}'_{k_3}(\hat{\xi}_s)\delta_{j_1,k_1}\delta_{j_2,k_2}. \tag{3.22}$$

The non-diagonal terms provide:

$$I_4 = \sum_{p=1}^{r+1}\sum_{q=1}^{r+1}\sum_{s=1}^{r+1} \hat{\omega}_p\hat{\omega}_q\hat{\omega}_s B_{12}(\hat{\xi}_p, \hat{\xi}_q, \hat{\xi}_s)$$
$$\times \hat{\varphi}_{j_1}(\hat{\xi}_p)\hat{\varphi}'_{j_2}(\hat{\xi}_q)\hat{\varphi}_{j_3}(\hat{\xi}_s)\hat{\varphi}'_{k_1}(\hat{\xi}_p)\hat{\varphi}_{k_2}(\hat{\xi}_q)\hat{\varphi}_{k_3}(\hat{\xi}_s)$$
$$= \hat{\omega}_{j_1}\hat{\omega}_{k_2}\hat{\omega}_{k_3} B_{12}(\hat{\xi}_{j1}, \hat{\xi}_{k_2}, \hat{\xi}_{k_3})\hat{\varphi}'_{j_2}(\hat{\xi}_{k_2})\hat{\varphi}'_{k_1}(\hat{\xi}_{j_1})\delta_{j_3,k_3}. \tag{3.23}$$

$$I_5 = \hat{\omega}_{j_1}\hat{\omega}_{k_2}\hat{\omega}_{k_3} B_{13}(\hat{\xi}_{j_1}, \hat{\xi}_{k_2}, \hat{\xi}_{k_3})\hat{\varphi}'_{j_3}(\hat{\xi}_{k_3})\hat{\varphi}'_{k_1}(\hat{\xi}_{j_1})\delta_{j_2,k_2}. \tag{3.24}$$

$$I_6 = \hat{\omega}_{k_1}\hat{\omega}_{j_2}\hat{\omega}_{k_3} B_{12}(\hat{\xi}_{k_1}, \hat{\xi}_{j_2}, \hat{\xi}_{k_3})\hat{\varphi}'_{j_1}(\hat{\xi}_{k_1})\hat{\varphi}'_{k_2}(\hat{\xi}_{j_2})\delta_{j_3,k_3}. \tag{3.25}$$

$$I_7 = \hat{\omega}_{k_1}\hat{\omega}_{j_2}\hat{\omega}_{k_3} B_{23}(\hat{\xi}_{k_1}, \hat{\xi}_{j_2}, \hat{\xi}_{k_3})\hat{\varphi}'_{j_3}(\hat{\xi}_{k_3})\hat{\varphi}'_{k_2}(\hat{\xi}_{j_2})\delta_{j_1,k_1}. \tag{3.26}$$

$$I_8 = \hat{\omega}_{k_1}\hat{\omega}_{k_2}\hat{\omega}_{j_3} B_{13}(\hat{\xi}_{k_1}, \hat{\xi}_{k_2}, \hat{\xi}_{j_3})\hat{\varphi}'_{j_1}(\hat{\xi}_{k_1})\hat{\varphi}'_{k_3}(\hat{\xi}_{j_3})\delta_{j_2,k_2}. \tag{3.27}$$

$$I_9 = \hat{\omega}_{k_1}\hat{\omega}_{k_2}\hat{\omega}_{j_3} B_{23}(\hat{\xi}_{k_1}, \hat{\xi}_{k_2}, \hat{\xi}_{j_3})\hat{\varphi}'_{j_2}(\hat{\xi}_{k_2})\hat{\varphi}'_{k_3}(\hat{\xi}_{j_3})\delta_{j_1,k_1}. \tag{3.28}$$

Remark: Equation (3.18) shows that, in 2D, the test function interacts with all the functions whose support intersects its support,which leads to a full interaction. In 3D, the presence of a Kronecker's symbol in all the components shows that only the functions for which a plane containing the degree of freedom corresponding to the test function and parallel to two directions of space interact. This dramatically reduces the number of interactions (which is in $O(r^2)$ instead of $O(r^3)$, as we shall see below).

3.2 First-Order Formulation of the Acoustics Equation

Another way to solve the acoustics equation is to use the original formulation derived from the Euler's equations, which leads to the following first-order system:
Find $u : \Omega \times [0, T] \rightarrow \mathbb{R}$, $\underline{v} : \Omega \times [0, T] \rightarrow \mathbb{R}^d$ such that

$$\lambda(\underline{x})\,\frac{\partial u}{\partial t}(\underline{x}, t) - \nabla \cdot \underline{v}(\underline{x}, t) = F(\underline{x}, t), \tag{3.29a}$$

$$\underline{\underline{\mu}}^{-1}(\underline{x})\,\frac{\partial \underline{v}}{\partial t}(\underline{x}, t) - \nabla u(\underline{x}, t) = 0, \tag{3.29b}$$

with the boundary and initial conditions

$$\underline{v} \cdot \underline{n}(\underline{x}, t) = 0 \text{ on } \partial\Omega,$$

$$u(\underline{x}, 0) = u_0(\underline{x}),\ \underline{v}(\underline{x}, 0) = \underline{v}_0(\underline{x}).$$

and where

$$\frac{\partial F}{\partial t} = f.$$

Remark: Another formulation (in which $\underline{v}$ has no physical meaning) of (3.1) is

$$\lambda(\underline{x}) \frac{\partial^2 u}{\partial^2 t}(\underline{x}, t) - \nabla \cdot \underline{v}(\underline{x}, t) = f(\underline{x}, t), \tag{3.30a}$$

$$\underline{\underline{\mu}}^{-1}(\underline{x})\, \underline{v}(\underline{x}, t) - \nabla u(\underline{x}, t) = 0. \tag{3.30b}$$

This formulation can be of some interest, as we shall see in Sect. 3.3.

3.2.1 *The Mixed Formulation*

Equation (3.29a) and (3.29b) is actually a mixed formulation of (3.1). A classical way to get a variational formulation of this system is to set $u(.,t) \in L^2(\Omega)$ and $\underline{v}(.,t) \in H(div, \Omega)$[2] [1]. However, this approach is not convenient to get the discrete system because of the normal continuity required by $H(div, \Omega)$. So, we set $u(.,t) \in H^1(\Omega)$ and $\underline{v}(.,t) \in [L^2(\Omega)]^d$. This provides the variational formulation:
Find $u(.,t) \in H^1(\Omega)$ and $\underline{v}(.,t) \in [L^2(\Omega)]^d$ such that

$$\frac{\mathrm{d}}{\mathrm{d}t} \int_\Omega \lambda\, u\, \varphi\, \mathrm{d}\underline{x} + \int_\Omega \underline{v} \cdot \nabla\varphi\, \mathrm{d}\underline{x} = \int_\Omega F \varphi\, \mathrm{d}\underline{x},\ \forall \varphi \in H^1(\Omega), \tag{3.31a}$$

$$\frac{\mathrm{d}}{\mathrm{d}t} \int_\Omega \underline{\underline{\mu}}^{-1}\, \underline{v} \cdot \underline{\psi}\, \mathrm{d}\underline{x} - \int_\Omega \nabla u \cdot \underline{\psi}\, \mathrm{d}\underline{x} = 0,\ \forall \underline{\psi} \in [L^2(\Omega)]^d, \tag{3.31b}$$

$$u(\underline{x}, 0) = u_0(\underline{x}),\ \underline{v}(\underline{x}, 0) = \underline{v}_0(\underline{x}). \tag{3.31c}$$

The approximate problem derived from this variational formulation reads:
Find $u_h(.,t) \in U_h^r$ and $\underline{v}_h(.,t) \in \underline{V}_h^r$ such that

---

[2] Actually, the physical point of view would suggest to set $u(.,t) \in H^1(\Omega)$ and $\underline{v}(.,t) \in H(div, \Omega)$, but too much regularity induces a loss of accuracy.

$$\frac{\mathrm{d}}{\mathrm{d}t}\int_{\Omega}\lambda\, u_h\, \varphi_h\, \mathrm{d}\underline{x} + \int_{\Omega}\underline{v}_h\cdot\nabla\varphi_h\, \mathrm{d}\underline{x} = \int_{\Omega} F\varphi_h\, \mathrm{d}\underline{x},\ \forall\varphi_h\in U_h^r, \tag{3.32a}$$

$$\frac{\mathrm{d}}{\mathrm{d}t}\int_{\Omega}\underline{\underline{\mu}}^{-1}\underline{v}_h\cdot\underline{\psi}_h\, \mathrm{d}\underline{x} - \int_{\Omega}\nabla u_h\cdot\underline{\psi}_h\, \mathrm{d}\underline{x} = 0,\ \forall\underline{\psi}_h\in\underline{V}_h^r, \tag{3.32b}$$

$$u_h(\underline{x},0) = u_0(\underline{x}),\ \underline{v}_h(\underline{x},0) = \underline{v}_0(\underline{x}), \tag{3.32c}$$

where U_h^r is defined in (3.5) and $\underline{V}_h^r \subset [L^2(\Omega)]^d$ is to be fixed.

A natural (and simple) definition of $\underline{V}_h^r$ would be

$$\underline{V}_h^r = \left\{\underline{w}_h \in [L^2(\Omega)]^d \text{ such that } \underline{w}_{h|K_\ell}\circ\underline{F}_\ell \in [Q_r]^d\right\},$$

where K_ℓ is an element of $\mathscr{T}_h$ defined in (3.4).

However, for the sake of simplification (which will appear later), we set:

$$\underline{V}_h^r = \left\{\underline{w}_h \in [L^2(\Omega)]^d \text{ such that } |J_\ell| DF_\ell^{-1}\underline{w}_{h|K_\ell}\circ\underline{F}_\ell \in [Q_r]^d\right\}, \tag{3.33}$$

where $J_\ell = \det DF_\ell$.

3.2.2 The Mass Matrices

Obviously, the mass integral of (3.32a) is computed as in Sect. 3.1. Let us compute the second mass integral. We have:

$$\begin{aligned}\int_{\Omega}\underline{\underline{\mu}}^{-1}\underline{v}_h\cdot\underline{\psi}_h\, \mathrm{d}\underline{x} &= \sum_{\ell=1}^{N_e}\left\{\int_{K_\ell}\underline{\underline{\mu}}^{-1}\underline{v}_h\cdot\underline{\psi}_h\, \mathrm{d}\underline{x}\right\}\\ &= \sum_{\ell=1}^{N_e}\left\{\int_{\widehat{K}}|J_\ell|\,\underline{\underline{\mu}}^{-1}\circ\underline{F}_\ell\,\underline{v}_h\circ\underline{F}_\ell\cdot\underline{\psi}_h\circ\underline{F}_\ell\, \mathrm{d}\hat{\underline{x}}\right\}.\end{aligned} \tag{3.34}$$

Since $\underline{V}_h^r \subset [L^2(\Omega)]^d$, functions of $\underline{V}_h^r$ have no continuity on ∂K_ℓ. So, the basis functions of this space have their support in only one element of $\mathscr{T}_h$. Let us now define the basis functions on $\widehat{K}$ by $\hat{\varphi}_n\,\underline{e}_p$, $n = 1\ldots(r+1)^d$, $p = 1\ldots d$, $\hat{\varphi}_n$ defined as in (3.13a) and (3.13b) and $\underline{e}_p$ being a vector of the canonical basis of $\mathbb{R}^d$. Taking into account the definition of $\underline{V}_h^r$, we have, for any basis function $\underline{\psi}_{\ell,n}^p$ of K_ℓ,

$$\underline{\psi}_{\ell,n}^p\circ\underline{F}_\ell = \frac{1}{|J_\ell|}DF_\ell\,\hat{\varphi}_n\,\underline{e}_p. \tag{3.35}$$

Since

$$\underline{v}_{h|K_\ell} = \sum_{p=1}^{d} \sum_{i=1}^{(r+1)^d} v^p_{\ell,i}\, \underline{\psi}^p_{\ell,i}, \tag{3.36}$$

by setting $\underline{\psi}_h = \underline{\psi}^q_{\ell,j}$ and taking into account (3.35), we get:

$$\begin{aligned} I_M &= \int_{\widehat{K}} |J_\ell|\, \underline{\underline{\mu}}^{-1} \circ \underline{F}_\ell\, \underline{v}_h \circ \underline{F}_\ell \cdot \underline{\psi}_h \circ \underline{F}_\ell\, \mathrm{d}\underline{\hat{x}} \\ &= \sum_{p=1}^{d} \sum_{i=1}^{(r+1)^d} v^p_{\ell,i} \int_{\widehat{K}} \hat{\varphi}_i\, \hat{\varphi}_j\, \frac{1}{|J_\ell|} DF_\ell^T\, \underline{\underline{\hat{\mu}}}^{-1} DF_\ell\, \underline{e}_p \cdot \underline{e}_q \mathrm{d}\underline{\hat{x}}, \end{aligned} \tag{3.37}$$

with $\underline{\underline{\hat{\mu}}}^{-1} = \underline{\underline{\mu}}^{-1} \circ \underline{F}_\ell$.

By setting $C_\ell = |J_\ell|^{-1} DF_\ell^T\, \underline{\underline{\hat{\mu}}}^{-1} DF_\ell = (c^\ell_{p,q})_{p=1\ldots d, q=1\ldots d}$, one can easily check that $C_\ell\, \underline{e}_p \cdot \underline{e}_q = c^\ell_{p,q}$. On the other hand, by computing the last integral of (3.37) by using the Gauss–Lobatto rule and by taking into account notations (3.11)–(3.13b), we get:

$$\begin{aligned} I_M &= \sum_{p=1}^{d} \sum_{i=1}^{(r+1)^d} v^p_{\ell,i} \int_{\widehat{K}} \hat{\varphi}_i\, \hat{\varphi}_j\, c^\ell_{p,q} \mathrm{d}\underline{\hat{x}}, \\ &\simeq \sum_{p=1}^{d} \sum_{i=1}^{(r+1)^d} v^p_{\ell,i} \sum_{n=1}^{(r+1)^d} \hat{\omega}_n \hat{\varphi}_i(\hat{\underline{\xi}}_n)\, \hat{\varphi}_j(\hat{\underline{\xi}}_n)\, c^\ell_{p,q} = \sum_{p=1}^{d} \hat{\omega}_j\, c^\ell_{p,q}\, v^p_{\ell,i}, \end{aligned} \tag{3.38}$$

$\hat{\omega}_n = \hat{\omega}_{n_1} \hat{\omega}_{n_2}$, $\hat{\underline{\xi}}_n = \left(\hat{\xi}_{n_1}, \hat{\xi}_{n_2}\right)$ for $d = 2$ and $\hat{\omega}_n = \hat{\omega}_{n_1} \hat{\omega}_{n_2} \hat{\omega}_{n_3}$, $\hat{\underline{\xi}}_n = \left(\hat{\xi}_{n_1}, \hat{\xi}_{n_2}, \hat{\xi}_{n_2}\right)$ for $d = 3$ being the Gauss–Lobatto quadrature weights and points.

Equation (3.38) shows that we get a $d \times d$ symmetric mass matrix whose terms are $\hat{\omega}_j\, c^\ell_{p,q}$ for each point $\underline{F}_\ell(\hat{\underline{\xi}}_j)$ of interpolation of K_ℓ. In other words, the mass matrix is a $d \times d$ block-diagonal symmetric matrix whose blocks are $\hat{\omega}_j\, C_\ell$.

3.2.3 *The Stiffness Matrices*

Let us now compute the stiffness integrals. First, it is easy to check that the two stiffness integrals of (3.32a) and (3.32b) are symmetrical to each other. We choose to compute the first one. By using (3.36) and setting $\varphi_h = \varphi^\ell_j = v_{m(\ell,j)|K_\ell}$, $v_{m(\ell,j)}$ being defined as in (3.7), we can write:

$$\int_{\Omega} \underline{v}_h \cdot \nabla \varphi_h \, d\underline{x} = \sum_{\ell=1}^{N_e} \sum_{p=1}^{d} \sum_{i=1}^{r+1} v_{\ell,i}^p \left\{ \int_{K_\ell} \underline{\psi}_{\ell,i}^p \cdot \nabla \varphi_j^\ell \, d\underline{x} \right\} \tag{3.39}$$

Now, by using (3.14) and (3.33), we have:

$$\begin{aligned} I_S' &= \int_{K_\ell} \underline{\psi}_{\ell,i}^p \cdot \nabla \varphi_j^\ell \, d\underline{x} = \int_{\widehat{K}} |J_\ell| \, \underline{\psi}_{\ell,i}^p \circ \underline{F}_\ell \cdot (\nabla \varphi_j^\ell) \circ \underline{F}_\ell \, d\hat{\underline{x}} \\ &= \int_{\widehat{K}} |J_\ell| \frac{1}{|J_\ell|} DF_\ell \, \hat{\varphi}_i \, \underline{e}_p \cdot DF_\ell^{-T} \hat{\nabla} \hat{\varphi}_j \, d\hat{\underline{x}} \\ &= \int_{\widehat{K}} \hat{\varphi}_i \, \underline{e}_p \cdot \hat{\nabla} \hat{\varphi}_j \, d\hat{\underline{x}}. \end{aligned} \tag{3.40}$$

Equation (3.40) shows that, thanks to the Piola's transform which appears in the definition of $\underline{V}_h^r$,[3] the elementary contributions of the stiffness matrix in any element K_ℓ does not depend on this element. That means that, if we assembly this matrix at each time-step,[4] we do not have to store it. On the other hand, $\underline{\psi}_h$ being discontinuous, the second stiffness matrix requires no assembly.

Now, let us use the the Gauss–Lobatto rule to compute the stiffness matrix in K_ℓ. Using (3.13a) and (3.13b), we get:

$$\begin{aligned} I_{S,2}'' &\simeq \sum_{p=1}^{d} \sum_{i=1}^{r+1} v_{\ell,i}^p \sum_{n=1}^{(r+1)^d} \hat{\omega}_n \hat{\varphi}_i(\hat{\underline{\xi}}_n) \, \underline{e}_p \cdot \hat{\nabla} \hat{\varphi}_j(\hat{\underline{\xi}}_n) \\ &\simeq \sum_{p=1}^{d} \sum_{i=1}^{r+1} v_{\ell,i}^p \sum_{n=1}^{(r+1)^d} \hat{\omega}_n \hat{\varphi}_i(\hat{\underline{\xi}}_n) \frac{\partial \hat{\varphi}_j}{\partial \hat{x}_p}(\hat{\underline{\xi}}_n) \end{aligned} \tag{3.41}$$

Let us now set $d = 2$. Using (3.11) and (3.13a), we get, for $I_{S,2}''$,

$$\begin{aligned} &\sum_{i=1}^{(r+1)^2} v_{\ell,i}^1 \sum_{n_1=1}^{r+1} \sum_{n_2=1}^{r+1} \hat{\omega}_{n_1} \hat{\omega}_{n_2} \hat{\varphi}_{i_1}(\hat{\underline{\xi}}_{n_1}) \hat{\varphi}_{i_2}(\hat{\underline{\xi}}_{n_2}) \hat{\varphi}_{j_1}'(\hat{\underline{\xi}}_{n_1}) \hat{\varphi}_{j_2}(\hat{\underline{\xi}}_{n_2}) \\ &+ \sum_{i=1}^{(r+1)^d} v_{\ell,i}^2 \sum_{n_1=1}^{r+1} \sum_{n_2=1}^{r+1} \hat{\omega}_{n_1} \hat{\omega}_{n_2} \hat{\varphi}_{i_1}(\hat{\underline{\xi}}_{n_1}) \hat{\varphi}_{i_2}(\hat{\underline{\xi}}_{n_2}) \hat{\varphi}_{j_1}(\hat{\underline{\xi}}_{n_1}) \hat{\varphi}_{j_2}'(\hat{\underline{\xi}}_{n_2}), \end{aligned} \tag{3.42}$$

[3] We recall that this transform ensures the normal continuity of the solution, which is useless for $\underline{v}_h$ which is discontinuous on the faces of the elements. One can also understand why we took the absolute value of J_ℓ in our definition.

[4] Whose (cheap) cost decreases with the order of approximation since it only involves the degrees of freedom on the boundary of K_ℓ.

which finally provides:

$$
\begin{aligned}
I''_{S,2} \simeq & \sum_{i_1=1}^{r+1} v^1_{\ell,(i_1,j_2)} \hat{\omega}_{i_1} \hat{\omega}_{j_2} \hat{\varphi}'_{j_1}(\hat{\underline{\xi}}_{i_1}) \delta_{i_2,j_2} \\
& + \sum_{i_2=1}^{r+1} v^2_{\ell,(j_1,i_2)} \hat{\omega}_{j_1} \hat{\omega}_{i_2} \hat{\varphi}'_{j_2}(\hat{\underline{\xi}}_{i_2}) \delta_{i_1,j_1},
\end{aligned} \tag{3.43}
$$

where

$$
\begin{aligned}
v^1_{\ell,(i_1,j_2)} &= v^1_{\ell,(r+1)(j_2-1)+i_1}, \\
v^2_{\ell,(j_1,i_2)} &= v^2_{\ell,(r+1)(i_2-1)+j_1}.
\end{aligned}
$$

For $d = 3$, a similar computation leads to

$$
\begin{aligned}
I''_{S,3} \simeq & \sum_{i_1=1}^{r+1} v^1_{\ell,(i_1,j_2,j_3)} \hat{\omega}_{i_1} \hat{\omega}_{j_2} \hat{\omega}_{j_3} \hat{\varphi}'_{j_1}(\hat{\underline{\xi}}_{i_1}) \delta_{i_2,j_2} \delta_{i_3,j_3} \\
& + \sum_{i_2=1}^{r+1} v^2_{\ell,(j_1,i_2,j_3)} \hat{\omega}_{j_1} \hat{\omega}_{i_2} \hat{\omega}_{j_3} \hat{\varphi}'_{j_2}(\hat{\underline{\xi}}_{i_2}) \delta_{i_1,j_1} \delta_{i_3,j_3} \\
& + \sum_{i_3=1}^{r+1} v^2_{\ell,(j_1,j_2,i_3)} \hat{\omega}_{j_1} \hat{\omega}_{j_2} \hat{\omega}_{i_3} \hat{\varphi}'_{j_3}(\hat{\underline{\xi}}_{i_3}) \delta_{i_1,j_1} \delta_{i_2,j_2}
\end{aligned} \tag{3.44}
$$

where

$$
\begin{aligned}
v^1_{\ell,(i_1,j_2,j_3)} &= v^1_{\ell,(r+1)[(r+1)(j_3-1)+j_2-1]+i_1}, \\
v^2_{\ell,(j_1,i_2,j_3)} &= v^2_{\ell,(r+1)[(r+1)(j_3-1)+i_2-1]+j_1}, \\
v^3_{\ell,(j_1,j_2,i_3)} &= v^3_{\ell,(r+1)[(r+1)(i_3-1)+j_2-1]+j_1}.
\end{aligned}
$$

Equations (3.43) and (3.44) show that in dimensions 2 or 3, only one line of basis functions of $\underline{V}^r_h$ interacts with φ^ℓ_j in each direction (Fig. 3.1), which provides a very sparse stiffness matrix. A similar computation shows that, for the second stiffness matrix (which is transposed to the first one), a vector-valued test function interacts with the line of scalar degrees of freedom in its direction.

Remark: In both approximations, the matrix character of $\underline{\underline{\mu}}$ does not induce an additional cost in the method.

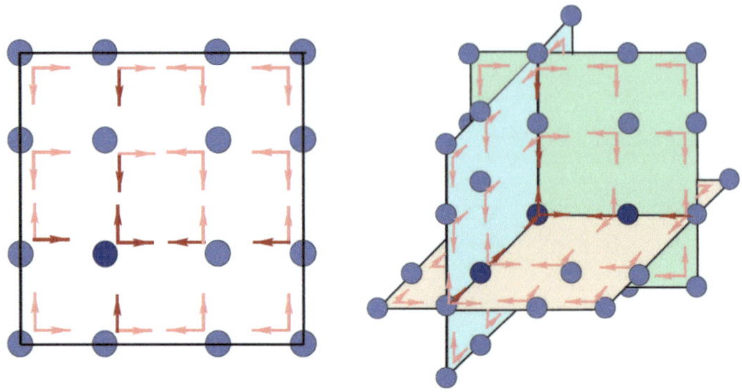

Fig. 3.1 The interactions for the first stiffness matrix of the mixed formulation in 2D (*left*) and 3D (*right*)

3.3 Comparison of the Methods

3.3.1 Matrix Formulation

Let us summarize the results of the previous sections in a matrix formulation of the second-order and first-order problems. The matrix form of (3.3) reads:

$$\frac{d^2}{d^2t} D_h \widetilde{U}_h + K_h \widetilde{U}_h = F_h'. \tag{3.45}$$

The matrix form of (3.29a) and (3.29b) reads:

$$\frac{d}{dt} D_h U_h + R_h V_h = F_h, \tag{3.46a}$$

$$\frac{d}{dt} B_h V_h - R_h^T U_h = 0. \tag{3.46b}$$

In (3.45) and (3.46a) and (3.46b), $\widetilde{U}_h \in \mathbb{R}^{n_d} \times [0, T]$, $U_h \in \mathbb{R}^{n_d} \times [0, T]$, $F_h \in \mathbb{R}^{n_d} \times [0, T]$ (F_h' being its derivative in time), n_d defined as in (3.7), $V_h \in \mathbb{R}^{3N_e(r+1)^d} \times [0, T]$, D_h is a r-order diagonal matrix, B_h is a 3×3 block-diagonal matrix (containing $N(r+1)^d$ blocks), K_h is the matrix derived from (3.18) or (3.19) and R_h is the matrix derived from (3.43) or (3.44) which only needs a storage on $\widehat{K}$ (provided the assembly of R_h).

Now, from (3.46b), we get:

$$\frac{d}{dt} V_h = B_h^{-1} R_h^T U_h. \tag{3.47}$$

By plugging (3.46b) in the derivative in time of (3.46a), we finally obtain:

$$\frac{\mathrm{d}^2}{\mathrm{d}^2 t} D_h U_h + R_h B_h^{-1} R_h^T U_h = F_h'. \tag{3.48}$$

It is therefore legitimate to ask if there is any relation between $\widetilde{U}_h$ and U_h and K_h and $R_h B_h^{-1} R_h^T$. This will be fixed in the next section.

3.3.2 A Theorem of Equivalence

Theorem 4 *Let*

$$\frac{\mathrm{d}^2}{\mathrm{d}t^2} \int_\Omega^{GL} \lambda\, \tilde{u}_h\, v_h \,\mathrm{d}\underline{x} + \int_\Omega^{GL} \underline{\underline{\mu}}\, \nabla \tilde{u}_h \cdot \nabla v_h \,\mathrm{d}\underline{x} = \int_\Omega^{GL} f\, v_h \,\mathrm{d}\underline{x}, \qquad \forall v_h \in \mathscr{B}_U, \tag{3.49}$$

be the discrete form of (3.3) and

$$\frac{\mathrm{d}}{\mathrm{d}t} \int_\Omega^{GL} \lambda\, u_h\, \varphi_h \,\mathrm{d}\underline{x} + \int_\Omega^{GL} \underline{v}_h \cdot \nabla \varphi_h \,\mathrm{d}\underline{x} = \int_\Omega^{GL} F \varphi_h \,\mathrm{d}\underline{x}, \ \forall \varphi_h \in \mathscr{B}_U, \tag{3.50a}$$

$$\frac{\mathrm{d}}{\mathrm{d}t} \int_\Omega^{GL} \underline{\underline{\mu}}^{-1}\, \underline{v}_h \cdot \underline{\psi}_h \,\mathrm{d}\underline{x} - \int_\Omega^{GL} \nabla u_h \cdot \underline{\psi}_h \,\mathrm{d}\underline{x} = 0, \ \forall \underline{\psi}_h \in \underline{V}_h^r, \tag{3.50b}$$

the discrete form of (3.32a) and (3.32b), where $\int_\Omega^{GL}$ *is the approximation of* $\int_\Omega$ *by a Gauss–Lobatto quadrature rule and* $\mathscr{B}_U$ *is the set of basis functions of* U_h^r.[5] *We have:*

$$\tilde{u}_h = u_h. \tag{3.51}$$

Proof If $v_h \in \mathscr{B}_U$, it is obvious that $\nabla v_h \notin \underline{V}_h^r$. However, $\forall K_\ell \in \mathscr{T}_h$, it exists a function $\underline{\psi}_h \in \underline{V}_h^r$ such that

$$\underline{\psi}_h \circ \underline{F}_\ell(\hat{\underline{\xi}}_n) = \underline{\underline{\mu}} \circ \underline{F}_\ell(\hat{\underline{\xi}}_n)(\nabla v_h) \circ \underline{F}_\ell(\hat{\underline{\xi}}_n), \ \forall n \in (r+1)^d. \tag{3.52}$$

[5] It would be equivalent to set $v_h \in U_h^r$ and $\varphi_h \in U_h^r$, but this formulation is more convenient for our proof.

Let us write explicitly (3.50b):

$$\begin{aligned}
&\frac{\mathrm{d}}{\mathrm{d}t}\int_{\Omega}^{GL} \underline{\underline{\mu}}^{-1}\,\underline{v}_h\cdot\underline{\psi}_h\,\mathrm{d}\underline{x} - \int_{\Omega}^{GL} \nabla u_h\cdot\underline{\psi}_h\,\mathrm{d}\underline{x},\\
&= \frac{\mathrm{d}}{\mathrm{d}t}\sum_{K_\ell\in\mathscr{T}_h}\sum_{n=1}^{(r+1)^d} \hat{\omega}_n |J_\ell|(\hat{\underline{\xi}}_n)\underline{\underline{\mu}}^{-1}\circ\underline{F}_\ell(\hat{\underline{\xi}}_n)\,\underline{v}_h\circ\underline{F}_\ell(\hat{\underline{\xi}}_n)\cdot\underline{\psi}_h\circ\underline{F}_\ell(\hat{\underline{\xi}}_n)\\
&\quad - \sum_{K_\ell\in\mathscr{T}_h}\sum_{n=1}^{(r+1)^d} \hat{\omega}_n |J_\ell|(\hat{\underline{\xi}}_n)(\nabla v_h)\circ\underline{F}_\ell(\hat{\underline{\xi}}_n)\cdot\underline{\psi}_h\circ\underline{F}_\ell(\hat{\underline{\xi}}_n).
\end{aligned} \tag{3.53}$$

By inserting (3.52) in (3.53) and coming back the Gauss–Lobatto integral formulation, we get:

$$\frac{\mathrm{d}}{\mathrm{d}t}\int_{\Omega}^{GL} \underline{v}_h\cdot\nabla v_h\,\mathrm{d}\underline{x} = \int_{\Omega}^{GL} \underline{\underline{\mu}}\,\nabla u_h\cdot\nabla v_h\,\mathrm{d}\underline{x}. \tag{3.54}$$

Now, if we derivate in time (3.50a) and we take into account relation (3.54), we obtain:

$$\begin{aligned}
&\frac{\mathrm{d}^2}{\mathrm{d}t^2}\int_{\Omega}^{GL} \lambda\, u_h\, v_h\,\mathrm{d}\underline{x} + \int_{\Omega}^{GL} \underline{\underline{\mu}}\,\nabla u_h\cdot\nabla v_h\,\mathrm{d}\underline{x} = \int_{\Omega}^{GL} f\, v_h\,\mathrm{d}\underline{x},\\
&\forall v_h\in\mathscr{B}_U.
\end{aligned} \tag{3.55}$$

Comparing (3.55) with (3.49) and using the unicity of the discrete solution leads to (3.51). ◇

An immediate application of Theorem 5 is

Corollary 2 *Problems (3.45) and (3.46a) and (3.46b) are equivalent. We actually have:*

$$K_h = R_h B_h^{-1} R_h^T \tag{3.56}$$

which ensures that

$$\widetilde{U}_h = U_h. \tag{3.57}$$

Remarks:

1. This proof can be extended to the Maxwell's equations and even to other linear hyperbolic systems.
2. This equivalence theorem is formulated as Corollary 2 in [2, 3] and is extended to the linear elastodynamics system in [4] but its proof, based on matrix analysis, is much longer and less elegant than the proof given here which is due to S. Pernet.

3.3.3 Comparison of the Costs

As we showed in Sects. 3.1 and 3.2, both approximations lead to mass-lumping and sparse matrices (in particular, in 3D). It would be therefore interesting to compare the costs of both approximations in terms of storage and number of computations.

In order to compare the methods, we use the discrete problem in time and space using a leapfrog approximation in time with a time-step equal to Δt. Since $\widetilde{U}_h = U_h$, (3.45) and (3.46a) and (3.46b) become

$$D_h \frac{U_h^{n+1} - 2U_h^n + U_h^{n-1}}{\Delta t^2} + K_h U_h^n = F_h'^n \tag{3.58}$$

and

$$D_h \frac{U_h^{n+1} - U_h^n}{\Delta t} + R_h V_h^{n+\frac{1}{2}} = F_h^{n+\frac{1}{2}}, \tag{3.59a}$$

$$B_h \frac{V_h^{n+\frac{1}{2}} - V_h^{n-\frac{1}{2}}}{\Delta t} - R_h^T U_h^n = 0, \tag{3.59b}$$

with $U_h^n = U_h(n\Delta t)$ and $V_h^{n+\frac{1}{2}} = V_h((n+1/2)\Delta t)$.[6]

Another formulation of (3.59a) and (3.59b), derived from (3.30a) and (3.30b) reads:

$$D_h \frac{U_h^{n+1} - 2U_h^n + U_h^{n-1}}{\Delta t^2} + R_h V_h^n = F_h^n, \tag{3.60a}$$

$$B_h V_h^n = R_h^T U_h^n. \tag{3.60b}$$

This formulation needs a lower storage than (3.59a) and (3.59b) since the storage of $V_h^{n-\frac{1}{2}}$ (whose dimension is $3N_e(r+1)^d$) is replaced by this of U_h^n (whose dimension is $N_e r^d$). For this reason, we use (3.60a) and (3.60b) to compare the costs.

We compare the storage and the number of multiplications for both approaches. We just take into account multiplications since the number of additions is roughly the same. Moreover, we assimilate $N_e + 1$ to N_e (which is supposed to be large) in our evaluation.

The storage of (3.58) in 2D (S_2^1) is composed of $2N_e r^2$ components of U_h, $N_e r^2$ coefficients of D_h and the stiffness matrix which can be decomposed as follows:

- $N_e(r-1)^2 \times (r+1)^2$ coefficients for the interior points of the elements,
- $2N_e(r-1) \times (r+1)(2r+1)$ coefficients for the points on the edges,
- $N_e \times (2r+1)^2$ coefficients for the summits.

[6]From a rigorous point of view, U_h and V_h are not the same in (3.45) and (3.46a) and (3.46b) and in (3.58) and (3.59a) and (3.59b).

Since the stiffness matrix is symmetric, we divide the number of its coefficients by 2 and we add its half diagonal, i.e. $N_e r^2/2$ coefficients. We finally obtain

$$S_2^1 = N_e \left(\frac{1}{2} r^4 + \frac{11}{2} r^2 + 2r^3 \right).$$

The storage of (3.60a) and (3.60b) in 2D (S_2^2) is composed of $2N_e r^2$ components of U_h, $N_e r^2$ coefficients of D_h $3N_e r^2$ coefficients of B_h (which is symmetric) and a mapping array of length $N_e r^2/2$, which provides

$$S_2^2 = N_e \frac{13}{2} r^2.$$

The number of multiplications of (3.58) in 2D (C_2^1) is equal to the number of coefficients of the stiffness matrix to which we add the multiplication by D_h. This leads to

$$C_2^1 = N_e \left(r^4 + 4r^3 + 5r^2 \right).$$

The number of multiplications of (3.60a) and (3.60b) in 2D (C_2^2) is equal to the number of coefficients of D_h plus $6N_e(r+1)^2$ multiplications for B_h plus $(r+1)^2 \times 2(r+1)$ multiplications for each stiffness matrix, i.e.

$$C_2^2 = N_e \left(4r^3 + 19r^2 + 24r + 10 \right).$$

These results are summarized in Fig. 3.2.

Figure 3.2 shows that (3.60a) and (3.60b) is much more efficient than (3.58) in terms of storage but becomes faster than (3.58) from the fourth order. However,

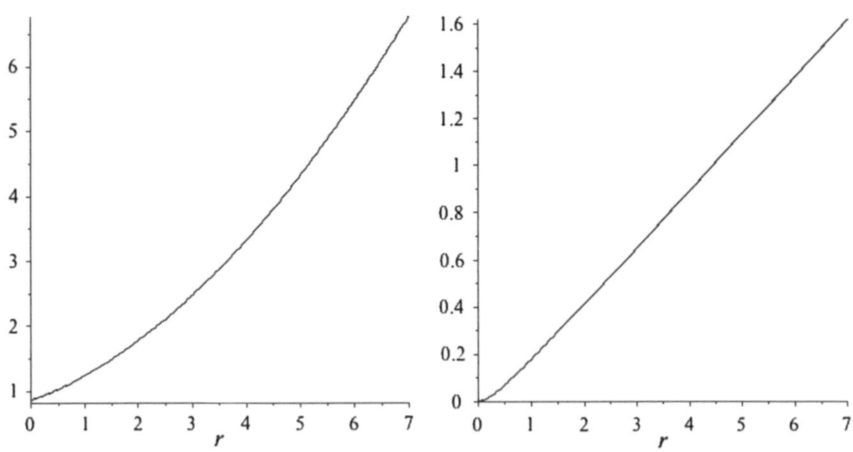

Fig. 3.2 S_2^1/S_2^2 (*left*) and C_2^1/C_2^2 (*right*)

numerical results show that C_2^2 is under-evaluated. So, it should be worth using (3.60a) and (3.60b) from the third order.

The storage of (3.58) in 3D (S_3^1) is composed of $2N_e r^3$ components of U_h, $N_e r^3$ coefficients of D_h and the stiffness matrix which can be decomposed as follows:

- $3N_e(r-1)^3 \times (r+1)^2$ coefficients for the interior points of the elements,
- $3N_e(r-1)^2 \times \left(2(r+1)(2r+1)+(r+1)^2\right)$ coefficients for the points on the faces,
- $3N_e(r-1) \times \left(2(r+1)(2r+1)+(2r+1)^2\right)$ coefficients for the points on the edges,
- $N_e \times 3(2r+1)^2$ coefficients for the summits.

We finally obtain

$$S_3^1 = N_e\left(\frac{3}{2}r^5 + 6r^4 + \frac{19}{2}r^3\right).$$

The storage of (3.60a) and (3.60b) in 3D (S_3^2) is composed of $2N_e r^3$ components of U_h, $N_e r^3$ coefficients of D_h $6N_e r^2$ coefficients of B_h and a mapping array of length $N_e r^3/2$, which provides

$$S_3^2 = N_e \frac{19}{2} r^3.$$

The number of multiplications of (3.58) in 3D (C_3^1) is equal to the number of coefficients of the stiffness matrix to which we add the multiplication by D_h. This leads to

$$C_3^1 = N_e\left(3r^5 + 12r^4 + 13r^3\right).$$

The number of multiplications of (3.60a) and (3.60b) in 3D (C_3^2) is equal to the number of coefficients of D_h plus $9N_e(r+1)^2$ multiplications for B_h plus $(r+1)^3 \times 3(r+1)$ multiplications for each stiffness matrix, i.e.

$$C_2^2 = N_e\left(6r^4 + 34r^3 + 63r^2 + 51r + 15\right).$$

These results are summarized in Fig. 3.3.

Figure 3.3 seems to show that it is worth to use (3.60a) and (3.60b) from the third order but numerical results seem to indicate that it could be used from order 2.

Another interesting comparison is the storage required by the discretization of (3.58) without using a Gauss–Lobatto rule (i.e. without mass-lumping)[7] ($\widetilde{S}_3^1$) versus S_3^2. The result, given in Fig. 3.4, shows the important gain induced by the use of a Gauss–Lobatto rule.

[7] Of course, the number of multiplications has no meaning in this case since the mass matrix obtained would be n-diagonal and would require an iterative algorithm of resolution to be inverted at each time-step.

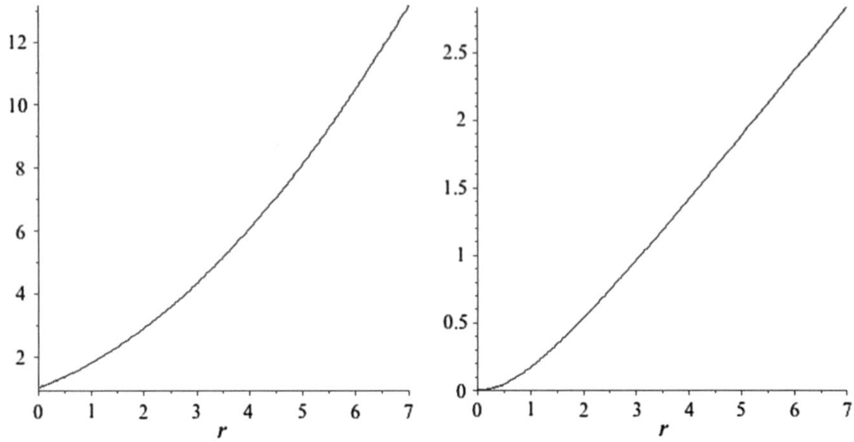

Fig. 3.3 S_3^1/S_3^2 (*left*) and C_3^1/C_3^2 (*right*)

Fig. 3.4 $\widetilde{S}_3^1/S_3^2$

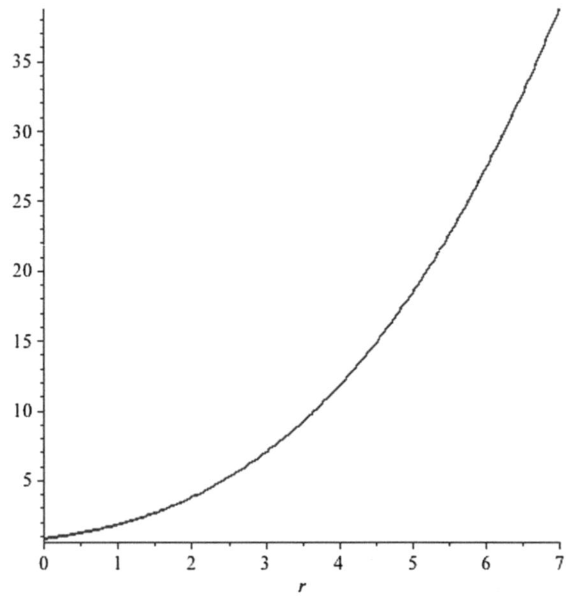

3.4 Dispersion Relation

The solution of the wave equation can be expressed as a sum of plane waves [2]. This remark justifies the use of plane waves to analyze continuous models. In [5], this use is extended to discrete equations. The purpose of this section is to show how to find dispersion relations of the method described in Sect. 3.1. As we shall see, this relation is an important source of information on the stability and accuracy properties of the numerical models. This chapter mainly addresses the 1D case, which is simple and can be extended to higher dimensions. A last point: plane wave analysis does not take into account the polynomial definition of the solution for finite element methods. It just considers the discrete equations as difference equations whose solutions are exponential functions.

3.4.1 The Continuous Equation

Let us set $\Omega = \mathbb{R}^d$, $\lambda = \lambda_0$, $\underline{\underline{\mu}} = \mu_0 I_d$ in (3.1), where λ_0 and μ_0 are real positive constants and I_d is the identity matrix of $\mathbb{R}^d$. We get:

$$\frac{\partial^2 u}{\partial t^2} - c^2 \Delta u = 0 \tag{3.61}$$

with $c^2 = \mu_0/\lambda_0$.

Now, let

$$u = \mathrm{e}^{\mathrm{i}(\omega t - \underline{k}\cdot\underline{x})} \tag{3.62}$$

be a plane wave solution of (3.61). By inserting (3.62) in (3.61) and dividing by u, we get:

$$\omega^2 = c^2 |\underline{k}|^2. \tag{3.63}$$

Equivalently, one can write:

$$c^2 = \frac{\omega^2}{|\underline{k}|^2}, \tag{3.64}$$

which shows that $\omega/|\underline{k}|$ is constant.[8]

3.4.2 A Didactic Case: The P_1 Approximation

Plane wave analysis is only possible when $\Omega = \mathbb{R}^d$ and is divided into cells of same shape which can be deduced from each other by a translation. So, we set here $d = 1$ and

$$\Omega = \mathbb{R} = \bigcup_{\ell \in \mathbb{Z}} [x_\ell, x_{\ell+1}], \tag{3.65}$$

where $x_\ell = \ell h$, $h \in \mathbb{R}^{+*}$.

[8]This constant character of $\omega/|\underline{k}|$ shows that the wave equation is non dispersive, i.e. that the velocity of the waves does not depend on $|\underline{k}|$ or, equivalently, on the frequency. As we shall see, this is not the case of its approximation.

With this notation, we define $V_h^1(\mathbb{R})$ as in (2.3) and discretize (3.61) by using a first-order Gauss–Lobatto rule (which actually is the trapezoidal rule).

Any basis function φ_ℓ of $V_h^1(\mathbb{R})$ can be written as

$$\varphi_\ell(x) = \frac{1}{h}(x - x_{\ell-1})\,\chi_{\ell-1} + \frac{1}{h}(x - x_\ell)\,\chi_\ell,$$

where χ_ℓ is the characteristic function of $[x_\ell, x_{\ell+1}]$. Now, if φ_i and φ_j are two basis functions, one can easily check that

$$\int_{\mathbb{R}}^{GL} \varphi_i\,\varphi_j\,\mathrm{d}x = \frac{h}{2}\delta_{i,j}. \tag{3.66}$$

So, the mass matrix is diagonal with all its terms equal to $h/2$.

On the other hand, we easily obtain:

$$\int_{\mathbb{R}}^{GL} \frac{\partial \varphi_i}{\partial x}\frac{\partial \varphi_j}{\partial x}\,\mathrm{d}x = \begin{cases} \dfrac{1}{h} \text{ if } j = i, \\ \dfrac{1}{2h} \text{ if } |j - i| = 1, \\ 0 \text{ otherwise.} \end{cases} \tag{3.67}$$

Equations (3.66) and (3.67) provide the following approximation of (3.61):

$$\frac{\mathrm{d}^2 u_\ell}{\mathrm{d}t^2} = \frac{c^2}{h^2}(u_{\ell+1}(t) - 2u_\ell(t) + u_{\ell-1}(t)), \tag{3.68}$$

where $u_\ell(t) = u_h(x_\ell, t)$.

Equation (3.68) actually corresponds to a second-order centered finite difference approximation in space of (3.61) when $d = 1$ [2].

Now, let us insert a plane wave solution defined by $u_\ell = \mathrm{e}^{\mathrm{i}(\omega_h t - \ell k h)}$ in (3.68). After dividing by u_ℓ, we get:

$$-\omega_h^2 = \frac{c^2}{h^2}\left(\mathrm{e}^{-\mathrm{i}k h} - 2 + \mathrm{e}^{\mathrm{i}k h}\right) = \frac{c^2}{h^2}\,(2\cos kh - 2)\,,$$

which finally provides:

$$\omega_h^2 = \frac{4c^2}{h^2}\sin^2\frac{k h}{2}. \tag{3.69}$$

Equations (3.63) and (3.69) are called the dispersion relations of (3.61) and (3.68) respectively. One can easily check that (3.69) is an approximation of (3.63) when $d = 1$, which reads:

$$\omega^2 = c^2 k^2. \tag{3.70}$$

In the same way, one can see that ω_h/k is an approximation of $\omega/k = c$. So, one can set $c_h = \omega_h/k$, which is the velocity of the solution of (3.68) which, unlike c, is not constant. We then define the a-dimensional quantity q_h by

$$q_h = \frac{c_h}{c} = \frac{\omega_h}{\omega}, \tag{3.71}$$

called the dispersion coefficient and which measures the error committed on the velocity for the approximation. In our case, we have:

$$q_h = \frac{2}{k\,h} \sin \frac{k\,h}{2}. \tag{3.72}$$

We obviously have $\lim\limits_{h \to 0} q_h = 1$. More precisely, we have:

$$q_h = 1 - \frac{k^2}{24} h^2 + \frac{k^4}{1920} h^4 + O\left(h^6\right), \tag{3.73}$$

which shows that we have a second-order error on the velocity.

Remark: Equations (3.63) and (3.69) can be obtained by using the (continuous and discrete) Fourier transform of (3.61) and (3.68) [2].

A convenient way to visualize the error committed on the velocity is to draw the dispersion curves. These curves generally depend on the quantity $K = k\,h/2\pi$ whose meaning is the following:

A plane wave with a pulsation ω and a velocity c can be characterized by

- its frequency: $f = \dfrac{\omega}{2\pi}$,
- its wavelength: : $\lambda = \dfrac{c}{f} = \dfrac{2\pi c}{\omega}$.

By using (3.70), we can write, for $\omega > 0$:

$$\lambda = \frac{2\pi}{k}. \tag{3.74}$$

Let N_λ be the number of points of a mesh per wavelength. We have

$$N_\lambda = \frac{\lambda}{h} = \frac{2\pi}{k\,h} = \frac{1}{K}. \tag{3.75}$$

So, K is the inverse of the number of points per wavelength, which is a quantity independent of the wavelength of the wave and which tends to infinity when $h \to 0$. With these notations, (3.72) becomes

$$q_h = \frac{1}{\pi K} \sin \pi K. \tag{3.76}$$

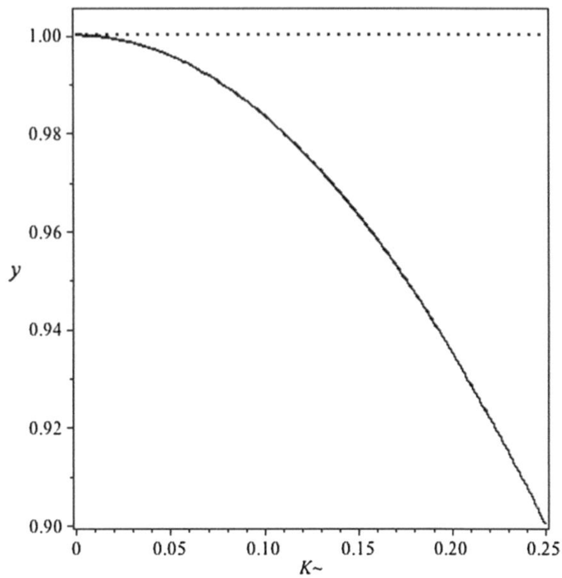

Fig. 3.5 The dispersion curve corresponding to the P_1 approximation of the 1D wave equation. The abscissa is K

Remarks:

1. For P_1, the number of elements per wavelength coincides with he number of points per wavelength. For higher orders, since h represents the size of a cell, which contains r points (the second end belonging to the next cell, as shown in Figs. 3.7 and 3.9), we have $N_\lambda = r\lambda/h$ and then $kh = 2r\pi K$.
2. These definitions hold for $d > 1$ if we replace k by $|\underline{k}|$.

The graphical representation of (3.76) is given in Fig. 3.5.

3.4.3 *The Concept of Numerical Dispersion*

Besides the error committed on the velocity, q_h measures the numerical dispersion of the scheme which produces parasitic waves around the solution.

In fact, these parasitic waves arise from the following process: As we know, the continuous velocity c in a homogeneous medium is a constant, whereas the discrete velocity c_h is a function of $\underline{k}$ and, therefore, of the frequency of the wave. So, when the wave is polychromatic, the dependency on the frequency implies that each monochromatic component of the wave moves with a different velocity. When c_h depends significantly on the frequency, some of these components leave the physical wave and produce a sequence of parasitic waves which have no physical meaning. Of course, such a phenomenon can seriously damage the results obtained by a numerical model.

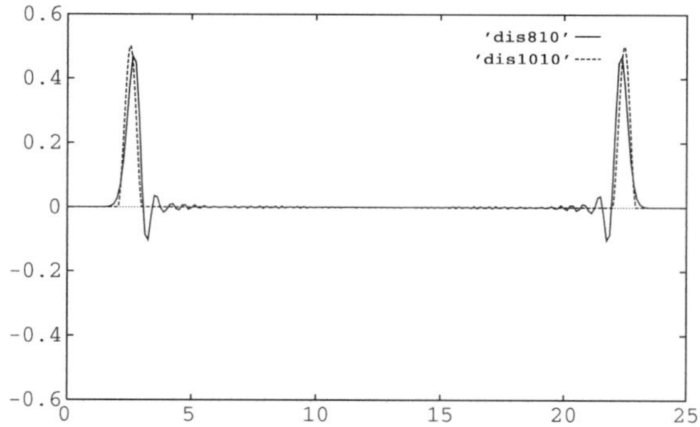

Fig. 3.6 Comparison of the solution obtained with a dispersive scheme (in *continuous line*) and the exact solution (in *dotted line*)

This can happen in two ways: either the time-step is too large for the high frequency components of the wave or the coefficient in front of the leading power of h in the Taylor's equation is too large.

Schemes with large coefficients can develop such parasitic waves even for small space-steps. They are called *dispersive* schemes. So, two schemes of the same order can be more or less dispersive because of this coefficient. The knowledge of this coefficient actually justifies the computation of the dispersion relation.

We give a numerical illustration of this phenomenon in Fig. 3.6.

3.4.4 P_2 Approximation

The basis functions of $V_h^2(\mathbb{R})$ are of two kinds:

$$\varphi_\ell(x) = \frac{2}{h^2}(x - x_{\ell-1})(x - x_{\ell-1,1})\,\chi_{\ell-1} + \frac{2}{h^2}(x - x_{\ell+1})(x - x_{\ell,1})\,\chi_\ell \qquad (3.77)$$

corresponding to x_ℓ, where $x_{\ell,1} = (x_\ell + x_{\ell+1})/2$,

$$\varphi_{\ell,1}(x) = -\frac{4}{h^2}(x - x_{\ell+1})(x - x_\ell)\,\chi_\ell \qquad (3.78)$$

corresponding to $x_{\ell,1}$.

By computing the mass and stiffness integrals by using the second-order Gauss–Lobatto rule (which is the Simpson's rule in this case), we get the two following classes of equations:

$$\begin{aligned}\frac{d^2u_\ell}{dt^2}(t) = -\frac{c^2}{h^2}(14u_\ell(t) - 8(u_{\ell,1}(t) + u_{\ell-1,1}(t))\\ + u_{\ell+1}(t) + u_{\ell-1}(t)), \quad \forall \ell \in \mathbb{Z},\end{aligned} \qquad (3.79a)$$

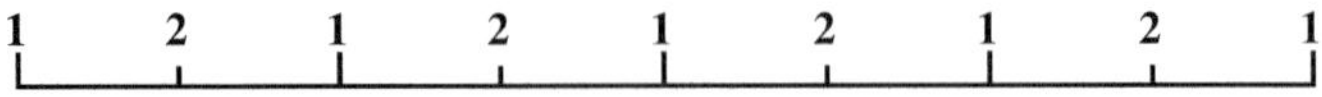

Fig. 3.7 The two classes of degrees of freedom for P_2

$$\frac{d^2 u_{\ell,1}}{dt^2}(t) = \frac{4c^2}{h^2}(u_\ell(t) - 2u_{\ell,1}(t) + u_{\ell+1}(t)), \quad \forall \ell \in \mathbb{Z} \tag{3.79b}$$

obtained by taking (3.77) and (3.78) as test functions respectively, where $u_{\ell,1}(t) = u_h(x_{\ell,1}, t)$.

We have here two classes of degrees of freedom invariable by translation (Fig. 3.7), the solution is composed of two types of plane waves:

$$u_\ell = \alpha_1 \mathrm{e}^{\mathrm{i}(\ell k h - \omega_h t)}, \tag{3.80a}$$

$$u_{\ell,1} = \alpha_2 \mathrm{e}^{\mathrm{i}((\ell+\frac{1}{2})k h - \omega_h t)}, \tag{3.80b}$$

where $(\alpha_1, \alpha_2)^T \in \mathbb{C}^2$.

By inserting (3.80a) and (3.80b) into (3.79a) and (3.79b) and after dividing by u_ℓ and $u_{\ell,1}$, we get:

$$\omega_h^2 \alpha_1 = \frac{c^2}{h^2}(14\alpha_1 - 8\alpha_2(\mathrm{e}^{\mathrm{i}kh/2} + \mathrm{e}^{-\mathrm{i}k h/2}) + \alpha_1(\mathrm{e}^{\mathrm{i}k h} + \mathrm{e}^{-\mathrm{i}k h})), \tag{3.81a}$$

$$\omega_h^2 \alpha_2 = -\frac{4c^2}{h^2}(\alpha_1 \mathrm{e}^{-\mathrm{i}k h/2} - 2\alpha_2 + \alpha_1 \mathrm{e}^{\mathrm{i}k h/2}). \tag{3.81b}$$

which can be written, after simplification, as the following generalized eigenvalue problem:

$$\widehat{N}_{1,2}\, \underline{U}_\alpha = \omega_h^2\, \underline{U}_\alpha, \tag{3.82}$$

where

$$\underline{U}_\alpha = \begin{pmatrix} \alpha_1 \\ \alpha_2 \end{pmatrix}, \quad \widehat{N}_{1,2}(k h) = \frac{c^2}{h^2} \begin{pmatrix} 14 + 2\cos kh & -16\cos \dfrac{kh}{2} \\ -8\cos \dfrac{kh}{2} & 8 \end{pmatrix},$$

After setting $w = \sin k h/2$, the characteristic equation of (3.82) reads:

$$h^4\, \omega_h^4 + 4c^2 h^2 (w^2 - 6)\, \omega_h^2 + 96 c^4 w^2 = 0. \tag{3.83}$$

The two zeroes of this quartic equation:

$$\omega_{h,1}^2 = \frac{2c^2}{h^2}(6 - w^2 - \sqrt{36 - 36w^2 + w^4}), \tag{3.84a}$$

$$\omega_{h,2}^2 = \frac{2c^2}{h^2}(6 - w^2 + \sqrt{36 - 36w^2 + w^4}), \tag{3.84b}$$

are the two dispersion relations of the problem.

The Taylor expansions of the two relations lead to

$$\omega_{h,1}^2 = c^2k^2\left(1 - \frac{k^4h^4}{1440} - \frac{k^6h^6}{48\,384} + O(k^8h^8)\right), \tag{3.85a}$$

$$\omega_{h,2}^2 = c^2k^2\left(\frac{24}{k^2h^2} - 2 + \frac{k^2h^2}{12} - \frac{k^4h^4}{480} + O(k^6h^6)\right). \tag{3.85b}$$

Equation (3.85a) is a fourth-order approximation of the dispersion relation of the continuous wave equation whereas (3.85b) reveals the presence of a parasitic wave whose velocity tends to infinity when h tends to zero. The amplitude of this wave is in $O(h^4)$ [2], which attenuates its effect when h is small enough. However, these parasitic waves somehow weaken the efficiency of the finite element approximations.

The graphical representation of (3.84a) is given in Fig. 3.8.

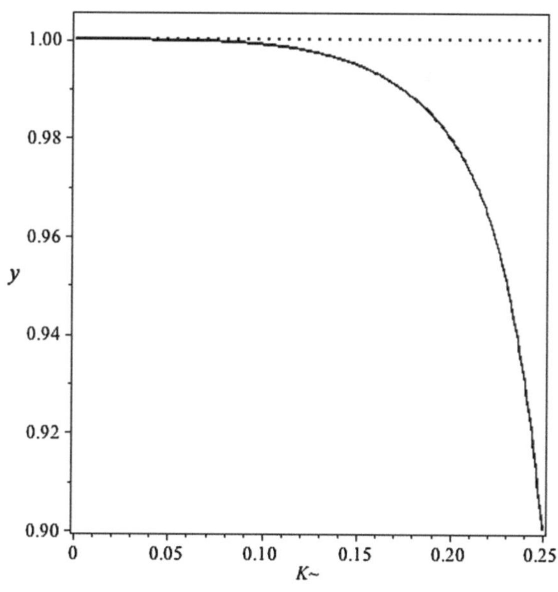

Fig. 3.8 The dispersion curve corresponding to the P_2 approximation of the 1D wave equation. The abscissa is K

3.4.5 P_3 and Higher-Order Approximations

The basis functions of $V_h^3(\mathbb{R})$ are of three kinds:

$$\begin{aligned}\varphi_\ell(x) = {} & \frac{5}{h^3}(x - x_{\ell-1})(x - x_{\ell-1,1})(x - x_{\ell-1,2})\,\chi_{\ell-1} \\ & + \frac{5}{h^3}(x - x_{\ell+1})(x - x_{\ell,1})(x - x_{\ell,2})\,\chi_\ell \end{aligned} \tag{3.86}$$

corresponding to x_ℓ, where $x_{\ell,1} = x_\ell + h\mu$, $x_{\ell,2} = x_\ell + h\nu$, $\mu = (5 - \sqrt{5})/10$, $\nu = 1 - \mu = (5 + \sqrt{5})/10$,

$$\varphi_{\ell,1}(x) = \frac{5\sqrt{5}}{h^3}(x - x_{\ell+1})(x - x_{\ell,2})(x - x_\ell)\,\chi_\ell \tag{3.87}$$

corresponding to $x_{\ell,1}$,

$$\varphi_{\ell,2}(x) = -\frac{5\sqrt{5}}{h^3}(x - x_{\ell+1})(x - x_{\ell,1})(x - x_\ell)\,\chi_\ell \tag{3.88}$$

corresponding to $x_{\ell,2}$.

To these three classes of functions correspond the following three classes of equations:

$$\begin{aligned}\frac{d^2u_\ell}{dt^2}(t) + \frac{c^2}{h^2}\Big[& 52u_\ell(t) - \frac{5}{2}(5 + 3\sqrt{5})(u_{\ell-1,2}(t) + u_{\ell,1}(t)) \\ & + \frac{5}{2}(3\sqrt{5} - 5)(u_{\ell-1,1}(t) + u_{\ell,2}(t)) \\ & -(u_{\ell-1}(t) + u_{\ell+1}(t))\Big] = 0, \ \forall \ell \in \mathbb{Z}, \end{aligned} \tag{3.89a}$$

$$\begin{aligned}\frac{d^2u_{\ell,1}}{dt^2}(t) + \frac{c^2}{h^2}[& -(5 + 3\sqrt{5})u_\ell(t) + 20u_{\ell,1}(t) - 10u_{\ell,2}(t) \\ & + (3\sqrt{5} - 5)u_{\ell+1}(t)] = 0, \ \forall \ell \in \mathbb{Z}, \end{aligned} \tag{3.89b}$$

$$\begin{aligned}\frac{d^2u_{\ell,2}}{dt^2}(t) + \frac{c^2}{h^2}[& (3\sqrt{5} - 5)u_\ell(t) - 10u_{\ell,1}(t) + 20u_{\ell,2}(t) \\ & - (5 + 3\sqrt{5})u_{\ell+1}(t)] = 0, \ \forall \ell \in \mathbb{Z}. \end{aligned} \tag{3.89c}$$

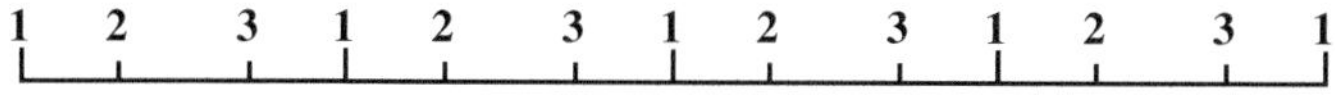

Fig. 3.9 The three classes of degrees of freedom for P_3

Here, as shown if Fig. 3.9, the plane wave solution is subdivided into three classes represented by

$$u_\ell = \alpha_1 \mathrm{e}^{\mathrm{i}(\ell k\,h - \omega_h t)}, \tag{3.90a}$$

$$u_{\ell,1} = \alpha_2 \mathrm{e}^{\mathrm{i}((\ell+\mu)k\,h - \omega_h t)}, \tag{3.90b}$$

$$u_{\ell,2} = \alpha_3 \mathrm{e}^{\mathrm{i}((\ell+\nu)k\,h - \omega_h t)}, \tag{3.90c}$$

where $(\alpha_1, \alpha_2, \alpha_3)^T \in \mathbb{C}^3$.

By inserting (3.90a)–(3.90c) into (3.89a)–(3.89c), we obtain, after simplification:

$$\begin{aligned}\alpha_1\omega_h^2 + \frac{c^2}{h^2}\Big[52\alpha_1 &- \frac{5}{2}(5+3\sqrt{5})(\alpha_3\mathrm{e}^{-\mathrm{i}\mu k\,h} + \alpha_2\mathrm{e}^{\mathrm{i}\mu k\,h})\\ &+ \frac{5}{2}(3\sqrt{5}-5)(\alpha_2\mathrm{e}^{-\mathrm{i}\nu k\,h} + \alpha_3\mathrm{e}^{\mathrm{i}\nu k\,h})\\ &-\alpha_1(\mathrm{e}^{-\mathrm{i}k\,h} + \mathrm{e}^{\mathrm{i}k\,h})\Big] = 0,\end{aligned} \tag{3.91a}$$

$$\begin{aligned}\alpha_2\omega_h^2 + \frac{c^2}{h^2}[&-(5+3\sqrt{5})\alpha_1\mathrm{e}^{-\mathrm{i}\mu k\,h} + 20\alpha_2 - 10\alpha_3\mathrm{e}^{\mathrm{i}\sqrt{5}k\,h}\\ &- (5+3\sqrt{5})\alpha_1\mathrm{e}^{\mathrm{i}\nu k\,h}] = 0,\end{aligned} \tag{3.91b}$$

$$\begin{aligned}\alpha_3\omega_h^2 + \frac{c^2}{h^2}[&(3\sqrt{5}-5)\alpha_1\mathrm{e}^{-\mathrm{i}\nu k\,h} - 10\alpha_2\mathrm{e}^{-\mathrm{i}\sqrt{5}k\,h} + 20\alpha_3\\ &- (5+3\sqrt{5})\alpha_1\mathrm{e}^{\mathrm{i}\nu k\,h}] = 0,\end{aligned} \tag{3.91c}$$

which provides the following three-dimensional eigenvalue problem

$$\widehat{N}_{1,3}\,\underline{U}_\alpha = \omega_h^2\,\underline{U}_\alpha, \tag{3.92}$$

where

$$\underline{U}_\alpha = \begin{pmatrix}\alpha_1\\ \alpha_2\\ \alpha_3\end{pmatrix},$$

$$\frac{h^2}{c^2}\widehat{K}_{1,3} =$$

$$\begin{bmatrix} 52 - 2\cos k\,h & \dfrac{-5\,a\mathrm{e}^{\mathrm{i}k\,h\mu} + 5\,b\mathrm{e}^{-\mathrm{i}k\,h\nu}}{2} & \dfrac{-5\,a\mathrm{e}^{-\mathrm{i}k\,h\mu} + 5\,b\mathrm{e}^{\mathrm{i}k\,h\nu}}{2} \\ -a\mathrm{e}^{-\mathrm{i}k\,h\mu} + b\mathrm{e}^{\mathrm{i}k\,h\nu} & 20 & -10\,\mathrm{e}^{\mathrm{i}\frac{\sqrt{5}}{5}k\,h} \\ \dfrac{-5a\mathrm{e}^{\mathrm{i}k\,h\mu} + 5b\mathrm{e}^{-\mathrm{i}k\,h\nu}}{4} & \dfrac{-25\mathrm{e}^{-\mathrm{i}\frac{\sqrt{5}}{5}k\,h}}{2} & 25 \end{bmatrix},$$

with $a = 5 + 3\sqrt{5}$ and $b = -5 + 3\sqrt{5}$. Here also, $\widehat{M}_{1,3}$ and $\widehat{K}_{1,3}$ are the Fourier transforms of the mass and the stiffness matrix.

By setting $w = \cos k\,h$, the characteristic polynomial of (3.92) can be written as

$$h^6\omega_h^6 + 2c^2h^4(w-46)\omega_h^4 + 120c^4h^2(w+14)\omega_h^2 + 3600c^6(w-1) = 0. \quad (3.93)$$

By using the trigonometric method for solving (3.93), we obtain the following three roots: of this equation

$$\omega_{h,1}^2 = \frac{2}{3}\sqrt{a_1}\cos\left(\frac{1}{3}\arccos\left(\frac{27\,a_2}{a_1^{3/2}}\right) + \frac{\pi}{3}\right) - a_3, \quad (3.94a)$$

$$\omega_{h,2}^2 = \frac{2}{3}\sqrt{a_1}\cos\left(\frac{1}{3}\arccos\left(\frac{27\,a_2}{a_1^{3/2}}\right) - \frac{\pi}{3}\right) - a_3, \quad (3.94b)$$

$$\omega_{h,3}^2 = -\frac{2}{3}\sqrt{a_1}\cos\left(\frac{1}{3}\arccos\left(\frac{27\,a_2}{a_1^{3/2}}\right)\right) - a_3, \quad (3.94c)$$

where

$$a_1 = \frac{4c^4\left(w^2 - 182\,w + 856\right)}{h^4},$$

$$a_2 = \frac{8}{27}\,\frac{c^6\left(w^3 - 273\,w^2 + 16743\,w - 16471\right)}{h^6},$$

$$a_3 = \frac{2}{3}\,\frac{c^2\,(w-46)}{h^2},$$

which are all real and their Taylor expansions are

$$\omega_{h,1}^2 = c^2k^2\left(\frac{60}{k^2h^2} + 5 - \frac{7}{6}k^2h^2 + \frac{23}{60}k^4h^4 - \frac{1899}{11\,200}k^6h^6 + O(h^8)\right), \quad (3.95a)$$

$$\omega_{h,2}^2 = c^2k^2\left(\frac{30}{k^2h^2} - 5 + \frac{13}{12}k^2h^2 - \frac{137}{360}k^4h^4 + \frac{51\,259}{302\,400}k^6h^6 + O(h^8)\right), \tag{3.95b}$$

$$\omega_{h,3}^2 = c^2k^2\left(1 - \frac{1}{302\,400}k^6h^6 - \frac{1}{427\,680\,000}k^{10}h^{10} + O(h^{12})\right). \tag{3.95c}$$

Here also, we obtain one physical dispersion relations and two other corresponding to parasitic waves whose amplitudes are in $O(h^6)$ or $O(h^5)$ [2, 6].

The graphical representation of (3.94c) and the comparison of dispersion curves provided by (3.76), (3.84a) and (3.94c) are given in Figs. 3.10 and 3.11.

Of course, such an approach, which provides the explicit form of the eigenvalues, is difficult to apply to higher-order methods since the degree of the characteristic polynomial increases with the order of the method. However, the characteristic equation (or its Taylor expansion) can be obtained for rather high orders. In that case, the Taylor expansions of the eigenvalues can be computed by a recursive algorithm. As an example, we are going to compute, in the following, the Taylor expansions of the eigenvalues (3.94a)–(3.94c) by this algorithm.

We first look for a sixth-order Taylor expansion around $h = 0$ of the characteristic polynomial (3.93), in which we set $\lambda = \omega_h^2$. We get:

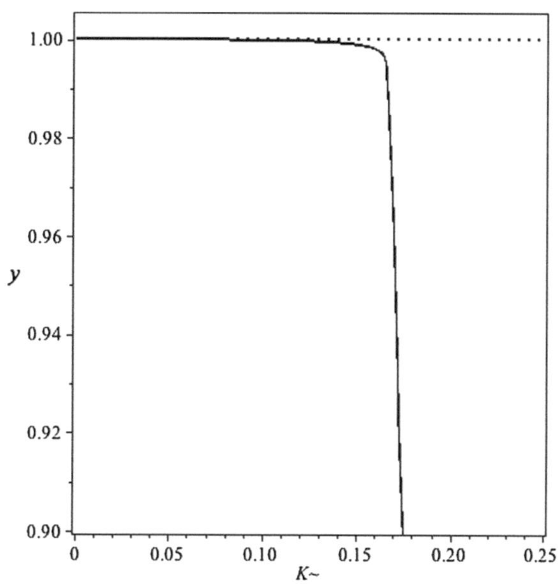

Fig. 3.10 The dispersion curve corresponding to the P_3 approximation of the 1D wave equation. The abscissa is K. One can notice the quick degeneration of the accuracy

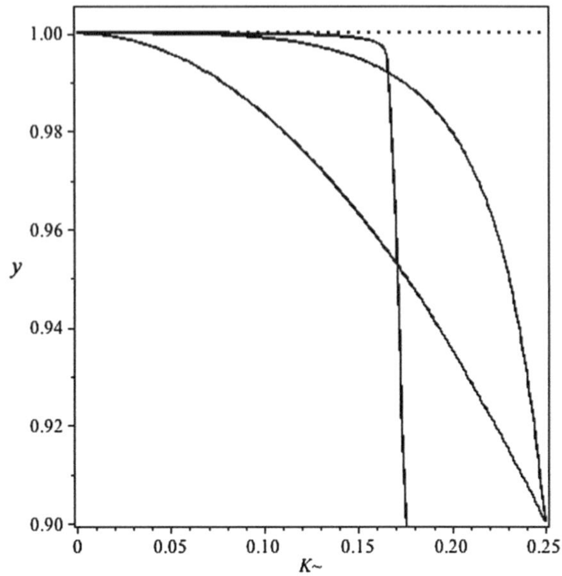

Fig. 3.11 The dispersion curves from P_1 (*lower*) to P_3 (*upper*) approximations in space of the 1D wave equation. The abscissa is K. One can notice that the gain of accuracy decreases with the order

$$\begin{aligned}
&\frac{1800\,(c^6k^2 - c^4\lambda)}{h^4} + \frac{-150\,c^6k^4 + 90\,c^2\lambda^2 + 60\,c^4k^2\lambda}{h^2} \\
&\quad - \lambda^3 + c^2k^2\lambda^2 - 5\,c^4k^4\lambda + 5\,c^6k^6 \\
&\quad + \left(-\frac{1}{12}\,c^2k^4\lambda^2 + \frac{1}{6}\,c^4k^6\lambda - \frac{5}{56}\,c^6k^8\right)h^2 \\
&\quad + \left(\frac{1}{360}\,c^2k^6\lambda^2 - \frac{1}{336}\,c^4k^8\lambda + \frac{1}{1008}\,c^6k^{10}\right)h^4 \\
&\quad + \left(-\frac{1}{20\,160}\,c^2k^8\lambda^2 + \frac{1}{30\,240}\,c^4k^{10}\lambda - \frac{1}{133\,056}\,c^6k^{12}\right)h^6 = 0. \qquad (3.96)
\end{aligned}$$

Then, we look for a solution of this Taylor expansion of the form:

$$\lambda = c^2\frac{\gamma}{h^2}, \qquad (3.97)$$

that we plug into (3.96).

The lowest-order term of the result is

$$\frac{c^6}{h^6}(-\gamma^3 + 90\gamma^2 - 1800\gamma), \qquad (3.98)$$

which has three roots equal to 0, 30 and 60. Of course, the root equal to 0 corresponds to the physical solution and the two other roots to the parasitic waves. We take $\gamma = 30$. Then, we look for a solution of (3.96) of the form:

$$\lambda = c^2 \left(\frac{30}{h^2} + \gamma k^2 \right). \tag{3.99}$$

The lowest-order term of (3.96) becomes

$$\frac{c^6}{h^4}(900\gamma k^2 + 4500k^2), \tag{3.100}$$

whose unique root is $\gamma = -5$.

Once more we look for a solution of the form:

$$\lambda = c^2 \left(\frac{30}{h^2} - 5k^2 + \gamma k^4 h^2 \right). \tag{3.101}$$

After inserting (3.101) into (3.96), we obtain its lowest-order term:

$$\frac{c^6}{h^2}(900k^4\gamma - 975k^4). \tag{3.102}$$

Its root is $\gamma = 13/12$.

So, let us take

$$\lambda = c^2 \left(\frac{30}{h^2} - 5k^2 + \frac{13}{12}k^4h^2 + \gamma k^6 h^4 \right) \tag{3.103}$$

and plug it into (3.96). Then, its lowest-order term is

$$c^6 \left(900k^6\gamma - \frac{685}{2}k^6 \right), \tag{3.104}$$

whose unique root is $\gamma = -137/360$.

Finally, we look for an eigenvalue of the form:

$$\lambda = c^2 \left(\frac{30}{h^2} - 5k^2 + \frac{13}{12}k^4h^2 - \frac{137}{360}k^6h^4 + \gamma k^8 h^6 \right). \tag{3.105}$$

The lowest-order term of (3.96), into which we inserted (3.105), is

$$c^6h^2 \left(900k^8\gamma - \frac{51\,259}{336}k^8 \right) \tag{3.106}$$

and its unique root is $\gamma = 51\,259/302\,400$.

We can iterate the process as long as we want, but we have reached the maximal accuracy for a sixth-order Taylor expansion of the characteristic polynomial. Higher accuracy requires adding to this expansion as many terms as the additional terms of the eigenvalue.

The rest of the Taylor expansion of (3.93) in which we plug the sixth-order expansion of λ is in $O(h^4)$, which means that we solved (3.93) up to the fourth-order.

Of course, we would obtain (3.95a) or (3.95b) by starting with $\gamma = 60$ or $\gamma = 0$.

If we did not know the form of the first term, we could start with any (even) negative power of h and we would find that the lowest term is proportional to γ, which would imply that $\gamma = 0$.

We could also take the unknown term of λ just proportional to h^{2r} without taking into account the term in k^{2r+2}. Then, we would obtain γ in terms of k^{2r+2}.

So, we have given a general algorithm which enables us to obtain the Taylor expansions of the dispersion relations as soon as we know how to compute a sufficiently accurate Taylor expansion of the characteristic polynomial of the eigenvalue problem derived from the plane wave solution of the problem.

3.4.6 Extension to Higher Dimensions

The construction of 2D and 3D elements based on the 1D elements enables us to derive, for regular meshes, the eigenvalues and eigenvectors in higher dimensions from those obtained in 1D in the plane wave analysis. This result is based on the following lemma[9]:

Lemma 1 *Let $N_{d,r} = D_h^{-1}K_h$, D_h and K_h being defined in (3.45), and $N_{d,r}(p,q)$ the term located at the pth line and qth column of this matrix ($1 \le p \le \nu^d$ and $1 \le q \le \nu^d$, $d = 1\ldots3$). Then, for a regular mesh composed of squares, we have:*

$$N_{2,r}((\ell_2-1)\nu+\ell_1,(m_2-1)\nu+m_1) = N_{1,r}(\ell_1,m_1)\delta_{\ell_2,m_2} + N_{1,r}(\ell_2,m_2)\delta_{\ell_1,m_1}, \tag{3.107}$$

$$\begin{aligned}N_{3,r}(((\ell_3-1)\nu+\ell_2-1)\nu+\ell_1),((m_3-1)\nu+m_2-1)\nu+m_1) \\ = N_{1,r}(\ell_1,m_1)\delta_{\ell_2,m_2}\delta_{\ell_3,m_3} + N_{1,r}(\ell_2,m_2)\delta_{\ell_1,m_1}\delta_{\ell_3,m_3} \\ + N_{1,r}(\ell_3,m_3)\delta_{\ell_1,m_1}\delta_{\ell_2,m_2},\end{aligned} \tag{3.108}$$

where $\delta_{\ell,m}$ is the Kronecker symbol.

Proof We give the proof in details in the 2D case and then extend it to the 3D case. This proof is based on the fact that any basis function $\varphi_{p,q}$ has a rectangular support $[\alpha_1,\alpha_2]\times[\beta_1,\beta_2]$ and we have:

$$\varphi_{p,q}(x_1,x_2) = \varphi_p(x_1)\varphi_q(x_2), \tag{3.109}$$

[9]The results of this section were given by N. Tordjman in her thesis [6] for Q_2 and Q_3 in 2D. Their generalization to any order and to the 3D case is a part of S. Fauqueux's thesis [7].

where φ_p and φ_q are two 1D basis functions so that $[\alpha_1, \alpha_2]$ is the support of φ_p and $[\beta_1, \beta_2]$ is that of φ_q.

On the other hand, the intersection of the supports of two 2D basis functions is always a rectangle that we denote $\mathscr{S} = [a_1, a_2] \times [b_1, b_2]$ in the following. With this notation, the terms of the mass-matrix can be written as

$$\int_{\mathscr{S}} \varphi_{\ell_1,\ell_2} \varphi_{m_1,m_2} \mathrm{d}\underline{x}. \tag{3.110}$$

Now, by using (3.109), we obtain:

$$\begin{aligned}
&\int_{\mathscr{S}} \varphi_{\ell_1,\ell_2} \varphi_{m_1,m_2} \mathrm{d}\underline{x} \\
&\quad = \int_{a_1}^{a_2} \int_{b_1}^{b_2} \varphi_{\ell_1}(x_1) \varphi_{\ell_2}(x_2) \varphi_{m_1}(x_1) \varphi_{m_2}(x_2)\, \mathrm{d}x_1 \mathrm{d}x_2 \\
&\quad = \int_{a_1}^{a_2} \varphi_{\ell_1}(x_1) \varphi_{m_1}(x_1)\, \mathrm{d}x_1 \int_{b_1}^{b_2} \varphi_{\ell_2}(x_2) \varphi_{m_2}(x_2)\, \mathrm{d}x_2.
\end{aligned} \tag{3.111}$$

So, if $D_{d,r}(p, q)$ denotes the term located at the pth line and qth column of the diagonal mass matrix in dimension d obtained by computing the above integrals by a Gauss–Lobatto rule, we obtain the following relation:

$$\begin{aligned}
&D_{2,r}((\ell_2 - 1)\nu + \ell_1, (m_2 - 1)\nu + m_1) \\
&\quad = D_{1,r}(\ell_1, m_1) \times D_{1,r}(\ell_2, m_2).
\end{aligned} \tag{3.112}$$

In the same way, the terms of the stiffness matrix are given by the integrals:

$$\int_{\mathscr{S}} \nabla \varphi_{\ell_1,\ell_2} \cdot \nabla \varphi_{m_1,m_2} \mathrm{d}\underline{x}. \tag{3.113}$$

Since

$$\nabla \varphi_{\ell_1,\ell_2}(x_1, x_2) = \begin{pmatrix} \dfrac{\mathrm{d}\varphi_{\ell_1}}{\mathrm{d}x_1}(x_1)\, \varphi_{\ell_2}(x_2) \\ \\ \varphi_{\ell_1}(x_1)\, \dfrac{\mathrm{d}\varphi_{\ell_2}}{\mathrm{d}x_2}(x_2) \end{pmatrix}, \tag{3.114}$$

we get

$$\begin{aligned}
&\int_{\mathscr{S}} \nabla \varphi_{\ell_1,\ell_2} \cdot \nabla \varphi_{m_1,m_2} \mathrm{d}\underline{x} \\
&\quad = \int_{\mathscr{S}} \left[\frac{\mathrm{d}\varphi_{\ell_1}}{\mathrm{d}x_1}(x_1)\, \frac{\mathrm{d}\varphi_{m_1}}{\mathrm{d}x_1}(x_1)\, \varphi_{\ell_2}(x_2)\, \varphi_{m_2}(x_2) \right.
\end{aligned}$$

$$
\begin{aligned}
&\quad + \varphi_{\ell_1}(x_1)\,\varphi_{m_1}(x_1)\,\frac{\mathrm{d}\varphi_{\ell_2}}{\mathrm{d}x_2}(x_2)\,\frac{\mathrm{d}\varphi_{m_2}}{\mathrm{d}x_2}(x_2)\Big]\,\mathrm{d}x_1\mathrm{d}x_2 \\
&= \int_{a_1}^{a_2} \frac{\mathrm{d}\varphi_{\ell_1}}{\mathrm{d}x_1}(x_1)\frac{\mathrm{d}\varphi_{m_1}}{\mathrm{d}x_1}(x_1)\,\mathrm{d}x_1 \int_{b_1}^{b_2} \varphi_{\ell_2}(x_2)\varphi_{m_2}(x_2)\,\mathrm{d}x_2 \\
&+ \int_{a_1}^{a_2} \varphi_{\ell_1}(x_1)\varphi_{m_1}(x_1)\,\mathrm{d}x_1 \int_{b_1}^{b_2} \frac{\mathrm{d}\varphi_{\ell_2}}{\mathrm{d}x_2}(x_2)\frac{d\varphi_{m_2}}{\mathrm{d}x_2}(x_2)\,\mathrm{d}x_2. \qquad (3.115)
\end{aligned}
$$

So, if we define $K_{d,r}(p,q)$ for the stiffness matrix as for the mass matrix, we have:

$$
\begin{aligned}
&K_{2,r}((\ell_2-1)\nu+\ell_1,(m_2-1)\nu+m_1) \\
&\quad = K_{1,r}(\ell_1,m_1)D_{1,r}(\ell_2,m_2) + K_{1,r}(\ell_2,m_2)D_{1,r}(\ell_1,m_1). \qquad (3.116)
\end{aligned}
$$

Since the mass matrices are all diagonal, we can write:

$$
\begin{aligned}
&N_{2,r}((\ell_2-1)\nu+\ell_1,(m_2-1)\nu+m_1) \\
&\quad = D_{2,r}^{-1}((\ell_2-1)\nu+\ell_1,(\ell_2-1)\nu+\ell_1) \\
&\quad\quad \times K_{2,r}((\ell_2-1)\nu+\ell_1,(m_2-1)\nu+m_1) \\
&\quad = \frac{1}{D_{1,r}(\ell_1,\ell_1)}\frac{1}{D_{1,r}(\ell_2,\ell_2)}\big[K_{1,r}(\ell_1,m_1)D_{1,r}(\ell_2,m_2) \\
&\quad\quad + K_{1,r}(\ell_2,m_2)D_{1,r}(\ell_1,m_1)\big] \\
&\quad = K_{1,r}(\ell_1,m_1)\frac{D_{1,r}(\ell_1,m_1)}{D_{1,r}(\ell_1,\ell_1)} + K_{1,r}(\ell_2,m_2)\frac{D_{1,r}(\ell_2,m_2)}{D_{1,r}(\ell_2,\ell_2)}. \qquad (3.117)
\end{aligned}
$$

So, by taking into account the fact that $D_{1,r}$ is diagonal, we obtain (3.107).

In order to obtain (3.108), one proves, by a similar process, that

$$
\begin{aligned}
&D_{3,r}((((\ell_3-1)\nu+\ell_2-1)\nu+\ell_1),((m_3-1)\nu+m_2-1)\nu+m_1) \\
&\quad = D_{1,r}(\ell_1,m_1)D_{1,r}(\ell_2,m_2)D_{1,r}(\ell_3,m_3) \qquad (3.118)
\end{aligned}
$$

and

$$
\begin{aligned}
&K_{3,r}((((\ell_3-1)\nu+\ell_2-1)\nu+\ell_1),((m_3-1)\nu+m_2-1)\nu+m_1) \\
&\quad = K_{1,r}(\ell_1,m_1)D_{1,r}(\ell_2,m_2)D_{1,r}(\ell_3,m_3) \\
&\quad\quad + K_{1,r}(\ell_2,m_2)D_{1,r}(\ell_1,m_1)D_{1,r}(\ell_3,m_3) \\
&\quad\quad + K_{1,r}(\ell_3,m_3)D_{1,r}(\ell_1,m_1)D_{1,r}(\ell_2,m_2). \qquad (3.119)
\end{aligned}
$$

3.4.6.1 Decomposition of the Eigenvalues and Eigenvectors

Now, we search for a plane wave solution of the discrete wave equation on the regular mesh:

$$\frac{\mathrm{d}^2}{\mathrm{d}t^2}\underline{U} + N_{d,r}\underline{U} = 0. \tag{3.120}$$

For this purpose, we suppose that our regular mesh is unbounded in all directions and we replace, in the following, the one-dimensional indexes by d-dimensional indexes. For instance, $(\ell_2 - 1)\nu + \ell_1$ is replaced by $(\ell_1, \ell_2) \in \mathbb{Z}^2$ and $((\ell_3 - 1)\nu + \ell_2 - 1)\nu + \ell_1$ by $(\ell_1, \ell_2, \ell_3) \in \mathbb{Z}^3$. Here, we have r^d degrees of freedom invariable by translation which define r^d classes. Each of these classes can be considered as the Cartesian product of r classes in 1D, as represented in Fig. 3.12.

With this notation, the dispersion relation is given by the following eigenvalue problem:

$$\widehat{N}_{2,r}\,\underline{U}_\alpha = \omega_h^2\,\underline{U}_\alpha, \tag{3.121}$$

where, if $\mathscr{I}_d = \{0, \ldots, r-1\}^d$, $\underline{U}_\alpha = \left(\bar{\alpha}_{\underline{p}}\right)_{\underline{p}\in\mathscr{I}_d}$ with $\underline{p} = p$ in 1D, $\underline{p} = (p_1, p_2)$ in 2D and $\underline{p} = (p_1, p_2, p_3)$ in 3D.

In the following, we suppose that $d = 2$. The plane wave solution of (3.120) is defined as

$$\underline{U} = \left(\alpha_{p,q}\mathrm{e}^{\mathrm{i}(\omega_h t - k_1\tilde{x}_p - k_2\tilde{y}_q)}\right)_{(p,q)\in\mathbb{Z}^2}. \tag{3.122}$$

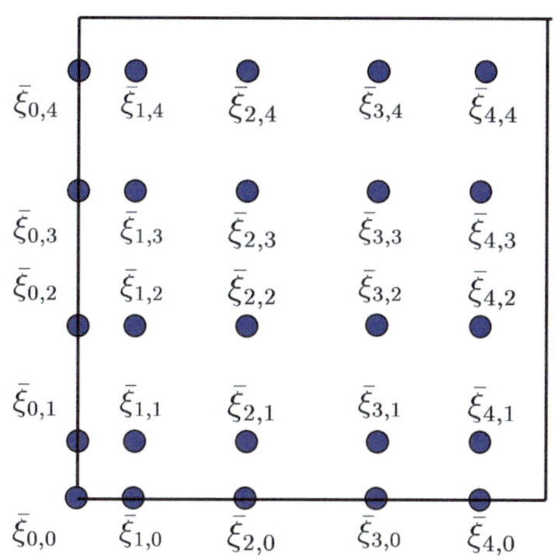

Fig. 3.12 The r^d classes of degrees of freedom in 2D (here $r = 5$ and $d = 2$). The corresponding 1D classes are those located on the edges

So, for $(p, q) = (\ell_1, \ell_2)$, after inserting (3.122) into (3.120), we obtain:

$$\omega_h^2 \alpha_{\ell_1,\ell_2} e^{i(\omega_h t - k_1 \tilde{x}_{\ell_1} - k_2 \tilde{y}_{\ell_2})} = \sum_{(m_1,m_2)\in\mathbb{Z}^2} N_{2,r}((\ell_1,\ell_2),(m_1,m_2))\alpha_{m_1,m_2} e^{i(\omega_h t - k_1 \tilde{x}_{m_1} - k_2 \tilde{y}_{m_2})}. \tag{3.123}$$

Now, by taking into account the fact that we have r^2 degrees of freedom invariable by translation, one can write:

$$\alpha_{\ell_1,\ell_2} = \alpha_{rp_1+q_1, rp_2+q_2} = \bar{\alpha}_{q_1,q_2}, \tag{3.124a}$$

$$\alpha_{m_1,m_2} = \alpha_{rp_1'+q_1', rp_2'+q_2'} = \bar{\alpha}_{q_1',q_2'}, \tag{3.124b}$$

where $1 \le q_j \le r$, $1 \le q_j' \le r$, $p_j \in \mathbb{Z}$, $p_j' \in \mathbb{Z}$, $j = 1, 2$.

With this notation, we obtain from (3.123), after simplification,

$$\begin{aligned}\omega_h^2 \bar{\alpha}_{q_1,q_2} &= \sum_{(q_1',q_2')\in\mathcal{I}_2} \bar{\alpha}_{q_1',q_2'} \sum_{(p_1',p_2')\in\mathbb{Z}^2} N_{2,r}((\ell_1,\ell_2),(rp_1'+q_1', rp_2'+q_2')) \\ &\quad \times e^{i(k_1(\tilde{x}_{rp_1'+q_1'} - \tilde{x}_{\ell_1}) - k_2(\tilde{y}_{rp_2'+q_2'} - \tilde{y}_{\ell_2}))} \\ &= \sum_{(q_1',q_2')\in\mathcal{I}_2} \widehat{N}_{2,r}[k_1,k_2]((q_1,q_2),(q_1',q_2'))\, \bar{\alpha}_{q_1',q_2'}.\end{aligned} \tag{3.125}$$

Now, by taking (3.107) into account, (3.123) can be rewritten as

$$\begin{aligned}&\omega_h^2 \alpha_{\ell_1,\ell_2} e^{i(\omega_h t - k_1 \tilde{x}_{\ell_1} - k_2 \tilde{y}_{\ell_2})} \\ &\quad = \sum_{(m_1,m_2)\in\mathbb{Z}^2} N_{1,r}(\ell_1,m_1)\delta_{\ell_2,m_2}\alpha_{m_1,m_2} e^{i(\omega_h t - k_1 \tilde{x}_{m_1} - k_2 \tilde{y}_{m_2})} \\ &\quad + \sum_{(m_1,m_2)\in\mathbb{Z}^2} N_{1,r}(\ell_2,m_2)\delta_{\ell_1,m_1}\alpha_{m_1,m_2} e^{i(\omega_h t - k_1 \tilde{x}_{m_1} - k_2 \tilde{y}_{m_2})} \\ &\quad = \sum_{m_1\in\mathbb{Z}} N_{1,r}(\ell_1,m_1)\alpha_{m_1,\ell_2} e^{i(\omega_h t - k_1 \tilde{x}_{m_1} - k_2 \tilde{y}_{\ell_2})} \\ &\quad + \sum_{m_2\in\mathbb{Z}} N_{1,r}(\ell_2,m_2)\alpha_{\ell_1,m_2} e^{i(\omega_h t - k_1 \tilde{x}_{\ell_1} - k_2 \tilde{y}_{m_2})},\end{aligned} \tag{3.126}$$

which provides, after simplification,

$$\begin{aligned}\omega_h^2 \alpha_{\ell_1,\ell_2} &= \sum_{m_1\in\mathbb{Z}} N_{1,r}(\ell_1,m_1)\alpha_{m_1,\ell_2} e^{ik_1(\tilde{x}_{m_1} - \tilde{x}_{\ell_1})} \\ &\quad + \sum_{m_2\in\mathbb{Z}} N_{1,r}(\ell_2,m_2)\alpha_{\ell_1,m_2} e^{ik_2(\tilde{y}_{m_2} - \tilde{y}_{\ell_2})}.\end{aligned} \tag{3.127}$$

So, by writing

$$\alpha_{m_1,\ell_2} = \alpha_{rp'_1+q'_1,rp_2+q_2} = \bar{\alpha}_{q'_1,q_2}, \tag{3.128a}$$

$$\alpha_{\ell_1,m_2} = \alpha_{rp_1+q_1,rp'_2+q'_2} = \bar{\alpha}_{q_1,q'_2}, \tag{3.128b}$$

we get from (3.127)

$$\begin{aligned}\omega_h^2\bar{\alpha}_{q_1,q_2} &= \sum_{q'_1=0}^{r-1}\bar{\alpha}_{q'_1,q_2}\sum_{p'_1\in\mathbb{Z}}N_{1,r}(\ell_1, rp'_1+q'_1)\mathrm{e}^{\mathrm{i}k_1(\tilde{x}_{rp_1+q_1}-\tilde{x}_{\ell_1})}\\ &+\sum_{q'_2=0}^{r-1}\bar{\alpha}_{q_1,q'_2}\sum_{p'_2\in\mathbb{Z}}N_{1,r}(\ell_2, rp'_2+q'_2)\mathrm{e}^{\mathrm{i}k_2(\tilde{y}_{rp'_2+q'_2}-\tilde{y}_{\ell_1})},\end{aligned} \tag{3.129}$$

which can be written as

$$\omega_h^2\bar{\alpha}_{q_1,q_2} = \sum_{q'_1=0}^{r-1}\widehat{N}_{1,r}[k_1](q'_1, q_2)\,\bar{\alpha}_{q'_1,q_2} + \sum_{q'_2=0}^{r-1}\widehat{N}_{1,r}[k_2](q_1, q'_2)\,\bar{\alpha}_{q_1,q'_2}. \tag{3.130}$$

By comparing (3.125) and (3.130), we obtain the following relation between $\widehat{N}_{2,r}[k_1, k_2]$, $\widehat{N}_{1,r}[k_1]$ and $\widehat{N}_{1,r}[k_2]$:

$$\begin{aligned}\widehat{N}_{2,r}[k_1, k_2]((q_1, q_2), (q'_1, q'_2)) = \widehat{N}_{1,r}[k_1](q_1, q'_1)\delta_{q_2,q'_2}\\ + \widehat{N}_{1,r}[k_2](q_2, q'_2)\delta_{q_1,q'_1}.\end{aligned} \tag{3.131}$$

Let us recall the tensor product of two vectors $\underline{X}$ and $\underline{Y}$ of $\mathbb{R}^r$:

$$\underline{X}\otimes\underline{Y} = (X_1Y_1, X_2Y_1, \ldots, X_rY_1, X_1Y_2, \ldots, X_rY_2, \ldots, X_1Y_r, \ldots, X_rY_r). \tag{3.132}$$

Now, let $\underline{U}_\lambda[k_1] = (U_1, \ldots, U_r)^T$ and $\underline{U}_{\lambda'}[k_2] = (U'_1, \ldots, U'_r)^T$ be two eigenvectors corresponding to two eigenvalues $\lambda[k_1]$ and $\lambda'[k_2]$ of $\widehat{N}_{1,r}[k_1]$ and $\widehat{N}_{1,r}[k_2]$. We have:

$$\begin{aligned}&\left(\widehat{N}_{2,r}[k_1, k_2]\,(\underline{U}_\lambda[k_1]\otimes\underline{U}_{\lambda'}[k_2])\right)_{q_1,q_2}\\ &= \sum_{(q'_1,q'_2)\in\mathscr{I}_2}\widehat{N}_{2,r}[k_1, k_2]((q_1, q_2), (q'_1, q'_2))U_{q'_1}U'_{q'_2}\\ &= \sum_{(q'_1,q'_2)\in\mathscr{I}_2}\widehat{N}_{1,r}[k_1](q_1, q'_1)\delta_{q_2,q'_2}U_{q'_1}U'_{q'_2}\\ &\quad+\sum_{(q'_1,q'_2)\in\mathscr{I}_2}\widehat{N}_{1,r}[k_2](q_2, q'_2)\delta_{q_1,q'_1}U_{q'_1}U'_{q'_2}\\ &= \sum_{q'_1\in\mathscr{I}_1}\widehat{N}_{1,r}[k_1](q_1, q'_1)U_{q'_1}U'_{q_2} + \sum_{q'_2\in\mathscr{I}_1}\widehat{N}_{1,r}[k_2](q_2, q'_2)U_{q_1}U'_{q'_2}\end{aligned}$$

$$
\begin{aligned}
&= \lambda[k_1]U_{q_1}U'_{q_2} + \lambda'[k_2]U_{q_1}U'_{q_2} \\
&= (\lambda[k_1] + \lambda'[k_2])U_{q_1}U'_{q_2}, \forall (q_1, q_2) \in \mathscr{I}_2. \qquad (3.133)
\end{aligned}
$$

This shows that, for any eigenvalue $\lambda[k_1]$ of $\widehat{N}_{1,r}[k_1]$ and for any eigenvalue $\lambda'[k_2]$ of $\widehat{N}_{1,r}[k_2]$, $\underline{U}_\lambda[k_1] \otimes \underline{U}_{\lambda'}[k_2]$ is the eigenvector of $\widehat{N}_{2,r}[k_1, k_2]$ associated to the eigenvalue $\lambda[k_1] + \lambda'[k_2]$. This important result can be summarized in

Theorem 5 *Let* $\{\lambda_j\}_{j=1}^r$ *be the eigenvalues and* $\left\{\underline{W}_j\right\}_{j=1}^r$ *the eigenvectors of* $\widehat{N}_{r,1}$, $\{T\lambda_j\}_{j=1}^{r^2}$ *the eigenvalues and* $\left\{\underline{TW}_j\right\}_{j=1}^{r^2}$ *the eigenvectors of* $\widehat{N}_{r,2}$ *and* ℓ *an integer between 1 and* r^2 *such that* $\ell = rp + q$, $0 \le p \le r - 1$, $1 \le q \le r$. *Then, we have:*

$$
\begin{cases}
T\lambda_\ell = \lambda_{p+1} + \lambda_q \\
\\
\underline{TW}_\ell = \underline{W}_{p+1} \otimes \underline{W}_q
\end{cases}
$$

where $\underline{W}_{q+1} \otimes \underline{W}_r$ *is defined as in (3.132).*

In 3D, a similar process leads to the relation:

$$
\begin{aligned}
&\widehat{N}_{2,r}[k_1, k_2, k_3]((q_1, q_2, q_3), (q'_1, q'_2, q'_3)) \\
&\quad = \widehat{N}_{1,r}[k_1](q_1, q'_1)\delta_{q_2,q'_2}\delta_{q_3,q'_3} + \widehat{N}_{1,r}[k_2](q_2, q'_2)\delta_{q_1,q'_1}\delta_{q_3,q'_3} \\
&\widehat{N}_{1,r}[k_2](q_3, q'_3)\delta_{q_1,q'_1}\delta_{q_2,q'_2}. \qquad (3.134)
\end{aligned}
$$

By denoting

$$
\underline{X} \otimes \underline{Y} \otimes \underline{Z} = (X_1Y_1Z_1, \ldots, X_rY_1Z_1, \ldots, X_1Y_rZ_r, \ldots, X_rY_rZ_r). \qquad (3.135)
$$

one shows, in the same way, that if $\underline{U}_\lambda[k_1]$ and $\underline{U}_{\lambda'}[k_2]$ and $\underline{U}_{\lambda''}[k_3]$ are three eigenvectors corresponding to three eigenvalues $\lambda[k_1]$, $\lambda'[k_2]$ and $\lambda''[k_3]$ of $\widehat{N}_{1,r}[k_1]$, $\widehat{N}_{1,r}[k_2]$ and $\widehat{N}_{1,r}[k_3]$, then, $\underline{U}_\lambda[k_1] \otimes \underline{U}_{\lambda'}[k_2] \otimes \underline{U}_{\lambda''}[k_3]$ is an eigenvector of $\widehat{N}_{2,r}[k_1, k_2, k_3]$ corresponding to the eigenvalue $\lambda[k_1] + \lambda'[k_2] + \lambda''[k_3]$.

These results imply, in particular, that the stability condition in dimension d is equal to the stability condition in 1D divided by $\sqrt{d}$.

On the other hand, the expressions of the eigenvectors and the eigenvalues in 2D enable us to obtain the errors committed on the amplitudes on a regular mesh [6].

Remarks

1. The infinite sums involved in (3.125) and (3.129) are actually finite.
2. In 2D, relation (3.131) can be written as

$$
\widehat{N}_{2,r}[k_1, k_2] = \widehat{N}_{1,r}[k_1] \otimes Id_r + Id_r \otimes \widehat{N}_{1,r}[k_2], \qquad (3.136)
$$

where Id_r is the identity matrix of order r and, for any $r \times r$ matrix N:

$$N \otimes Id_r = \begin{pmatrix} N & 0 & \dots & 0 & 0 \\ 0 & N & \dots & 0 & 0 \\ \vdots & \vdots & \ddots & \vdots & \vdots \\ 0 & 0 & \dots & N & 0 \\ 0 & 0 & \dots & 0 & N \end{pmatrix}.$$

$$Id_r \otimes N = $$

$$\begin{pmatrix} N_{1,1} & \dots & 0 & N_{1,2} & \dots & 0 & & N_{1,r} & \dots & 0 \\ \vdots & \ddots & \vdots & \vdots & \ddots & \vdots & \dots & \vdots & \ddots & \vdots \\ 0 & \dots & N_{1,1} & 0 & \dots & N_{1,2} & & 0 & \dots & N_{1,r} \\ & & & & & & & & & \\ N_{2,1} & \dots & 0 & N_{2,2} & \dots & 0 & & N_{2,r} & \dots & 0 \\ \vdots & \ddots & \vdots & \vdots & \ddots & \vdots & \dots & \vdots & \ddots & \vdots \\ 0 & \dots & N_{2,1} & 0 & \dots & N_{2,2} & & 0 & \dots & N_{2,r} \\ & & & & & & & & & \\ & \vdots & & & \vdots & & \ddots & & \vdots & \\ & & & & & & & & & \\ N_{r,1} & \dots & 0 & N_{r,2} & \dots & 0 & & N_{r,r} & \dots & 0 \\ \vdots & \ddots & \vdots & \vdots & \ddots & \vdots & \dots & \vdots & \ddots & \vdots \\ 0 & \dots & N_{r,1} & 0 & \dots & N_{r,2} & & 0 & \dots & N_{r,r} \end{pmatrix},$$

This notation provides a more straightforward proof of Theorem 5 by using the fact that $N \otimes Id_r(\underline{X} \otimes \underline{Y}) = (N\underline{X}) \otimes \underline{Y}$ and $Id_r \otimes N(\underline{X} \otimes \underline{Y}) = \underline{X} \otimes (N\underline{Y})$ but it is not easy to generalize these formulas to the 3D case.

3.4.6.2 Some Analysis of Non-regular Meshes in 2D

In this section, we analyze, first by plane waves then numerically, the effect of the distortion of an element on the accuracy and the stability of the method. In order to control this distortion, the analysis is made on periodic meshes.

A dispersion analysis is based on plane wave analysis which must be performed on a periodic infinite mesh. For this purpose, we define a periodic mesh of $\mathbb{R}^2$ composed of square cells of size $2h$ divided into four quadrilaterals. If $\mathscr{P} = \{P_i\}_{i=1\dots4}$ are the

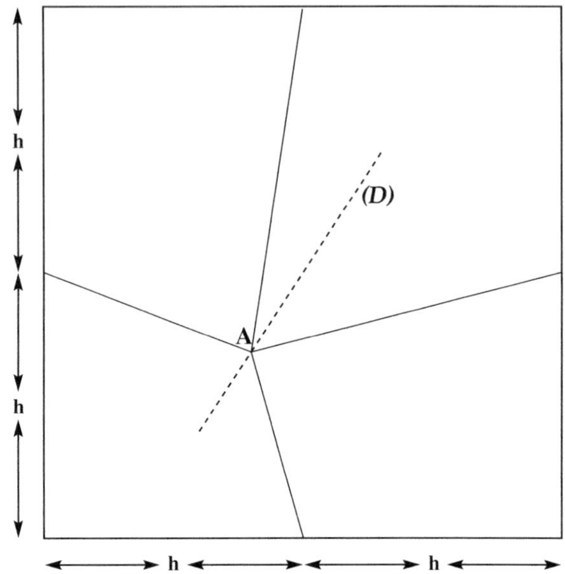

Fig. 3.13 The square cell. In the analysis and the experiments, $A = (ah, bh)$ is moved along the straight line (D) of equation $b = (3a - 1)/2$ $(0.6 \leq a \leq 1.4)$. Reprinted from G. Cohen, S. Fauqueux, Mixed finite elements with mass-lumping for the transient wave equation, J. of Comp. Acous., vol. 8 (1), pp. 171–188, Copyright © 2000 World Scientific Publishing. Reprinted with permission

vertices of the square, $\mathscr{C} = \{C_i\}_{i=1...4}$, the midpoints of its edges and A an interior point, each quadrilateral has two vertices in $\mathscr{C}$, one vertex in $\mathscr{P}$ and A as fourth vertex (Fig. 3.13).

For a Q_3 approximation, this periodic structure contains 36 classes of equations instead of 9 for a regular mesh (3 in 1D). As for a regular mesh in 1D, we substitute in these 36 equations a 36-dimensional vector valued plane wave solution which leads to a 36-dimensional eigenvalue problem. The matrix of this problem is constructed with the help of Maple and its eigenvalues are then computed (by a double precision FORTRAN program) numerically. Of course, we obtain one physical eigenvalue and 35 parasitic ones.

In Fig. 3.14, we give the dispersion curves, i.e. the ratio between the numerical velocity c_h of the physical wave and the exact velocity c. One can see that the loss of accuracy remains reasonable, even for significant distortions.

We conclude this section by the three following remarks

1. This kind of plane wave analysis which introduces different classes of the solution is a numerical version of the computation of Bloch waves in a crystal [8].
2. The Taylor expansions of the approximated velocities derived from (3.85a) and (3.95a) are of the same order as the dispersion relations, i.e. of fourth-order for P_2 and of sixth-order for P_3. This reveals a superconvergence phenomenon versus error estimates obtained below.
3. Ainsworth developed a general computation of the physical dispersion which are expressed as Padé's approximants of trigonometric functions [9]. This study was also extended to discontinuous Galerkin methods (DGM) [10] and Maxwell's equations [11].

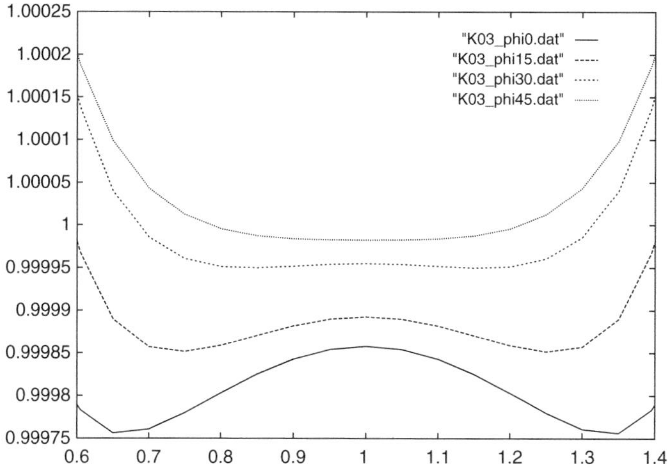

Fig. 3.14 Dispersion curves (c_h/c) versus a for some angles of propagation (0° (*lower curve*), 15°, 30°, 45° (*upper curve*)) and 3 elements per wavelength. $a = 1$ corresponds to a regular orthogonal mesh $a = 0.6$ or $a = 1.4$ correspond to degenerations into triangles. Reprinted from G. Cohen, S. Fauqueux, Mixed finite elements with mass-lumping for the transient wave equation, vol. 8 (1), pp. 171–188, Copyright © 2000 World Scientific Publishing. Reprinted with permission

3.5 Reflection-Transmission by a Discontinuous Interface

The purpose of this section is to study the effect of a discontinuity on the accuracy of the method. This effect is measured by a plane wave analysis on a two-layer medium.

3.5.1 *The Continuous Problem*

Let us consider the following two-layer medium:

$$(\mu(x), \lambda(x)) = \begin{cases} (\mu_1, \lambda_1) & \text{for } x < 0, \\ (\mu_2, \lambda_2) & \text{for } x > 0. \end{cases} \tag{3.137}$$

The velocity of a wave is $c_1 = \sqrt{\mu_1/\lambda_1}$ in the first layer and $c_2 = \sqrt{\mu_2/\lambda_2}$ in the second one.

In such a medium, an incident plane wave of amplitude 1 which propagates in the first layer is decomposed, after crossing the interface between the two layers, into a transmitted wave of amplitude T and a reflected wave of amplitude R. Then, the solution can be written as

$$u(x,t) = \chi_{\mathbb{R}^-}(x)\left(\mathrm{e}^{\mathrm{i}(\omega t - K_1 x)} + R\mathrm{e}^{\mathrm{i}(\omega t + K_1 x)}\right) + T\chi_{\mathbb{R}^+}(x)\mathrm{e}^{\mathrm{i}(\omega t - K_2 x)}, \tag{3.138}$$

where $\chi_{\mathbb{R}^+}$ and $\chi_{\mathbb{R}^-}$ are the characteristic functions of the sets $\mathbb{R}^+$ and $\mathbb{R}^-$. On the other hand, K_1 and K_2 are defined by the dispersion relations in each layer:

$$K_1 = \frac{\omega}{c_1}, \tag{3.139a}$$

$$K_2 = \frac{\omega}{c_2}. \tag{3.139b}$$

Since u and its normal derivative $\mu\partial u/\partial n$ are continuous at the interface, we can write:

$$u(0^-) = \mathrm{e}^{\mathrm{i}\omega t} + R\,\mathrm{e}^{\mathrm{i}\omega t} = u(0^+) = T\,\mathrm{e}^{\mathrm{i}\omega t}, \tag{3.140a}$$

$$\mu_1 u'(0^-) = -\mathrm{i}\mu_1 K_1 \mathrm{e}^{\mathrm{i}\omega t} + R\mathrm{i}\mu_1 K_1 \mathrm{e}^{\mathrm{i}\omega t} = \mu_2 u'(0^+) = -T\mathrm{i}\mu_2 K_2 \mathrm{e}^{\mathrm{i}\omega t}. \tag{3.140b}$$

After simplification, we obtain:

$$1 + R = T, \tag{3.141a}$$

$$K_1\mu_1 R + K_2\,\mu_2\,T = K_1\mu_1. \tag{3.141b}$$

By taking into account (3.139a) and (3.139b), this can be rewritten as

$$R = \frac{\sigma_1 - \sigma_2}{\sigma_1 + \sigma_2}, \tag{3.142a}$$

$$T = \frac{2\sigma_1}{\sigma_1 + \sigma_2}, \tag{3.142b}$$

where $\sigma_j = \sqrt{\lambda_j \mu_j}$, $j = 1, 2$ is the *acoustic impedance* of each layer.

We now give guidelines for formulating the equations satisfied by the transmitted and reflected waves for the P_2 approximation. Then, we simply give the order of the error in this case and in the case of higher-order approximations.

3.5.2 FEM Approximation of the Heterogeneous Wave Equation

The variational formulation of the 1D heterogeneous wave equation:

$$\lambda(x)\frac{\partial^2 u}{\partial t^2} - \frac{\partial}{\partial x}\left(\mu(x)\frac{\partial u}{\partial x}\right) = 0, \tag{3.143}$$

is obviously

$$\frac{d^2}{dt^2}\int_{\mathbb{R}} \lambda\, uv dx \; + \; \int_{\mathbb{R}} \mu \frac{\partial u}{\partial x}\frac{\partial v}{\partial x} dx = 0 \quad \forall v \in H^1(\mathbb{R}). \tag{3.144}$$

In this case, it is necessary to compute the mass integral and the stiffness integral by using a Gauss–Lobatto quadrature rule, because of the presence of λ and μ. This process leads, for P_2, to the following semi-discrete in space equations on a regular mesh:

$$\begin{aligned}\lambda_p \frac{d^2 u_p}{dt^2}(t) = \frac{1}{h^2}\Bigg[&-\frac{1}{2}(3\mu_{p-1} - 4\mu_{p-1,1} + 3\mu_p)u_{p-1}(t) \\ &+ 2(\mu_{p-1} + 3\mu_p)u_{p-1,1}(t) \\ &- \frac{1}{2}(\mu_{p-1} + 4\mu_{p-1,1} + 18\mu_p + 4\mu_{p,1} + \mu_{p+1})u_p(t) \\ &+ 2(\mu_p + 3\mu_{p+1})u_{p,1}(t) \\ &- \frac{1}{2}(3\mu_p - 4\mu_{p,1} + 3\mu_{p+1})u_{p+1}(t)\Bigg], \quad \forall p \in \mathbb{Z}, s \end{aligned} \tag{3.145a}$$

$$\begin{aligned}\lambda_{p,1}\frac{d^2 u_{p,1}}{dt^2}(t) = \frac{1}{h^2}\big[&(3\mu_p + \mu_{p+1})u_p(t) - 4(\mu_p + \mu_{p+1})u_{p,1}(t) \\ &+ (\mu_p + 3\mu_{p+1})u_{p+1}(t)\big], \quad \forall p \in \mathbb{Z}, \end{aligned} \tag{3.145b}$$

where $\lambda_p = \lambda(x_p)$, $\lambda_{p,1} = \lambda(x_{p,1})$, $\mu_p = \mu(x_p)$ and $\mu_{p,1} = \mu(x_{p,1})$.

3.5.3 *Taylor Expansion of the Wavenumber*

We look for an expression of k in terms of ωh in the first step of our study. This expression can be derived from the characteristic polynomial of (3.82) which can be written as

$$4\left(24 + \frac{\omega^2 h^2}{c^2}\right)\sin^2\frac{kh}{2} - \frac{\omega^2 h^2}{c^2}\left(24 - \frac{\omega^2 h^2}{c^2}\right) = 0, \tag{3.146}$$

from which one can easily derive

$$k = \pm\frac{2}{h}\arcsin\left(\frac{\omega h}{2c}\sqrt{\left(24 - \frac{\omega^2 h^2}{c^2}\right)\left(24 + \frac{\omega^2 h^2}{c^2}\right)^{-1}}\right). \tag{3.147}$$

The Taylor expansion of (3.147) is

$$k = \pm\frac{\omega}{c}\left(1 + \frac{\omega^4 h^4}{2880\, c^4} + \frac{\omega^6 h^6}{96\,768\, c^6} + O(h^8)\right). \tag{3.148}$$

The equation giving k versus ωh has two solutions of opposite sign although the inverse equation has four solutions. This implies, in particular, that we have no parasitic wave coming from the interface, which can occur in finite difference methods [2].

Now, let us consider a two-layer infinite medium such that $\lambda = \lambda_1$ and $\mu = \mu_1$ for $x < 0$ and $\lambda = \lambda_2$ and $\mu = \mu_2$ for $x > 0$. In each layer, we write:

$$k_j = \pm\frac{\omega}{c_j}\left(1 + \frac{\omega^4 h^4}{2880\, {c_j}^4} + \frac{\omega^6 h^6}{96768\, {c_j}^6} + O(h^8)\right), \quad j = 1, 2. \tag{3.149}$$

3.5.4 Interface Between Two Elements

Our first study will be carried out for the case of an interface located between two elements. In this case, the solution can be written as

$$u_p = \begin{cases} \alpha_1 (\mathrm{e}^{\mathrm{i}(\omega t - p k_1 h)} + R_h\, \mathrm{e}^{\mathrm{i}(\omega t + p k_1 h)}) \text{ for } p \le 0, \\ \alpha_2\, T_h\, \mathrm{e}^{\mathrm{i}(\omega t - p k_2 h)} \text{ for } p \ge 0, \end{cases} \tag{3.150a}$$

$$u_{p,1} = \begin{cases} \beta_1 (\mathrm{e}^{\mathrm{i}(\omega t - (p + \frac{1}{2}) k_1 h)} + R_h\, \mathrm{e}^{\mathrm{i}(\omega t + (p + \frac{1}{2}) k_1 h)}) \text{ for } p \le -1, \\ \beta_2\, T_h\, \mathrm{e}^{\mathrm{i}(\omega t - (p + \frac{1}{2}) k_2 h)} \text{ for } p \ge 0, \end{cases} \tag{3.150b}$$

where $(\alpha_j, \beta_j)^T$, $j = 1, 2$ is the eigenvector corresponding to the physical solution of the discrete system given in (3.83) (which is the correct one for $\omega h/c < 2$), in which k is replaced by k_j.

The reflection-transmission coefficients R_h and T_h are determined by

- the condition of continuity at $x = 0$:

$$\alpha_1(1 + R_h) = \alpha_2 T_h, \tag{3.151}$$

- the equation at this same point:

$$\begin{aligned} \omega^2 \frac{\lambda_1 + \lambda_2}{2} u_0 = \frac{1}{h^2}&(7(\mu_1 + \mu_2)u_0 - 8(\mu_2 u_{0,1} + \mu_1 u_{-1,1}) \\ &+ \mu_2 u_1 + \mu_1 u_{-1}), \end{aligned} \tag{3.152}$$

in which we plug (3.150a) and (3.150b).

An elementary computation (aided by Maple) leads to the following result:

$$R_h = \frac{\sigma_1 - \sigma_2}{\sigma_1 + \sigma_2} + \frac{1}{288} \frac{\sigma_1 \sigma_2}{(\sigma_1 + \sigma_2)^2} \left[\frac{1}{c_1^4} - \frac{1}{c_2^4} \right] \omega^4 h^4 + O(h^6), \tag{3.153a}$$

$$T_h = \frac{2\sigma_1}{\sigma_1 + \sigma_2} - \frac{1}{288} \frac{\sigma_1^2}{(\sigma_1 + \sigma_1)^2} \left[\frac{1}{c_1^4} - \frac{1}{c_2^4} \right] \omega^4 h^4 + O(h^6), \tag{3.153b}$$

where

$$\sigma_j = \sqrt{\lambda_j \mu_j}, \quad j = 1, 2. \tag{3.154}$$

Equations (3.153a) and (3.153b) show that the error committed on the reflection-transmission coefficients is, in this case, of the order of the dispersion.

3.5.5 *Interface at an Interior Point*

When the interface is located inside the element whose ends are 0 and h, three equations (instead of one) mix the two layers together: those at $x = 0$, at $x = h/2$ and at $x = h$. In order to have as many equations as unknowns, we consider the value $u_{0,1}$ as an unknown that we add to R and T.

The three equations are:

$$\omega^2 \lambda_1 u_0 = \frac{1}{h^2} \left(\frac{25\mu_1 + 3\mu_2}{2} u_0 - (6\mu_1 + 2\mu_2) u_{0,1} \right.$$
$$\left. -8\mu_1 u_{-1,1} + \frac{\mu_1 + \mu_2}{2} u_1 + \mu_1 u_{-1} \right), \tag{3.155a}$$

$$\omega^2 \frac{\lambda_1 + \lambda_2}{2} u_{0,1} = -\frac{1}{h^2} ((3\mu_1 + \mu_2) u_0 + (3\mu_2 + \mu_1) u_1$$
$$- 4(\mu_1 + \mu_2) u_{0,1}), \tag{3.155b}$$

$$\omega^2 \lambda_2 u_1 = \frac{1}{h^2} \left(\frac{\mu_1 + \mu_2}{2} u_0 - (6\mu_2 + 2\mu_1) u_{0,1} \right.$$
$$\left. -8\mu_2 u_{1,1} + \frac{25\mu_2 + 3\mu_1}{2} u_1 + \mu_2 u_2 \right). \tag{3.155c}$$

By inserting the plane wave solution

$$u_p = \begin{cases} \alpha_1 (\mathrm{e}^{\mathrm{i}(\omega t - p k_1 h)} + R_h \, \mathrm{e}^{\mathrm{i}(\omega t + p k_1 h)}) \text{ for } p \leq 0, \\ \alpha_2 T_h \mathrm{e}^{\mathrm{i}(\omega t - p k_2 h)} \text{ for } p \geq 1, \end{cases} \tag{3.156a}$$

$$u_{p,1} = \begin{cases} \beta_1 (\mathrm{e}^{\mathrm{i}(\omega t - (p+\frac{1}{2}) k_1 h)} + R_h \, \mathrm{e}^{\mathrm{i}(\omega t + (p+\frac{1}{2}) k_1 h)}) \text{ for } p \leq -1, \\ \beta_2 \, T_h \, \mathrm{e}^{\mathrm{i}(\omega t - (p+\frac{1}{2}) k_2 h)} \text{ for } p \geq 1, \end{cases} \tag{3.156b}$$

into (3.155a)–(3.155c), we obtain the following orders for R_h and T_h

$$R_h = \frac{\sigma_1 - \sigma_2}{\sigma_1 + \sigma_2} + O(h), \tag{3.157a}$$

$$T_h = \frac{2\sigma_1}{\sigma_1 + \sigma_2} + O(h). \tag{3.157b}$$

This result shows that, if one locates the interface of discontinuity at an interior point of the element, one loses the accuracy provided by the use of a higher-order method.

Remark: The condition of continuity is contained in Eq. (3.155a) and (3.155b) since the solution is not expressed as an explicit plane wave in the interval]0, h[.

3.5.6 *Extension to Higher-Order Approximations*

We carried out the computations up to a P_5 approximation. For all the orders, the characteristic polynomial provided only two solutions for k in terms of ωh, which means that we never have parasitic waves generated by a discontinuity. So, for an interface between two elements, the equations remain the same i.e., the condition of continuity at $x = 0$ and the equation at the same point. We obtained the following results [7] summarized in Table 3.1.

- P_3 elements:

$$R_h = \frac{\sigma_1 - \sigma_2}{\sigma_1 + \sigma_2} + \frac{1}{1800} \frac{\sigma_1(\sigma_2 - \sigma_1)}{c_1^4 (\sigma_1 + \sigma_2)^2} \omega^4 h^4 + O(h^6), \tag{3.158a}$$

$$T_h = \frac{2\sigma_1}{\sigma_1 + \sigma_2} + \frac{1}{1800} \frac{\sigma_1(\sigma_2 - \sigma_1)}{c_1^4 (\sigma_1 + \sigma_2)^2} \omega^4 h^4 + O(h^6). \tag{3.158b}$$

- P_4 elements:

$$R_h = \frac{\sigma_1 - \sigma_2}{\sigma_1 + \sigma_2} + \frac{1}{470\,400} \frac{\sigma_1(\sigma_2 - \sigma_1)}{c_1^6 (\sigma_1 + \sigma_2)^2} \omega^6 h^6 + O(h^8), \tag{3.159a}$$

Table 3.1 Comparison of the errors for different orders of approximation

	L^2 error estimates	Reflection-transmission	Numerical dispersion
P_2	h^3	h^4	h^4
P_3	h^4	h^4	h^6
P_4	h^5	h^6	h^8
P_5	h^6	h^6	h^{10}

$$T_h = \frac{2\sigma_1}{\sigma_1+\sigma_2} + \frac{1}{470\,400}\frac{\sigma_1(\sigma_2-\sigma_1)}{c_1^6(\sigma_1+\sigma_2)^2}\omega^6 h^6 + O(h^8). \tag{3.159b}$$

- P_5 elements:

$$R_h = \frac{\sigma_1-\sigma_2}{\sigma_1+\sigma_2} - \frac{1}{1\,587\,600}\frac{\sigma_1(\sigma_2-\sigma_1)}{c_1^6(\sigma_1+\sigma_2)^2}\omega^6 h^6 + O(h^8), \tag{3.160a}$$

$$T_h = \frac{2\sigma_1}{\sigma_1+\sigma_2} - \frac{1}{1\,587\,600}\frac{\sigma_1(\sigma_2-\sigma_1)}{c_1^6(\sigma_1+\sigma_2)^2}\omega^6 h^6 + O(h^8). \tag{3.160b}$$

These results show that the order of the error induced by a discontinuity increases by two orders of approximation. However, one can notice that the coefficient in front of the powers of h decreases with the order of the scheme. The different orders given in Table 3.1 show that the reflection-transmission is roughly of the same order as the global error estimates whereas the numerical dispersion always has a superconvergence phenomenon.

For all the orders of approximation, we get an error in $O(h)$ when the interface crosses an element. This result shows that one must carefully treat the discontinuities when one uses finite elements. These results hold in 2D and 3D. So, for multidimensional meshes, the mesh must follow the interfaces of discontinuities in order to maintain a good accuracy. In particular, for very discontinuous media (as in seismic simulations, for instance), the accuracy of the method can be completely destroyed if the mesh does not follow the interfaces. This poses a difficult problem in 3D heterogenous media if one wants to adapt the space-step to the wavelength on an hexahedral mesh.

3.5.7 A Two-Layer Experiment

In this section, we carry out a numerical study of a reflection-transmission process in 2D in order to check if the theoretical results found in 1D are confirmed in a 2D configuration.

Let us consider a quasi-rectangular domain $\Omega \subset \mathbb{R}^2$ of dimensions 4200 m × 3000 m whose upper boundary is a curve and such that $\Omega = \Omega_1 \cup \Omega_2$. Moreover,

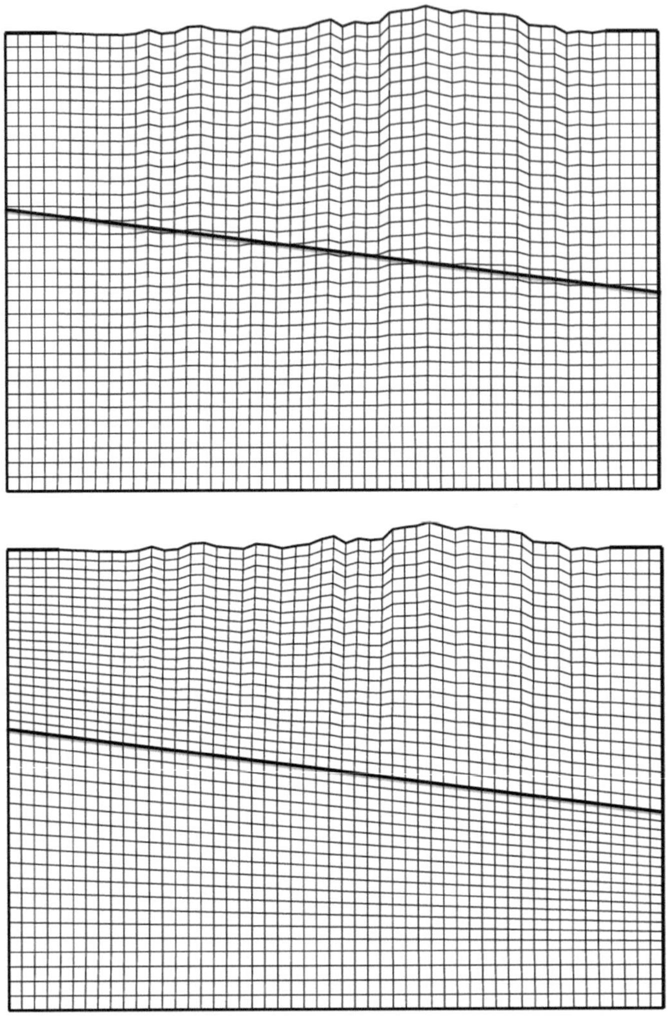

Fig. 3.15 The meshes $\mathscr{M}_1$ (*above*) and $\mathscr{M}_2$ (*below*). The *bold line* marks the interface

$\bar{\Omega}_1 \cap \bar{\Omega}_2 = D$, where D is a straight line with a negative slope (Fig. 3.15). We solve the wave equation:

$$\frac{1}{c^2(\underline{x})}\frac{\partial^2 u}{\partial t^2}(\underline{x}, t) - \Delta u(\underline{x}, t) = f(\underline{x}, t) \text{ in } \Omega. \tag{3.161}$$

We have an homogeneous Dirichlet condition on the upper boundary and the three other boundaries are open. We set $c = c_1$ in Ω_1 and $c = c_2$ in Ω_2. In this domain,

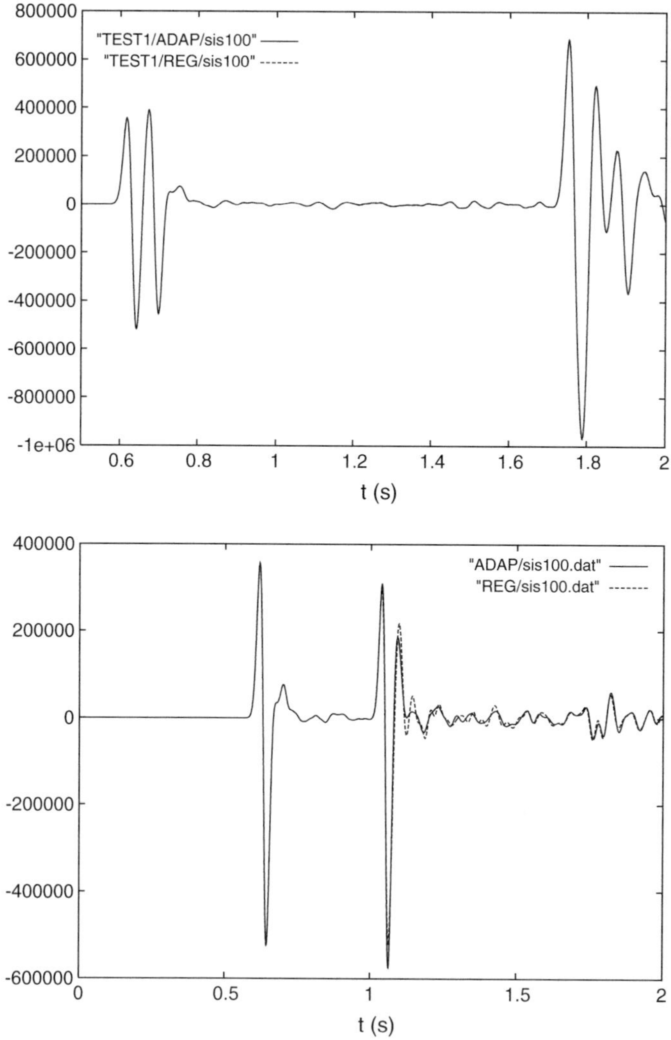

Fig. 3.16 The seismogram at the point of abscissa 2100, 100 m below the upper boundary obtained for the two meshes when $c_1 = c_2 = 3500$ m/s (*above*) and when $c_1 = 3500$ m/s in the higher part of the domain and $c_2 = 5500$ m/s in the lower part (*below*)

we propagate a pulse with a frequency of 17 Hz located at the middle of the upper boundary. The domain is meshed in two ways:

- By a regular mesh $\mathscr{M}_1$ (Fig. 3.15)
- By a mesh $\mathscr{M}_2$ which follows the interface between Ω_1 and Ω_2 (Fig. 3.15).

In the first experiment, we set $c_1 = c_2 = 3500$ m/s and we draw the seismograms (at the midpoint in the x direction and 100 m below the upper boundary) obtained

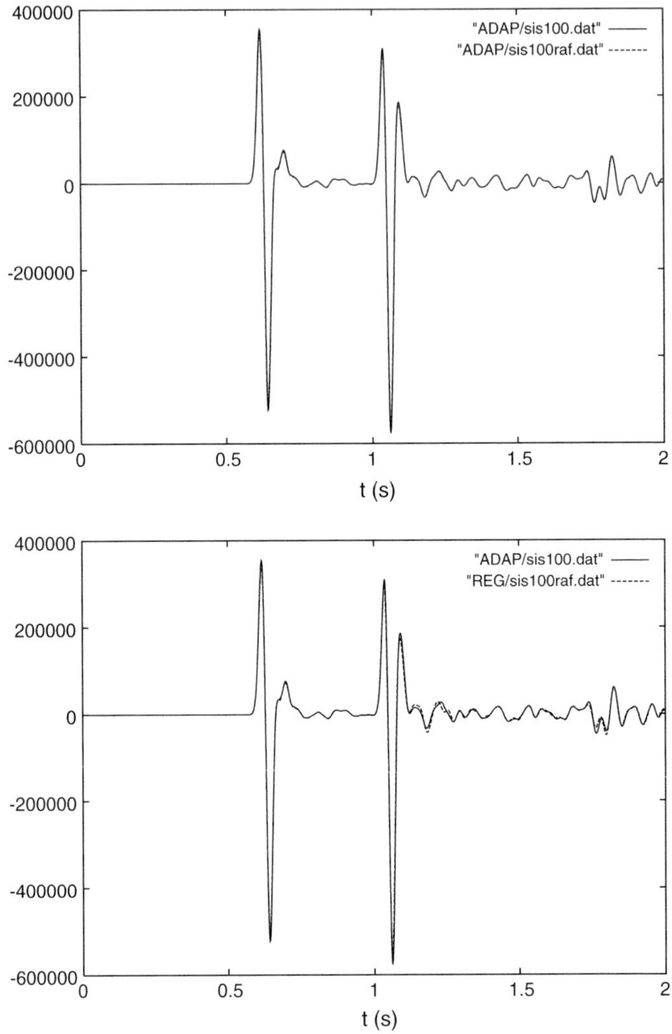

Fig. 3.17 The seismogram at the point of abscissa 2100, 100 m below the upper boundary obtained for the refined adapted mesh (*above*) and the refined regular mesh (*below*) when $c_1 = 3500\,\text{m/s}$ in the higher part of the domain and $c_2 = 5500\,\text{m/s}$ in the lower part. The two seismograms are compared to the seismogram obtained by the coarse adapted mesh

for the two meshes on the same figure. One can notice the perfect fitting of the two curves (Fig. 3.16) which proves that the solution is not at all affected by the mesh (the number of elements is taken in order to obtain an accurate solution).

In a second experiment, we set $c_1 = 3500\,\text{m/s}$ in the higher part of the domain and $c_2 = 5500\,\text{m/s}$ in the lower part. The seismogram in Fig. 3.16 shows that the two curves do not agree as soon as the wave crosses the interface. This indicates the loss of accuracy predicted in Sect. 3.5 when the discretization in space does not follow the interface between two media. This loss of accuracy is confirmed by the use of refined regular and adapted meshes in which each quadrilateral is divided by 4. In Fig. 3.17, we see that the refinement of mesh provides the same solution for the adapted mesh whereas the solution provided by the refined regular mesh is different from that given by the coarse mesh but closer to that given by the adapted mesh.

Remarks:

1. Higher-dimensional plane reflection-transmission analysis leads to implicit equations which seem impossible to solve.
2. One could use averaging techniques used for finite difference methods in order to increase the order of the approximation inside a cell [12, 13], but these are not obvious to apply to finite element methods.
3. Plane wave analysis can also enable us to compute the error committed on the amplitudes of physical and parasitic waves [2, 6]. However, this computation (which is rather tedious) is not very significant since the results only hold on regular meshes for which parasitic waves are not really troublesome. On the other hand, the amplitudes of the approximate physical waves can be derived from error estimates. For these reason, we do not develop this technique here.

3.6 *hp-a priori* Error Estimates

In this section, we present a *hp*-convergence analysis of the finite element approximation (3.49) of the second order acoustic equation (3.1). We underline the spectral convergence of the method in the sense that, when the solutions and the data are very smooth, the discretization is of unlimited order. Equivalence Theorem 5 directly implies some convergence results about the mixed approximation (3.50a) and (3.50b).

The main difficulties of this study are:

1. Unlike the tetrahedral finite element, the transfomation F_K is not affine and the functions $v_K = \hat{v}_K \circ F_K$ with $\hat{v}_K \in Q_r$, are not necessarily polynomial. For this reason, specific tools must be developed for the hexahedral case (see Sects. 3.6.1 and 3.6.2).
2. We consider that the integrals are not exactly computed and we must estimate the numerical integration error (see Sect. 3.6.3). Actually, we study the influence of the Gauss–Lobatto integration rule on the convergence of the finite element method.
3. We want to obtain the h and r dependences in the error estimates.

Several classical steps will be used to estimate the error $e_h = u - u_h$. We first introduce a projection I_h^r on the finite element space U_h^r and next, we decompose e_h into two contributions: $e_h = (u - I_h^r u) + (I_h^r u - u_h) = e_h^P + e_h^I$ which are called the projection and the interpolation error respectively. In Sect. 3.6.2, we propose an appropriate projection I_h^r for our error analysis and we derive some error estimates about this latter. In Sect. 3.6.4, we show that the interpolation error e_h^I can to be controlled by the projection error e_h^P and some terms which come from the inexact integration and, so we obtain the error estimate for e_h.

Finally, in this section, we assume that

1. The physical parameters λ and $\underline{\underline{\mu}}$ are piecewise constants,
2. The convergence is studied for a regular family of meshes (see Definition 3.1).

Remark: A h-version of this study can be found in [14].

3.6.1 Some Properties of Meshes

We assume that all quadrilaterals or hexahedra K are convex in order to ensure the existence of the diffeomorphism $\underline{F}_K \in [Q_1(\hat{K})]^d$ with $d = 2$ or 3 (such that $\underline{F}_K(\widehat{K}) = K$).

Now, let us give some definitions and properties on the quadrilateral or hexahedral finite elements (for more details see [15]) and on the transformation $\underline{F}_K$: to characterize an element $K \in \mathscr{T}_h$, we define:

$$h_K = \text{diameter of K}, \tag{3.162a}$$

$$\sigma_K = \frac{h_K}{\rho_K} = \text{regularity parameter}, \tag{3.162b}$$

where $\rho_K := \| J_{\underline{F}_K^{-1}} \|_{\infty,K}^{\frac{1}{d}}$ with $J_{\underline{F}_K^{-1}}$ is the determinant of the Jacobian matrix of $\underline{F}_K^{-1}$.

Remarks:

1. The diameter h_K is defined as the greatest distance between two points of the cell K. One can choose another measure which is the greatest distance between two vertices of K. Indeed, this distance is naturally compatible with the Q_1 transformation F_K.
2. In two dimensions, we can give a geometric characterization of ρ_K (see [16]). Actually, in this case, ρ_K is the minimum of the diameters of the inscribed circles in the four triangles possibly built with the nodes of the quadrangle K.

We denote:

$$|\underline{F}_K|_{m,\infty,\hat{K}} = \sup_{\hat{x}\in\hat{K}} \| D^m \underline{F}_K(\hat{x}) \|_{\mathscr{L}_m(\mathbb{R}^d,\mathbb{R}^d)}, \tag{3.163a}$$

$$|\underline{F}_K^{-1}|_{m,\infty,K} = \sup_{x\in K} \| D^m \underline{F}_K^{-1}(x) \|_{\mathscr{L}_m(\mathbb{R}^d,\mathbb{R}^d)}, \tag{3.163b}$$

where $\mathscr{L}_m(\mathbb{R}^d, \mathbb{R}^d)$ is the set of the m-linear applications of $\mathbb{R}^d$ into $\mathbb{R}^d$, $D^m \underline{F}_K(\hat{x})$ and $D^m \underline{F}_K^{-1}(x)$ are respectively the mth-derivatives of $\underline{F}_K$ and $\underline{F}_K^{-1}$ at the points $\hat{x}$ and x.

We will use the following estimates:

$$|\underline{F}_K|_{1,\infty,\hat{K}} \leq Ch_K, \ \|J_K\|_{\infty,\hat{K}} \leq Ch_K^d, \tag{3.164a}$$

$$|\underline{F}_K^{-1}|_{1,\infty,K} \leq C\frac{h_K^{d-1}}{\rho_K^d}, \tag{3.164b}$$

$$|\underline{F}_K|_{2,\infty,\widehat{K}} \leq Ch_K, \ |\underline{F}_K|_{2,\infty,\widehat{K}} \leq Ch_K^2 \text{ if } K \text{ is almost a parallelepiped} \tag{3.164c}$$

where $C > 0$ is independent of K and r.

Remarks:

1. By the expression "almost a parallelepiped", we mean a small deformation of a parallelepipedic cell. In this case, the second derivatives of $\underline{F}_K$ are zero.
2. We have, by definition,

$$D(\underline{F}_K^{-1})(x) = \left(D\underline{F}_K(F^{-1}(x))\right)^{-1}, \tag{3.165a}$$

$$J_{\underline{F}_K^{-1}} o \underline{F}_K = \frac{1}{J_K}, \tag{3.165b}$$

$\underline{\text{v}}$ where $D(\underline{F}_K^{-1})$ is the Jacobian matrix of $\underline{F}_K^{-1}$.

Using properties (3.164a)–(3.164c), it is easy to deduce the following proposition

Proposition 3.1 *We have the following estimates :* $\forall \hat{x} \in \widehat{K}$,

$$\lambda((D\underline{F}_K D\underline{F}_K^T)(\hat{x})) \leq Ch_K^2, \tag{3.166a}$$

$$\lambda((D\underline{F}_K^{-1} D\underline{F}_K^{-T})(\hat{x})) \leq C\frac{h_K^{2(d-1)}}{\rho_K^{2d}}, \tag{3.166b}$$

where $\lambda(A)$ *belongs to the spectrum of* A *and* $C > 0$ *is independent of* K *and* r.

Proof Let $\hat{x} \in \widehat{K}$. As $(D\underline{F}_K D\underline{F}_K^T)(\hat{x})$ and $(D\underline{F}_K^{-1} D\underline{F}_K^{-T})(\hat{x})$ are symmetrical matrices, we can write

$$\begin{aligned}\rho((D\underline{F}_K D\underline{F}_K^T)(\hat{x})) &= \sup_{\underline{v}\in\mathbb{R}^{*d}} \frac{\|(D\underline{F}_K D\underline{F}_K^T)(\hat{x})\underline{v}\|}{\|\underline{v}\|} = \|(D\underline{F}_K D\underline{F}_K^T)(\hat{x})\| \\ &\leq \|(D\underline{F}_K)(\hat{x})\| \|(D\underline{F}_K^T)(\hat{x})\| \leq |\underline{F}_K|_{1,\infty,\widehat{K}}^2, \end{aligned} \tag{3.167}$$

$\rho(A)$ being the spectral radius of A. Using (3.164a), we immediately obtain (3.166a). A similar process leads to (3.166b).

We finally define the regularity of a mesh.

Definition 3.1 A family $(\mathscr{T}_h)_{h>0}$ of triangulation of Ω is regular when h tends toward 0, if it exists a number $\sigma > 0$, independent of h, such that

$$\sigma_K \leq \sigma, \ \forall K \in \mathscr{T}_h. \tag{3.168}$$

3.6.2 Some Interpolation Errors for Quadrilaterals and Hexahedra

We first introduce the projector $\hat{I}_r$ defined as a Lagrange interpolation at the Gauss–Lobatto points on $Q_r(\hat{K})$.

Definition 3.2 Let $\hat{v} \in H^s(\widehat{K})$ with $s > d/2$ and $r \geq 0$. The Lagrange interpolation of $\hat{v}$ at the Gauss–Lobatto points $(\hat{\xi}_{l_1}, \ldots, \hat{\xi}_{l_d})_{l_1,\ldots,l_d=1,\ldots,r+1}$ on $Q_r(\hat{K})$ is defined by

$$(\hat{I}_r\hat{v})(\hat{x}_1, \ldots, \hat{x}_d) = \sum_{l_1,\ldots,l_d=1}^{r+1} \hat{v}(\hat{\xi}_{l_1}, \ldots, \hat{\xi}_{l_d})\hat{\varphi}_{l_1}(\hat{x}_1) \ldots \hat{\varphi}_{l_d}(\hat{x}_d) \tag{3.169}$$

Remark: The condition $s > d/2$ is used to ensure the injection of the space $H^s(\widehat{K})$ into the space of the continuous functions and hence, $\hat{I}_r$ is well-defined.

We then define a projector I_h^r on U_h^r as follows:
Let $v \in H^s(\Omega)$ with $s > d/2$. For each cell $K \in \mathscr{T}_h$, we have:

$$\left(I_{h|K}^r v\right) \circ \underline{F}_K = \hat{I}_r\hat{v} \tag{3.170}$$

where $\hat{v} = v \circ \underline{F}_K$.

To study the projection error introduced by I_h^r, we use the bracket semi-norm. Let $u \in W^{m,p}(\widehat{K})$,

$$[u]_{m,p,\widehat{K}}^2 = \sum_{i=1}^{d} \left\| \frac{\partial^m u}{\partial \hat{x}_i^m} \right\|_{p,\widehat{K}}^2 \tag{3.171}$$

and the Bramble–Hilbert lemma adapted to Q_r (see [15, 16]).

Lemma 3.1 (Bramble–Hilbert) *Let p, q be two numbers such that $1 \leq p, q \leq \infty$ and let r, m be two integers such that $r \geq 0$ and $m \leq r + 1$,*

$$W^{r+1,p}(\widehat{K}) \hookrightarrow W^{m,q}(\widehat{K}). \tag{3.172}$$

Let $\Pi \in \mathscr{L}(W^{r+1,p}(\widehat{K}); W^{m,q}(\widehat{K}))$ be an operator which verifies:

$$\forall p \in Q_r, \ \Pi p = p, \tag{3.173}$$

then it exists C dependent on K and r such that

$$\forall v \in W^{r+1,p}(\widehat{K}),\ |v - \Pi v|_{m,q,\widehat{K}} \leq C[v]_{r+1,p,\widehat{K}}. \tag{3.174}$$

In (3.174), $|\cdot|_{m,q,\widehat{K}}$ is the semi-norm defined by: $\forall v \in W^{m,q}(\widehat{K})$

$$|v|_{m,q,\widehat{K}} = \left(\sum_{|\alpha|=m} \int_{\widehat{K}} \left| \frac{\partial^{|\alpha|}}{\partial \hat{x}^{\alpha}} v \right|^q d\hat{x} \right)^{\frac{1}{q}}.$$

Moreover, for $p = 2$, $m, q, \widehat{K}$ is replace by $m, \widehat{K}$ for all the semi norms of norms.

We now derive two hp-error estimates for the projector $\hat{I}_r$.

Theorem 6 *If $v \in H^{r+1}(\widehat{K})$, then for all $r \geq 1$, it exists a positive constant C independent of r such that*

$$|\hat{v} - \hat{I}_r \hat{v}|_{m,\widehat{K}} \leq \frac{C}{r^{r+1-m}} [v]_{r+1,\widehat{K}},\ m = 0, 1. \tag{3.175}$$

Proof The Bramble–Hilbert lemma applied to $\hat{I}_r$ immediately leads to: for $r \geq 1$ and $m \leq r+1$, it exists C dependent on K and r such that

$$\forall \hat{v} \in H^{r+1}(\widehat{K}),\ |\hat{v} - \hat{I}_r \hat{v}|_{m,\widehat{K}} \leq C[\hat{v}]_{r+1,\widehat{K}}. \tag{3.176}$$

In order to derive the hp-projection error estimates for I_h^r, we must specify the exact r-dependence of constant C of (3.176). To do so, we come back to the proof of the Bramble–Hilbert lemma by directly considering $\hat{I}_r$.

The first step, to prove this type of result, is to write (see [15]): $\forall \hat{v} \in H^{r+1}(\widehat{K})$,

$$\begin{aligned} |\hat{v} - \hat{I}_r \hat{v}|_{m,\widehat{K}} &\leq \|I - \hat{I}_r\|_{\mathscr{L}(H^{r+1}(\widehat{K}), H^m(\widehat{K}))} \inf_{\hat{p} \in Q_r(\widehat{K})} \|\hat{v} + \hat{p}\|_{r+1,\widehat{K}} \\ &\leq C_1 \|I - \hat{I}_r\|_{\mathscr{L}(H^{r+1}(\widehat{K}), H^m(\widehat{K}))} [v]_{r+1,\widehat{K}} \end{aligned} \tag{3.177}$$

where C_1 is independent of r and $\mathscr{L}(H^{r+1}(\widehat{K}), H^m(\widehat{K}))$ is the vector space of continuous linear mapping from $H^{r+1}(\widehat{K})$ to $H^m(\widehat{K})$.

The second step is to use the classical spectral error estimates [17, 18] for the interpolation operator $\hat{I}_r$: let $\hat{v} \in H^s(\hat{K})$ with $s > d/2$, then it exists a constant $C > 0$ independent of r such that

$$\|\hat{v} - \hat{I}_r \hat{v}\|_{0,\hat{K}} \leq C r^{-s} \|\hat{v}\|_{s,\hat{K}}, \tag{3.178a}$$

$$\|\hat{v} - \hat{I}_r \hat{v}\|_{1,\hat{K}} \leq C r^{1-s} \|\hat{v}\|_{s,\hat{K}}. \tag{3.178b}$$

By using (3.178a) and (3.178b) and setting $s = r + 1$, (3.177) immediately provides:

$$|\hat{v} - \hat{I}_r \hat{v}|_{m,\widehat{K}} \le \frac{C_2(\widehat{K})}{r^{r+1-m}} [\hat{v}]_{r+1,\widehat{K}}, \ m = 0, 1. \tag{3.179}$$

In order to determine the projector errors, we will the following estimate.

Lemma 3.2 *Let $K \in \mathscr{T}_h$ and $v \in W^{m,p}(K)$. We have the estimate:*

$$[v \circ \underline{F}_K]_{m,p,\widehat{K}} \le C \frac{h_K^m}{\rho_K^{\frac{d}{p}}} |v|_{m,p,K}. \tag{3.180}$$

If $\mathscr{T}_h$ belongs to a regular family of triangulation, we have:

$$[v \circ \underline{F}_K]_{m,p,\widehat{K}} \le C \sigma^{\frac{3}{p}} h_K^{m-\frac{d}{p}} |v|_{m,p,K} \tag{3.181}$$

where $C > 0$ independent of K and r.

Proof We denote $\underline{F}_K = (F_K^1, \dots, F_K^d)$. To prove this lemma, we use the property:

$$\partial^2_{\hat{x}_k^2} \underline{F}^i_K = 0 \text{ for } i = 1, \dots, d, \tag{3.182}$$

since $F^i_K \in Q_1(\widehat{K})$.

Lemma 3.3 *Let $v \in H^{r+1}(K)$, $r \ge 1$. It exists C a positive constant independent of K and r such that*

$$\|v - I_K^r v\|_{0,K} \le C \frac{h_K^{\frac{d}{2}}}{r^{r+1}} [\hat{v}]_{r+1,\widehat{K}}, \tag{3.183a}$$

$$|v - I_K^r v|_{1,K} \le C \frac{h_K^{\frac{d}{2}-1}}{r^r} [\hat{v}]_{r+1,\widehat{K}}. \tag{3.183b}$$

Proof We only prove the second inequality. The same process provides the first one. Let $w = v - I_h^r v$, we have:

$$|w|^2_{1,K} = \sum_{l=1}^{d} \int_{\widehat{K}} |J_K| |(\partial_{x_l} w) o \underline{F}_K|^2 d\,\hat{x}, \tag{3.184}$$

where $\partial_{x_l} = \dfrac{\partial}{\partial x_l}$.

By definition, we have: $w = \hat{w} o \underline{F}_K^{-1}$ where $\hat{w} = \hat{v} - \hat{\pi}_r^1 \hat{v}$ and by using the classical rule of derivation, we have:

$$|w|_{1,K^2} = \sum_{l=1}^{d} \int_{\widehat{K}} |J_K| \left| \sum_{k=1}^{d} \partial_{\hat{x}_k} \hat{w} \partial_{x_l} \hat{x}_k \circ \underline{F}_K \right|^2 d\hat{x}. \tag{3.185}$$

On the other hand, estimates (3.164a)–(3.164c) imply: $|\partial_{x_l} \hat{x}_k o \underline{F}_K| \leq C/h_K$ and $|J_K| \leq C h_K^d$. This immediately leads to

$$|w|_{1,K}^2 \leq C h_K^{d-2} \sum_{k=1}^{d} \int_{\widehat{K}} |\partial_{\hat{x}_k} \hat{w}|^2 \, d\hat{x} \tag{3.186}$$

where $C > 0$ is independent of K and r. (3.175) finally provides (3.183b).

Now, by grouping (3.183a) and (3.183b) and (3.181), we obtain the following error estimates.

Theorem 7 *Let $v \in H^{r+1}(K)$ with $r \geq 1$, then it exists a positive constant C independent of K and r such that*

$$\|v - I_h^r v\|_{0,K} \leq C \frac{h_K^{r+1}}{r^{r+1}} |v|_{r+1,K}, \tag{3.187a}$$

$$|v - I_h^r v|_{1,K} \leq C \frac{h_K^r}{r^r} |v|_{r+1,K}. \tag{3.187b}$$

By using the interpolation Theorem 1.4 of [16], we extend the result to real exponents.

Theorem 8 *Let $v \in H^{s+1}(K)$, $s \geq 1$ real. It exists C independent of K and r such that*

$$\|v - I_h^r v\|_{0,K} \leq C \frac{h_K^{\min(s+1,r+1)}}{r^{s+1}} \|v\|_{s+1,K}, \tag{3.188a}$$

$$|v - I_h^r v|_{1,K} \leq C \frac{h_K^{\min(s,r)}}{r^s} \|v\|_{s+1,K}. \tag{3.188b}$$

Proof Let $r_1 < r_2$ be two positive integers and $\theta \in [0, 1]$. Assume that $I_{h|K}^r \in \mathscr{L}(H^{r_1+1}(K), H^m(K)) \cap \mathscr{L}(H^{r_2+1}(K), H^m(K))$ for $m = 0, 1$.

Then we have:

$$\begin{aligned} &\|I - I_{h|K}^r\|_{\mathscr{L}(H^{\theta r_1 + (1-\theta) r_2 + 1}(K), H^m(K))} \\ &\leq C \|I - I_{h|K}^r\|_{\mathscr{L}(H^{r_1+1}(K), H^m(K))}^{\theta} \|I - I_{h|K}^r\|_{\mathscr{L}(H^{r_2+1}(K), H^m(K))}^{1-\theta}. \end{aligned} \tag{3.189}$$

Inequalities (3.187a) and (3.187b) lead to

$$\|I - I^r_{h|K}\|_{\mathscr{L}(H^{r_1+1}(K),H^m(K))} \leq C\frac{h_K^{r_1+1-m}}{r^{r_1+1-m}}, \tag{3.190a}$$

$$\|I - I^r_{h|K}\|_{\mathscr{L}(H^{r_2+1}(K),H^m(K))} \leq C\frac{h_K^{r_2+1-m}}{r^{r_2+1-m}}. \tag{3.190b}$$

So, we obtain:

$$\|I - I^r_{h|K}\|_{\mathscr{L}(H^{\theta r_1+(1-\theta)r_2+1}(K),H^m(K))} \leq C\frac{h_K^{\theta r_1+(1-\theta)r_2-m+1}}{r^{\theta r_1+(1-\theta)r_2+1-m}}. \tag{3.191}$$

Finally, by taking $r_1 = 0$, $r_2 = r$ and $s = (1-\theta)$, we can write the inequality:

$$\begin{aligned}\|v - I^r_{h|K}v\|_{m,K} &\leq \|I - I^r_{h|K}\|_{\mathscr{L}(H^{s+1}(K),H^m(K))}\|v\|_{s+1,K}\\ &\leq C\frac{h_K^{s-m}}{r^{s+1-m}}\|v\|_{s+1,K}.\end{aligned} \tag{3.192}$$

3.6.3 *hp*-Estimation of Numerical Integration Errors

In this section, we establish some estimations of numerical integration errors due the use of the Gauss–Lobatto rule. The choice of the quantities to be estimated directly comes from the next part which deals with hp a priori error analysis of the finite element discretization of the problem (3.1).

We begin by giving some technical intermediate results.

Proposition 1 *There exists two positive constants α and β independent of r such that $\forall \hat{v} \in Q_r(\hat{K})$,*

$$\alpha\|\hat{v}\|^2_{0,\hat{K}} \leq \int_{\hat{K}}^{GL} \hat{v}^2\,dx \leq \beta\|\hat{v}\|^2_{0,\hat{K}}, \tag{3.193}$$

Moreover, it exists four positive constants α_1, β_1 and α_2, β_2 independent of h and r such that $\forall v_h \in U^r_h$ and $\forall K \in \mathscr{T}_h$,

$$\alpha_1\|v_h\|^2_{0,K} \leq \int_K^{GL} v_h^2\,dx \leq \beta_1\|v_h\|^2_{0,K}, \tag{3.194a}$$

$$\alpha_2\|v_h\|^2_{1,K} \leq \int_K^{GL} \underline{\underline{\mu}}|\nabla v_h|^2\,dx \leq \beta_2\|v_h\|^2_{1,K}. \tag{3.194b}$$

Proof We start to prove the first inequality i.e we prove the equivalence of norms $\hat{v} \in Q_r(\hat{K}) \to \left(\int_{\hat{K}}^{GL} \hat{v}^2 \, d\hat{x} \right)^{1/2}$ and $\| \cdot \|_{0,\hat{K}}$ with constants which are independent of r. For that, we consider the basis of $Q_r(\hat{K})$ composed of products of Legendre polynomials defined on $[0, 1]$: a basis function is written $\hat{w} := P_{l_1}(\hat{x}_1) \dots P_{l_d}(\hat{x}_d)$ for $l_i = 0, \dots, r$. Since the Gauss–Lobatto rule with $(r+1)^d$ points is exact for the functions in $Q_{2r-1}(\hat{K})$, the only product of basis functions for which the integration by the Gauss–Lobatto is not exact, is $\hat{v} = P_{l_1}(\hat{x}_1) \dots P_{l_d}(\hat{x}_d)$ with at least one l_i equal to r. In this case, we have $\int_{[0,1]}^{GL} P_r^2(\hat{x}) \, d\hat{x} = \sum_{i=1}^{r+1} \hat{\omega}_i P_r^2(\hat{x}_i)$. Now, by using the expression of the weights $\hat{\omega}_i = 1/(r(r+1)[P_r(\hat{x}_i)]^2)$ for $i = 2, \dots, r$ and $\hat{\omega}_1 = \hat{\omega}_{r+1} = 1/(r(r+1))$, we immediately have $\int_{[0,1]}^{GL} P_r^2(\hat{x}) \, d\hat{x} = 1/r$. Finally, we recall $\int_{[0,1]} P_r^2(\hat{x}) \, d\hat{x} = 2/(2r+1)$ and we have $\int_{[0,1]}^{GL} P_r^2(\hat{x}) \, d\hat{x} = (1 + 1/(2r)) \int_{[0,1]} P_r^2(\hat{x}) \, d\hat{x}$. So, the equivalence constant are independent of the polynomial order r.

Now, we derive (3.194a). First, we have

$$\int_K^{GL} v_h^2 \, dx = \int_{\hat{K}}^{GL} |J_K| \hat{v}_h^2 \, d\hat{x}. \tag{3.195}$$

By using estimates (3.164a)–(3.164c), we can immediately write the double inequality:

$$C_1 h_K^d \int_{\widehat{K}}^{GL} \hat{v}_h^2 \, d\hat{x} \le \int_K^{GL} v_h^2 \, dx \le C_2 h_K^d \int_{\widehat{K}}^{GL} \hat{v}_h^2 \, d\hat{x} \tag{3.196}$$

where $C_1, C_2 > 0$ are independent of K and r.

From (3.193), we can now rewrite (3.196) in this way:

$$C_1 h_K^d \int_{\widehat{K}} \hat{v}_h^2 \, d\hat{x} \le \int_K^{GL} v_h^2 \, dx \le C_2 h_K^d \int_{\widehat{K}} \hat{v}_h^2 \, d\hat{x} \tag{3.197}$$

where $C_1, C_2 > 0$ are independent of K and r.

Returning to the cell K of the mesh, we get:

$$C_1 h_K^d \int_K \frac{1}{|J_K| \circ \underline{F}_K^{-1}} v_h^2 \, dx \le \int_K^{GL} v_h^2 \, dx \le C_2 h_K^d \int_K \frac{1}{|J_K| \circ \underline{F}_K^{-1}} v_h^2 \, dx, \tag{3.198}$$

Equations (3.166a) and (3.166b) then yield (3.194a).

Equation (3.194b) is derived by the same technique.

The main results are contained in the following proposition:

Proposition 2 *$\forall K \in \mathscr{T}_h$, let E_K^I be the integration error defined by*

$$E_I^K(g,h) = \left| \int_K^{GL} g\,h\,dx - \int_K g\,h\,dx \right| \tag{3.199}$$

Let $r \in \mathbb{N}^*$ and $s \in \mathbb{R}$ such that $s \geq 1$. Moreover, we assume $0 < h_K \leq 1$. For all $g \in H^{s+1}(K)$ and $\forall w_h \in U_h^r$, we have:

$$E_I^K(g,w_h) \leq C \frac{h_K^{\min(g^d(s),\,g^d(r))}}{r^s} \|g\|_{s+1,K}\, \|w_h\|_{0,K} \tag{3.200}$$

where $g^d(x) = x - \min(x, d-1)$.

Proof We begin the proof by assuming $g \in H^{r+1}(K)$. By using the transformation $\underline{F}_K$, $E_I^K(g,w_h)$ can be rewritten as follows:

$$E_I^K(g,w_h) = \left| \int_{\widehat{K}}^{GL} |J_K| \hat{g}_K\, \hat{w}_K\, d\hat{x} - \int_{\widehat{K}} |J_K| \hat{g}_K\, \hat{w}_K\, d\hat{x} \right|, \tag{3.201}$$

with $g_{|K} \circ \underline{F}_K = \widehat{g}_K$ and $w_{h|K} \circ \underline{F}_K = \widehat{w}_K$.

Let $\hat{z}_K = |J_K| \hat{g}_K$.

First, by using the interpolation operator $\hat{I}_r$ associated to the Gauss–Lobatto quadrature rule of order r, we can rewrite (3.201) in this way:

$$E_I^K(g,w_h) = \left| \int_{\widehat{K}}^{GL} \hat{I}_r(\widehat{z}_K)\, \hat{w}_K\, d\hat{x} - \int_{\widehat{K}} \widehat{z}_K\, \widehat{w}_K\, d\hat{x} \right| \tag{3.202}$$

Secondly, we know that the Gauss–Lobatto integration rule of order r is exact for polynomials in $Q_{2r-1}(\widehat{K})$. In particular, we have:

$$\int_{\widehat{K}}^{GL} \hat{I}_{r-1}(\widehat{z}_K)\, \hat{w}_K\, d\hat{x} = \int_{\widehat{K}} \hat{I}_{r-1}(\widehat{z}_K)\, \hat{w}_K\, d\hat{x}. \tag{3.203}$$

Finally, by combining (3.202) and (3.203), we can write:

$$E_I^K(g,w_h) = \left| \int_{\widehat{K}}^{GL} \left(\hat{I}_r(\widehat{z}_K) - \hat{I}_{r-1}(\widehat{z}_K) \right) \hat{w}_K\, d\hat{x} \right.$$
$$\left. + \int_{\widehat{K}} \left(\hat{I}_{r-1}(\widehat{z}_K) - \widehat{z}_K \right) \widehat{w}_K\, d\hat{x} \right|. \tag{3.204}$$

Now, the interpolation error estimate (3.175) and (3.193) implies:

$$E_I^K(g,w_h)$$
$$\leq C\big(\|\widehat{z}_K - \hat{I}_{r-1}(\widehat{z}_K)\|_{0,\widehat{K}} + \|\widehat{z}_K - \hat{I}_r(\widehat{z}_K)\|_{0,\widehat{K}} \big) \|\widehat{w}_K\|_{0,\widehat{K}}$$

$$\leq C\left(\frac{1}{r^r}[\hat{z}_K]_{r,\widehat{K}} + \frac{1}{r^{r+1}}[\hat{z}_K]_{r+1,\widehat{K}}\right)\|\widehat{w}_K\|_{0,\widehat{K}}. \tag{3.205}$$

We are going to estimate the term $[\hat{z}_K]_{m,\widehat{K}}$ for $m = r,\ r+1$: the definition of $\hat{z}_K$ leads to

$$[\hat{z}_K]_{m,\widehat{K}} = [|J_K| g \circ \underline{F}_K]_{m,\widehat{K}}. \tag{3.206}$$

Developing (3.206), we get:

$$\begin{aligned}
[|J_K| g \circ \underline{F}_K]^2_{m,\widehat{K}} &= \sum_{l=1}^{d} \int_{\widehat{K}} \left| \frac{\partial^m}{\partial \hat{x}_l^m} \big(|J_K| g \circ \underline{F}_K\big) \right|^2 d\hat{x} \\
&\leq \sum_{l=1}^{d} \int_{\widehat{K}} \left| \sum_{n=0}^{m} \binom{n}{m} \frac{\partial^n}{\partial \hat{x}_l^n}(|J_K|) \frac{\partial^{m-n}}{\partial \hat{x}_l^{m-n}} (g \circ \underline{F}_K) \right|^2 d\hat{x}.
\end{aligned} \tag{3.207}$$

It is easy to see that $\forall n \geq d,\ \dfrac{\partial^n}{\partial \hat{x}_k^n}\big(|J_K|\big) = 0$ since $J_K \in Q_{d-1}(\widehat{K})$ and that $0 \leq n \leq d-1,\ \left|\dfrac{\partial^n}{\partial \hat{x}_k^n}\big(|J_K|\big)\right| \leq C h_K^d$.

We finally have:

$$[\hat{z}_K]^2_{m,\widehat{K}} \leq C h_K^{2d} \sum_{n=0}^{\min(m,d-1)} [g \circ \underline{F}_K]^2_{m-n,\widehat{K}}. \tag{3.208}$$

Now, using (3.181) and a classical scaling argument, we deduce the two following estimates: for $n \leq \min(m, d-1)$,

$$[g \circ \underline{F}_K]_{m-n,\widehat{K}} \leq C h_K^{m-n-\frac{d}{2}} |g|_{m-n,K} \tag{3.209a}$$

$$\left\| w_h \circ \underline{F}_K \right\|_{0,\widehat{K}} \leq C h_K^{-\frac{d}{2}} \|w_h\|_{0,K} \tag{3.209b}$$

Injecting (3.209a) and (3.209b) in (3.208), we obtain:

$$[\hat{z}_K]_{r+1,\widehat{K}} \leq C h_K^{r+1+\frac{d}{2}} \sum_{n=0}^{d-1} h_K^{-n} |g|_{r+1-n,K}, \tag{3.210a}$$

$$[\hat{z}_K]_{r,\widehat{K}} \leq C h_K^{r+\frac{d}{2}} \sum_{n=0}^{\min(r,d-1)} h_K^{-n} |g|_{r-n,K}. \tag{3.210b}$$

From (3.205) and by using the assumption $0 < h_K \leq 1$, we can write $\forall r \geq 1$,

$$
\begin{aligned}
E_I^K(g, w_h) &\leq C\left(\frac{h_K^r}{r^r}\sum_{n=0}^{\min(r,d-1)} h_K^{-n}|g|_{r-n,K} \right.\\
&\qquad \left. + \frac{h_K^{r+1}}{r^{r+1}}\sum_{n=0}^{d-1} h_K^{-n}|g|_{r+1-n,K}\right)\|w_h\|_{0,K}\\
&\leq C\frac{h_K^{r-\min(r,d-1)}}{r^r}\|g\|_{r+1}\,\|w_h\|_{0,K}\,. \qquad (3.211)
\end{aligned}
$$

Let $r \in \mathbb{N}^*$ and $w_h \in U_h^r$. We define the linear operator A as follows: $k \in \mathbb{N}^*$,

$$
A : H^{k+1}(K) \to \mathbb{R}, \qquad (3.212a)
$$

$$
g \in H^{k+1}(K) \mapsto Ag = \int_K^{GL} g\, w_h\, dx - \int_K g\, w_h\, dx. \qquad (3.212b)
$$

By using the same way that the one used to derive (3.211), we immediately obtain the continuity of the linear form A: if $0 < h_K \leq 1$,

$$
\|A\|_{\mathscr{L}(H^{k+1}(K),\mathbb{R})} \leq C\frac{h_K^{\min(g^d(k)\,,\,g^d(r))}}{r^k}\|w_h\|_{0,K}\,, \qquad (3.213)
$$

where $g^d(x) = x - \min(x, d-1)$.

So, by using a standard interpolation to define the Sobolev space with real regularity, we have: $\forall s \geq 1$ (real),

$$
\|A\|_{\mathscr{L}(H^{s+1}(K),\mathbb{R})} \leq C\frac{h_K^{\min(g^d(s)\,,\,g^d(r))}}{r^s}\|w_h\|_{0,K} \qquad (3.214)
$$

and we obtain (3.200).

We finish this part by the following corollary of Proposition 2.

Corollary 3 *Let $w \in H^{s+1}(K)$ with $s \geq 2$ and $v_h \in U_h^r$. Moreover, we assume $0 < h_K \leq 1$. We have the following numerical integration error estimate:*

$$
E_I^K\left(\underline{\underline{\mu}}^{1/2}\nabla I_h^r w\,,\,\underline{\underline{\mu}}^{1/2}\nabla v_h\right) \leq C\frac{h_K^{\min(g^d(s-1)\,,\,g^d(r))}}{r^{s-1}}\|w\|_{s+1,K}\left\|\underline{\underline{\mu}}^{1/2}\nabla v_h\right\|_{0,K} \qquad (3.215)
$$

where $g^d(x) = x - \min(x, d-1)$ and $C >$ is a constant independent of h_K and r.

Proof Since we have assumed that $w \in H^{s+1}(K)$ with $s \geq 2$ then $I_h^r(\nabla w)$ is well-defined and we can write:

$$\int_K^{GL} \underline{\underline{\mu}} \nabla I_h^r w \cdot \nabla v_h \, dx = \int_K^{GL} \underline{\underline{\mu}} \, I_h^r (\nabla w) \cdot \nabla v_h \, dx$$

$$+ \int_K^{GL} \underline{\underline{\mu}} \left(\nabla I_h^r w - I_h^r (\nabla w) \right) \cdot \nabla v_h \, dx \tag{3.216a}$$

$$= \int_K^{GL} \underline{\underline{\mu}} \, \nabla w \cdot \nabla v_h \, dx + \int_K^{GL} \underline{\underline{\mu}} \left(\nabla I_h^r w - I_h^r (\nabla w) \right) \cdot \nabla v_h \, dx$$

$$\int_K \underline{\underline{\mu}} \nabla I_h^r w \cdot \nabla v_h \, dx = \int_K \underline{\underline{\mu}} \, \nabla w \cdot \nabla v_h \, dx + \int_K \underline{\underline{\mu}} \, \nabla \left(I_h^r w - w \right) \cdot \nabla v_h \, dx \tag{3.216b}$$

Equation (3.216a) and (3.216b) lead to

$$\begin{aligned} E_I^K \left(\underline{\underline{\mu}}^{1/2} \nabla I_h^r w \, , \, \underline{\underline{\mu}}^{1/2} \nabla v_h \right) &\leq E_I^K \left(\underline{\underline{\mu}}^{1/2} \nabla w \, , \, \underline{\underline{\mu}}^{1/2} \nabla v_h \right) \\ &+ \left\| \underline{\underline{\mu}}^{1/2} \left(\nabla I_h^r w - I_h^r (\nabla w) \right) \right\|_{0,K,GL} \left\| \underline{\underline{\mu}}^{1/2} \nabla v_h \right\|_{0,K,GL} \\ &+ \left\| \underline{\underline{\mu}}^{1/2} \nabla \left(I_h^r w - w \right) \right\|_{0,K} \left\| \underline{\underline{\mu}}^{1/2} \nabla v_h \right\|_{0,K} \end{aligned} \tag{3.217}$$

where $\| \cdot \|_{0,K,GL}$ is the L^2-norm computed by using the Gauss–Lobatto quadrature rule.

We now examine more precisely the term $\left\| \underline{\underline{\mu}}^{1/2} \left(\nabla I_h^r w - I_h^r (\nabla w) \right) \right\|_{0,K,GL}$:

$$\begin{aligned} &\left\| \underline{\underline{\mu}}^{1/2} \left(\nabla I_h^r w - I_h^r (\nabla w) \right) \right\|_{0,K,GL}^2 \\ &= \int_{\hat{K}}^{GL} |J_K| \left| DF_K^{*-1} \widehat{\nabla} (\hat{I}_r \hat{w}_K) - \hat{I}_r (DF_K^{-1} \widehat{\nabla} \hat{w}_K) \right|^2 \, d\hat{x} \\ &= \sum_{n=1}^{(r+1)^d} \hat{\omega}_n |J_K| (\widehat{\underline{\xi}_n}) \left| DF_K^{*-1} (\widehat{\underline{\xi}_n}) \left(\widehat{\nabla} (\hat{I}_r \hat{w}_K)(\widehat{\underline{\xi}_n}) - \widehat{\nabla} \hat{w}_K (\widehat{\underline{\xi}_n}) \right) \right|^2 \end{aligned} \tag{3.218}$$

By using (3.164a) and (3.164b), (3.219) becomes:

$$\begin{aligned} &\left\| \underline{\underline{\mu}}^{1/2} \left(\nabla I_h^r w - I_h^r (\nabla w) \right) \right\|_{0,K,GL}^2 \\ &\leq C h_K^{d-2} \sum_{n=1}^{(r+1)^d} \hat{\omega}_n \left| \widehat{\nabla} (\hat{I}_r \hat{w}_K)(\widehat{\underline{\xi}_n}) - \widehat{\nabla} \hat{w}_K (\widehat{\underline{\xi}_n}) \right|^2 \\ &= C h_K^{d-2} \int_{\hat{K}}^{GL} \left| \widehat{\nabla} (\hat{I}_r \hat{w}_K) - \hat{I}_r \left(\widehat{\nabla} \hat{w}_K \right) \right|^2 \, d\hat{x} \end{aligned} \tag{3.219}$$

The inequalities of (3.193) lead to

$$\left\|\underline{\underline{\mu}}^{1/2}\left(\nabla I_h^r w - I_h^r\left(\nabla w\right)\right)\right\|_{0,K,GL}^2 \leq C h_K^{d-2}\left\|\widehat{\nabla}(\hat{I}_r \hat{w}_K) - \hat{I}_r\left(\widehat{\nabla}\hat{w}_K\right)\right\|_{0,\hat{K}}^2$$
$$\leq C h_K^{d-2}\left(\left\|\widehat{\nabla}(\hat{I}_r \hat{w}_K) - \widehat{\nabla}\hat{w}_K\right\|_{0,\hat{K}}^2 + \left\|\widehat{\nabla}\hat{w}_K - \hat{I}_r\left(\widehat{\nabla}\hat{w}_K\right)\right\|_{0,\hat{K}}^2\right) \quad (3.220)$$

Now we come back on the cell K of the mesh. First, we have:

$$\left\|\widehat{\nabla}(\hat{I}_r \hat{w}_K) - \widehat{\nabla}\hat{w}_K\right\|_{0,\hat{K}}^2 = \int_{\hat{K}} \frac{|J_K|}{|J_K|}\left|DF_K^*\left(\nabla\left(I_h^r w\right)\circ F_K - \nabla\left(w\right)\circ F_K\right)\right|^2 d\,\hat{x}$$
$$\leq C\, h_K^{2-d}\int_{\hat{K}} |J_K|\left|\nabla\left(I_h^r w\right)\circ F_K - \nabla\left(w\right)\circ F_K\right|^2 d\,\hat{x}$$
$$= C\, h_K^{2-d}\left\|\nabla(I_h^r\, w) - \nabla\, w\right\|_{0,K}^2$$
$$= C\, h_K^{2-d}\left|I_h^r\, w - w\right|_{1,K}^2$$
$$\leq C\frac{h_K^{\min(2(s+1),2(r+1))-d}}{r^{2s}}\|w\|_{s+1,K}^2 \text{ by using (3.189b).} \quad (3.221)$$

To study $\left\|\widehat{\nabla}\hat{w}_K - \hat{I}_r\left(\widehat{\nabla}\hat{w}_K\right)\right\|_{0,\hat{K}}^2$, we first assume that $\nabla\, w \in H^{r+1}(K)^d$. In this case, (3.179) leads to

$$\left\|\widehat{\nabla}\hat{w}_K - \hat{I}_r\left(\widehat{\nabla}\hat{w}_K\right)\right\|_{0,\hat{K}}^2 \leq \frac{C}{r^{2(r+1)}}\left[\widehat{\nabla}\hat{w}_K\right]_{r+1,\hat{K}}^2 \quad (3.222)$$

where the bracket semi-norm is taken componentwise

$$\left[\widehat{\nabla}\hat{w}_K\right]_{m,\hat{K}}^2 = \sum_{i=1}^{d}\sum_{j=1}^{d}\left\|\frac{\partial^m}{\partial\hat{x}_j^m}\left(\frac{\partial\hat{w}_K}{\partial\hat{x}_i}\right)\right\|_{0,\hat{K}}^2 \quad (3.223)$$

Now, if we define the Jacobian matrix DF_K by $(J_{i,j})_{1\leq i,j\leq d}$, by using the property $\dfrac{\partial^2 J_{i,j}}{\partial\hat{x}_k^2} = 0,\ \forall k = 1,\ldots,d$ and (3.164c), (3.223) becomes:

$$\left[\widehat{\nabla}\hat{w}_K\right]_{m,\hat{K}}^2 = \sum_{i=1}^{d}\sum_{j=1}^{d}\left\|\sum_{k=1}^{d}\frac{\partial^m}{\partial\hat{x}_j^m}\left(J_{k,i}\frac{\partial w}{\partial x_k}\circ F_K\right)\right\|_{0,\hat{K}}^2$$
$$= \sum_{i=1}^{d}\sum_{j=1}^{d}\left\|\sum_{k=1}^{d}\left(m\frac{\partial J_{k,i}}{\partial\hat{x}_j}\frac{\partial^{m-1}}{\partial\hat{x}_j^{m-1}}\left(\frac{\partial w}{\partial x_k}\circ F_K\right) + J_{k,i}\frac{\partial^m}{\partial\hat{x}_j^m}\left(\frac{\partial w}{\partial x_k}\circ F_K\right)\right)\right\|_{0,\hat{K}}^2$$

$$\leq Ch_K^2 \sum_{j=1}^{d}\sum_{k=1}^{d}\left(\left\|\frac{\partial^{m-1}}{\partial \hat{x}_j^{m-1}}\left(\frac{\partial w}{\partial x_k}\circ F_K\right)\right\|_{0,\hat{K}}^2 + \left\|\frac{\partial^{m}}{\partial \hat{x}_j^{m}}\left(\frac{\partial w}{\partial x_k}\circ F_K\right)\right\|_{0,\hat{K}}^2\right)$$
$$\leq Ch_K^2\left([\nabla w\circ F_K]_{m-1,\hat{K}}^2 + [\nabla w\circ F_K]_{m,\hat{K}}^2\right) \tag{3.224}$$

Finally, (3.181) gives:

$$\left[\widehat{\nabla}\hat{w}_K\right]_{r+1,\hat{K}}^2 \leq Ch_K^{2r+2-d}\|\nabla w\|_{r+1,K}^2 \tag{3.225}$$

and (3.222) gives:

$$\left\|\widehat{\nabla}\hat{w}_K - \hat{I}_r\left(\widehat{\nabla}\hat{w}_K\right)\right\|_{0,\hat{K}}^2 \leq C\frac{h_K^{2(r+1)-d}}{r^{2(r+1)}}\|\nabla w\|_{r+1,K}^2 \tag{3.226}$$

Now, by using a standard interpolation (see previous subsection) to define the Sobolev space with real regularity, we have: $\forall s \geq 2$ (real) and $\forall w \in H^{s+1}(K)$,

$$\left\|\widehat{\nabla}\hat{w}_K - \hat{I}_r\left(\widehat{\nabla}\hat{w}_K\right)\right\|_{0,\hat{K}} \leq C\frac{h_K^{\min(s,r+1)-\frac{d}{2}}}{r^s}\|w\|_{s+1,K} \tag{3.227}$$

By using (3.221) and (3.227), (3.220) becomes:

$$\left\|\underline{\underline{\mu}}^{1/2}\left(\nabla I_h^r w - I_h^r(\nabla w)\right)\right\|_{0,K,GL}^2$$
$$\leq Ch_K^{d-2}\left(\left\|\widehat{\nabla}(\hat{I}_r\hat{w}_K) - \widehat{\nabla}\hat{w}_K\right\|_{0,\hat{K}}^2 + \left\|\widehat{\nabla}\hat{w}_K - \hat{I}_r\left(\widehat{\nabla}\hat{w}_K\right)\right\|_{0,\hat{K}}^2\right)$$
$$\leq C\frac{h_K^{\min(2(s-1),2r)}}{r^{2s}}\|w\|_{s+1,K}^2 \tag{3.228}$$

Finally, by using (3.228) and by applying (3.200) componentwise with $g = \underline{\underline{\mu}}^{1/2}\nabla w \in H^s(K)^d$, we obtain the result.

3.6.4 *hp a priori Error Estimate for the Semi-discrete Approximation*

In this part, we assume that the data of the problem (3.1) lead to the following regularity of the weak solution (see Sect. 1.2.4):

$$u \in C^0\left([0,T],\ H^1(\Omega)\right) \tag{3.229a}$$

$$\frac{\partial u}{\partial t} \in C^0\left([0, T], L^2(\Omega)\right) \tag{3.229b}$$

So, in this case, we have $e_h = u - u_h \in C^0\left([0, T], H^1(\Omega)\right) \cap C^1([0, T], L^2(\Omega))$.

Moreover, we add some regularity assumptions to derive the error estimates:

$$\frac{\partial^i u}{\partial t^i} \in L^\infty\left(]0, T[, H^{s+1}(\Omega)\right) \text{ for all } i \in [\![0, 2]\!] \text{ with } s \geq r + 1, \tag{3.230a}$$

$$f \in L^\infty\left(]0, T[, H^{s'+1}(\Omega)\right) \text{ for all } i \in [\![0, 1]\!] \text{ with } s' \geq r \tag{3.230b}$$

with the convention $\dfrac{\partial^0 u}{\partial t^0} = u$.

Remark: Assumptions (3.230a) and (3.230b) mean that we have sufficient regularity on f and u such that

$$\min(g^d(s \text{ or } s'), g^d(r)) = g^d(r), \tag{3.231a}$$

$$\min(g^d(s-1), g^d(r)) = g^d(r), \tag{3.231b}$$

where $g_1^d(x) = x + 1 - min(x, d-1)$ and $g_2^d(x) = x - min(x, d-1)$ are used in Sect. 3.6.3 in the estimations of numerical integration errors.

We introduce the continuous and the discrete energy norms on the space

$$\|v\|^2 = \left\|\lambda^{1/2}\frac{\partial v}{\partial t}\right\|^2_{0,\Omega} + \left\|\underline{\underline{\mu}}^{1/2}\nabla v\right\|^2_{0,\Omega}, \tag{3.232a}$$

$$\|v\|_h^2 = \int_\Omega^{GL} \lambda\left(\frac{\partial v}{\partial t}\right)^2 dx + \int_\Omega^{GL} \underline{\underline{\mu}}\,|\nabla v|^2\, dx. \tag{3.232b}$$

We recall that the solution of (3.1) verifies the identity

$$\int_\Omega \lambda\frac{\partial^2 u}{\partial t^2} v\, dx + \int_\Omega \underline{\underline{\mu}}\nabla u \cdot \nabla v\, dx = \int_\Omega f\, v dx, \;\forall v \in H^1(\Omega) \tag{3.233}$$

and the finite element approximation of this problem is defined by: $u_h(t,\cdot) \in U_h^r$ such that

$$\int_\Omega^{GL} \lambda\frac{\partial^2 u_h}{\partial t^2} v_h\, dx + \int_\Omega^{GL} \underline{\underline{\mu}}\nabla u_h \cdot \nabla v_h\, dx = \int_\Omega^{GL} f\, v_h dx, \;\forall v_h \in U_h^r \tag{3.234}$$

Let $I_h^r u$ be the interpolation of the solution u on the finite space U_h^r (see Sect. 3.6.2). We can write:

$$\begin{aligned}
&\int_\Omega^{GL} \lambda \frac{\partial^2 (u_h - I_h^r u)}{\partial t^2} v_h \, dx + \int_\Omega^{GL} \underline{\underline{\mu}} \nabla (u_h - I_h^r u) \cdot \nabla v_h \, dx \\
&= \int_\Omega f v_h \, dx - \int_\Omega \lambda \frac{\partial^2 I_h^r u}{\partial t^2} v_h \, dx - \int_\Omega \underline{\underline{\mu}} \nabla I_h^r u \cdot \nabla v_h \, dx \\
&+ \left[\int_\Omega^{GL} f v_h \, dx - \int_\Omega f v_h \, dx \right] \\
&+ \left[\int_\Omega \lambda \frac{\partial^2 I_h^r u}{\partial t^2} v_h \, dx - \int_\Omega^{GL} \lambda \frac{\partial^2 I_h^r u}{\partial t^2} v_h \, dx \right] \\
&+ \left[\int_\Omega \underline{\underline{\mu}} \nabla I_h^r u \cdot \nabla v_h \, dx - \int_\Omega^{GL} \underline{\underline{\mu}} \nabla I_h^r u \cdot \nabla v_h \, dx \right], \quad \forall v_h \in U_h^r.
\end{aligned} \tag{3.235}$$

By using (3.233), (3.235) becomes:

$$\begin{aligned}
&\int_\Omega^{GL} \lambda \frac{\partial^2 (u_h - I_h^r u)}{\partial t^2} v_h \, dx + \int_\Omega^{GL} \underline{\underline{\mu}} \nabla (u_h - I_h^r u) \cdot \nabla v_h \, dx \\
&= \int_\Omega \lambda \frac{\partial^2 (u - I_h^r u)}{\partial t^2} v_h \, dx + \int_\Omega \underline{\underline{\mu}} \nabla (u - I_h^r u) \cdot \nabla v_h \, dx \\
&+ \left[\int_\Omega^{GL} f v_h \, dx - \int_\Omega f v_h \, dx \right] \\
&+ \left[\int_\Omega \lambda \frac{\partial^2 I_h^r u}{\partial t^2} v_h \, dx - \int_\Omega^{GL} \lambda \frac{\partial^2 I_h^r u}{\partial t^2} v_h \, dx \right] \\
&+ \left[\int_\Omega \underline{\underline{\mu}} \nabla I_h^r u \cdot \nabla v_h \, dx - \int_\Omega^{GL} \underline{\underline{\mu}} \nabla I_h^r u \cdot \nabla v_h \, dx \right], \quad \forall v_h \in U_h^r.
\end{aligned} \tag{3.236}$$

Now, we define the interpolation $e_h^I = u_h - I_h^r u$ and the projection $e_h^P(u) = u - I_h^r u$ errors and we take the test-function $v_h = \partial e_h^I / \partial t$. So, (3.236) yields:

$$\begin{aligned}
\frac{1}{2} \frac{d}{dt} \left\| e_h^I \right\|_h^2 &= \int_\Omega \lambda \frac{\partial^2 e_h^P(u)}{\partial t^2} \frac{\partial e_h^I}{\partial t} \, dx + \int_\Omega \underline{\underline{\mu}} \nabla e_h^P(u) \cdot \nabla \frac{\partial e_h^I}{\partial t} \, dx \\
&+ \left[\int_\Omega^{GL} f \frac{\partial e_h^I}{\partial t} \, dx - \int_\Omega f \frac{\partial e_h^I}{\partial t} \, dx \right] \\
&+ \left[\int_\Omega \lambda \frac{\partial^2 I_h^r u}{\partial t^2} \frac{\partial e_h^I}{\partial t} \, dx - \int_\Omega^{GL} \lambda \frac{\partial^2 I_h^r u}{\partial t^2} \frac{\partial e_h^I}{\partial t} \, dx \right] \\
&+ \left[\int_\Omega \underline{\underline{\mu}} \nabla I_h^r u \cdot \nabla \frac{\partial e_h^I}{\partial t} \, dx - \int_\Omega^{GL} \underline{\underline{\mu}} \nabla I_h^r u \cdot \nabla \frac{\partial e_h^I}{\partial t} \, dx \right].
\end{aligned} \tag{3.237}$$

The time regularity assumptions yield the identities:

$$\int_{\Omega} \underline{\underline{\mu}} \nabla e_h^P(u) \cdot \nabla \frac{\partial e_h^I}{\partial t}\, dx = \frac{d}{dt} \int_{\Omega} \underline{\underline{\mu}} \nabla e_h^P(u) \cdot \nabla e_h^I\, dx - \int_{\Omega} \underline{\underline{\mu}} \nabla e_h^P\left(\frac{\partial u}{\partial t}\right) \cdot \nabla e_h^I\, dx \tag{3.238a}$$

$$\int_{\Omega} \underline{\underline{\mu}} \nabla I_h^r u \cdot \nabla \frac{\partial e_h^I}{\partial t}\, dx = \frac{d}{dt} \int_{\Omega} \underline{\underline{\mu}} \nabla I_h^r u \cdot \nabla e_h^I\, dx - \int_{\Omega} \underline{\underline{\mu}} \nabla I_h^r\left(\frac{\partial u}{\partial t}\right) \cdot \nabla e_h^I\, dx \tag{3.238b}$$

$$\int_{\Omega}^{GL} \underline{\underline{\mu}} \nabla I_h^r u \cdot \nabla \frac{\partial e_h^I}{\partial t}\, dx = \frac{d}{dt} \int_{\Omega}^{GL} \underline{\underline{\mu}} \nabla I_h^r u \cdot \nabla e_h^I\, dx - \int_{\Omega}^{GL} \underline{\underline{\mu}} \nabla I_h^r\left(\frac{\partial u}{\partial t}\right) \cdot \nabla e_h^I\, dx \tag{3.238c}$$

We now establish the following lemma.

Lemma 2 *We have the estimate*

$$\begin{aligned}
&\int_{\Omega} \lambda \frac{\partial^2 I_h^r u}{\partial t^2} \frac{\partial e_h^I}{\partial t}\, dx - \int_{\Omega}^{GL} \lambda \frac{\partial^2 I_h^r u}{\partial t^2} \frac{\partial e_h^I}{\partial t}\, dx \\
&\leq \int_{\Omega} \lambda \frac{\partial^2 u}{\partial t^2} \frac{\partial e_h^I}{\partial t}\, dx - \int_{\Omega}^{GL} \lambda \frac{\partial^2 u}{\partial t^2} \frac{\partial e_h^I}{\partial t}\, dx \\
&\quad + C \left\| I_h^r\left(\frac{\partial^2 u}{\partial t^2}\right) - \frac{\partial^2 u}{\partial t^2} \right\|_0 \left\| \lambda^{1/2} \frac{\partial e_h^I}{\partial t} \right\|_0
\end{aligned} \tag{3.239}$$

with C is a positive constant independent of r and h.

Proof First, the regularity assumptions on the exact solution u imply:

$$\frac{\partial^2 I_h^r u}{\partial t^2} = I_h^r\left(\frac{\partial^2 u}{\partial t^2}\right). \tag{3.240}$$

So, we can write:

$$\int_{\Omega}^{GL} \lambda \frac{\partial^2 I_h^r u}{\partial t^2} \frac{\partial e_h^I}{\partial t}\, dx = \int_{\Omega}^{GL} \lambda\, I_h^r\left(\frac{\partial^2 u}{\partial t^2}\right) \frac{\partial e_h^I}{\partial t}\, dx = \int_{\Omega}^{GL} \lambda \frac{\partial^2 u}{\partial t^2} \frac{\partial e_h^I}{\partial t}\, dx \tag{3.241a}$$

$$\int_{\Omega} \lambda \frac{\partial^2 I_h^r u}{\partial t^2} \frac{\partial e_h^I}{\partial t} \, dx = \int_{\Omega} \lambda \, I_h^r \left(\frac{\partial^2 u}{\partial t^2} \right) \frac{\partial e_h^I}{\partial t} \, dx$$

$$= \int_{\Omega} \lambda \frac{\partial^2 u}{\partial t^2} \frac{\partial e_h^I}{\partial t} \, dx + \int_{\Omega} \lambda \left(I_h^r \left(\frac{\partial^2 u}{\partial t^2} \right) - \frac{\partial^2 u}{\partial t^2} \right) \frac{\partial e_h^I}{\partial t} \, dx \tag{3.241b}$$

We obtain (3.239) from (3.241a) and (3.241b) by using the Cauchy–Schwarz inequality in L^2.

Now, by using identities (3.238a)–(3.238c) and Lemma 2, (3.237) becomes:

$$\begin{aligned}
\frac{1}{2} \frac{d}{dt} \|e_h^I\|_h^2 \leq & \int_{\Omega} \lambda e_h^P \left(\frac{\partial^2 u}{\partial t^2} \right) \frac{\partial e_h^I}{\partial t} \, dx + \frac{d}{dt} \int_{\Omega} \underline{\underline{\mu}} \nabla e_h^P(u) \cdot \nabla e_h^I \, dx \\
& - \int_{\Omega} \underline{\underline{\mu}} \nabla e_h^P \left(\frac{\partial u}{\partial t} \right) \cdot \nabla e_h^I \, dx + \left[\int_{\Omega}^{GL} f \frac{\partial e_h^I}{\partial t} \, dx - \int_{\Omega} f \frac{\partial e_h^I}{\partial t} \, dx \right] \\
& + \left[\int_{\Omega} \lambda \frac{\partial^2 u}{\partial t^2} \frac{\partial e_h^I}{\partial t} \, dx - \int_{\Omega}^{GL} \lambda \frac{\partial^2 u}{\partial t^2} \frac{\partial e_h^I}{\partial t} \, dx \right] \\
& - \left[\int_{\Omega} \underline{\underline{\mu}} \nabla I_h^r \left(\frac{\partial u}{\partial t} \right) \cdot \nabla e_h^I \, dx - \int_{\Omega}^{GL} \underline{\underline{\mu}} \nabla I_h^r \left(\frac{\partial u}{\partial t} \right) \cdot \nabla e_h^I \, dx \right] \\
& + \frac{d}{dt} \left[\int_{\Omega} \underline{\underline{\mu}} \nabla I_h^r u \cdot \nabla e_h^I \, dx - \int_{\Omega}^{GL} \underline{\underline{\mu}} \nabla I_h^r u \cdot \nabla e_h^I \, dx \right] \\
& + C \left\| I_h^r \left(\frac{\partial^2 u}{\partial t^2} \right) - \frac{\partial^2 u}{\partial t^2} \right\|_0 \left\| \lambda^{1/2} \frac{\partial e_h^I}{\partial t} \right\|_0
\end{aligned} \tag{3.242}$$

where C is a positive constant independent of r and h.

The Cauchy–Schwarz inequality enables us to rewrite (3.242) as follows:

$$\begin{aligned}
\frac{1}{2} \frac{d}{dt} \|e_h^I\|_h^2 \leq & \left\| \lambda^{1/2} e_h^P \left(\frac{\partial^2 u}{\partial t^2} \right) \right\|_0 \left\| \lambda^{1/2} \frac{\partial e_h^I}{\partial t} \right\|_0 \\
& + \left\| \underline{\underline{\mu}}^{1/2} \nabla e_h^P \left(\frac{\partial u}{\partial t} \right) \right\|_0 \left\| \underline{\underline{\mu}}^{1/2} \nabla e_h^I \right\|_0 + E_I \left(f, \frac{\partial e_h^I}{\partial t} \right) \\
& + E_I \left(\lambda \frac{\partial^2 u}{\partial t^2}, \frac{\partial e_h^I}{\partial t} \right) + E_I \left(\underline{\underline{\mu}}^{1/2} \nabla I_h^r \left(\frac{\partial u}{\partial t} \right), \underline{\underline{\mu}}^{1/2} \nabla e_h^I \right) \\
& + \frac{d}{dt} \left[\int_{\Omega} \underline{\underline{\mu}} \nabla I_h^r u \cdot \nabla e_h^I \, dx - \int_{\Omega}^{GL} \underline{\underline{\mu}} \nabla I_h^r u \cdot \nabla e_h^I \, dx \right] \\
& + \frac{d}{dt} \int_{\Omega} \underline{\underline{\mu}} \nabla e_h^P(u) \cdot \nabla e_h^I \, dx + C \left\| I_h^r \left(\frac{\partial^2 u}{\partial t^2} \right) - \frac{\partial^2 u}{\partial t^2} \right\|_0 \left\| \lambda^{1/2} \frac{\partial e_h^I}{\partial t} \right\|_0,
\end{aligned} \tag{3.243}$$

where

$$E_I(g, h) = \left| \int_{\Omega}^{GL} g \, h \, dx - \int_{\Omega} g \, h \, dx \right|. \tag{3.244}$$

Integration over $[0, \zeta]$ yields $\forall \zeta \in \bar{J} = [0, T]$,

$$\begin{aligned}
\frac{1}{2}\left(\left\|e_h^I\right\|_h^2(\zeta) - \left\|e_h^I\right\|_h^2(0)\right) \leq \int_0^\zeta \Bigg(& \left\|\lambda^{1/2} e_h^P\left(\frac{\partial^2 u}{\partial t^2}\right)\right\|_0 \left\|\lambda^{1/2}\frac{\partial e_h^I}{\partial t}\right\|_0 \\
& + \left\|\underline{\underline{\mu}}^{1/2}\nabla e_h^P\left(\frac{\partial u}{\partial t}\right)\right\|_0 \left\|\underline{\underline{\mu}}^{1/2}\nabla e_h^I\right\|_0\Bigg) dt \\
& + \int_0^\zeta \Bigg[E_I\left(f, \frac{\partial e_h^I}{\partial t}\right) + E_I\left(\lambda\frac{\partial^2 u}{\partial t^2}, \frac{\partial e_h^I}{\partial t}\right) \\
& + E_I\left(\underline{\underline{\mu}}^{1/2}\nabla I_h^r\left(\frac{\partial u}{\partial t}\right), \underline{\underline{\mu}}^{1/2}\nabla e_h^I\right)\Bigg] dt \\
& + \left\|\underline{\underline{\mu}}^{1/2}\nabla e_h^P(u)\right\|_0^2(\zeta) \left\|\underline{\underline{\mu}}^{1/2}\nabla e_h^I\right\|_0^2(\zeta) \\
& + \left\|\underline{\underline{\mu}}^{1/2}\nabla e_h^P(u)\right\|_0^2(0) \left\|\underline{\underline{\mu}}^{1/2}\nabla e_h^I\right\|_0^2(0) \\
& + E_I\left(\underline{\underline{\mu}}^{1/2}\nabla I_h^r u, \underline{\underline{\mu}}^{1/2}\nabla e_h^I\right)(\zeta) + E_I\left(\underline{\underline{\mu}}^{1/2}\nabla I_h^r u, \underline{\underline{\mu}}^{1/2}\nabla e_h^I\right)(0) \\
& + C\int_0^\zeta \left\| I_h^r\left(\frac{\partial^2 u}{\partial t^2}\right) - \frac{\partial^2 u}{\partial t^2}\right\|_0 \left\|\lambda^{1/2}\frac{\partial e_h^I}{\partial t}\right\|_0 dt. \qquad (3.245)
\end{aligned}$$

The results of the previous section and, in particular, the results of Sect. 3.6.2, Proposition 2 and Corollary 3 lead to the following error estimates: for $0 < h \leq 1$,

$$\left\|\underline{\underline{\mu}}^{1/2}\nabla e_h^P\left(\frac{\partial u}{\partial t}\right)\right\|_0 \leq C\frac{h^r}{r^s}\left\|\frac{\partial u}{\partial t}\right\|_{L^\infty(J, H^{s+1}(\Omega))}, \qquad (3.246a)$$

$$E_I\left(f, \frac{\partial e_h^I}{\partial t}\right) \leq C\frac{h^{g^d(r)}}{r^{s'}}\|f\|_{L^\infty(J, H^{s'+1}(\Omega))}\left\|\lambda^{1/2}\frac{\partial e_h^I}{\partial t}\right\|_{0,\Omega}, \qquad (3.246b)$$

$$E_I\left(\lambda\frac{\partial^2 u}{\partial t^2}, \frac{\partial e_h^I}{\partial t}\right) \leq C\frac{h^{g^d(r)}}{r^s}\left\|\frac{\partial^2 u}{\partial t^2}\right\|_{L^\infty(J, H^{s+1}(\Omega))}\left\|\lambda^{1/2}\frac{\partial e_h^I}{\partial t}\right\|_{0,\Omega}, \qquad (3.246c)$$

$$\begin{aligned}
& E_I\left(\underline{\underline{\mu}}^{1/2}\nabla I_h^r\left(\frac{\partial u}{\partial t}\right), \underline{\underline{\mu}}^{1/2}\nabla e_h^I\right) \\
& \quad \leq C\frac{h^{g^d(r)}}{r^{s-1}}\left\|\frac{\partial u}{\partial t}\right\|_{L^\infty(J, H^{s+1}(\Omega))}\left\|\underline{\underline{\mu}}^{1/2}\nabla e_h^I\right\|_{0,\Omega}, \qquad (3.246d)
\end{aligned}$$

$$E_I\left(\underline{\underline{\mu}}^{1/2}\nabla I_h^r u, \underline{\underline{\mu}}^{1/2}\nabla e_h^I\right) \leq C\frac{h^{g^d(r)}}{r^{s-1}}\|u\|_{L^\infty(J, H^{s+1}(\Omega))}\left\|\underline{\underline{\mu}}^{1/2}\nabla e_h^I\right\|_{0,\Omega}, \quad (3.246e)$$

$$\left\| I_h^r\left(\frac{\partial^2 u}{\partial t^2}\right) - \frac{\partial^2 u}{\partial t^2}\right\|_0 \leq C\frac{h^{r+1}}{r^{s+1}}\left\|\frac{\partial^2 u}{\partial t^2}\right\|_{L^\infty(J,H^{s+1}(\Omega))} \tag{3.246f}$$

where $C > 0$ is a constant independent of $h = \max_{K\in\mathcal{T}_h} h_K$ and r, $g^d(x) = x - \min(x, d-1)$.

So, by choosing $u_h(x,0) = I_h^r u_0$ and $\frac{\partial u_h}{\partial t}(\cdot,0) = I_h^r v_0$ (ie $e_h^I(\cdot,0) = 0$ and $\frac{\partial e_h^I}{\partial t}(\cdot,0) = 0$), estimate (3.245) and the Proposition 1 yield: $\forall s \in \bar{J}$ and $0 < h \leq 1$,

$$\begin{aligned}\left\|e_h^I\right\|^2(\zeta) \leq CT&\left(\frac{h^{r+1}}{r^{s+1}}\left\|\frac{\partial^2 u}{\partial t^2}\right\|_{L^\infty(J,H^{s+1}(\Omega))} + \frac{h^{g^d(r)}}{r^{s'}}\|f\|_{L^\infty(J,H^{s'+1}(\Omega))}\right.\\ &\left.+\frac{h^{g^d(r)}}{r^s}\left\|\frac{\partial^2 u}{\partial t^2}\right\|_{L^\infty(J,H^{s+1}(\Omega))}\right)\left\|\lambda^{1/2}\frac{\partial e_h^I}{\partial t}\right\|_{L^\infty(J,L^2(\Omega))}\\ +CT&\frac{h^{g^d(r)}}{r^{s-1}}\left\|\frac{\partial u}{\partial t}\right\|_{L^\infty(J,H^{s+1}(\Omega))}\left\|\underline{\underline{\mu}}^{1/2}\nabla e_h^I\right\|_{L^\infty(J,L^2(\Omega))}\\ +C&\frac{h^{g^d(r)}}{r^{s-1}}\left(\|u\|_{L^\infty(J,H^{s+1}(\Omega))}\right.\\ &\left.+\left\|\frac{\partial u}{\partial t}\right\|_{L^\infty(J,H^{s+1}(\Omega))}\right)\left\|\underline{\underline{\mu}}^{1/2}\nabla e_h^I\right\|_{L^\infty(J,L^2(\Omega))}\end{aligned} \tag{3.247}$$

where C is a positive constant independent of h and r.

Finally, we have the following semi-discrete error estimates:

Theorem 9 (a priori error estimate with numerical integration) *Let* $r \in \mathbb{N}^*$ *and* $J =]0, T[$. *Assume that* $0 < h \leq 1$ *and the following regularity assumptions hold*

$$u,\ \frac{\partial u}{\partial t},\ \frac{\partial^2 u}{\partial t^2} \in L^\infty\left(J\ ,\ H^{s+1}(\Omega)\right),\ s \geq r+1, \tag{3.248a}$$

$$f \in L^\infty\left(J\ ,\ H^{s'+1}(\Omega)\right),\ s' \geq r. \tag{3.248b}$$

So, we have the a priori error estimate:

$$\sup_{\zeta\in J}\|e_h(\zeta)\| \leq \frac{h^{r-min(r,d-1)}}{r^{min(s-1,s')}}\left(C_1(u,f) + T\,C_2(u,f)\right) \tag{3.249}$$

where the positive constants $C_1(u, f)$ *and* $C_2(u, f)$ *depend on the norm in* $L^\infty(J, H^{s+1}\Omega))$ *or* $L^\infty(J, H^{s'+1}(\Omega))$ *of* u, $\frac{\partial u}{\partial t}$, $\frac{\partial^2 u}{\partial t^2}$, f *but are independent of* h *and* r.

Theorem 10 (a priori error estimate with exact integration) *Let* $r \in \mathbb{N}^*$ *and* $J =]0, T[$. *Assume that* $0 < h \leq 1$ *and the following regularity assumptions hold:*

$$u, \ \frac{\partial u}{\partial t}, \ \frac{\partial^2 u}{\partial t^2} \in L^\infty \left(J \, , \ H^{s+1}(\Omega)\right), \ s \geq r \tag{3.250}$$

So, we have the a priori error estimate:

$$\sup_{\zeta \in J} \|e_h(\zeta)\| \leq \frac{h^r}{r^s} \left(C_1(u) + T \, C_2(u)\right), \tag{3.251}$$

where the positive constants $C_1(u, f)$ *and* $C_2(u, f)$ *depend on the norm in* $L^\infty(J, H^{s+1}(\Omega))$ *or* $L^\infty(J, H^{s'+1}(\Omega))$ *of* $u, \ \dfrac{\partial u}{\partial t}, \ \dfrac{\partial^2 u}{\partial t^2}$ *but are independent of* h *and* r.

Remarks:

1. The a priori error estimates (3.249) and (3.251) are obtained by considering a general regular (see Definition 3.1) family of the unstructured meshes. The h-convergence can be improved on cartesian or almost cartesian meshes [14].
2. The estimates can be also enhanced by using a Gauss quadrature rule instead of the Gauss–Lobatto one to integrate the the rigidity term in the discrete weak formulation [14].

3.7 The Linear Elastodynamics System

This section addresses the linear elastodynamics system in its classical formulation and in a new form which enables to get a more efficient way of resolution.

3.7.1 Second Order Formulation

Let us recall the classical formulation with a free surface conditions, which reads as the second-order system:
Find $\underline{v} : \Omega \times [0, T] \rightarrow \mathbb{R}^d$ such that

$$\rho(\underline{x}) \frac{\partial^2 \underline{v}}{\partial t^2}(\underline{x}, t) - \mathbf{div} \underline{\underline{\tau}}(\underline{x}, t) = \underline{f}(\underline{x}, t) \text{ in } \Omega, \tag{3.252a}$$

$$\underline{\underline{\tau}}(\underline{x}, t) = \underline{\underline{C}}(\underline{x}) \underline{\underline{\varepsilon}}(\underline{v})(\underline{x}, t) \text{ in } \Omega, \tag{3.252b}$$

$$\underline{\tau}_k \cdot \underline{n} = 0, \quad k = 1 \ldots d, \text{ on } \partial\Omega, \tag{3.252c}$$

+ initial conditions, $\underline{n}$ denoting the exterior normal to $\partial\Omega$ and **div** defined by (1.26c).

After inserting (3.252b) into (3.252a), multiplying by $\underline{\varphi} \in \left[H^1(\Omega)\right]^d$ and integrating by part, we get the variational formulation:
Find $\underline{v} \in \left[H^1(\Omega)\right]^d$ such that

$$\frac{\mathrm{d}^2}{\mathrm{d}t^2}\int_\Omega \rho\,\underline{v}\cdot\underline{\varphi}\,\mathrm{d}\underline{x} + \int_\Omega \underline{\underline{C}}\,\underline{\underline{\varepsilon}}(\underline{v}) : \underline{\underline{\varepsilon}}(\underline{\varphi})\,\mathrm{d}\underline{x} = \int_\Omega \underline{f}\cdot\underline{\varphi}\,\mathrm{d}\underline{x},\ \forall\underline{\varphi}\in\left[H^1(\Omega)\right]^d, \tag{3.253}$$

where, for two d-dimensional tensors $\underline{\underline{u}} = (\underline{u}_1, \ldots, \underline{u}_d)^T$ with $\underline{u}_j = (u_{j,1}, \ldots, u_{j,d})^T$ and $\underline{\underline{v}} = (\underline{v}_1, \ldots, \underline{v}_d)^T$ with $\underline{v}_j = (v_{j,1}, \ldots, v_{j,d})^T$, we denote

$$\underline{\underline{u}} : \underline{\underline{v}} = \sum_{j=1}^{d} \underline{u}_j \cdot \underline{v}_j = \sum_{j=1}^{d}\sum_{k=1}^{d} u_{j,k} v_{j,k}. \tag{3.254}$$

Equation (3.253) provides the straightforward finite element approximation of (3.252a)–(3.252c):
Find $\underline{v}_h \in \left[U_h^r\right]^d$ such that

$$\frac{\mathrm{d}^2}{\mathrm{d}t^2}\int_\Omega \rho\,\underline{v}_h\cdot\underline{\varphi}_h\,\mathrm{d}\underline{x} + \int_\Omega \underline{\underline{C}}\,\underline{\underline{\varepsilon}}(\underline{v}_h) : \underline{\underline{\varepsilon}}(\underline{\varphi}_h)\,\mathrm{d}\underline{x} = \int_\Omega \underline{f}\cdot\underline{\varphi}_h\,\mathrm{d}\underline{x},\ \forall\underline{\varphi}_h\in\left[U_h^r\right]^d. \tag{3.255}$$

3.7.2 First-Order Formulation

3.7.2.1 Reformulation of the Continuous Problem

We have

$$\begin{aligned}\tau_{ik} &= \frac{1}{2}\sum_{j=1}^{d}\sum_{l=1}^{d} c_{ikjl}\left(\frac{\partial v_j}{\partial x_l} + \frac{\partial v_l}{\partial x_j}\right)\\ &= \frac{1}{2}\sum_{j=1}^{d}\sum_{l=1}^{d} c_{iklj}\frac{\partial v_j}{\partial x_l} + \frac{1}{2}\sum_{j=1}^{d}\sum_{l=1}^{d} c_{ikjl}\frac{\partial v_j}{\partial x_l}.\end{aligned}$$

Since $c_{ikjl} = c_{iklj}$, we finally get:

$$\tau_{ik} = \sum_{j=1}^{d}\sum_{l=1}^{d} c_{ikjl}\frac{\partial v_j}{\partial x_l}. \tag{3.256}$$

Using (3.256), we get:

$$\nabla \cdot \underline{\tau}]_i = \sum_{k=1}^{d} \frac{\partial \tau_{ik}}{\partial x_k} = \sum_{k=1}^{d} \frac{\partial}{\partial x_k} \left(\sum_{j=1}^{d} \sum_{l=1}^{d} c_{ikjl} \frac{\partial v_j}{\partial x_l} \right)$$

$$= \sum_{j=1}^{d} \left[\sum_{k=1}^{d} \frac{\partial}{\partial x_k} \left(\sum_{l=1}^{d} c_{ikjl} \frac{\partial v_j}{\partial x_l} \right) \right],$$

which can be rewritten as

$$\nabla \cdot \underline{\tau_i} = \sum_{j=1}^{d} \nabla \cdot [A_{ij} \nabla v_j], \tag{3.257}$$

where $A_{ij} = \begin{pmatrix} c_{i1j1} & c_{i1j2} \\ c_{i2j1} & c_{i2j2} \end{pmatrix}$ for $d = 2$ and $A_{ij} = \begin{pmatrix} c_{i1j1} & c_{i1j2} & c_{i1j3} \\ c_{i2j1} & c_{i2j2} & c_{i2j3} \\ c_{i3j1} & c_{i3j2} & c_{i3j3} \end{pmatrix}$ for $d = 3$.

Now, let us introduce the variables $\underline{\gamma}_{ij} = (\gamma_{ij}^{\ell})_{\ell=1\ldots d} \in \mathbb{R}^d$ and $\underline{\gamma}_i = (\gamma_i^{\ell})_{\ell=1\ldots d} \in \mathbb{R}^d$ such that

$$\frac{\partial \underline{\gamma}_i}{\partial t} = \nabla v_j, \ \forall i = 1 \ldots d \text{ and } \underline{\gamma}_{ij} = A_{ij} \underline{\gamma}_i, \ \forall i, j = 1 \ldots d.$$

We obviously have:

$$\tau_{ij} = \sum_{k=1}^{d} \frac{\partial \underline{\gamma}_{ik}^{j}}{\partial t}. \tag{3.258}$$

With these notations, the free surface condition can be rewritten as follows:

$$\underline{\underline{\tau}} \underline{n} = 0 \Longleftarrow \sum_{j=1}^{d} \underline{\gamma}_{ij} \cdot \underline{n} = 0 \ \forall i = 1 \ldots d. \tag{3.259}$$

Remarks:

1. The symmetry properties of $\underline{\underline{C}}$ show that $A_{ij} = A_{ij}^T, \ \forall i, j = 1 \ldots d$.
2. The definition of $\underline{\gamma}_{ij}$ and (3.258) imply that symmetry of $\underline{\underline{\tau}}$ is kept.

We can now reformulate the system given by (3.252a)–(3.252c) as the following first order system:

Find $\underline{v} : \Omega \times [0, T] \to \mathbb{R}^d$, $\underline{\gamma}_{ij} : \Omega \times [0, T] \to \mathbb{R}^d, \ i, j = 1 \ldots d$ and $\underline{\gamma}_i : \Omega \times [0, T] \to \mathbb{R}^d, \ i = 1 \ldots d$ such that

$$\rho(\underline{x})\frac{\partial v_i}{\partial t}(\underline{x},t) - \sum_{j=1}^{d} \nabla \cdot \underline{\gamma}_{ij}(\underline{x},t) = F_i(\underline{x},t), \quad \forall i = 1\ldots d, \text{ in } \Omega \times [0,T], \tag{3.260a}$$

$$\frac{\partial \underline{\gamma}_i}{\partial t}(\underline{x},t) = \nabla v_i(\underline{x},t), \quad \forall i = 1\ldots d, \text{ in } \Omega \times [0,T], \tag{3.260b}$$

$$\underline{\gamma}_{ij}(\underline{x},t) = A_{ij}(\underline{x})\underline{\gamma}_i(\underline{x},t), \quad \forall i,j = 1\ldots d, \text{ in } \Omega \times [0,T], \tag{3.260c}$$

$$\sum_{j=1}^{d} \underline{\gamma}_{ij}(\underline{x},t) \cdot \underline{n} = 0, \quad \forall i = 1\ldots d, \text{ on } \partial\Omega, \tag{3.260d}$$

+ initial conditions, where

$$\frac{\partial F_i(\underline{x},t)}{\partial t} = f_i(\underline{x},t), \text{ with } \underline{f} = (f_i)_{i=1\ldots d}\,.$$

3.7.2.2 Variational and Approximate Formulations

For the variational formulation, we set $\underline{v}(.,t) \in \left[H^1(\Omega)\right]^d$, $\underline{\gamma}_i(.,t) \in \left[L^2(\Omega)\right]^d$ and $\underline{\gamma}_{ij}(.,t) \in \left[L^2(\Omega)\right]^d$. After integrating by part and taking (3.260d) into account, this choice leads to the following formulation:
Find $\underline{v}(.,t) \in \left[H^1(\Omega)\right]^d$, $\underline{\gamma}_i(.,t) \in \left[L^2(\Omega)\right]^d$ and $\underline{\gamma}_{ij}(.,t) \in \left[L^2(\Omega)\right]^d$ such that

$$\frac{\mathrm{d}}{\mathrm{d}t}\int_\Omega \rho\, v_i\, \varphi\, \mathrm{d}\underline{x} + \sum_{j=1}^{d}\int_\Omega \underline{\gamma}_{ij} \cdot \nabla\varphi\, \mathrm{d}\underline{x} = \int_\Omega F_i\, \varphi\, \mathrm{d}\underline{x}, \ \forall \varphi \in H^1(\Omega), \ \forall i = 1\ldots d, \tag{3.261a}$$

$$\frac{\mathrm{d}}{\mathrm{d}t}\int_\Omega \underline{\gamma}_i \cdot \underline{\psi}_1 \mathrm{d}\underline{x} = \int_\Omega \nabla v_i \cdot \underline{\psi}_1 \mathrm{d}\underline{x}, \quad \forall \underline{\psi}_1 \in \left[L^2(\Omega)\right]^d, \ \forall i = 1\ldots d, \tag{3.261b}$$

$$\int_\Omega \underline{\gamma}_{ij} \cdot \underline{\psi}_2 \mathrm{d}\underline{x} = \int_\Omega A_{ij}\underline{\gamma}_i \cdot \underline{\psi}_2 \mathrm{d}\underline{x}, \quad \forall \underline{\psi}_2 \in \left[L^2(\Omega)\right]^d, \ \forall i,j = 1\ldots d, \tag{3.261c}$$

+ initial conditions.

As for acoustics and for the same reasons, the approximate problem reads:
Find $\underline{v}_h(.,t) \in \left[U_h^r\right]^d$, $\underline{\gamma}_{ih}(.,t) \in \underline{V}_h^r$ and $\underline{\gamma}_{ijh}(.,t) \in \underline{V}_h^r$ such that

$$\frac{\mathrm{d}}{\mathrm{d}t}\int_\Omega \rho\, v_{ih}\, \varphi_h\, \mathrm{d}\underline{x} + \sum_{j=1}^{d}\int_\Omega \underline{\gamma}_{ijh} \cdot \nabla\varphi_h\, \mathrm{d}\underline{x} = \int_\Omega F_i\, \varphi_h\, \mathrm{d}\underline{x}, \ \forall \varphi_h \in U_h^r, \ \forall i = 1\ldots d, \tag{3.262a}$$

$$\frac{\mathrm{d}}{\mathrm{d}t}\int_{\Omega}\underline{\gamma}_{ih}\cdot\underline{\psi}_{1h}\mathrm{d}\underline{x} = \int_{\Omega}\nabla v_{ih}\cdot\underline{\psi}_{1h}\mathrm{d}\underline{x}, \quad \forall\underline{\psi}_{1h}\in\underline{V}_h^r, \ \forall i = 1\ldots d, \tag{3.262b}$$

$$\int_{\Omega}\underline{\gamma}_{ijh}\cdot\underline{\psi}_{2h}\mathrm{d}\underline{x} = \int_{\Omega}A_{ijh}\underline{\gamma}_{ih}\cdot\underline{\psi}_{2h}\mathrm{d}\underline{x}, \quad \forall\underline{\psi}_{2h}\in\underline{V}_h^r, \ \forall i,j = 1\ldots d, \tag{3.262c}$$

+ initial conditions.

3.7.3 Comparison of the Two Approaches

3.7.3.1 Equivalence Theorem

As we said in Sect. 3.3, we have an equivalence theorem for the approximations defined by (3.255) and by (3.262a)–(3.262c). In this section, we provide this theorem without proof (which can be found in [4]). A first step is the following remark: with the above notations, Eq. (3.252a) can be written as

$$\rho(\underline{x})\frac{\partial^2 v_i}{\partial t^2}(\underline{x},t) - \sum_{j=1}^{d}\nabla\cdot(A_{ij}\nabla v_j)(\underline{x},t) = f_i(\underline{x},t), \quad \forall i = 1\ldots d. \tag{3.263}$$

A finite element approximation of (3.263) in U_h^r leads to the matrix problem:

$$\frac{\mathrm{d}^2}{\mathrm{d}^2 t}D_h\widetilde{V}_{ih} + \sum_{j=1}^{d}K_{ijh}\widetilde{V}_{jh} = F'_{ih}, \quad \forall i = 1\ldots d. \tag{3.264}$$

On the other hand, the matrix formulation of (3.262a)–(3.262c) reads:

$$\frac{\mathrm{d}}{\mathrm{d}t}D_h V_{ih} + R_h\sum_{j=1}^{d}\Gamma_{ijh} = F_{ih}, \quad \forall i = 1\ldots d, \tag{3.265a}$$

$$\frac{\mathrm{d}}{\mathrm{d}t}B_h\,\Gamma_{ih} - R_h^T\,V_{ih} = 0, \quad \forall i = 1\ldots d, \tag{3.265b}$$

$$B_h\,\Gamma_{ijh} - A_{ijh}\,\Gamma_{ih} = 0, \quad \forall i,j = 1\ldots d. \tag{3.265c}$$

By combining (3.265b) and (3.265c) and setting $B_{ijh} = A_{ijh}\,B_h^{-1}$, we get:

$$\frac{\mathrm{d}}{\mathrm{d}t}D_h V_{ih} + R_h\sum_{j=1}^{d}\Gamma_{ijh} = F_{ih}, \quad \forall i = 1\ldots d, \tag{3.266a}$$

$$\frac{\mathrm{d}}{\mathrm{d}t}B_h\,\Gamma_{ijh} - B_{ijh}\,R_h^T\,V_{ih} = 0, \quad \forall i,j = 1\ldots d. \tag{3.266b}$$

D_h, B_h and R_h are defined as in (3.45) and (3.46a) and (3.46b). On the other hand, $\widetilde{V}_{ih} \in \mathbb{R}^{n_d} \times [0, T]$, $V_{ih} \in \mathbb{R}^{n_d} \times [0, T]$, $F_{ih} \in \mathbb{R}^{n_d} \times [0, T]$ (F'_{ih} being its derivative in time), n_d defined as in (3.7), $\Gamma_{ih} \in \mathbb{R}^{3N_e(r+1)^d} \times [0, T]$ and $\Gamma_{ijh} \in \mathbb{R}^{3N_e(r+1)^d} \times [0, T]$. K_{ijh} is the stiffness matrix derived from (3.263) and A_{ijh} is a $d \times d$ block-diagonal matrix such that $A_{jih} = A^T_{ijh}$.

We can now state the equivalence theorem:

Theorem 11 *Problems (3.264) and (3.266a) and (3.266b) are equivalent, i.e.*

$$\widetilde{V}_{ih} = V_{ih}.$$

Moreover, we have

$$K_{ijh} = R_h\, B_{ijh}\, R_h^T = R_h\, B_h^{-1}\, A_{ijh}\, B_h^{-1}\, R_h^T, \quad \forall i = 1 \ldots d. \tag{3.267}$$

Remark: B_h being symmetric, (3.267) shows that $K_{jih} = K^T_{ijh}$.

3.7.3.2 Three Algorithms of Resolution

So, (3.264) and (3.266a) and (3.266b) provide two different ways to compute V_{ih}. Actually, as for acoustics, (3.266a) and (3.266b) can be written as

$$\frac{\mathrm{d}^2}{\mathrm{d}t^2} D_h V_{ih} + R_h \sum_{j=1}^{d} \Gamma_{ijh} = F_{ih}, \quad \forall i = 1 \ldots d, \tag{3.268a}$$

$$B_h\, \Gamma_{ijh} - B_{ijh}\, R_h^T\, V_{ih} = 0, \quad \forall i, j = 1 \ldots d, \tag{3.268b}$$

which provides a third way to compute V_{ih}.

As for acoustics, we discretize these three systems by using a leapfrog scheme in time. Obviously, for this discretization, (3.268a) and (3.268b) require a lower storage than (3.266a) and (3.266b) since Γ_{ijh} can be used as local variables on each element. A priori, (3.264) seems to be the most expensive in terms of storage, since it requires the storage of K_{jih}. However, Maday et al. [19] proposed a low storage algorithm which consists into a reconstruction of K_{jih} at each time-step by using the Jacobian J_ℓ, the Jacobian matrix DF_ℓ and the tensor $\underline{\underline{C}}$ (which is generally constant or defined by zone (in layered media) and requires a negligible storage, whereas B_{ijh} must be stored (taking into account its symmetry properties) at each point of interpolation of each cell).

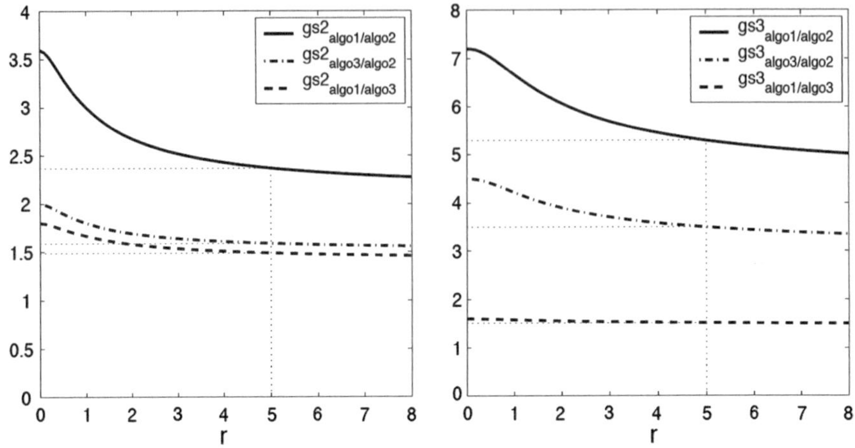

Fig. 3.18 Comparison of the storage required by the three algorithms in 2D (*left*) and 3D (*right*). For $r = 5$, *Algo2* requires about 3.5 times less storage than *Algo3* and 5 times less storage than *Algo1* in 3D

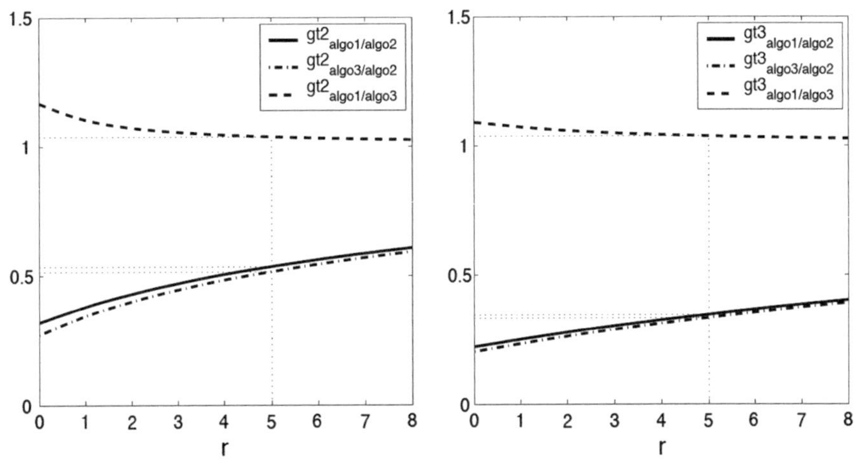

Fig. 3.19 Comparison of the computational time required by the three algorithms in 2D (*left*) and 3D (*right*). For $r = 5$, *Algo2* requires about 3.5 times more time than *Algo3* and *Algo1* in 3D. In practice, this difference seems to be even larger

Let us denote *Algo1* the algorithm derived from (3.266a) and (3.266b), *Algo2* this derived from (3.264) and *Algo3* this derived from (3.268a) and (3.268b). In Figs. 3.18 and 3.19, we give the comparison of these three algorithms in terms of storage and computational time.

References

1. Raviart, P.A., Thomas, J.M.: A mixed finite element method for 2nd order elliptic problems. In: Dold, A., Eckmann, B. (eds.) Mathematical Aspects of Finite Element Methods. Lecture Notes in Mathematics, vol. 606. Springer, New York (1977)
2. Cohen, G.: High Order Numerical Methods for Transient Wave Equations. Scientific Computation. Springer, New York (2001)
3. Cohen, G., Fauqueux, S.: Mixed finite elements with mass-lumping for the transient wave equation. J. Comput. Acoust. **8**(1), 171–188 (2000)
4. Cohen, G., Fauqueux, S.: Mixed spectral finite elements for the linear elasticity system in unbounded domains. SIAM J. Sci. Comput. **26**(3), 864–884 (2005)
5. Trefethen, L.N.: Group velocity in finite difference schemes. SIAM Rev. **24**(2), 113–136 (1982)
6. Tordjman, N.: Eléments finis d'ordre élevé avec condensation de masse pour l'équation des ondes, thèse de doctorat, U. Paris IX-Dauphine (1995)
7. Fauqueux, S.: Modélisation de la propagation d'ondes en milieu élastique par éléments finis mixtes avec condensation de masse, thèse de doctorat, U. de Paris IX-Dauphine (2003)
8. Keller, J.B., Odeh, F.: Partial differential equations with periodic coefficients and Bloch waves in crystals. J. Math. Phys. **5**, 1499–1504 (1964)
9. Ainsworth, M.: Discrete dispersion relation for hp-version finite element approximation at high wave number. SIAM J. Numer. Anal. **42**(2), 553–575 (2004)
10. Ainsworth, M.: Dispersive and dissipative behaviour of high order discontinuous Galerkin finite element methods. J. Comput. Phys. **198**(1), 106–130 (2004)
11. Ainsworth, M.: Dispersive properties of high order Nédélec/edge element approximation of the time-harmonic Maxwell equations. Philos. Trans. R. Soc. Ser. A **362**(1816), 471–491 (2004)
12. Moczo, P., Lucka, M., Kristek, J., Kristekova, M.: 3D displacement finite differences and a combined memory optimization. Bull. Seismol. Soc. Am. **89**(1), 69–79 (1999)
13. Zahradnik, J., O'Leary, P., Sochacki, J.: Finite-difference schemes for elastic waves based on the integration approach. Geophysics **59**(6), 928–937 (1994)
14. Duruflé, M., Grob, P.: Joly, influence of Gauss and Gauss-Lobatto quadrature rules on the accuracy of a quadrilateral finite element method in the time domain. Numer. Methods Partial Differ. Equ. **25**(3), 526–551 (2009)
15. Ciarlet, P.G.: The Finite Element Method for Elliptic Problems. North-Holland, Amsterdam (2002)
16. Girault, V., Raviart, P.-A.: Finite Element Methods for Navier-Stokes Equations. Sringer, New York (1986)
17. Bernardi, C., Maday, Y.: Approximations spectrales de problèmes aux limites elliptiques. Mathématiques et Applications, vol. 10. SMAI, Springer, Paris (1992)
18. Bergot, M.: Eléments finis d'ordre élevé pour maillages hybrides. Application à la résolution de systèmes hyperboliques linéaires en régimes harmonique et temporel, thèse de doctorat, U. de Paris-Dauphine (Paris IX) (2010)
19. Maday, Y., Ronquist, E., Patera, A.: Optimal error analysis of spectral methods with emphasis on non-constant coefficients and deformed geometries. Comput. Methods Appl. Mech. Eng. **80**(1–3), 91–115 (1990)

Chapter 4
Discontinuous Galerkin Methods

Abstract This chapter is devoted to the construction, based on discrete energy control, of discontinuous Galerkin methods (DGM) which are well-adapted to the solution of wave problems. In a first part, these methods are described by using an abstract framework for first-order linear hyperbolic problems which covers, in particular, all the transient wave equations considered in the first chapter of this book. In a next part, an explicit description of DGM approximations by triangles, tetrahedra, quadrilaterals and hexahedra is presented in details. A comparison between the different DGM approaches is then proposed for Maxwell's equations. The penultimate part deals with analyzing the numerical dispersion and dissipation introduced by the DGM approximation by using a plane wave analysis. Finally, some DGM formulations adapted to the second-order wave equations and some of their properties are given.

Discontinuous Galerkin methods (DGM) can be seen as an extension of finite volume methods and were first introduced by Reed and Hill [1] in 1973 for neutron transport. Their first mathematical analysis was made by Lesaint and Raviart [2] in 1974. This method was then used and analyzed for non-linear problems in a first step, and applied to linear hyperbolic problems in a second step. Numerous papers were devoted to this method. A wide description can be found in [3] and a presentation of the method for linear hyperbolic problems as well as an update of the bibliography in [4].

4.1 General Formulation for Linear Hyperbolic Problems

4.1.1 The Discontinuous Galerkin Formulation

4.1.1.1 A Preliminary Result

Let A be a linear first-order differential operator such that $D(A) \subset \left[L^2(\Omega)\right]^q$ and $Im(A) \subset \left[L^2(\Omega)\right]^p$, Ω being an open set of $\mathbb{R}^d$ and A^* its adjoint operator. One can decompose A as follows:

G. Cohen and S. Pernet, *Finite Element and Discontinuous Galerkin Methods for Transient Wave Equations*, Scientific Computation,
DOI 10.1007/978-94-017-7761-2_4

$$A = \sum_{i=1}^{d} A_i \frac{\partial}{\partial x_i}, \tag{4.1}$$

where A_i is a $p \times q$ real matrix.

Let us take $\underline{\psi} \in \mathscr{D}(\Omega)^q$ and $\underline{\psi}' \in \mathscr{D}(\Omega)^p$ (where $\mathscr{D}(\Omega)$ is the space of functions which are C^∞ and compactly supported in Ω). By integrating by parts in each direction we have:

$$\begin{aligned}\int_\Omega A\underline{\psi} \cdot \underline{\psi}' \, \mathrm{d}\underline{x} &= \sum_{i=1}^{d} \int_\Omega A_i \frac{\partial \underline{\psi}}{\partial x_i} \cdot \underline{\psi}' \, \mathrm{d}\underline{x} = -\sum_{i=1}^{d} \int_\Omega \underline{\psi} \cdot \frac{\partial (A_i^T \underline{\psi}')}{\partial x_i} \, \mathrm{d}\underline{x} \\ &= -\int_\Omega \underline{\psi} \cdot \sum_{i=1}^{d} \left(\frac{\partial A_i^T}{\partial x_i} \underline{\psi}' + A_i^T \frac{\partial \underline{\psi}'}{\partial x_i} \right) \mathrm{d}\underline{x} = \int_\Omega \underline{\psi} \cdot A^* \underline{\psi}' \, \mathrm{d}\underline{x},\end{aligned} \tag{4.2}$$

where A_i^T is the transposed matrix of A_i.

In particular, when A_i does not depend on space, we get:

$$A^* = -\sum_{i=1}^{d} A_i^T \frac{\partial}{\partial x_i}. \tag{4.3}$$

4.1.1.2 The Continuous Problem

Let us formulate the discontinuous Galerkin method for the formal linear hyperbolic problem on Ω:

$$\underline{\underline{\lambda}}(\underline{x}) \frac{\partial \underline{u}}{\partial t}(\underline{x}, t) + A\underline{v}(\underline{x}, t) = \underline{f}(\underline{x}, t) \text{ in } \Omega, \tag{4.4a}$$

$$\underline{\underline{\mu}}(\underline{x}) \frac{\partial \underline{v}}{\partial t}(\underline{x}, t) - A^* \underline{u}(\underline{x}, t) = 0 \text{ in } \Omega, \tag{4.4b}$$

$$A^*(n)\, \underline{u}(\underline{x}, t) = 0 \text{ on } \partial\Omega, \tag{4.4c}$$

$$\underline{u}(\underline{x}, 0) = \underline{u}_0(\underline{x}), \;\; \underline{v}(\underline{x}, 0) = \underline{v}_0(\underline{x}) \text{ in } \Omega, \tag{4.4d}$$

where $\underline{u}(\underline{x}, t) \in \mathbb{R}^p$, $\underline{v}(\underline{x}, t) \in \mathbb{R}^q$, $\underline{f}(\underline{x}, t) \in \mathbb{R}^p$, $\underline{\underline{\lambda}}(\underline{x}) \in \mathbb{R}^p \times \mathbb{R}^p$, $\underline{\underline{\mu}}(\underline{x}) \in \mathbb{R}^q \times \mathbb{R}^q$ are symmetric, definite, positive matrices, $\underline{n} = (n_i)_{i=1}^d$ is the exterior normal to Ω, $A(\underline{n}) = \sum_{i=1}^{d} A_i n_i$, $A^*(\underline{n}) = -\sum_{i=1}^{d} A_i^T n_i$ and A does not depend on space.

4.1.1.3 Functional Framework and Notations

We can rewrite (4.4) in the Hille–Yosida framework defined in Chap. 1 i.e.

$$\frac{\partial \underline{w}}{\partial t} + B\,\underline{w} = F \tag{4.5}$$

where B is a maximal monotone operator defined by

$$B = \begin{bmatrix} 0 & \underline{\underline{\lambda}}^{-1}\,A \\ -\underline{\underline{\mu}}^{-1}\,A^* & 0 \end{bmatrix}$$

and the domain of the operator B is $D(B) = H_0(A^*,\,\Omega) \times H(A,\,\Omega) \subset H = L^2(\Omega)^{p+q}$ with

$$H(A,\,\Omega) = \{\underline{w} \in L^2(\Omega)^q\ :\ A\,\underline{w} \in L^2(\Omega)^p\},$$

$$H_0(A^*,\,\Omega) = \{\underline{w} \in L^2(\Omega)^p\ :\ A^*\,\underline{w} \in L^2(\Omega)^q \text{ and } A^*(\underline{n})\underline{w} = 0 \text{ on } \Gamma\}.$$

Moreover, we assume that the operator $I + B$ is surjective onto H and if $(\underline{u}_0, \underline{v}_0) \in D(B)$, we obtain that the unique solution of (4.4) has the following regularity properties:

$$\underline{u} \in C^1([0,T], L^2(\Omega)^p) \cap C^0([0,T], H_0(A^*,\,\Omega)), \tag{4.7a}$$

$$\underline{v} \in C^1([0,T], L^2(\Omega)^q) \cap C^0([0,T], H(A,\,\Omega)). \tag{4.7b}$$

Remarks:

1. The graph spaces $H(A,\,\Omega)$ and $H(A^*,\,\Omega)$ are Hilbert spaces for the scalar products
$$(\underline{v},\,\underline{w})_A = (\underline{v},\,\underline{w})_0 + (A\underline{v},\,A\underline{w})_0,$$
$$(\underline{v},\,\underline{w})_{A^*} = (\underline{v},\,\underline{w})_0 + (A^*\underline{v},\,A^*\underline{w})_0.$$
2. The trace operators
$$A(\underline{n})\ :\ H(A,\,\Omega) \to [H^{-1/2}(\partial\Omega)]^p$$
$$\underline{w} \mapsto A(\underline{n})\underline{w},$$
$$A^*(\underline{n})\ :\ H(A^*,\,\Omega) \to [H^{-1/2}(\partial\Omega)]^q$$
$$\underline{w} \mapsto A^*(\underline{n})\underline{w},$$
are linear continuous but not necessarily surjective [5].
3. We have the integration by parts formula: for all $\underline{v} \in H(A^*,\,\Omega)$ and $\underline{w} \in H^1(\Omega)^q$,
$$< A^*(\underline{n})\underline{v},\,\underline{w} >= \int_\Omega A^*\underline{v}\cdot\underline{w}\,\mathrm{d}\underline{x} - \int_\Omega \underline{v}\cdot A\underline{w}\,\mathrm{d}\underline{x},$$

where $<,>$ denotes the duality bracket between $H(A^*, \Omega)$ and $[H^{-1/2}(\partial\Omega)]^q$. This can be extended to $H(A^*, \Omega) \times H(A, \Omega)$.

Let

$$\mathscr{T}_h = \bigcup_{\ell=1}^{N_e} K_\ell, \tag{4.10}$$

be a mesh of an open (polyhedral) set Ω, where K_ℓ is an element of any shape.

The principle of DGM is to search for a solution on this mesh which has no continuity requirements on the faces of the mesh. So, basically, the solution should be sought in L^2. However, the lack of continuity across the faces is compensated by the introduction of jumps of the traces of the unknowns on the boundary. For sake of simplicity, we shall assume an extra regularity (see first remark below) of the exact solution of (4.4) in order to ensure that the traces $A^*(\underline{n}_\Gamma)\underline{u}$ and $A(\underline{n}_\Gamma)\underline{v}$ belong to L^2 for all the faces Γ of the mesh $\mathscr{T}_h$, $\underline{n}_\Gamma$ being an unit normal vector to Γ. For this purpose, we define the following space:

$$H^s(\mathscr{T}_h) = \left\{ u \in L^2(\Omega) \text{ such that, } u_{|K_\ell} \in H^s(K_\ell) \right\}, \tag{4.11}$$

where $H^s(K_\ell)$ is the Sobolev space of order $s > \frac{1}{2}$. $H^s(\mathscr{T}_h)$ is equipped with the norm

$$||u||_{s,h} = \left(\sum_{K_\ell \in \mathscr{T}_h} ||u||^2_{s,\ell} \right)^{\frac{1}{2}}, \tag{4.12}$$

where $||u||_{s,\ell}$ is the usual norm of $H^s(K_\ell)$.

For a n-dimensional vector-valued function $\underline{v} \in \underline{H}^s_n(\mathscr{T}_h) = [H^s(\mathscr{T}_h)]^n$, we define the norm:

$$||\underline{v}||_{s,h} = \left(\sum_{i=1}^{n} ||v_i||^2_{s,h} \right)^{\frac{1}{2}}. \tag{4.13}$$

So, we assume that $\exists s > \dfrac{1}{2}$,

$$\underline{u}(\cdot, t) \in H_0(A^*, \Omega) \cap \underline{H}^s_p(\mathscr{T}_h), \ \forall t, \tag{4.14a}$$

$$\underline{v}(\cdot, t) \in H(A, \Omega) \cap \underline{H}^s_q(\mathscr{T}_h), \ \forall t. \tag{4.14b}$$

Now, let us call $\mathscr{F}_I$ and $\mathscr{F}_B$ the sets of interior and boundary faces (edges in 2D) of the mesh $\mathscr{T}_h$. On a face $\Gamma_{\ell,m} = K_\ell \cap K_m \in \mathscr{F}_I$, we denote:

$$[\underline{v}]^{K_\ell}_{\Gamma_{\ell,m}} = (\underline{v}_{|K_m})_{|\Gamma_{\ell,m}} - (\underline{v}_{|K_\ell})_{|\Gamma_{\ell,m}} \tag{4.15}$$

the jump of $\underline{v}$ across $\Gamma_{\ell,m}$, $(\underline{v}_{|K_\ell})_{|\Gamma_{\ell,m}}$ being the trace of $\underline{v}_{|K_\ell}$ on $\Gamma_{\ell,m}$ (which exists in our functional framework).

Moreover, on a face $\Gamma = \partial K \cap \partial \Omega \in \mathscr{F}_B$, we denote:

$$[\underline{v}]_{\Gamma}^{K} = -(\underline{v}_{|K})_{|\Gamma}. \tag{4.16}$$

Remarks:

1. Actually, in practice, $\underline{v}_{|K_\ell}$ belongs to a polynomial space[1] and is locally much more regular than functions of $\underline{H}^s(\mathscr{T}_h)$. However, this space is convenient for error estimates.
2. DGM enable us the use of non-conforming meshes (one must however avoid hanging nodes [6]). So $\Gamma_{\ell,q}$ can be a subset of a face of K_ℓ or K_q.

4.1.1.4 DGM Formulation

With the above notations, we introduce the following problem:

Find $\underline{u}(.,t) \in U_p$, $\underline{v}(.,t) \in U_q$ such that $\forall K_\ell \in \mathscr{T}_h$ and $\forall \underline{\varphi} \in U_p$, $\forall \underline{\psi} \in U_q$,

$$\begin{aligned}\frac{\mathrm{d}}{\mathrm{d}t}\int_{K_\ell} \underline{\underline{\lambda}}\underline{u}_\ell \cdot \underline{\varphi}_\ell \,\mathrm{d}\underline{x} = &-\int_{K_\ell} A\underline{v}_\ell \cdot \underline{\varphi}_\ell \,\mathrm{d}\underline{x} + \int_{\partial K_\ell} \alpha\, A(\underline{n})[\underline{v}]_{\partial K_\ell}^{K_\ell} \cdot \underline{\varphi}_\ell \,\mathrm{d}\sigma \\ &+ \int_{\partial K_\ell} \gamma\, C[\underline{u}]_{\partial K_\ell}^{K_\ell} \cdot \underline{\varphi}_\ell \,\mathrm{d}\sigma + \int_{K_\ell} \underline{f} \cdot \underline{\varphi}_\ell \,\mathrm{d}\underline{x},\end{aligned} \tag{4.17a}$$

$$\begin{aligned}\frac{\mathrm{d}}{\mathrm{d}t}\int_{K_\ell} \underline{\underline{\mu}}\underline{v}_\ell \cdot \underline{\psi}_\ell \,\mathrm{d}\underline{x} = &\int_{K_\ell} A^*\underline{u}_\ell \cdot \underline{\psi}_\ell \,\mathrm{d}\underline{x} + \int_{\partial K_\ell} \beta\, A^*(\underline{n})[\underline{u}]_{\partial K_\ell}^{K_\ell} \cdot \underline{\psi}_\ell \,\mathrm{d}\sigma \\ &+ \int_{\partial K_\ell} \delta\, C'[\underline{v}]_{\partial K_\ell}^{K_\ell} \cdot \underline{\psi}_\ell \,\mathrm{d}\sigma,\end{aligned} \tag{4.17b}$$

$$\underline{u}(\underline{x},0) = \underline{u}_0(\underline{x}),\ \ \underline{v}(\underline{x},0) = \underline{v}_0(\underline{x}) \text{ in } \Omega, \tag{4.17c}$$

where $\underline{u}_\ell = \underline{u}_{|K_\ell}$, $\underline{v}_\ell = \underline{v}_{|K_\ell}$, $\underline{\varphi}_\ell = \underline{\varphi}_{|K_\ell}$, $\underline{\psi}_\ell = \underline{\psi}_{|K_\ell}$. Moreover, $C = A(\underline{n})\, A^*(\underline{n})$ and $C' = A^*(\underline{n})\, A(\underline{n})$ are symmetric positive matrices, α, β, γ, δ are real piecewise constants on each face of $\mathscr{T}_h$ and U_p and U_q are functional spaces to be defined.

Remark: In (4.17a) and (4.17b), the integrals containing C and C' can be viewed as penalization terms which impose the transmission conditions $A(\underline{n})[\underline{v}]_{\partial K_\ell}^{K_\ell} = 0$ and $A^*(\underline{n})[\underline{u}]_{\partial K_\ell}^{K_\ell} = 0$ in a least-square sense. In particular, this term are consistent with the exact solution since they are null for this solution.

Before defining the approximate problem, we first prove that formulation (4.17a)–(4.17c) is consistent with the continuous problem. Actually, we have

[1] It can also be a rational function if K_ℓ is a pyramid.

Proposition 3 *For $U_p = \underline{H}^s_p(\mathscr{T}_h)$, $U_q = \underline{H}^s_q(\mathscr{T}_h)$ and if α, β, γ, δ satisfy*

1. γ, $\beta \neq 0$ *on* $\mathscr{F}_I$,
2. $\alpha = \delta = 0$ *and* γ *or* $\beta \neq 0$ *on* $\mathscr{F}_B$,

problems (4.4a)–(4.4d) and (4.17a)–(4.17c) are equivalent.

Proof Let $\underline{u}$ and $\underline{v}$ be the solutions of (4.4a)–(4.4d). Regularity properties (4.14a)–(4.14b) of this solution immediately lead to

1. $A^*(\underline{n})[\underline{u}]_\Gamma = 0$ for all $\Gamma \in \mathscr{F}_B \cup \mathscr{F}_I$,
2. $A(\underline{n})[\underline{v}]_\Gamma = 0$ for all $\Gamma \in \mathscr{F}_I$.

Consequently, the hypothesis on the parameters α, β, γ, δ show that $(\underline{u},\ \underline{v})$ verifies (4.17a)–(4.17c).

Now, let $(\underline{u},\ \underline{v})$ be a solution of (4.17a)–(4.17c).

First, $\forall K \in \mathscr{T}_h$ and by taking $\underline{\varphi} \in \mathscr{D}(\overset{\circ}{K})^p$ and $\underline{\psi} \in \mathscr{D}(\overset{\circ}{K})^q$, (4.17a) and (4.17b) can be rewritten as regular distributions.

$$\frac{\mathrm{d}}{\mathrm{d}t} < \underline{\underline{\lambda}}\underline{u},\ \underline{\varphi} >_{\mathscr{D}',\mathscr{D}} = - < A\underline{v},\ \underline{\varphi} >_{\mathscr{D}',\mathscr{D}} + < \underline{f},\ \underline{\varphi} >_{\mathscr{D}',\mathscr{D}}, \tag{4.18a}$$

$$\frac{\mathrm{d}}{\mathrm{d}t} < \underline{\underline{\mu}}\underline{v},\ \underline{\psi} >_{\mathscr{D}',\mathscr{D}} = < A^*\underline{u},\ \underline{\psi} >_{\mathscr{D}',\mathscr{D}} . \tag{4.18b}$$

The H^s-regularity of $(\underline{u},\ \underline{v})$ implies:

$$\underline{\underline{\lambda}}\frac{\partial \underline{u}}{\partial t} = -A\underline{v} + \underline{f}, \quad \text{a.e. in } K, \tag{4.19a}$$

$$\underline{\underline{\mu}}\frac{\partial \underline{v}}{\partial t} = A^*\underline{u}, \quad \text{a.e. in } K. \tag{4.19b}$$

Then, by taking $\underline{\varphi} \in \mathscr{D}(\overset{\circ}{\overbrace{K_\ell \cup K_m}})^p$ and $\underline{\psi} \in \mathscr{D}(\overset{\circ}{\overbrace{K_\ell \cup K_m}})^q$ and by using (4.19a)–(4.19b), we have by looking at (4.17a)–(4.17c) for K_ℓ and K_m,

$$\alpha \int_{\Gamma_{\ell,m}} A(\underline{n})[\underline{v}]^{K_\ell}_{\Gamma_{\ell,m}} \cdot \underline{\varphi}\, \mathrm{d}\sigma + \gamma \int_{\Gamma_{\ell,m}} C[\underline{u}]^{K_\ell}_{\Gamma_{\ell,m}} \cdot \underline{\varphi}\, \mathrm{d}\sigma = 0, \tag{4.20a}$$

$$\alpha \int_{\Gamma_{\ell,m}} A(\underline{n})[\underline{v}]^{K_\ell}_{\Gamma_{\ell,m}} \cdot \underline{\varphi}\, \mathrm{d}\sigma - \gamma \int_{\Gamma_{\ell,m}} C[\underline{u}]^{K_\ell}_{\Gamma_{\ell,m}} \cdot \underline{\varphi}\, \mathrm{d}\sigma = 0, \tag{4.20b}$$

$$\beta \int_{\Gamma_{\ell,m}} A^*(\underline{n})[\underline{u}]^{K_\ell}_{\Gamma_{\ell,m}} \cdot \underline{\psi}\, \mathrm{d}\sigma + \delta \int_{\Gamma_{\ell,m}} C'[\underline{v}]^{K_\ell}_{\Gamma_{\ell,m}} \cdot \underline{\psi}\, \mathrm{d}\sigma = 0, \tag{4.20c}$$

$$\beta \int_{\Gamma_{\ell,m}} A^*(\underline{n})[\underline{u}]_{\Gamma_{\ell,m}}^{K_\ell} \cdot \underline{\psi} \, d\sigma - \delta \int_{\Gamma_{\ell,m}} C'[\underline{v}]_{\Gamma_{\ell,m}}^{K_\ell} \cdot \underline{\psi} \, d\sigma = 0. \tag{4.20d}$$

These four equations hold for all $\underline{\psi} \in \mathscr{D}(\mathring{\Gamma}_{\ell,m})^q$ and $\underline{\varphi} \in \mathscr{D}(\mathring{\Gamma}_{\ell,m})^p$ and the L^2 regularity of the traces yields:

$$\alpha A(\underline{n})[\underline{v}]_{\Gamma_{\ell,m}}^{K_\ell} + \gamma C[\underline{u}]_{\Gamma_{\ell,m}}^{K_\ell} = 0, \quad \text{a.e. on } \Gamma_{\ell,m}, \tag{4.21a}$$

$$\alpha A(\underline{n})[\underline{v}]_{\Gamma_{\ell,m}}^{K_\ell} - \gamma C[\underline{u}]_{\Gamma_{\ell,m}}^{K_\ell} = 0, \quad \text{a.e. on } \Gamma_{\ell,m}, \tag{4.21b}$$

$$\beta A^*(\underline{n})[\underline{u}]_{\Gamma_{\ell,m}}^{K_\ell} + \delta C'[\underline{v}]_{\Gamma_{\ell,m}}^{K_\ell} \sigma = 0, \quad \text{a.e. on } \Gamma_{\ell,m}, \tag{4.21c}$$

$$\beta A^*(\underline{n})[\underline{u}]_{\Gamma_{\ell,m}}^{K_\ell} - \delta C'[\underline{v}]_{\Gamma_{\ell,m}}^{K_\ell} = 0, \quad \text{a.e. on } \Gamma_{\ell,m}. \tag{4.21d}$$

By adding (4.21a) and (4.21b) and next (4.21c) and (4.21d), we obtain $\forall \Gamma_{\ell,m} \in \mathscr{F}_I$:

$$A(\underline{n})[\underline{v}]_{\Gamma_{\ell,m}}^{K_\ell} = A(\underline{n})[\underline{v}]_{\Gamma_{\ell,m}}^{K_m} = 0, \tag{4.22a}$$

$$A^*(\underline{n})[\underline{u}]_{\Gamma_{\ell,m}}^{K_\ell} = A^*(\underline{n})[\underline{u}]_{\Gamma_{\ell,m}}^{K_m} = 0. \tag{4.22b}$$

Equations (4.22a)–(4.22b) implies $\underline{u} \in H(A^*, \Omega)$ and $\underline{v} \in H(A, \Omega)$ and $(\underline{u}, \underline{v})$ satisfies (4.4a) and (4.4b).

In the same way, if $\Gamma \in \mathscr{F}_B$ then since $\alpha = \delta = 0$ we obtain:

$$\gamma C \underline{u}_{|\Gamma} = \gamma A(\underline{n}) A^*(\underline{n}) \underline{u}_{|\Gamma} = 0 \Rightarrow \; A^*(\underline{n}) \underline{u}_{|\Gamma} = 0 \text{ if } \gamma \neq 0, \tag{4.23a}$$

$$\beta A^*(\underline{n}) \underline{u}_{|\Gamma} = 0 \Rightarrow \; A^*(\underline{n}) \underline{u}_{|\Gamma} = 0 \text{ if } \beta \neq 0. \tag{4.23b}$$

So, we obtain boundary condition (4.4c) and the proof is finished.

The DG approximation of (4.17a)–(4.17c) is defined, in its functional form, by setting $U_p = \underline{H}_p^s(\mathscr{T}_h)$ and $U_q = \underline{H}_q^s(\mathscr{T}_h)$. However, as we said above, its actual discretization is made in the finite dimensional approximate spaces $\underline{V}_{h,n} \subset \underline{H}_n^s(\mathscr{T}_h)$ with $n = p, q$ which will be defined below.

4.1.2 Energy Identity

A necessary condition of the well-posedness of problem (4.17a)–(4.17c) is the existence of a non increasing discrete energy of the system. Let us derive this energy from the equations. For this purpose, we first integrate by parts the volumic stiffness term of (4.17a). We have

$$\int_{K_\ell} A\underline{v}_\ell \cdot \underline{\varphi}_\ell \,\mathrm{d}\underline{x} = \int_{K_\ell} \sum_{i=1}^{d} A_i \frac{\partial \underline{v}_\ell}{\partial x_i} \cdot \underline{\varphi}_\ell \,\mathrm{d}\underline{x}$$

By using Green's theorem and the constant character of A_i, we get:

$$\int_{K_\ell} A\underline{v}_\ell \cdot \underline{\varphi}_\ell \,\mathrm{d}\underline{x} = -\int_{K_\ell} \underline{v}_\ell \cdot \sum_{i=1}^{d} A_i^T \frac{\partial \underline{\varphi}_\ell}{\partial x_i} \,\mathrm{d}\underline{x} + \int_{\partial K_\ell} \sum_{i=1}^{d} n_i\, A_i \underline{v}_\ell \cdot \underline{\varphi}_\ell \,\mathrm{d}\sigma. \tag{4.24}$$

Now, by setting $\underline{f} = 0$ and replacing $\underline{\varphi}_\ell$ by $\underline{u}_\ell$ in (4.17a) and $\underline{\psi}_\ell$ by $\underline{v}_\ell$ in (4.17b) and by using (4.24), we obtain:

$$\begin{aligned}\int_{K_\ell} \underline{\underline{\lambda}} \frac{\partial \underline{u}_\ell}{\partial t} \cdot \underline{u}_\ell \,\mathrm{d}\underline{x} &= -\int_{K_\ell} \underline{v}_\ell \cdot A^* \underline{u}_\ell \,\mathrm{d}\underline{x} - \int_{\partial K_\ell} A(\underline{n}) \underline{v}_\ell \cdot \underline{u}_\ell \,\mathrm{d}\sigma \\ &+ \int_{\partial K_\ell} \alpha\, A(\underline{n}) [\underline{v}_h]_{\partial K_\ell}^{K_\ell} \cdot \underline{u}_\ell \,\mathrm{d}\sigma + \int_{\partial K_\ell} \gamma\, C [\underline{u}_h]_{\partial K_\ell}^{K_\ell} \cdot \underline{u}_\ell \,\mathrm{d}\sigma,\end{aligned} \tag{4.25a}$$

$$\begin{aligned}\int_{K_\ell} \underline{\underline{\mu}} \frac{\partial \underline{v}_\ell}{\partial t} \cdot \underline{v}_\ell \,\mathrm{d}\underline{x} &= \int_{K_\ell} A^* \underline{u}_\ell \cdot \underline{v}_\ell \,\mathrm{d}\underline{x} + \int_{\partial K_\ell} \beta\, A^*(\underline{n}) [\underline{u}_h]_{\partial K_\ell}^{K_\ell} \cdot \underline{v}_\ell \,\mathrm{d}\sigma \\ &+ \int_{\partial K_\ell} \delta\, C' [\underline{v}_h]_{\partial K_\ell}^{K_\ell} \cdot \underline{v}_\ell \,\mathrm{d}\sigma.\end{aligned} \tag{4.25b}$$

$\underline{\underline{\lambda}}$ and $\underline{\underline{\mu}}$ being symmetric, definite, positive, one can write $\underline{\underline{\lambda}} = \underline{\underline{\tilde{\lambda}}}^T \underline{\underline{\tilde{\lambda}}}$ and $\underline{\underline{\mu}} = \underline{\underline{\tilde{\mu}}}^T \underline{\underline{\tilde{\mu}}}$. Then

$$\underline{\underline{\lambda}} \frac{\partial \underline{u}_\ell}{\partial t} \cdot \underline{u}_\ell = \frac{\partial (\underline{\underline{\tilde{\lambda}}} \underline{u}_\ell)}{\partial t} \cdot \underline{\underline{\tilde{\lambda}}} \underline{u}_\ell = \frac{1}{2} \frac{\partial || \underline{\underline{\tilde{\lambda}}} \underline{u}_\ell ||^2}{\partial t} \tag{4.26}$$

and, in a same way,

$$\underline{\underline{\mu}} \frac{\partial \underline{v}_\ell}{\partial t} \cdot \underline{v}_\ell = \frac{1}{2} \frac{\partial || \underline{\underline{\tilde{\mu}}} \underline{v}_\ell ||^2}{\partial t} \tag{4.27}$$

By adding (4.25a) to (4.25b) and taking into account (4.26) and (4.27), we get the discrete energy $\mathscr{E}_h^\ell$ on K_ℓ of the system such that

$$\begin{aligned}\frac{\mathrm{d}}{\mathrm{d}t} \mathscr{E}_h^\ell &= \frac{1}{2} \frac{\mathrm{d}}{\mathrm{d}t} \left\{ \int_{K_\ell} || \underline{\underline{\tilde{\lambda}}} \underline{u}_\ell ||^2 \mathrm{d}\underline{x} + \int_{K_\ell} || \underline{\underline{\tilde{\mu}}} \underline{v}_\ell ||^2 \mathrm{d}\underline{x} \right\} = -\int_{\partial K_\ell} A(\underline{n}) \underline{v}_\ell \cdot \underline{u}_\ell \,\mathrm{d}\sigma \\ &+ \int_{\partial K_\ell} \alpha\, A(\underline{n}) [\underline{v}_h]_{\partial K_\ell}^{K_\ell} \cdot \underline{u}_\ell \,\mathrm{d}\sigma + \int_{\partial K_\ell} \beta\, A^*(\underline{n}) [\underline{u}_h]_{\partial K_\ell}^{K_\ell} \cdot \underline{v}_\ell \,\mathrm{d}\sigma \\ &+ \int_{\partial K_\ell} \gamma\, C [\underline{u}_h]_{\partial K_\ell}^{K_\ell} \cdot \underline{u}_\ell \,\mathrm{d}\sigma + \int_{\partial K_\ell} \delta\, C' [\underline{v}_h]_{\partial K_\ell}^{K_\ell} \cdot \underline{v}_\ell \,\mathrm{d}\sigma,\end{aligned} \tag{4.28}$$

Let us now write explicitly the right hand side R_E^ℓ of (4.28) on the face $\Gamma_{\ell,m}$ by taking into account (4.15) and using the definitions of $A(\underline{n})$ and $A^*(\underline{n})$ (which imply that $(A^*(\underline{n}))^T = -A(\underline{n})$). We obtain:

$$\begin{aligned} R_E^\ell = &-\int_{\Gamma_{\ell,m}} A(\underline{n})\underline{v}_\ell \cdot \underline{u}_\ell \, \mathrm{d}\sigma \\ &+\alpha\int_{\Gamma_{\ell,m}} A(\underline{n})(\underline{v}_m - \underline{v}_\ell)\cdot \underline{u}_\ell \, \mathrm{d}\sigma - \beta\int_{\Gamma_{\ell,m}} (\underline{u}_m - \underline{u}_\ell)\cdot A(\underline{n})\underline{v}_\ell \, \mathrm{d}\sigma \\ &+\gamma\int_{\Gamma_{\ell,m}} C(\underline{u}_m - \underline{u}_\ell)\cdot \underline{u}_\ell \, \mathrm{d}\sigma + \delta\int_{\Gamma_{\ell,m}} C'(\underline{v}_m - \underline{v}_\ell)\cdot \underline{v}_\ell \, \mathrm{d}\sigma. \end{aligned} \tag{4.29}$$

By noticing that the exterior normal of K_m is the opposite of that of K_ℓ and that $A(-\underline{n}) = -A(\underline{n})$, we can write, in the same way:

$$\begin{aligned} R_E^m = &\int_{\Gamma_{\ell,m}} A(\underline{n})\underline{v}_m \cdot \underline{u}_m \, \mathrm{d}\sigma \\ &+\alpha\int_{\Gamma_{\ell,m}} A(\underline{n})(\underline{v}_m - \underline{v}_\ell)\cdot \underline{u}_m \, \mathrm{d}\sigma - \beta\int_{\Gamma_{\ell,m}} (\underline{u}_m - \underline{u}_\ell)\cdot A(\underline{n})\underline{v}_m \, \mathrm{d}\sigma \\ &-\gamma\int_{\Gamma_{\ell,m}} C(\underline{u}_m - \underline{u}_\ell)\cdot \underline{u}_m \, \mathrm{d}\sigma - \delta\int_{\Gamma_{\ell,m}} C'(\underline{v}_m - \underline{v}_\ell)\cdot \underline{v}_m \, \mathrm{d}\sigma. \end{aligned} \tag{4.30}$$

By using $C = A(\underline{n})A^*(\underline{n})$ and $C' = A^*(\underline{n})A(\underline{n})$ and after a little algebra, we get:

$$\begin{aligned} R_E^\ell + R_E^m = &(1+\alpha-\beta)\int_{\Gamma_{\ell,m}} A(\underline{n})(\underline{v}_m \cdot \underline{u}_m - \underline{v}_\ell \cdot \underline{u}_\ell)\, \mathrm{d}\sigma \\ &+(\alpha+\beta)\int_{\Gamma_{\ell,m}} A(\underline{n})(\underline{v}_m \cdot \underline{u}_l - \underline{v}_\ell \cdot \underline{u}_m)\, \mathrm{d}\sigma \\ &-\gamma\int_{\Gamma_{\ell,m}} ||A^*(\underline{n})(\underline{u}_m - \underline{u}_\ell)||^2 \, \mathrm{d}\sigma - \delta\int_{\Gamma_{\ell,m}} ||A(\underline{n})(\underline{v}_m - \underline{v}_\ell)||^2 \, \mathrm{d}\sigma. \end{aligned} \tag{4.31}$$

Since $\alpha = 0$ for $\Gamma_\ell \in \mathscr{F}_B$, we get:

$$R_E^\ell = (1-\beta)\int_{\Gamma_\ell} A(\underline{n})\underline{v}_m \cdot \underline{u}_m \, \mathrm{d}\sigma - \gamma\int_{\Gamma_\ell} ||A^*(\underline{n})\underline{u}_m||^2 \, \mathrm{d}\sigma - \delta\int_{\Gamma_\ell} ||A(\underline{n})\underline{v}_m||^2 \, \mathrm{d}\sigma. \tag{4.32}$$

Finally, by summing on all the faces (or edges) of the mesh, we get the energy $\mathscr{E}_h$ of the system such that

$$\frac{\mathrm{d}}{\mathrm{d}t}\mathscr{E}_h = \sum_{\Gamma_{\ell,m}\in\mathscr{F}_I} (R_E^\ell + R_E^m) + \sum_{\Gamma_\ell\in\mathscr{F}_B} R_E^\ell. \tag{4.33}$$

Equations (4.31) and (4.32) imply that, in order to have $\mathrm{d}\mathscr{E}_h/\mathrm{d}t \le 0$, we must set:

1. $\alpha = -\beta = -1/2$ for $\Gamma_\ell \in \mathscr{F}_I$,
2. $\beta = 1$ for $\Gamma_\ell \in \mathscr{F}_B$,
3. If $\gamma > 0$ and $\delta > 0$, we have $\mathrm{d}\mathscr{E}_h/\mathrm{d}t < 0$ and the energy decreases. The scheme is then *dissipative*.
4. If $\gamma = 0$ and $\delta = 0$, we have $\mathrm{d}\mathscr{E}_h/\mathrm{d}t = 0$ and the energy is constant. The scheme is then *conservative*. We have an energy conservation.

Remarks:

1. The dissipative term is also called "penalty" or "damping" term.
2. It seems natural to set $\gamma = 0$ and $\delta = 0$ in order to avoid dissipation which induces a damping of the amplitude of the solution. However, we shall see later that this choice can be troublesome and it is safer to have some dissipation in some cases.
3. Another way to formulate (4.17a) is to integrate by parts and combine the boundary term with the jump. This provides:

$$\frac{\mathrm{d}}{\mathrm{d}t}\int_{K_\ell} \underline{\underline{\lambda}} \underline{u}_\ell \cdot \underline{\varphi}_\ell \, \mathrm{d}\underline{x} = -\int_{K_\ell} \underline{v}_\ell \cdot A^* \underline{\varphi}_\ell \, \mathrm{d}\underline{x} - \int_{\partial K_\ell} A(\underline{n}) \{\underline{v}\}_{\partial K_\ell}^{K_\ell} \cdot \underline{\varphi}_\ell \, \mathrm{d}\sigma$$
$$+ \gamma \int_{\partial K_\ell} C[\underline{u}_h]_{\partial K_\ell}^{K_\ell} \cdot \underline{\varphi}_\ell \, \mathrm{d}\sigma + \int_{K_\ell} \underline{f} \cdot \underline{\varphi}_\ell \, \mathrm{d}\underline{x}, \tag{4.34}$$

where

$$\{\underline{v}\}_{\Gamma_{\ell,m}}^{K_\ell} = \frac{(\underline{v}_{|K_m})_{|\Gamma_{\ell,m}} + (\underline{v}_{|K_\ell})_{|\Gamma_{\ell,m}}}{2} \tag{4.35}$$

is the mean value of the solution on the boundary. One can transform (4.17b) in the same way.

4.1.3 Application to Some Wave Equations

4.1.3.1 The Acoustics System

For the acoustics system defined by (3.29a)–(3.29b), we have $p = 1$, $q = d$, $\underline{u} = u$, $\underline{v} = \underline{v}$, $A = -\mathrm{div}$, $A^* = \mathrm{grad}$, $A(\underline{n}) = -(n_1\ n_2\ n_3)$ and $A^*(\underline{n}) = (n_1\ n_2\ n_3)^T$ for $d = 3$ ($A(\underline{n}) = -(n_1\ n_2)$ and $A^*(\underline{n}) = (n_1\ n_2)^T$ for $d = 2$). On the other hand, the dissipative jumps are useless for these equations. So, with the notations of (4.17a)–(4.17c), the discontinuous Galerkin formulation of this system reads:

Find $u_h(.,t) \in V_h \subset H^s(\mathscr{T}_h)$, $\underline{v}_h(.,t) \in \underline{V}_{h,d} \subset \underline{H}_d^s(\mathscr{T}_h)$ such that $\forall K_\ell \in \mathscr{T}_h$ and $\forall \varphi_h \in V_h$, $\forall \underline{\psi}_h \in \underline{V}_{h,d}$,

$$\frac{\mathrm{d}}{\mathrm{d}t}\int_{K_\ell}\lambda\, u_\ell\, \varphi_\ell\, \mathrm{d}\underline{x} = \int_{K_\ell}\nabla\cdot\underline{v}_\ell\,\varphi_\ell\,\mathrm{d}\underline{x} + \int_{\partial K_\ell}\alpha\,\underline{n}\cdot[\underline{v}_h]^{K_\ell}_{\partial K_\ell}\varphi_\ell\,\mathrm{d}\sigma + \int_{K_\ell} f\,\varphi_\ell\,\mathrm{d}\underline{x}, \tag{4.36a}$$

$$\frac{\mathrm{d}}{\mathrm{d}t}\int_{K_\ell}\underline{\underline{\mu}}\underline{v}_\ell\cdot\underline{\psi}_\ell\,\mathrm{d}\underline{x} = \int_{K_\ell}\nabla u_\ell\cdot\underline{\psi}_\ell\,\mathrm{d}\underline{x} + \int_{\partial K_\ell}\beta\,\underline{n}[u]^{K_\ell}_{\partial K_\ell}\cdot\underline{\psi}_\ell\,\mathrm{d}\sigma, \tag{4.36b}$$

$$u_h(\underline{x},0) = u_{0h}(\underline{x}),\ \underline{v}_h(\underline{x},0) = \underline{v}_{0h}(\underline{x}) \text{ in } \Omega, \tag{4.36c}$$

with

1. $\alpha = -\beta = -1/2$ for $\Gamma_\ell \in \mathscr{F}_I$,
2. $\beta = 1$ for $\Gamma_\ell \in \mathscr{F}_B$.

Remark: The above DG formulation corresponds to the approximation of an acoustic problem with a homogeneous Dirichlet boundary condition. We can consider a Neumann boundary condition by taking $A = \nabla$ and $A^* = -\nabla\cdot$.

4.1.3.2 The Maxwell's System

For the Maxwell's system defined by (1.8a)–(1.8b), we have $p = 3$, $q = 3$, $\underline{u} = \underline{E}$, $\underline{v} = \underline{H}$, $A = A^* = \underline{curl}$,

$$A(\underline{n}) = A^*(\underline{n}) = \begin{pmatrix} 0 & n_3 & -n_2 \\ -n_3 & 0 & n_1 \\ n_2 & -n_1 & 0 \end{pmatrix}.$$

Moreover, we set:

$$C = C' = \begin{pmatrix} n_2^2+n_3^2 & -n_1n_2 & -n_1n_3 \\ -n_1n_2 & n_1^2+n_3^2 & -n_2n_3 \\ -n_1n_3 & -n_2n_3 & n_1^2+n_2^2 \end{pmatrix}.$$

Actually if $\underline{V} \in \mathbb{R}^3$, we have $A(\underline{n})\underline{V} = \underline{V}\times\underline{n}$ and $C\,\underline{V} = \underline{n}\times\underline{V}\times\underline{n}$. These definitions lead to the following DG formulation:

Find $\underline{E}_h(.,t) \in \underline{V}_{h,3} \subset \underline{H}^s_3(\mathscr{T}_h)$, $\underline{H}_h(.,t) \in \underline{V}_{h,3}$ such that $\forall K_\ell \in \mathscr{T}_h$ and $\forall \underline{\varphi}_h \in \underline{V}_{h,3}$, $\forall \underline{\psi}_h \in \underline{V}_{h,3}$,

$$\frac{\mathrm{d}}{\mathrm{d}t}\int_{K_\ell}\underline{\underline{\varepsilon}}\underline{E}_\ell\cdot\underline{\varphi}_\ell\,\mathrm{d}\underline{x} = \int_{K_\ell}\nabla\times\underline{H}_\ell\cdot\underline{\varphi}_\ell\,\mathrm{d}\underline{x} + \int_{\partial K_\ell}\alpha\,[\underline{H}\times\underline{n}]^{K_\ell}_{\partial K_\ell}\cdot\underline{\varphi}_\ell\,\mathrm{d}\sigma$$
$$+\int_{\partial K_\ell}\gamma\,[\underline{n}\times\underline{E}\times\underline{n}]^{K_\ell}_{\partial K_\ell}\cdot\underline{\varphi}_\ell\,\mathrm{d}\sigma - \int_{K_\ell}\underline{J}\cdot\underline{\varphi}_\ell\,\mathrm{d}\underline{x}, \tag{4.37a}$$

$$\frac{\mathrm{d}}{\mathrm{d}t}\int_{K_\ell} \underline{\underline{\mu}} \underline{H}_\ell \cdot \underline{\psi}_\ell \,\mathrm{d}\underline{x} = -\int_{K_\ell} \nabla \times \underline{E}_\ell \cdot \underline{\psi}_\ell \,\mathrm{d}\underline{x} + \int_{\partial K_\ell} \beta\,[\underline{E} \times \underline{n}]_{\partial K_\ell}^{K_\ell} \cdot \underline{\psi}_\ell \,\mathrm{d}\sigma$$
$$+ \int_{\partial K_\ell} \delta\,[\underline{n} \times \underline{H} \times \underline{n}]_{\partial K_\ell}^{K_\ell} \cdot \underline{\psi}_\ell \,\mathrm{d}\sigma, \tag{4.37b}$$

$$\underline{E}_h(\underline{x}, 0) = \underline{E}_{0h}(\underline{x}),\ \ \underline{H}_h(\underline{x}, 0) = \underline{H}_{0h}(\underline{x}) \text{ in } \Omega \tag{4.37c}$$

with

1. $\alpha = -\beta = -1/2$ and $\gamma,\ \delta \geq 0$ for $\Gamma_\ell \in \mathscr{F}_I$,
2. $\beta = 1, \alpha = \delta = 0$ and $\gamma \geq 0$ for $\Gamma_\ell \in \mathscr{F}_B$.

Remarks:

1. The above DG formulation treats a perfectly conducting boundary condition i.e. $\underline{E} \times \underline{n} = 0$ on $\partial\Omega$.
2. In practice, it is sufficient to set $\gamma > 0,\ \delta = 0$ or $\gamma = 0,\ \delta > 0$.

4.1.3.3 The Linear Elastodynamics System

For this system, the formulation can be easily derived from the acoustics equations. If we set $A = -\mathrm{div}$, $A^* = \mathrm{grad}$, $A(\underline{n}) = -(n_1\ n_2\ n_3)$ and $A^*(\underline{n}) = (n_1\ n_2\ n_3)^T$ and we apply the discontinuous Galerkin formulation of the acoustics system to each equation defined by (3.260a)–(3.260b), we get:

Find $\underline{v}_h(.,t) \in \underline{V}_{h,3} \subset \underline{H}_3^s(\mathscr{T}_h)$, $\underline{\gamma}_{ijh}(.,t) \in \underline{V}_{h,3}$, $i = 1 \ldots d$, $j = 1 \ldots d$, $\underline{\gamma}_{ih}(.,t) \in \underline{V}_{h,3}$, $i = 1 \ldots d$, such that $\forall K_\ell \in \mathscr{T}_h$ and $\forall \underline{\varphi}_i \in \underline{V}_{h,3}$, $\forall \underline{\psi}_i \in \underline{V}_{h,3}$, $\forall \underline{\theta}_i \in \underline{V}_{h,3}$, $i = 1 \ldots d$,

$$\frac{\mathrm{d}}{\mathrm{d}t}\int_{K_\ell} \rho\, v_{ih,\ell}\, \varphi_{i,\ell} \,\mathrm{d}\underline{x} = \sum_{j=1}^{d} \int_{K_\ell} \nabla \cdot \underline{\gamma}_{ijh,\ell}\, \varphi_{i,\ell} \,\mathrm{d}\underline{x}$$
$$+ \sum_{j=1}^{d} \int_{\partial K_\ell} \alpha \underline{n} \cdot [\underline{\gamma}_{ijh,\ell}]_{\partial K_\ell}^{K_\ell} \varphi_{i,\ell} \,\mathrm{d}\sigma + \int_{K_\ell} F_i\, \varphi_{i,\ell} \,\mathrm{d}\underline{x}, \tag{4.38a}$$

$$\frac{\mathrm{d}}{\mathrm{d}t}\int_{K_\ell} \underline{\gamma}_{ih,\ell} \cdot \underline{\psi}_{i,\ell} \,\mathrm{d}\underline{x} = \int_{K_\ell} \nabla v_{ih,\ell} \cdot \underline{\psi}_{i,\ell} \,\mathrm{d}\underline{x} + \int_{\partial K_\ell} \beta \underline{n} [v_{ih}]_{\partial K_\ell}^{K_\ell} \cdot \underline{\psi}_{i,\ell} \,\mathrm{d}\sigma, \tag{4.38b}$$

$$\int_{K_\ell} \underline{\gamma}_{ijh} \cdot \underline{\theta}_{i,\ell} \,\mathrm{d}\underline{x} = \int_{K_\ell} A_{ij}(\underline{x}) \underline{\gamma}_i \cdot \underline{\theta}_{i,\ell} \,\mathrm{d}\underline{x}, \tag{4.38c}$$

+ initial conditions.

As for acoustics, the dissipative jump is here useless.

The drawback of this formulation is that it does not fit to the above general formulation and requires therefore to reformulate the discrete energy of the system. In order to avoid this drawback, we have to rewrite (3.260a)–(3.260d).

Let us set $\underline{\tilde{\gamma}}_1 \in \mathbb{R}^{d^3}$ and $\underline{\tilde{\gamma}}_2 \in \mathbb{R}^{d^3}$ such that $\underline{\tilde{\gamma}}_1 = (\underline{\gamma}_{11}, \ldots, \underline{\gamma}_{1d}, \ldots, \underline{\gamma}_{d1}, \ldots, \underline{\gamma}_{dd})$ and $\underline{\tilde{\gamma}}_2 = (\underline{\gamma}_1, \underline{\gamma}_1, \underline{\gamma}_1, \underline{\gamma}_2, \underline{\gamma}_2, \underline{\gamma}_2, \underline{\gamma}_3, \underline{\gamma}_3, \underline{\gamma}_3)$ when $d = 3$ and $\underline{\gamma}_2 = (\underline{\gamma}_1, \underline{\gamma}_1, \underline{\gamma}_2, \underline{\gamma}_2)$ when $d = 2$. Then, A^* and $A^*(\underline{n})$ can be defined as the following block-matrices:

- $$A^* = \begin{pmatrix} D_3 & O_3 & O_3 \\ D_3 & O_3 & O_3 \\ D_3 & O_3 & O_3 \\ O_3 & D_3 & O_3 \\ O_3 & D_3 & O_3 \\ O_3 & D_3 & O_3 \\ O_3 & O_3 & D_3 \\ O_3 & O_3 & D_3 \\ O_3 & O_3 & D_3 \end{pmatrix}, \quad A^*(\underline{n}) = \begin{pmatrix} N_3 & O_3 & O_3 \\ N_3 & O_3 & O_3 \\ N_3 & O_3 & O_3 \\ O_3 & N_3 & O_3 \\ O_3 & N_3 & O_3 \\ O_3 & N_3 & O_3 \\ O_3 & O_3 & N_3 \\ O_3 & O_3 & N_3 \\ O_3 & O_3 & N_3 \end{pmatrix},$$

  where $D_3 = (\partial/\partial x_1 \; \partial/\partial x_2 \; \partial/\partial x_3)^T$, $N_3 = (n_1 \; n_2 \; n_3)^T$ and $O_3 = (0\;0\;0)^T$ for $d = 3$.

- $$A^* = \begin{pmatrix} D_2 & O_2 \\ D_2 & O_2 \\ O_2 & D_2 \\ O_2 & D_2 \end{pmatrix}, \quad A^*(\underline{n}) = \begin{pmatrix} N_2 & O_2 \\ N_2 & O_2 \\ O_2 & N_2 \\ O_2 & N_2 \end{pmatrix},$$

  where $D_2 = (\partial/\partial x_1 \; \partial/\partial x_2)^T$, $N_2 = (n_1 \; n_2)^T$ and $O_2 = (0\;0)^T$ for $d = 2$.

Of course, for any d, we have $A = -(A^*)^T$ and $A(\underline{n}) = -(A^*(\underline{n}))^T$.

With the above notations, (3.260a)–(3.260d) can be rewritten as Find $\underline{v} : \Omega \times [0, T] \to \mathbb{R}^d$, $\underline{\tilde{\gamma}}_1 : \Omega \times [0, T] \to \mathbb{R}^{d^3}$ and $\underline{\tilde{\gamma}}_2 : \Omega \times [0, T] \to \mathbb{R}^{d^3}$ such that

$$\rho(\underline{x})\frac{\partial \underline{v}}{\partial t}(\underline{x}, t) + A\underline{\tilde{\gamma}}_1(\underline{x}, t) = \underline{F}(\underline{x}, t) \;\text{ in } \Omega \times [0, T], \tag{4.39a}$$

$$\frac{\partial \underline{\tilde{\gamma}}_2}{\partial t}(\underline{x}, t) - A^*\underline{v}(\underline{x}, t) \;\text{ in } \Omega \times [0, T], \tag{4.39b}$$

$$\underline{\tilde{\gamma}}_1(\underline{x}, t) = \check{C}(\underline{x})\underline{\tilde{\gamma}}_2(\underline{x}, t) \;\text{ in } \Omega \times [0, T], \tag{4.39c}$$

$$\sum_{j=1}^{d} \underline{\gamma}_{ij}(\underline{x}, t) \cdot \underline{n} = 0, \quad \forall i = 1 \ldots d, \;\text{ on } \partial\Omega, \tag{4.39d}$$

where $\check{C}$ is the symmetric, definite, positive block-matrix defined as

$$\begin{pmatrix} A_{11} & A_{12} & A_{13} \\ A_{21} & A_{22} & A_{23} \\ A_{31} & A_{32} & A_{33} \end{pmatrix}. \tag{4.40}$$

We can now derive the discontinuous Galerkin formulation of (4.39a)–(4.39d):

Find $\underline{v}_h(.,t) \in \underline{V}_{h,d}$, $\underline{\tilde{\gamma}}_{1h}(.,t) \in \underline{V}_{h,d^3}$, $\underline{\tilde{\gamma}}_{2h}(.,t) \in \underline{V}_{h,d^3}$ such that $\forall K_\ell \in \mathscr{T}_h$ and $\forall \underline{\varphi}_h \in \underline{V}_{h,d}$, $\forall \underline{\psi}_h \in \underline{V}_{h,d^3}$, $\forall \underline{\theta}_h \in \underline{V}_{h,d^3}$

$$\frac{\mathrm{d}}{\mathrm{d}t}\int_{K_\ell} \rho \underline{v}_\ell \cdot \underline{\varphi}_\ell \,\mathrm{d}\underline{x} = -\int_{K_\ell} A\underline{\tilde{\gamma}}_{1,\ell} \cdot \underline{\varphi}_\ell \,\mathrm{d}\underline{x}$$
$$- \int_{\partial K_\ell} \alpha A(\underline{n})[\underline{\tilde{\gamma}}_{1h}]^{K_\ell}_{\partial K_\ell} \cdot \underline{\varphi}_\ell \,\mathrm{d}\sigma + \int_{K_\ell} \underline{F} \cdot \underline{\varphi}_\ell \,\mathrm{d}\underline{x}, \tag{4.41a}$$

$$\frac{\mathrm{d}}{\mathrm{d}t}\int_{K_\ell} \underline{\tilde{\gamma}}_{2,\ell} \cdot \underline{\psi}_\ell \,\mathrm{d}\underline{x} = \int_{K_\ell} A^* \underline{v}_\ell \cdot \underline{\psi}_\ell \,\mathrm{d}\underline{x} + \int_{\partial K_\ell} \beta A^*(\underline{n})[\underline{v}_h]^{K_\ell}_{\partial K_\ell} \cdot \underline{\psi}_\ell \,\mathrm{d}\sigma \tag{4.41b}$$

$$\frac{\mathrm{d}}{\mathrm{d}t}\int_{K_\ell} \underline{\tilde{\gamma}}_{1,\ell} \cdot \underline{\theta}_\ell \,\mathrm{d}\underline{x}, = \frac{\mathrm{d}}{\mathrm{d}t}\int_{K_\ell} \check{C}\,\underline{\tilde{\gamma}}_{2,\ell} \cdot \underline{\theta}_\ell \,\mathrm{d}\underline{x}, \tag{4.41c}$$

with

1. $\alpha = -\beta = -1/2$ for $\Gamma_\ell \in \mathscr{F}_I$,
2. $\beta = 1$ for $\Gamma_\ell \in \mathscr{F}_B$.

In order to obtain the energy of (4.41a)–(4.41c), we just have to set $\underline{\varphi}_\ell = \underline{v}_\ell$, $\underline{\psi}_\ell = \underline{\tilde{\gamma}}_{1,\ell}$ and $\underline{\theta}_\ell = \underline{\tilde{\gamma}}_{2,\ell}$. Then, by plugging (4.41c) in (4.41b), we get:

$$\frac{\mathrm{d}}{\mathrm{d}t}\mathscr{E}_h^\ell = \frac{\mathrm{d}}{\mathrm{d}t}\left\{\int_{K_\ell} ||\rho \underline{v}_\ell||^2 \mathrm{d}\underline{x} + \int_{K_\ell} \check{C}\,\underline{\tilde{\gamma}}_{2,\ell} \cdot \underline{\tilde{\gamma}}_{2,\ell} \mathrm{d}\underline{x}\right\}. \tag{4.42}$$

$\mathscr{E}_h^\ell$ is obviously positive because of the definition of $\check{C}$. On the other hand, one can easily check the equivalence between (4.38a)–(4.38c) and (4.41a)–(4.41c).

4.2 Approximation by Triangles and Tetrahedra

A first polynomial approximation of (4.17a)–(4.17c) is provided by using triangles and tetrahedra defined in Sect. 2.3.1. Unlike continuous elements, this approximation is here possible since, as we shall see, the local character of the basis functions leads to a block-diagonal mass matrix. So, in this section, K_ℓ denotes a triangle or a tetrahedron of a mesh $\mathscr{T}_h$ and the solution is sought in the spaces:

$$\underline{V}^r_{h,n} = \left\{ \underline{w}_h \in [L^2(\Omega)]^n \text{ such that } \underline{w}_{h|K_\ell} \in [P_r]^n \right\}, \tag{4.43}$$

where $n = p$ or q.

All the integrals are then computed on the unit element $\hat{T}_d$, $d = 2, 3$ defined in (2.35) and (2.37) by using polynomials defined in (2.36) and (2.38).

4.2.1 The Mass Integrals

Let us first write a decomposition of the solution $\underline{u}_\ell$ on the polynomial basis $\left(\underline{\varphi}^m_{\ell,s} = \varphi_{\ell,s}\, \underline{e}_m\right)_{s=1\ldots N_r,\; m=1\ldots p}$ related to K_ℓ. We have:

$$\underline{u}_\ell = \sum_{m=1}^{p} \sum_{s=1}^{N_r} u^m_{\ell,s}\, \underline{\varphi}^m_{\ell,s}, \tag{4.44}$$

where N_r is the number of interpolation points in K_ℓ ($N_r = (r+1)(r+2)/2$ in 2D and $N_r = (r+1)(r+2)(r+3)/6$ in 3D).

Taking into account (4.44) and setting $\underline{\varphi}_\ell = \underline{\varphi}^n_{\ell,t}$ in the left-hand side of (4.17a) leads to

$$\int_{K_\ell} \underline{\underline{\lambda}}\, \underline{u}_\ell \cdot \underline{\varphi}_\ell \, \mathrm{d}\underline{x} = \underline{u}_\ell = \sum_{m=1}^{p} \sum_{s=1}^{N_r} u^m_{\ell,s} \int_{K_\ell} \underline{\underline{\lambda}}\, \underline{\varphi}^m_{\ell,s} \cdot \underline{\varphi}^n_{\ell,t} \, \mathrm{d}\underline{x}. \tag{4.45}$$

On the other hand, we have:

$$\begin{aligned} M^{m,n}_{s,t} = \int_{K_\ell} \underline{\underline{\lambda}}\, \underline{\varphi}^m_{\ell,s} \cdot \underline{\varphi}^n_{\ell,t} \, \mathrm{d}\underline{x} &= \int_{\hat{T}_d} |J_{T_\ell}| \underline{\underline{\lambda}} \circ \underline{F}_{T_\ell} \underline{\varphi}^m_{\ell,s} \circ \underline{F}_{T_\ell} \cdot \underline{\varphi}^n_{\ell,t} \circ \underline{F}_{T_\ell} \, \mathrm{d}\hat{\underline{x}} \\ &= \int_{\hat{T}_d} |J_{T_\ell}| \hat{\underline{\underline{\lambda}}}\, \hat{\underline{\varphi}}^m_s \cdot \hat{\underline{\varphi}}^n_t \, \mathrm{d}\hat{\underline{x}}, \end{aligned} \tag{4.46}$$

where $\underline{F}_{T_\ell}(\hat{x})$ is the mapping such that $\underline{F}_{T_\ell}(\hat{T}_d) = K_\ell$ $J_{T_\ell} = \det(DF_{T_\ell})$, DF_{T_ℓ} being the (constant) Jacobian matrix of $\underline{F}_{T_\ell}$ and $\hat{\underline{\underline{\lambda}}} = \underline{\underline{\lambda}} \circ \underline{F}_{T_\ell}$.

Now, according to Sect. 2.3.1.3, one can write, for any basis function $\hat{\varphi}_s$ such that $\hat{\underline{\varphi}}^m_s = \hat{\varphi}_s\, \underline{e}_m$ ($\underline{e}_m$ being a vector of the canonical basis of $\mathbb{R}^p$), when $d = 3$:

$$\hat{\varphi}_s \circ \underline{F}_c = \sum_{i=0}^{r} \sum_{j=0}^{r-i} \sum_{k=0}^{r-i-j} \alpha^s_{i,j,k}\, p_{i,j,k}, \tag{4.47}$$

where $\underline{F}_c(a, b, c)$ is the mapping such that $\underline{F}_c(C_3) = \hat{T}_3$, $C_3 = [-1, 1]^3$ and

$$p_{i,j,k}(a, b, c) = P_i^0(a)\, P_j^{2i+1}(b)\, P_k^{2i+2j+2}(c)(1-b)^i(1-c)^{i+j},$$

with $P_i^n = \widetilde{P}_i^{n,0}$, $\widetilde{P}_i^{n,0}$ being the Jacobi polynomial defined in (2.33).

In the same way, when $d = 2$:

$$\hat{\varphi}_s \circ \underline{F}_c = \sum_{i=0}^{r} \sum_{j=0}^{r-i} \alpha_{i,j}^s\, p_{i,j}, \tag{4.48}$$

where $\underline{F}_c(a, b)$ is the mapping such that $\underline{F}_c(C_2) = \hat{T}_2$, $C_2 = [-1, 1]^2$ and

$$p_{i,j}(a, b) = P_i^0(a)\, P_j^{2i+1}(b)(1-b)^i.$$

We set $d = 3$ in the following, the computations being similar when $d = 2$.

Let us denote $J_c = (1-b)(1-c)^2/4$ (≤ 0) the Jacobian of $\underline{F}_c$. By inserting (4.47) in the right-handside of (4.46) and making a change of variable, we get:

$$M_{s,t}^{m,n} = |J_{T_\ell}| \int_{C_3} J_c \left(\sum_{i=0}^{r} \sum_{j=0}^{r-i} \sum_{k=0}^{r-i-j} \alpha_{i,j,k}^s\, p_{i,j,k} \right) \times \left(\sum_{i'=0}^{r} \sum_{j'=0}^{r-i'} \sum_{k'=0}^{r-i'-j'} \alpha_{i',j',k'}^t\, p_{i',j',k'} \right) \underline{\underline{\lambda}} \underline{e}_m \cdot \underline{e}_n \mathrm{d}a\, \mathrm{d}b\, \mathrm{d}c. \tag{4.49}$$

Now, taking into account the orthogonality of $p_{i,j,k} \circ F_c^{-1}$ and $p_{i',j',k'} \circ F_c^{-1}$ (see Sect. 2.3.1.3), we finally obtain:

$$M_{s,t}^{m,n} = -|J_{T_\ell}| \sum_{i=0}^{r} \sum_{j=0}^{r-i} \sum_{k=0}^{r-i-j} \left(\alpha_{i,j,k}^s \alpha_{i,j,k}^t \int_{C_3} J_c |p_{i,j,k}|^2 \underline{\underline{\lambda}} \underline{e}_m \cdot \underline{e}_n \mathrm{d}a\, \mathrm{d}b\, \mathrm{d}c \right). \tag{4.50}$$

In particular, when $\underline{\underline{\lambda}} = \lambda I_p$, I_p being the unit matrix of $\mathbb{R}^p \times \mathbb{R}^p$, we have:

$$M_{s,t}^{m,n} = -|J_{T_\ell}| \sum_{i=0}^{r} \sum_{j=0}^{r-i} \sum_{k=0}^{r-i-j} \left(\alpha_{i,j,k}^s \alpha_{i,j,k}^t \int_{C_3} J_c \lambda |p_{i,j,k}|^2 \mathrm{d}a\, \mathrm{d}b\, \mathrm{d}c \right) \delta_{mn}. \tag{4.51}$$

Equation (4.51) shows that the mass matrix is a block-diagonal matrix containing p blocks in $\mathbb{R}^{N_r} \times \mathbb{R}^{N_r}$. If λ is constant per element, the definition of this matrix only requires the storage of J_{T_ℓ} and λ per element and the knowledge of the mass matrix on $\hat{T}_3$ to be defined.

Of course, the mass matrix derived from (4.17b) is obtained in the same way.

4.2.2 The Stiffness Integrals

The approximation of the stiffness integrals depends on the equation we have to solve. Approximations of the different equations described in Sect. 4.1.3 use the same change of variables, based on the following result: if $\underline{F}_{AB}$ denotes the mapping from a domain $A \in \mathbb{R}^d$ to a domain $B \in \mathbb{R}^d$ and DF_{AB} is its Jacobian matrix, we have:

$$\nabla \circ \underline{F}_{AB} = DF_{AB}^{-T} \hat{\nabla}, \tag{4.52}$$

where ∇ is the gradient on B, $\hat{\nabla}$ the gradient on A and DF_{AB}^{-T} is the inverse of the transposed matrix of DF_{AB}. However, the treatment of the stiffness integrals depends on the equations.

4.2.2.1 The Acoustics and Linear Elastodynamics Systems

As for the mass integral, we first decompose the variable $\underline{v}_\ell$ involved in (4.36a) on a basis of K_ℓ as follows:

$$\underline{v}_\ell = \sum_{m=1}^{d} \sum_{s=1}^{N_r} v_{\ell,s}^m \, \underline{\psi}_{\ell,s}^m. \tag{4.53}$$

By setting $\varphi_\ell = \varphi_{\ell,t}$, the stiffness integral of (4.36a) then becomes:

$$\int_{K_\ell} \nabla \cdot \underline{v}_\ell \, \varphi_\ell \, \mathrm{d}\underline{x} = \sum_{m=1}^{d} \sum_{s=1}^{N_r} v_{\ell,s}^m \int_{K_\ell} \nabla \cdot \underline{\psi}_{\ell,s}^m \, \varphi_{\ell,t} \, \mathrm{d}\underline{x}. \tag{4.54}$$

By setting $\underline{\psi}_{\ell,s}^m \circ \underline{F}_{T_\ell} = \hat{\psi}_s \, \underline{e}_m$, $\hat{\psi}_s \circ \underline{F}_c = \tilde{\psi}_s$, $\varphi_{\ell,t} \circ \underline{F}_{T_\ell} = \hat{\varphi}_t$, $\hat{\varphi}_t \circ \underline{F}_c = \tilde{\varphi}_t$ and by applying relation (4.52) to K_ℓ, $\hat{T}_d$ and C_d, we get:

$$\begin{aligned} S_{s,t}^m &= \int_{K_\ell} \nabla \cdot \underline{\psi}_{\ell,s}^m \, \varphi_{\ell,t} \, \mathrm{d}\underline{x} = -|J_{T_\ell}| \int_{C_d} J_c \, DF_{T_\ell}^{-T} \, DF_c^{-T} \widetilde{\nabla} \cdot (\tilde{\psi}_s \, \underline{e}_m) \, \tilde{\varphi}_t \, \mathrm{d}a \, \mathrm{d}b \, \mathrm{d}c \\ &= -s_\ell \int_{C_d} M_\ell \, M_c \widetilde{\nabla} \cdot (\tilde{\psi}_s \, \underline{e}_m) \, \tilde{\varphi}_t \, \mathrm{d}a \, \mathrm{d}b \, \mathrm{d}c, \end{aligned} \tag{4.55}$$

where s_ℓ is the sign of J_{T_ℓ}, $M_\ell = \mathrm{adj}(DF_{T_\ell})$ and $M_c = \mathrm{adj}(DF_c)$, $\mathrm{adj}(M)$ denoting the adjugate matrix of a matrix M and $\widetilde{\nabla}$ denoting the gradient operator on C_d.

Remark: To derive (4.55), we have used the identities $\nabla \cdot \underline{\psi}_{\ell,s}^m = \nabla(\underline{\psi}_{\ell,s}) \cdot \underline{e}_m$ and (4.52).

Now, by setting $\widetilde{\nabla} = (\delta_1 \ldots \delta_d)^T$, $M_c = (m^c_{ij})$ and $M_\ell = (m^\ell_{ij})$, since M_ℓ is constant, (4.55) reads:

$$S^m_{s,t} = -s_\ell \sum_{p=1}^{d} m^\ell_{mp} \left(\sum_{n=1}^{d} \int_{C_d} m^c_{pn} (\delta_n \tilde{\psi}_s)\, \tilde{\varphi}_t \, \mathrm{d}a \, \mathrm{d}b \, \mathrm{d}c \right). \tag{4.56}$$

Equation (4.56) shows that $S^m_{s,t}$ provides a $d\,N_r \times N_r$ stiffness matrix containing d blocks corresponding to the d directions.The blocks size is $N_r \times N_r$, each block being derived from the basis functions in one direction. The storage of this matrix can be done in two ways

1. Store all the coefficients $S^m_{s,t}$ in each element K_ℓ, which induces a storage of $d\,N_\ell\,N_r^2$ coefficients.
2. Store the $d\,N_r^2$ elementary integrals involved in (4.56) and the N_ℓ matrices M_ℓ and compute $S^m_{s,t}$ at each time-step. This point of view, proposed by Hesthaven [4] increases the computational time (in a reasonable way) but only requires the storage of $d\,N_r^2 + d^2\,N_\ell$ coefficients (the sign of J_{T_ℓ} is omitted), which leads to a substantial gain of storage, in particular for high-order methods and for large meshes. For instance, when $N_\ell = 400$ (which is quite reasonable), $d = 3$ and $r = 3$, the storage is divided by about 130!

The stiffness integral of (4.36b) is treated in the same way and leads to the adjoint matrix. On the other hand, the stiffness integrals of the linear elastodynamics system (4.38a)–(4.38c) using the gradient and divergence operators, their approximation can be easily deduced from the acoustics system.

4.2.2.2 The Maxwell's System

The two stiffness integrals appearing in (4.37a)–(4.37b) have obviously the same approximation. Let us approximate the first integral. $\underline{H}_\ell$ can decomposed on K_ℓ as follows:

$$\underline{H}_\ell = \sum_{m=1}^{3} \sum_{s=1}^{N_r} H^m_{\ell,s}\, \underline{\psi}^m_{\ell,s}. \tag{4.57}$$

If $\underline{\varphi}_\ell = \underline{\varphi}^n_{\ell,t}$, the stiffness integral of (4.37a) becomes:

$$\int_{K_\ell} \nabla \times \underline{H}_\ell \cdot \underline{\varphi}_\ell \, \mathrm{d}\underline{x} = \sum_{m=1}^{3} \sum_{s=1}^{N_r} H^m_{\ell,s} \int_{K_\ell} \nabla \times \underline{\psi}^m_{\ell,s} \cdot \underline{\varphi}^n_{\ell,t} \, \mathrm{d}\underline{x}. \tag{4.58}$$

Now, by using the same changes of variables as the previous section, we get:

$$
\begin{aligned}
S_{s,t}^{m,n} &= \int_{K_\ell} \nabla \times \underline{\psi}_{\ell,s}^{m} \cdot \underline{\varphi}_{\ell,t}^{n} \,\mathrm{d}\underline{x} \\
&= -|J_{T_\ell}| \int_{C_d} J_c \,(DF_{T_\ell}^{-T} DF_c^{-T} \widetilde{\nabla}) \times (\tilde{\psi}_s \, \underline{e}_m) \cdot \tilde{\varphi}_t \, \underline{e}_n \,\mathrm{d}a \,\mathrm{d}b \,\mathrm{d}c \\
&= -s_\ell \int_{C_d} (M_\ell \, M_c \widetilde{\nabla}) \times (\tilde{\psi}_s \, \underline{e}_m) \cdot \tilde{\varphi}_s \, \underline{e}_n \,\mathrm{d}a \,\mathrm{d}b \,\mathrm{d}c.
\end{aligned} \tag{4.59}
$$

After some algebra, (4.59) provides the following skew block matrix:

$$
\begin{pmatrix} 0 & B_{12} & B_{13} \\ -B_{12} & 0 & B_{23} \\ -B_{13} & -B_{23} & 0 \end{pmatrix}, \tag{4.60}
$$

where, using the notations of the previous section, $B_{mn} = (S_{s,t}^{m,n})$, with

$$
S_{s,t}^{m,n} = -s_\ell \sum_{p=1}^{3} m_{\kappa p}^{\ell} \left(\sum_{q=1}^{3} \int_{C_d} m_{pq}^{c} \,(\delta_q \tilde{\psi}_s) \, \tilde{\varphi}_t \,\mathrm{d}a \,\mathrm{d}b \,\mathrm{d}c \right),
$$

with[2] $\kappa = 6 - m - n$.

Of course, the two points of view for the storage given in the previous section hold for the Maxwell's equations.

4.2.3 The Jump Terms

As the stiffness integrals, the jump integrals cannot be treated in a general frame. These terms however all use the following relations. With the notations of Sect. 4.2.2, we have, for any unit normal $\underline{n}$ to ∂K_ℓ:

$$
\underline{n} \circ \underline{F}_{AB} = \frac{DF_{AB}^{-T} \hat{\underline{n}}}{||DF_{AB}^{-T} \hat{\underline{n}}||}. \tag{4.61}
$$

On the other hand, we have:

$$
\mathrm{d}\sigma = ||J_{AB} \, DF_{AB}^{-T} \hat{\underline{n}}|| \mathrm{d}\hat{\sigma}. \tag{4.62}
$$

[2] One could also set $\kappa = 6/(mn)$.

4.2.3.1 The Acoustics and Linear Elastodynamics Systems

Let us first set $\partial K_\ell = \bigcup_{p=1}^{d+1} \Gamma_{\ell,p}$, $\Gamma_{\ell,p}$ being a face (or an edge when $d = 2$) of ∂K_ℓ.
The jump term of (4.36a) can be rewritten as

$$\int_{\partial K_\ell} \underline{n} \cdot [\underline{v}_h]^{K_\ell}_{\partial K_\ell} \varphi_\ell \, \mathrm{d}\sigma = \sum_{p=1}^{d+1} \int_{\Gamma_{\ell,p}} \underline{n}^\ell_p \cdot [\underline{v}_h]^{K_\ell}_{\Gamma_{\ell,p}} \varphi_\ell \, \mathrm{d}\sigma, \tag{4.63}$$

where $\underline{n}^\ell_p$ is the exterior normal to K_ℓ on $\Gamma_{\ell,p}$.

Let us now consider $\Gamma_{\ell,p} = K_\ell \cap K_{\ell'}$. We have:

$$I = \int_{\Gamma_{\ell,p}} \underline{n}^\ell_p \cdot [\underline{v}_h]^{K_\ell}_{\Gamma_{\ell,p}} \varphi_\ell \, \mathrm{d}\sigma = \int_{\Gamma_{\ell,p}} \underline{n}^\ell_p \cdot (\underline{v}_{\ell'} - \underline{v}_\ell) \varphi_\ell \, \mathrm{d}\sigma. \tag{4.64}$$

By taking into account (4.53) and by setting $\varphi_\ell = \varphi_{\ell,t}$, we get:

$$I = \sum_{m=1}^{d} \left(\sum_{s=1}^{N^\partial_r} v^m_{\ell',s} \int_{\Gamma_{\ell,p}} \underline{n}^\ell_p \cdot \underline{\psi}^m_{\ell',s} \varphi_{\ell,t} \, \mathrm{d}\sigma - \sum_{s=1}^{N^\partial_r} v^m_{\ell,s} \int_{\Gamma_{\ell,p}} \underline{n}^\ell_p \cdot \underline{\psi}^m_{\ell,s} \varphi_{\ell,t} \, \mathrm{d}\sigma \right), \tag{4.65}$$

where N^∂_r ($= (r+1)(r+2)/2$ in 3D and $= r+1$ in 2D)[3] is the number of interpolation points on a face (an edge) of K_ℓ and s and s' are such that, if $\xi_{\ell,s}$ and $\xi_{\ell',s'}$ are the two interpolation points of K_ℓ and K'_ℓ such that $\underline{\psi}^m_{\ell,s}(\xi_{\ell,s}) = \underline{\psi}^m_{\ell',s'}(\xi_{\ell',s'}) = \underline{e}_m$, then $||\xi_{\ell',s'} - \xi_{\ell,s}|| = 0$.

So, we can write $s' = i(s)$. On the other hand, we have $\underline{\psi}^m_{\ell,s|\Gamma_{\ell,p}} = \underline{\psi}^m_{\ell',s'|\Gamma_{\ell,p}}$.
Equation (4.53) then becomes:

$$I = \sum_{m=1}^{d} \sum_{s=1}^{N^\partial_r} (v^m_{\ell',i(s)} - v^m_{\ell,s}) \int_{\Gamma_{\ell,p}} \underline{n}^\ell_p \cdot \underline{\psi}^m_{\ell,s} \varphi_{\ell,t} \, \mathrm{d}\sigma, \tag{4.66}$$

By using (4.61) and (4.62) for the changes of variables from ∂K_ℓ to $\hat{T}_d$ then to C_d, we get, with the notations of Sect. 4.2.2.1:

$$J^m_{s,t} = \int_{\Gamma_{\ell,p}} \underline{n}^\ell_p \cdot \underline{\psi}^m_{\ell,s} \varphi_{\ell,t} \, \mathrm{d}\sigma = -|J_{T_\ell}| \int_{\widetilde{\Gamma_p}} J_c \, DF^{-T}_{T_\ell} \, DF^{-T}_c \, \underline{\tilde{n}}_p \cdot \tilde{\psi}_s \, \underline{e}_m \, \tilde{\varphi}_t \, \mathrm{d}\tilde{\sigma}$$

---

[3] For sake of simplicity, we assume that the interpolation points of a face are sequentially numbered from 1 to N^∂_r.

$$= -s_\ell \int_{\widetilde{\Gamma}_p} M_\ell\, M_c\, \underline{\tilde{n}}_p \cdot \tilde{\psi}_s\, \underline{e}_m\, \tilde{\varphi}_t\, \mathrm{d}\tilde{\sigma}$$

$$= -s_\ell \sum_{q=1}^{d} m^\ell_{mq} \left(\sum_{n=1}^{d} \int_{\widetilde{\Gamma}_p} m^c_{qn} (\tilde{n}_n \tilde{\psi}_s)\, \tilde{\varphi}_t\, \mathrm{d}\tilde{\sigma} \right), \tag{4.67}$$

where $\widetilde{\Gamma}_p$ is a face (an edge in 2D) of C_d and $\underline{\tilde{n}}_p = (\tilde{n}_n)^d_{n=1}$ its exterior normal.
Since $\underline{\tilde{n}}_p$ is normal to the boundary of the unit cube (or square), we have:

$$n_n = \eta \delta_{pn}, \; \eta = \pm 1 \tag{4.68}$$

so that

$$J^m_{s,t} = -\eta\, s_\ell \sum_{q=1}^{d} m^\ell_{mq} \int_{\widetilde{\Gamma}_p} m^c_{qn} (\tilde{n}_p \tilde{\psi}_s)\, \tilde{\varphi}_t\, \mathrm{d}\tilde{\sigma}. \tag{4.69}$$

Remarks:

1. (4.67) shows that we get a jump matrix containing d blocks of $N^\partial_r \times N^\partial_r$ elements. The jump term of (4.36b) provides the transposed matrix. The jump terms of the linear elastodynamics system can be easily derived from the acoustics jump terms.
2. $i(s)$ provides the correspondence of the numbers of the interpolation points for one of the 4×6 possible permutations of the faces of a tetrahedron (3×2 for the edges of a triangle in 2D).

4.2.3.2 The Maxwell's System

With notations of Sects. 4.2.2.2 and 4.2.3.1, the non dissipative jump term of (4.37a) reads:

$$\int_{\partial K_\ell} [\underline{H} \times \underline{n}]^{K_\ell}_{\partial K_\ell} \cdot \underline{\varphi}_\ell\, \mathrm{d}\sigma = \sum_{p=1}^{d+1} \int_{\Gamma_{\ell,p}} [\underline{H} \times \underline{n}^\ell_p]^{K_\ell}_{\Gamma_{\ell,p}} \cdot \underline{\varphi}_\ell\, \mathrm{d}\sigma, \tag{4.70}$$

which provides, on $\Gamma_{\ell,p}$:

$$I = \int_{\Gamma_{\ell,p}} [\underline{H} \times \underline{n}^\ell_p]^{K_\ell}_{\Gamma_{\ell,p}} \cdot \underline{\varphi}_\ell\, \mathrm{d}\sigma, = \int_{\Gamma_{\ell,p}} (\underline{H}_{\ell'} - \underline{H}_\ell) \times \underline{n}^\ell_p \cdot \underline{\varphi}_\ell\, \mathrm{d}\sigma \tag{4.71}$$

Now, by taking into account (4.57) and by setting $\underline{\varphi}_\ell = \underline{\varphi}^n_{\ell,t}$, we get:

$$I = \sum_{m=1}^{d} \left(\sum_{s'=1}^{N^\partial_r} H^m_{\ell,s'} \int_{\Gamma_{\ell,p}} \underline{n}^\ell_p \times \underline{\psi}^m_{\ell,s'} \cdot \underline{\varphi}^n_{\ell,t}\, \mathrm{d}\sigma - \sum_{s=1}^{N^\partial_r} H^m_{\ell,s} \int_{\Gamma_{\ell,p}} \underline{n}^\ell_p \times \underline{\psi}^m_{\ell,s} \cdot \underline{\varphi}^n_{\ell,t}\, \mathrm{d}\sigma \right). \tag{4.72}$$

As for the acoustics equation and with notations of Sect. 4.2.2.2, the right term of I finally provides, taking into account (4.68):

$$\begin{aligned} J_{s,t}^{m,n} &= \int_{\Gamma_{\ell,p}} \underline{n}_p^{\ell} \times \underline{\psi}_{\ell,s}^{m} \cdot \underline{\varphi}_{\ell,t}^{n} \, \mathrm{d}\sigma \\ &= -\eta \, s_\ell \sum_{q=1}^{3} m_{\kappa q}^{\ell} \int_{\widetilde{\Gamma}_p} m_{qp}^{c} (n_p \tilde{\psi}_s) \, \tilde{\varphi}_t \, \mathrm{d}\widetilde{\sigma}, \end{aligned} \tag{4.73}$$

$\kappa = 6 - m - n$.

Remark: It is easy to see that (4.73) leads to a skew block matrix similar to (4.60).

Let us now consider the dissipative jump term of (4.37a). A similar process leads to the following elementary jump term:

$$J_{s,t}^{m,n} = \int_{K_\ell} (\underline{n}_p^{\ell} \times \underline{\psi}_{\ell,s}^{m}) \times \underline{n}_p^{\ell} \cdot \underline{\varphi}_{\ell,t}^{n} \, \mathrm{d}\sigma. \tag{4.74}$$

By applying the changes of variables from ∂K_ℓ to $\hat{T}_d$ then to C_d, we get, with the notations of Sect. 4.2.2.1:

$$J_{s,t}^{m,n} = -s_\ell \int_{\widetilde{\Gamma}_p} (M_\ell \, M_c \, \underline{\tilde{n}}_p \times \tilde{\psi}_s \, \underline{e}_m) \times M_\ell \, M_c \, \underline{\tilde{n}}_p \cdot \tilde{\varphi}_s \, \underline{e}_n \, \mathrm{d}\widetilde{\sigma}. \tag{4.75}$$

After some algebra, $J_{s,t}^{m,n}$ reads:

$$J_{s,t}^{m,n} = -s_\ell \int_{\widetilde{\Gamma}_p} [\delta_{mn} \sum_{q=1,\, q \neq n}^{3} P_{pq}^2 - (1 - \delta_{mn}) P_{pm} P_{pn}] \tilde{\psi}_s \tilde{\varphi}_s \, \mathrm{d}\widetilde{\sigma}, \tag{4.76}$$

with $P_{pq} = \sum_{i=1}^{3} m_{qi}^{\ell} m_{ip}^{c}$.

Equation (4.76) shows that $J_{s,t}^{m,n}$ uses products of coefficients of matrices M_ℓ and M_c. So, in order to avoid the storage of this term for each degree of freedom, one must first expand P_{pq}^2 and $P_{pm} P_{pn}$. In these expansions appear the products of the coefficients. In fact, one must then store the products of coefficients of M_c on the unit cube, which are, by taking into account symmetries, 45 different products. On the other hand, we store the products of coefficients of M_ℓ on each element, i.e., by taking into account symmetries, 18 coefficients. In the last step, one must recompute P_{pq}^2 and $P_{pm} P_{pn}$ from these coefficients, which induces an increase of the CPU time. However, this additional time is not substantial since only the boundary points of the elements are concerned.

4.3 Approximation by Quadrilaterals and Hexahedra

The approximation on quadrilaterals and hexahedra is based on mass-lumping and mixed formulation which, as we shall see, lead here also to a substantial gain of storage.

We consider the general system (4.17a)–(4.17c). As above, we first define a mesh $\mathscr{T}_h$ composed of quadrilaterals or hexahedra denoted K_ℓ and the unit element $\widehat{K} = [0, 1]^d$. $\underline{F}_\ell$ is the transform such that $K_\ell = \underline{F}_\ell(\widehat{K})$. Let us now introduce the two spaces of approximations. $\underline{u}_\ell$ is sought in

$$\underline{U}_h^r = \left\{\underline{w}_h \in [L^2(\Omega)]^p \text{ such that } M_{u,\ell}^{-1}\underline{w}_{h|K_\ell} \circ \underline{F}_\ell \in [Q_r]^p\right\} \tag{4.77}$$

and $\underline{v}_\ell$ is sought in

$$\underline{V}_h^r = \left\{\underline{w}_h \in [L^2(\Omega)]^q \text{ such that } M_{v,\ell}^{-1}\underline{w}_{h|K_\ell} \circ \underline{F}_\ell \in [Q_r]^q\right\} \tag{4.78}$$

where $M_{u,\ell}$ and $M_{v,\ell}$ are matrices such that

$$(A\underline{v}_\ell) \circ \underline{F}_\ell = (J_\ell^{-1} M_{u,\ell}^{-T}\, \hat{A})\, \hat{\underline{v}}_\ell,\ \forall \underline{v}_\ell \in \underline{V}_{h,}^r \tag{4.79a}$$

$$(A^*\underline{u}_\ell) \circ \underline{F}_\ell = (J_\ell^{-1} M_{v,\ell}^{-T}\, \hat{A}^*)\, \hat{\underline{u}}_\ell,\ \forall \underline{u}_\ell \in \underline{U}_{h,}^r \tag{4.79b}$$

where $\hat{A}$, $\hat{A}^*$, $\hat{\underline{u}}_\ell = M_{u,\ell}^{-1}\underline{u}_\ell$ and $\hat{\underline{v}}_\ell = M_{v,\ell}^{-1}\underline{v}_\ell$ are in $\widehat{K}$-coordinates and J_ℓ is the Jacobian of $\underline{F}_\ell$.

In the following, we only compute the integrals appearing in (4.17a), the integrals of (4.17b) being deduced in the same way. We suppose that the basis functions of $[Q_r]^s$ on the unit element are $\hat{\varphi}_i\, \underline{e}_n$, $i = 1 \ldots (r+1)^d$, $n = 1 \ldots s$, $s = p$ or q, $\hat{\varphi}_i$ being defined by (3.13a)–(3.13b). Moreover, if $\hat{\Xi} = (\hat{\xi}_j)_{j=1}^{r+1}$ denotes a set of quadrature points on [0, 1] whose weights are $(\hat{\omega}_k)_{k=1}^{r+1}$, we have:

$$\forall i = 1 \ldots (r+1)^d,\ \forall j = 1 \ldots (r+1)^d,\ \hat{\varphi}_i(\hat{\underline{\xi}}_j) = \delta_{ij}, \tag{4.80}$$

where $\hat{\underline{\xi}}_j \in \hat{\Xi}^d$.

4.3.1 The Mass Integrals

Let us compute the mass integral. As in Sect. 3.2.2, we have:

$$I_M = \int_{K_\ell} \underline{\underline{\lambda}}\, \underline{u}_\ell \cdot \underline{\varphi}_\ell\, \mathrm{d}\underline{x} = \int_{\widehat{K}} |J_\ell|\, \underline{\underline{\lambda}} \circ \underline{F}_\ell\, \underline{u}_\ell \circ \underline{F}_\ell \cdot \underline{\varphi}_\ell \circ \underline{F}_\ell\, \mathrm{d}\hat{\underline{x}}. \tag{4.81}$$

Taking into account the definition of $\underline{U}_h^r$, we can set $\underline{\varphi}_\ell = \underline{\varphi}_{\ell,j}^n$, where

$$\underline{\varphi}_{\ell,j}^n \circ \underline{F}_\ell = M_{u,\ell}\, \hat{\varphi}_j\, \underline{e}_n \tag{4.82}$$

Since, on the other hand,

$$\underline{u}_\ell = \sum_{m=1}^{p} \sum_{i=1}^{(r+1)^d} u_{\ell,i}^m\, \underline{\varphi}_{\ell,i}^m, \tag{4.83}$$

by applying (4.82) to $\underline{\varphi}_{\ell,i}^m$, we get

$$I_M = \sum_{m=1}^{p} \sum_{i=1}^{(r+1)^d} u_{\ell,i}^m \int_{\widehat{K}} |J_\ell|\, \hat{\varphi}_i\, \hat{\varphi}_j\, M_{u,\ell}^T \underline{\underline{\hat{\lambda}}}\, M_{u,\ell}\, \underline{e}_m \cdot \underline{e}_n \mathrm{d}\underline{\hat{x}}, \tag{4.84}$$

with $\underline{\underline{\hat{\lambda}}} = \underline{\underline{\lambda}} \circ \underline{F}_\ell$.

Now, by setting $C_\ell = |J_\ell| M_{u,\ell}^T \underline{\underline{\hat{\lambda}}}\, M_{u,\ell} = (c_{s,t}^\ell)_{s=1\ldots p,\, t=1\ldots p}$, one can easily check that $C_\ell\, \underline{e}_m \cdot \underline{e}_n = c_{m,n}^\ell$. On the other hand, by computing the last integral of (4.84) by using the quadrature rule defined above and taking into account notations (4.80), we get:

$$\begin{aligned} I_M &= \sum_{m=1}^{p} \sum_{i=1}^{(r+1)^d} u_{\ell,i}^m \int_{\widehat{K}} \hat{\varphi}_i\, \hat{\varphi}_j\, c_{m,n}^\ell \mathrm{d}\underline{\hat{x}} \\ &\simeq \sum_{m=1}^{p} \sum_{i=1}^{r+1} u_{\ell,i}^m \sum_{s=1}^{(r+1)^d} \hat{\omega}_s \hat{\varphi}_i(\underline{\hat{\xi}}_s)\, \hat{\varphi}_j(\underline{\hat{\xi}}_s)\, c_{m,n}^\ell = \sum_{m=1}^{p} \hat{\omega}_j\, c_{m,n}^\ell\, u_{\ell,j}^m. \end{aligned} \tag{4.85}$$

As in Sect. 3.2.2, (4.85) shows that we get a $p \times p$ symmetric mass matrix whose terms are $\hat{\omega}_j\, c_{m,n}^\ell$ for each point $\underline{F}_\ell(\underline{\hat{\xi}}_j)$ of interpolation of K_ℓ. In other words, the mass matrix is a $p \times p$ block-diagonal symmetric matrix whose blocks are $\hat{\omega}_j\, C_\ell$.

Remark: The components of the vector-valued basis functions are the functions defined by (3.13a)–(3.13b). So, for any basis function $\underline{\theta}_{\ell,j}^n \in \underline{U}_h^r$ or $\in \underline{V}_h^r$, one can write:

$$\underline{\theta}_{\ell,j}^n \circ \underline{F}_\ell = M_{u,\ell}\, \hat{\varphi}_j\, \underline{e}_n. \tag{4.86}$$

4.3.2 *The Stiffness Integrals*

As $\underline{u}_\ell, \underline{v}_\ell$ can be decomposed as follows:

$$\underline{v}_\ell = \sum_{m=1}^{q} \sum_{s=1}^{(r+1)^d} v^m_{\ell,i}\, \underline{\psi}^m_{\ell,s}. \tag{4.87}$$

By using (4.87) and by setting $\underline{\varphi}_\ell = \underline{\varphi}^n_{\ell,t}$, the stiffness integral of (4.17a) reads:

$$\int_{K_\ell} A\underline{v}_\ell \cdot \underline{\varphi}_\ell \,\mathrm{d}\underline{x} = \sum_{m=1}^{q} \sum_{s=1}^{N_r} v^m_{\ell,s} \int_{K_\ell} A\underline{\psi}^m_{\ell,s} \cdot \underline{\varphi}^n_{\ell,t} \,\mathrm{d}\underline{x}. \tag{4.88}$$

We have:

$$S^{m,n}_{s,t} = \int_{K_\ell} A\underline{\psi}^m_{\ell,s} \cdot \underline{\varphi}^n_{\ell,t} \,\mathrm{d}\underline{x} = \int_{\widehat{K}} |J_\ell| \left(A\underline{\psi}^m_{\ell,s}\right) \circ \underline{F}_\ell \cdot \underline{\varphi}^n_{\ell,t} \circ \underline{F}_\ell \,\mathrm{d}\hat{\underline{x}}. \tag{4.89}$$

Taking into account (4.77) and (4.79a), we get

$$\begin{aligned} S^{m,n}_{s,t} &= \int_{\widehat{K}} |J_\ell|\, (J_\ell^{-1} M^{-T}_{u,\ell}\, \hat{A}) \hat{\underline{\psi}}^m_s \cdot M_{u,\ell}\, \hat{\underline{\varphi}}^n_t \,\mathrm{d}\hat{\underline{x}} \\ &= s_\ell \int_{\widehat{K}} \hat{A} \hat{\underline{\psi}}^m_s \cdot \hat{\underline{\varphi}}^n_t \,\mathrm{d}\hat{\underline{x}}, \end{aligned} \tag{4.90}$$

where s_ℓ is the sign of J_ℓ.

Equation (4.90) shows that the stiffness matrix only needs a storage on the unit element. The discretization of the stiffness integral will be given below.

4.3.3 The Jump Terms

We set $\partial K_\ell = \bigcup_{s=1}^{2d} \Gamma_{\ell,i}$, $\Gamma_{\ell,i}$ being a face (or an edge when $d = 2$) of ∂K_ℓ. The non dissipative jump term of (4.17a) can then be rewritten as

$$\int_{\partial K_\ell} A(\underline{n})[\underline{v}_h]^{K_\ell}_{\partial K_\ell} \cdot \underline{\varphi}_\ell \,\mathrm{d}\sigma = \sum_{i=1}^{2d} \int_{\Gamma_{\ell,i}} A(\underline{n}^\ell_i)[\underline{v}_h]^{K_\ell}_{\Gamma_{\ell,i}} \cdot \underline{\varphi}_\ell \,\mathrm{d}\sigma, \tag{4.91}$$

where $\underline{n}^\ell_i$ is the exterior normal to the face $\Gamma_{\ell,i}$ of K_ℓ.

By setting $\Gamma_{\ell,i} = K_\ell \cap K_{\ell'}$, we have:

$$I = \int_{\Gamma_{\ell,i}} A(\underline{n}^\ell_i)[\underline{v}_h]^{K_\ell}_{\Gamma_{\ell,i}} \cdot \underline{\varphi}_\ell \,\mathrm{d}\sigma = \int_{\Gamma_{\ell,i}} A(\underline{n}^\ell_i) \cdot (\underline{v}_{\ell'} - \underline{v}_\ell)\varphi_\ell \,\mathrm{d}\sigma. \tag{4.92}$$

If $\underline{\varphi}_\ell = \underline{\varphi}^n_{\ell,t}$, we can write:

$$\begin{aligned} I = \sum_{m=1}^{q}\sum_{s'=1}^{N} v^m_{\ell',s'} \int_{\Gamma_{\ell,i}} A(\underline{n}^\ell_i)\underline{\psi}^m_{\ell',s'} \cdot \underline{\varphi}^n_{\ell,t}\, \mathrm{d}\sigma \\ - \sum_{m=1}^{q}\sum_{s=1}^{N} v^m_{\ell,s} \int_{\Gamma_{\ell,i}} A(\underline{n}^\ell_i)\underline{\psi}^m_{\ell,s} \cdot \underline{\varphi}^n_{\ell,t}\, \mathrm{d}\sigma = I_L - I_R, \end{aligned} \tag{4.93}$$

where $N \leq (r+1)^d$ will be defined later.

We now assume that

$$(A(\underline{n}^\ell_i)\underline{\psi}^m_{\ell,s}) \circ \underline{F}_\ell = \frac{(J^{-1}_\ell M^{-T}_{u,\ell}\, \hat{A}(\hat{\underline{n}}))}{||DF^{-T}_\ell \hat{\underline{n}}||} \hat{\varphi}_s\, \underline{e}_m, \tag{4.94}$$

where $\hat{\underline{n}}$ is the exterior normal to the face $\widehat{\Gamma}$ of $\widehat{K}$ such that $\underline{F}_\ell(\widehat{\Gamma}) = \Gamma_{\ell,i}$ and $\hat{A}(\hat{\underline{n}})$ is in $\widehat{K}$ coordinates.

Then, by change of variables and taking into account (4.82), the integral involved in I_R reads:

$$\begin{aligned} J^{m,n}_{s,t} &= \int_{\Gamma_{\ell,i}} A(\underline{n}^\ell_i)\underline{\psi}^m_{\ell,s} \cdot \underline{\varphi}^n_{\ell,t}\, \mathrm{d}\sigma \\ &= \int_{\widehat{\Gamma}} ||J_\ell\, DF^{-T}_\ell \hat{\underline{n}}|| \frac{(J^{-1}_\ell M^{-T}_{u,\ell}\, \hat{A}(\hat{\underline{n}}))}{||DF^{-T}_\ell \hat{\underline{n}}||} \hat{\varphi}_s\, \underline{e}_m \cdot M_{u,\ell}\, \hat{\varphi}_t\, \underline{e}_n\, \mathrm{d}\hat{\sigma} \\ &= s_\ell \int_{\widehat{\Gamma}} \hat{A}(\hat{\underline{n}})\, \hat{\varphi}_s\, \underline{e}_m \cdot \hat{\varphi}_t\, \underline{e}_n\, \mathrm{d}\hat{\sigma}. \end{aligned} \tag{4.95}$$

Note that $\underline{F}_{\ell'}(\widehat{\Gamma})$ can be different from $\Gamma_{\ell,i}$.[4] So, we denote $\underline{F}_{\ell'}(\widehat{\Gamma}') = \Gamma_{\ell,i}$ and the integral involved in I_L provides:

$$J^{m,n}_{s',t} = s_\ell \int_{\widehat{\Gamma}'} \hat{A}(\hat{\underline{n}})\, \hat{\varphi}_{s'}\, \underline{e}_m \cdot \hat{\varphi}_t\, \underline{e}_n\, \mathrm{d}\hat{\sigma}. \tag{4.96}$$

Equation (4.95) shows that, as the stiffness matrix, the non dissipative jump matrix needs only a storage on the unit element. The dissipative jump will be treated for the Maxwell's system below.

[4]This is due to the different orientations of $\widehat{K}$ which cannot ensure that the local faces $\underline{F}^{-1}_\ell(\Gamma_{\ell,i})$ and $\underline{F}^{-1}_{\ell'}(\Gamma_{\ell,i})$ are the same on both sides of $\Gamma_{\ell,i}$.

4.3.4 *Application to Wave Equations*

4.3.4.1 The Acoustics and Linear Elastodynamics Systems

We recall that, for the acoustics system defined by (3.29a)–(3.29b), we have $p = 1$, $q = d$, $\underline{u} = u$, $\underline{v} = \underline{v}$, $A = -\text{div}$, $A^* = \text{grad}$, $A(\underline{n}) = -(n_1\ n_2\ n_3)$ and $A^*(\underline{n}) = (n_1\ n_2\ n_3)^T$ for $d = 3$ ($A(\underline{n}) = -(n_1\ n_2)$ and $A^*(\underline{n}) = (n_1\ n_2)^T$ for $d = 2$). On the other hand, let us suppose that, for any $u \in \underline{U}^r_{h,}$, we have $\hat{u} = u \circ \underline{F}_\ell$. We get:

$$(\nabla u) \circ \underline{F}_\ell = DF_\ell^{-T} \hat{\nabla} \hat{u}, \tag{4.97}$$

where $\hat{\nabla}$ is the gradient in $\widehat{K}$ coordinates.

So, taking into account (4.78) and (4.79b), we get $M_{v,\ell} = J_\ell^{-1} DF_\ell$. On the other hand, this definition combined with (4.78) leads to the following identity, for any $\underline{v} \in \underline{V}^r_{h,}$:

$$\begin{aligned}
\int_{K_\ell} \nabla \cdot \underline{v} \varphi \, \mathrm{d}x &= -\int_{K_\ell} \underline{v} \cdot \nabla \varphi \, \mathrm{d}\underline{x} + \int_{\partial K_\ell} \underline{v} \cdot \underline{n} \, \varphi \, \mathrm{d}\sigma \\
&= -\int_{\widehat{K}} |J_\ell| \underline{v} \circ \underline{F}_\ell \cdot (\nabla \varphi) \circ \underline{F}_\ell \, \mathrm{d}\hat{\underline{x}} \\
&\quad + \int_{\partial \widehat{K}} ||J_\ell \, DF_\ell^{-T} \underline{\hat{n}}|| \underline{v} \circ \underline{F}_\ell \cdot \underline{n} \circ \underline{F}_\ell \, \varphi \circ \underline{F}_\ell \, \mathrm{d}\hat{\sigma} \\
&= -\int_{\widehat{K}} |J_\ell| J_\ell^{-1} DF_\ell \, \hat{\underline{v}} \cdot DF_\ell^{-T} \hat{\nabla} \hat{\varphi} \, \mathrm{d}\hat{\underline{x}} \\
&\quad + \int_{\partial \widehat{K}} ||J_\ell \, DF_\ell^{-T} \underline{\hat{n}}|| J_\ell^{-1} DF_\ell \, \hat{\underline{v}} \cdot \frac{(DF_\ell^{-T} \, \underline{\hat{n}})}{||DF_\ell^{-T} \underline{\hat{n}}||} \, \hat{\varphi} \, \mathrm{d}\hat{\sigma} \\
&= -s_\ell \int_{\widehat{K}} \hat{\underline{v}} \cdot \hat{\nabla} \hat{\varphi} \, \mathrm{d}\hat{\underline{x}} + s_\ell \int_{\partial \widehat{K}} \hat{\underline{v}} \cdot \underline{\hat{n}} \, \hat{\varphi} \, \mathrm{d}\hat{\sigma}.
\end{aligned} \tag{4.98}$$

By integrating by parts and using density arguments, we get:

$$(\nabla \cdot \underline{v}) \circ \underline{F}_\ell = J_\ell^{-1} \hat{\nabla} \cdot \hat{\underline{v}}, \tag{4.99}$$

which confirms our definition for $\hat{u}$. In order to avoid the sign s_ℓ, we replace in this case[5] $M_{v,\ell} = J_\ell^{-1} DF_\ell$ by $M_{v,\ell} = |J_\ell^{-1}| DF_\ell$.

The computation of the mass and stiffness matrices is the same as the computation matrices of continuous elements described in Sects. 3.2.2 and 3.2.3. Let us compute the jump term. according to the definition of $A(\underline{n})$ and $A^*(\underline{n})$, (4.94) holds for this system. So, we have to compute the integral:

---

[5]This will not be possible for the Maxwell's system, in which J_ℓ^{-1} appears in the change of variable and not in the definition of the space of approximation.

$$J_{s,t}^{m,n} = \int_{\widehat{\Gamma}} \underline{\hat{n}} \cdot \hat{\varphi}_s \, \underline{e}_m \, \hat{\varphi}_t \, \mathrm{d}\hat{\sigma} \tag{4.100}$$

by using the quadrature rule introduced in the beginning of Sect. 4.3. If we suppose that $d = 3$ and $\widehat{\Gamma} = \{\underline{\hat{x}} \in \widehat{K}$ such that $\hat{x}_1 = 0\}$, then $\underline{\hat{n}} = -\underline{e}_1$. So, we get:

$$J_{s,t}^{m,n} = -\delta_{1m} \int_0^1 \int_0^1 \hat{\varphi}_s(0, \hat{x}_2, \hat{x}_3)\, \hat{\varphi}_t(0, \hat{x}_2, \hat{x}_3)\, \mathrm{d}\hat{x}_2\, \mathrm{d}\hat{x}_3, \tag{4.101}$$

δ_{1m} being the Kronecker's symbol.

Now, by taking into account (3.11), (3.12) and (3.13a)–(3.13b) and the fact that $\hat{\varphi}_t$ is the test function, and by using the quadrature rule, we obtain:

$$\begin{aligned} J_{s,t}^{m,n} &\simeq -\delta_{1m} \hat{\varphi}_{s_1}(0)\, \hat{\varphi}_{t_1}(0) \sum_{n=1}^{r+1} \hat{\omega}_n\, \hat{\varphi}_{s_2}(\hat{\xi}_n)\, \hat{\varphi}_{t_2}(\hat{\xi}_n) \sum_{n=1}^{r+1} \hat{\omega}_n\, \hat{\varphi}_{s_3}(\hat{\xi}_n)\, \hat{\varphi}_{t_3}(\hat{\xi}_n) \\ &\simeq -\delta_{1m} \hat{\varphi}_{s_1}(0)\, \hat{\varphi}_{t_1}(0)\, \hat{\omega}_{t_2}\, \hat{\omega}_{t_3}. \end{aligned} \tag{4.102}$$

Let us now come back to the second term of (4.93). By using (4.102), this term reads, in our case:

$$I_R = -\delta_{1m}\, \hat{\varphi}_{t_1}(0)\, \hat{\omega}_{t_2}\, \hat{\omega}_{t_3} \sum_{s_1=1}^{r+1} v_{\ell,s}^m\, \hat{\varphi}_{s_1}(0), \tag{4.103}$$

with $s = (r+1)[(r+1)(t_3 - 1) + t_2 - 1] + s_1$.

To this point, we must define our quadrature rule. The discontinuous character of the degrees of freedom allows us to use either Gauss or Gauss–Lobatto rules. For the Gauss–Lobatto rule, we have $\hat{\xi}_1 = 0$ which implies that $s_1 = t_1 = 1$ and $\hat{\varphi}_{s_1}(0) = \hat{\psi}_{t_1}(0) = 1$, so that our jump becomes:

$$I_R = -\delta_{1m}\, \hat{\omega}_{t_2}\, \hat{\omega}_{t_3} v_{\ell,s}^m, \tag{4.104}$$

with $s = (r+1)[(r+1)(t_3 - 1) + t_2 - 1] + 1$.

By cons, none of the Gauss points is equal to 0. So, the jump is given by (4.103) and its computation involves $r + 1$ values of $v_{\ell,s}^m$ at the points $(\hat{\xi}_{s_1}, \hat{\xi}_{t_2}, \hat{\xi}_{t_3})$, $s_1 = 1 \dots r+1$. Moreover, this term exists for all the basis functions of the unit element, whereas the Gauss–Lobatto rule only involves functions on the boundary of this element. This seems to lead to a much more expensive approximation than Gauss–Lobatto points. However, we shall see later that this additional cost is balanced by the higher accuracy of the Gauss rule.

Now, since, for conforming meshes, we have $\underline{\psi}_{\ell',s'|\Gamma_{\ell,i}}^m = \underline{\psi}_{\ell,s|\Gamma_{\ell,i}}^m$ if $\underline{\psi}_{\ell',s'|\Gamma_{\ell,i}}^m(\xi) = \underline{\psi}_{\ell,s|\Gamma_{\ell,i}}^m(\xi) = 1$, ξ being an interpolation point of the face, one can check that (4.104) hold for I_L, by replacing s by $s' = i(s)$. As in Sect. 4.2.3.1, $i(s)$ provides the correspondence of the numbers of the interpolation points for one of the 6×8 possible permutations of the faces of an hexahedron (4×2 for the edges of a quadrilateral in 2D).

For Gauss points, (4.103) also holds for element $K_{\ell'}$ by replacing s by s'. However, one can no longer set $s' = i(s)$ since the points are not located on a face. The relation between s and s' can be expressed as follows: If $P(\underline{x})$ and $P'(\underline{x})$ denote the projection of a point $\underline{x}$ on the faces $\widehat{\Gamma}$ and $\widehat{\Gamma}'$ defined at the end of Sect. 4.3.3, then $\underline{F}_\ell(P(\underline{\xi}_s)) = \underline{F}_{\ell'}(P'(\underline{\xi}_{s'}))$. One can however notice that s' is not necessarily defined as s since the normals to $\widehat{\Gamma}$ and $\widehat{\Gamma}'$ are not necessarily in the same direction in local coordinates.

Equation (4.103) shows that the jump is non-zero only in the $\hat{x}_1$-direction. A similar computation leads to similar results for the other faces. In the 2D case, we get, for the edge defined by $\{\hat{x}_1 = 0\}$:

$$I_R = \begin{cases} -\delta_{1m}\,\hat{\omega}_{t_2} v^m_{\ell,s}, \quad s = (r+1)(t_2-1)+1 \text{ for the Gauss–Lobatto rule,} \\ -\delta_{1m}\,\hat{\varphi}_{t_1}(0)\,\hat{\omega}_{t_2} \displaystyle\sum_{s_1=1}^{r+1} v^m_{\ell,s}\,\hat{\varphi}_{s_1}(0), \quad s = (r+1)(t_2-1)+s_1 \text{ for the Gauss rule.} \end{cases}$$

Similar results are obtained for the other edges.

4.3.4.2 The Maxwell's System

For the Maxwell's system defined by (4.37a)–(4.37c), we have $p = 3, q = 3, \underline{u} = \underline{E}$, $\underline{v} = \underline{H}$, $A = A^* = \underline{curl}$, $A(\underline{n}) = A^*(\underline{n}) = \underline{n}\times$. Let us set, for any $\underline{u} \in \underline{U}^r_{h,}$, $\underline{u} \circ \underline{F}_\ell = DF_\ell^{-T}\,\hat{\underline{u}}$. This definition provides:

$$\begin{aligned} (\nabla \times \underline{u}) \circ \underline{F}_\ell &= (DF_\ell^{-T}\hat{\nabla}) \times (DF_\ell^{-T}\hat{\underline{u}}) \\ &= \operatorname{adj}(DF_\ell^{-T})(\hat{\nabla} \times \hat{\underline{u}}) \\ &= J_\ell^{-1} DF_\ell\,(\hat{\nabla} \times \hat{\underline{u}}), \end{aligned} \tag{4.105}$$

which confirms that $M_{u,\ell} = DF_\ell^{-T}$ [7].

In our case, we obviously have $\underline{U}^r_{h,} = \underline{V}^r_{h,}$ and $M_{u,\ell} = M_{v,\ell}$. On the other hand, we have:

$$\begin{aligned} (\underline{n} \times \underline{u}) \circ \underline{F}_\ell &= \frac{DF_\ell^{-T}\hat{\underline{n}}}{||DF_\ell^{-T}\hat{\underline{n}}||} \times (DF_\ell^{-T}\hat{\underline{u}}) \\ &= \frac{J_\ell^{-1} DF_\ell}{||DF_\ell^{-T}\hat{\underline{n}}||}(\hat{\underline{n}} \times \hat{\underline{u}}), \end{aligned} \tag{4.106}$$

which proves that (4.94) holds for Maxwell's equations.

So, by taking into account (4.105) and (4.90), (4.88) reads in our case:

$$
\begin{aligned}
I &= \int_{K_\ell} \nabla \times \underline{H}_\ell \cdot \underline{\varphi}_\ell \, \mathrm{d}\underline{x} = \sum_{m=1}^{3} \sum_{s=1}^{N_r} H_{\ell,s}^m \int_{K_\ell} \nabla \times \underline{\psi}_{\ell,s}^m \cdot \underline{\varphi}_{\ell,t}^n \, \mathrm{d}\underline{x} \\
&= s_\ell \sum_{m=1}^{3} \sum_{s=1}^{N_r} H_{\ell,s}^m \int_{\widehat{K}} \hat{\nabla} \times \underline{\hat{\psi}}_s^m \cdot \underline{\hat{\varphi}}_t^n \, \mathrm{d}\underline{\hat{x}} = s_\ell \sum_{m=1}^{3} \sum_{s=1}^{N_r} H_{\ell,s}^m \, S_{s,t}^{m,n}.
\end{aligned} \tag{4.107}
$$

Let us now compute $S_{s,t}^{m,n}$ by using the quadrature rule. We obviously have $S_{s,t}^{n,n} = 0$. For $m \neq n$, we get:

$$
\begin{aligned}
S_{s,t}^{m,n} &= \eta_{m,n} \int_{\widehat{K}} \frac{\partial \hat{\varphi}_s}{\partial \hat{x}_\kappa} \hat{\varphi}_t \, \mathrm{d}\underline{\hat{x}} \\
&= \eta_{m,n} \int_0^1 \hat{\varphi}'_{s_\kappa}(\hat{x}_\kappa)\, \hat{\varphi}_{t_\kappa}(\hat{x}_\kappa)\, \mathrm{d}\hat{x}_\kappa \int_0^1 \hat{\varphi}_{s_m}(\hat{x}_m)\, \hat{\varphi}_{t_m}(\hat{x}_m)\, \mathrm{d}\hat{x}_m \int_0^1 \hat{\varphi}_{s_n}(\hat{x}_n)\, \hat{\varphi}_{t_n}(\hat{x}_n)\, \mathrm{d}\hat{x}_n \\
&\simeq \eta_{m,n} \sum_{j=1}^{r+1} \hat{\omega}_j\, \hat{\varphi}'_{s_\kappa}(\hat{\xi}_j)\, \hat{\varphi}_{t_\kappa}(\hat{\xi}_j) \sum_{j=1}^{r+1} \hat{\omega}_j\, \hat{\varphi}_{s_m}(\hat{\xi}_j)\, \hat{\varphi}_{t_m}(\hat{\xi}_j) \sum_{j=1}^{r+1} \hat{\omega}_j\, \hat{\varphi}_{s_n}(\hat{\xi}_j)\, \hat{\varphi}_{t_n}(\hat{\xi}_j) \\
&= \eta_{m,n} \sum_{j=1}^{r+1} \hat{\omega}_j\, \hat{\varphi}'_{s_\kappa}(\hat{\xi}_j)\, \delta_{t_\kappa j} \sum_{j=1}^{r+1} \hat{\omega}_j\, \delta_{s_m j}\, \delta_{t_m j} \sum_{j=1}^{r+1} \hat{\omega}_j\, \delta_{s_n j}\, \delta_{t_n j} \\
&= \eta_{m,n} \hat{\omega}_{t_\kappa} \hat{\omega}_{t_m} \hat{\omega}_{t_n} \hat{\varphi}'_{s_\kappa}(\hat{\xi}_{t_\kappa}),
\end{aligned} \tag{4.108}
$$

with $\eta_{m,n} = \pm 1$ and $\kappa = 6 - m - n$.

Remark: $\eta_{m,n}$ represents the sign of the scalar product $\hat{\nabla} \times \underline{\hat{\psi}}_s^m \cdot \underline{\hat{\varphi}}_t^n$.

By reporting (4.108) in (4.107), we obtain:

$$
I = s_\ell\, \hat{\omega}_{t_n} \sum_{m=1, m\neq n}^{3} \hat{\omega}_{t_\kappa} \hat{\omega}_{t_m} \eta_{m,n} \sum_{s_\kappa=1}^{r+1} H_{\ell, s(s_\kappa)}^m\, \hat{\varphi}'_{s_\kappa}(\hat{\xi}_{t_\kappa}), \tag{4.109}
$$

where

$$
s(s_\kappa) = \begin{cases} (r+1)[(r+1)(s_\kappa - 1) + s_2 - 1] + s_1 & \text{for } m = 1, n = 2, \\ (r+1)[(r+1)(s_3 - 1) + s_\kappa - 1] + s_1 & \text{for } m = 1, n = 3, \\ (r+1)[(r+1)(s_3 - 1) + s_2 - 1] + s_\kappa & \text{for } m = 2, n = 3. \end{cases}
$$

Equation (4.109) shows that, as for the acoustics system, there are only $2(r+1)$ interactions instead of $2(r+1)^3$ with one basis function, which provides a very sparse stiffness matrix. Computations on $K_{\ell'}$ can be done in the same way and follow the same rules as the jumps for acoustics.

Let us now compute the non dissipative jump term. Equation (4.95) reads, in this case:

$$J_{s,t}^{m,n} = s_\ell \int_{\widehat{\Gamma}} \hat{\underline{n}} \times \hat{\varphi}_s \, \underline{e}_m \cdot \hat{\varphi}_t \, \underline{e}_n \, \mathrm{d}\hat{\sigma}, \tag{4.110}$$

$\hat{\underline{n}}$ being the exterior normal to a face of $\widehat{K}$, we can write $\hat{\underline{n}} = \eta \underline{e}_k$, $k \in \{1, 2, 3\}$, with $\eta = \pm 1$. On the other hand, $J_{s,t}^{m,n} \neq 0$ if and only if $(\underline{e}_k, \underline{e}_m, \underline{e}_n)$ form an orthonormal basis of $\mathbb{R}^3$. So, (4.110) becomes, by taking into account (3.13b):

$$\begin{aligned} J_{s,t}^{m,n} &= c_{kmn} \int_{\widehat{\Gamma}} \hat{\varphi}_s \, \hat{\varphi}_t \, \mathrm{d}\hat{\sigma} \\ &= c_{kmn} \hat{\varphi}_{s_k}(0) \, \hat{\varphi}_{t_k}(0) \int_0^1 \int_0^1 \hat{\varphi}_{s_m}(\hat{x}_m) \, \hat{\varphi}_{s_n}(\hat{x}_n) \, \hat{\varphi}_{t_m}(\hat{x}_m) \, \hat{\varphi}_{t_n}(\hat{x}_n) \, \mathrm{d}\hat{x}_m \, \mathrm{d}\hat{x}_n, \end{aligned} \tag{4.111}$$

with $c_{kmn} = \eta \, s_\ell \, \delta_{km} \, \delta_{kn} \, \delta_{mn}$.

Let us now compute $J_{s,t}^{m,n}$ by using the quadrature rule. We get:

$$\begin{aligned} J_{s,t}^{m,n} &\simeq c_{kmn} \hat{\varphi}_{s_k}(0) \, \hat{\varphi}_{t_k}(0) \sum_{j=1}^{r+1} \hat{\omega}_j \, \hat{\varphi}_{s_m}(\hat{\xi}_j) \, \hat{\varphi}_{t_m}(\hat{\xi}_j) \sum_{j=1}^{r+1} \hat{\omega}_j \, \hat{\varphi}_{s_n}(\hat{\xi}_j) \, \hat{\varphi}_{t_n}(\hat{\xi}_j) \\ &\simeq c_{kmn} \, \hat{\varphi}_{s_k}(0) \, \hat{\varphi}_{t_k}(0) \, \hat{\omega}_{t_m} \, \hat{\omega}_{t_n}. \end{aligned} \tag{4.112}$$

As for the acoustics equation, the second term of (4.93) provides:

$$I_R = c_{kmn} \, \hat{\varphi}_{t_k}(0) \, \hat{\omega}_{t_m} \, \hat{\omega}_{t_n} \sum_{s_k=1}^{r+1} H_{\ell,s}^m \, \hat{\varphi}_{s_k}(0) \tag{4.113}$$

and we finally get, s being defined as

$$I_R = \begin{cases} c_{kmn} \, \hat{\omega}_{t_m} \, \hat{\omega}_{t_n} \, H_{\ell,s}^m \text{ for the Gauss–Lobatto rule,} \\ c_{kmn} \, \hat{\varphi}_{t_k}(0) \, \hat{\omega}_{t_m} \, \hat{\omega}_{t_n} \displaystyle\sum_{s_k=1}^{r+1} H_{\ell,s}^m \, \hat{\varphi}_{s_k}(0) \text{ for the Gauss rule,} \end{cases}$$

with s defined as for (4.103).

We see that, here also, the jump matrix is very sparse and depends on the quadrature rule.

The last term which is useful for the Maxwell's system is the dissipative jump. With the above notations, this term reads:

$$I = \int_{\partial K_\ell} [\underline{n} \times \underline{E} \times \underline{n}]_{\partial K_\ell}^{K_\ell} \cdot \underline{\varphi}_\ell \, \mathrm{d}\sigma = \int_{\partial K_\ell} [\underline{n} \times \underline{E}]_{\partial K_\ell}^{K_\ell} \cdot (\underline{n} \times \underline{\varphi}_\ell) \, \mathrm{d}\sigma \tag{4.114}$$

which provides, on the face $\Gamma_{\ell,i}$:

$$
\begin{aligned}
I &= \int_{\Gamma_{\ell,i}} \underline{n} \times (\underline{E}_{\ell'} - \underline{E}_{\ell}) \cdot (\underline{n} \times \underline{\varphi}_{\ell}) \,\mathrm{d}\sigma \\
&= \sum_{m=1}^{3} \sum_{s=1}^{N_r} E^m_{\ell',s'} \int_{\Gamma_{\ell,i}} (\underline{n} \times \underline{\psi}^m_{\ell',s'}) \cdot (\underline{n} \times \underline{\varphi}^n_{\ell,t}) \,\mathrm{d}\sigma \\
&\quad - \sum_{m=1}^{3} \sum_{s=1}^{N_r} E^m_{\ell,s} \int_{\Gamma_{\ell,i}} (\underline{n} \times \underline{\psi}^m_{\ell,s}) \cdot (\underline{n} \times \underline{\varphi}^n_{\ell,t}) \,\mathrm{d}\sigma.
\end{aligned} \tag{4.115}
$$

By change of variables, the last integral of (4.115) becomes:

$$
\begin{aligned}
J^{m,n}_{s,t} &= \int_{\Gamma_{\ell,i}} (\underline{n} \times \underline{\psi}^m_{\ell,s}) \cdot (\underline{n} \times \underline{\varphi}^n_{\ell,t}) \,\mathrm{d}\sigma \\
&= \int_{\widehat{\Gamma}} \frac{DF_\ell^T \, DF_\ell}{||J_\ell \, DF_\ell^{-T} \underline{\hat{n}}||} (\underline{\hat{n}} \times \hat{\varphi}_s \, \underline{e}_m) \cdot (\underline{\hat{n}} \times \hat{\varphi}_t \, \underline{e}_n) \,\mathrm{d}\hat{\sigma}.
\end{aligned} \tag{4.116}
$$

Obviously, $J^{m,n}_{s,t} \neq 0$ if $\underline{e}_m \perp \underline{\hat{n}}$ and $\underline{e}_n \perp \underline{\hat{n}}$. In this case, by setting $\underline{\hat{n}} = \eta \underline{e}_k$, $\kappa_m = 6 - k - m$, $\kappa_n = 6 - k - n$ and $B_\ell = (b^\ell_{ij}) = ||J_\ell \, DF_\ell^{-T} \underline{\hat{n}}||^{-1} DF_\ell^T \, DF_\ell$, we get[6]:

$$
\begin{aligned}
J^{m,n}_{s,t} &= \int_{\widehat{\Gamma}} (B_\ell \, \underline{e}_{\kappa_m} \cdot \underline{e}_{\kappa_n}) \, \hat{\varphi}_s \, \hat{\varphi}_t \,\mathrm{d}\hat{\sigma} \\
&= \hat{\varphi}_{s_k}(0) \, \hat{\varphi}_{t_k}(0) \int_0^1 \int_0^1 b^\ell_{\kappa_n \kappa_m} \hat{\varphi}_{s_m}(\hat{x}_m) \, \hat{\varphi}_{s_n}(\hat{x}_n) \, \hat{\varphi}_{t_m}(\hat{x}_m) \, \hat{\varphi}_{t_n}(\hat{x}_n) \,\mathrm{d}\hat{x}_m \,\mathrm{d}\hat{x}_n
\end{aligned} \tag{4.117}
$$

By using the quadrature rule to compute $J^{m,n}_{s,t}$, we obtain:

$$
\begin{aligned}
J^{m,n}_{s,t} &\simeq \hat{\varphi}_{s_k}(0) \, \hat{\varphi}_{t_k}(0) \sum_{j=1}^{r+1} \sum_{l=1}^{r+1} \hat{\omega}_j \, \hat{\omega}_l \, b^\ell_{\kappa_n \kappa_m} (\hat{x}_k = 0, \hat{x}_m = \hat{\xi}_j, \hat{x}_n = \hat{\xi}_l) \\
&\quad \times \hat{\varphi}_{s_m}(\hat{\xi}_j) \, \hat{\varphi}_{s_n}(\hat{\xi}_l) \, \hat{\varphi}_{t_m}(\hat{\xi}_j) \, \hat{\varphi}_{t_n}(\hat{\xi}_l).
\end{aligned} \tag{4.118}
$$

This approximated integral is not equal to zero if and only if $l = t_n$, $s_m = t_n$ and $j = t_n$. We get, in this case:

$$
J^{m,n}_{s,t} \simeq \hat{\varphi}_{s_k}(0) \, \hat{\varphi}_{t_k}(0) \hat{\omega}_{t_n} \, \hat{\omega}_{t_n} \, b^\ell_{\kappa_n \kappa_m} (\hat{x}_k = 0, \hat{x}_m = \hat{\xi}_{t_n}, \hat{x}_n = \hat{\xi}_{t_n}). \tag{4.119}
$$

[6]Of course, B_ℓ is symmetric and depends on $\hat{x}_m$ and $\hat{x}_n$.

As for the non dissipative jump, we finally have:

$$I_R = \begin{cases} \tilde{b}^{\ell}_{\kappa_n \kappa_m} \, \hat{\omega}_{t_m} \, \hat{\omega}_{t_n} \, H^m_{\ell,s} \text{ for the Gauss–Lobatto rule,} \\ \tilde{b}^{\ell}_{\kappa_n \kappa_m} \, \hat{\varphi}_{t_k}(0) \, \hat{\omega}_{t_m} \, \hat{\omega}_{t_n} \displaystyle\sum_{s_k=1}^{r+1} H^m_{\ell,s} \, \hat{\varphi}_{s_k}(0) \text{ for the Gauss rule.} \end{cases}$$

where $\tilde{b}^{\ell}_{\kappa_n \kappa_m} = b^{\ell}_{\kappa_n \kappa_m}(\hat{x}_k = 0, \hat{x}_m = \hat{\xi}_{t_n}, \hat{x}_n = \hat{\xi}_{t_n})$ and s is defined as for (4.103).

Remarks:

1. $b^{\ell}_{\kappa_n \kappa_m}$ are the coefficients of a 2×2 matrix which must stored for each interpolation point of K_ℓ. Actually, this jump behaves as two-dimensional mass matrix.
2. The jump terms are orthogonal to $\underline{n}$ for the Maxwell's system and collinear to this normal for the acoustics and the linear elastodynamics systems.
3. All the jump terms are here computed for conforming meshes. Non conforming meshes require interpolations of the degrees of freedom. Several methods can be used for this purpose, but the order at the interface is not always obvious. However, as we said, such meshes provide stable schemes in time if there contain no hanging nodes, In the case of hanging nodes, mortar elements [8, 9] should be used. These elements present the important drawback of coupling the degrees of freedom in a global way.

4.4 Comparison of the DG Methods for Maxwell's Equations

We provide here a comparison of the different DG approaches described above for Maxwell's equations. Similar results can be obtained for other equations.

4.4.1 *Gauss or Gauss–Lobatto?*

4.4.1.1 Discrete Formulations and Their Properties

The discrete formulation obtained by Gauss–Lobatto quadrature reads:

$$B_\varepsilon \, \frac{\mathbf{E^{n+1}} - \mathbf{E^n}}{\Delta t} + R_h \, \mathbf{H^{n+1/2}} + \alpha \, D_h \, \mathbf{E^n} + \beta \, S_h \, \mathbf{H^{n+1/2}} + \mathbf{J^n} = 0, \tag{4.120a}$$

$$B_\mu \, \frac{\mathbf{H^{n+1/2}} - \mathbf{H^{n-1/2}}}{\Delta t} + R_h \, \mathbf{E^n} + \gamma \, S_h^* \, \mathbf{E^n} + \delta \, D_h^* \, \mathbf{H^{n-1/2}} = 0. \tag{4.120b}$$

Because of the complex definition of the jumps, (4.120a)–(4.120b) would provide an implicit scheme involving D_h and D_h^* which behave, for Gauss quadrature, as stiffness matrices. So, the resolution would imply the inversion of a matrix which is no longer block-diagonal. For this reason, we choose to uncenter the dissipative term. We have:

$$B_\varepsilon \frac{\mathbf{E}^{\mathbf{n}+1} - \mathbf{E}^{\mathbf{n}}}{\Delta t} + R_h\, \mathbf{H}^{\mathbf{n}+1/2} + \alpha\, D_h\, \mathbf{E}^{\mathbf{n}} + \beta\, S_h\, \mathbf{H}^{\mathbf{n}+1/2} + \mathbf{J}^{\mathbf{n}} = 0, \tag{4.121a}$$

$$B_\mu \frac{\mathbf{H}^{\mathbf{n}+1/2} - \mathbf{H}^{\mathbf{n}-1/2}}{\Delta t} + R_h\, \mathbf{E}^{\mathbf{n}} + \gamma\, S_h^*\, \mathbf{E}^{\mathbf{n}} + \delta\, D_h^*\, \mathbf{H}^{\mathbf{n}-1/2} = 0, \tag{4.121b}$$

where

- B_ε, B_μ are 3×3 block-diagonal symmetric mass matrices,
- R_h are very sparse matrix which needs local storage,
- S_h, S_h^* are jump block-diagonal[7] symmetric matrices which need local storage,
- D_h, ${D_h}^*$ are jump block-diagonal symmetric matrices which need a storage of symmetric 2×2 matrix all over the mesh, which is a reasonable additional storage.

The main advantages of the Gauss rules are:

1. They are exact for Q_{2r+1}, which implies in particular that the mass integrals can be exactly computed,
2. As we shall see below, they have very good dispersion properties,

whereas their main drawbacks are

1. The jumps are computed by a 1D extrapolation,
2. All the degrees of freedom produce jumps,
3. The uncentered treatment of the dissipative term induces a (reasonable) reduction of the stability condition (CFL).[8]

On the other hand, the main advantages of the Gauss–Lobatto rules are:

1. The jumps are computed without extrapolation (since there are degrees of freedom on the faces),
2. Only degrees of freedom on the faces produce jumps,
3. One can treat the dissipative terms in a centered way,

whereas their main drawbacks are

1. They are exact for Q_{2r-1}, which implies that the mass integrals are not exact,
2. As we shall see below, they have less good dispersion properties.

Actually, the gain in computational time obtained by the Gauss-Lobatto rules being balanced by the better accuracy of the Gauss scheme, it seems that both methods have equivalent performances.

[7] With larger blocks for Gauss quadrature than for Gauss–Lobatto rule.

[8] As well as the introduction of some numerical dissipation, which is not really a problem since this term is itself dissipative.

4.4.1.2 Cost of the Methods

Let us now evaluate the cost of both approaches by computing the number of operations in one element. In all the following (even for tetrahedra), we just address the number of multiplications, the number of addition being roughly the same. Moreover, we do not take into account the dissipative jump term. We have for both approaches:

- $9 \times (r+1)^3$ operations for the mass term,
- $6 \times (r+1)^3 \times (r+1)$ operations for the stiffness term,

Moreover we have:

- $6 \times 10 \times (r+1)^2 \times (r+1)$ operations for the jump term with Gauss rule,
- $6 \times 6 \times (r+1)^2$ operations for the jump term with Gauss–Lobatto rule.

We finally obtain:

- $N_G(r) = 6(r+1)^4 + 69(r+1)^3$ operations for the Gauss rule,
- $N_{GL}(r) = 6(r+1)^4 + 9(r+1)^3 + 36(r+1)^2$ operations for the Gauss–Lobatto rule.

In Fig. 4.1, we compare the costs of $N_G(r)$, $N_{GL}(r)$ and $N_{GL}(r+1)$ (which should be as accurate as $N_G(r)$). One can see that $N_{GL}(r)$ and $N_{GL}(r+1)$ have roughly the same cost, which suggests that the two approaches should be equivalent.

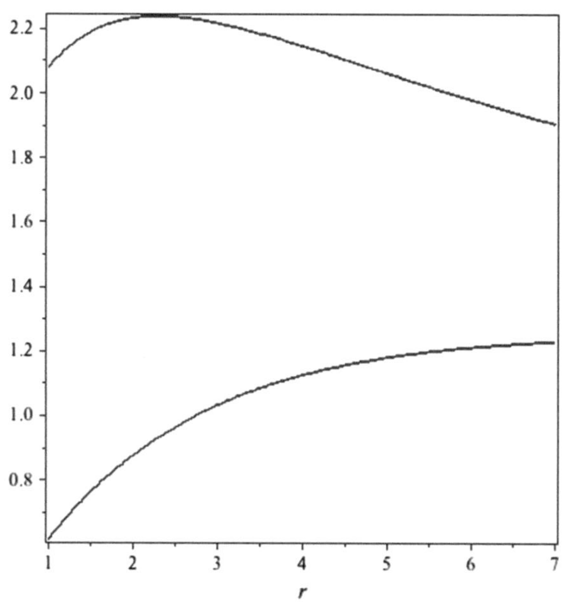

Fig. 4.1 $N_G(r)/N_{GL}(r)$ (in *red*) and $N_G(r)/N_{GL}(r+1)$ (in *blue*) for $r = 1$–7

4.4.2 *Tetrahedra with and Without Reconstruction of the Stiffness Matrix*

In the following, we denote $N_T^3 = (r+1)(r+2)(r+3)/6$ (number of points in a tetrahedron), $N_T^2 = (r+1)(r+2)/2$ (number of points on a face of a tetrahedron).

4.4.2.1 With Construction of the Stiffness Matrix at Each Time-Step

The number of multiplications for one tetrahedron are:

- $3(N_T^3)^2$ for the mass term,
- $(9+6)(N_T^3)^2$ for the stiffness term,
- $(4 \times (9(N_T^2)^2 + 36)$ for the jump term,

which provides $N_R(r) = 18(N_T^3)^2 + 36((N_T^2)^2 + 4)$ operations for this first approach.

4.4.2.2 With Storage of the Stiffness Matrix at Each Time-Step

The number of multiplications for one tetrahedron are:

- $3(N_T^3)^2$ for the mass term,
- $6(N_T^3)^2$ for the stiffness term,
- $4 \times 6(N_T^2)^2$ for the jump term,

which provides $N_S(r) = 9(N_T^3)^2 + 24(N_T^2)^2$ operations for this second approach.

4.4.2.3 Graphical Comparison

In Fig. 4.2, we compare the costs of $N_R(r)$ and $N_S(r)$. One can see that $N_R(r)$ is roughly twice more expensive than $N_S(r)$.

4.4.3 *Tetrahedra Versus Hexahedra*

4.4.3.1 Theoretical Costs

In order to compare the costs of hexahedra versus this of tetrahedra, one must first introduce the following normalizing coefficient which ensures that tetrahedral and hexahedral meshes have the same number of points:

$$n_{th} = \frac{6(r+1)^2}{(r+2)(r+3)}.$$

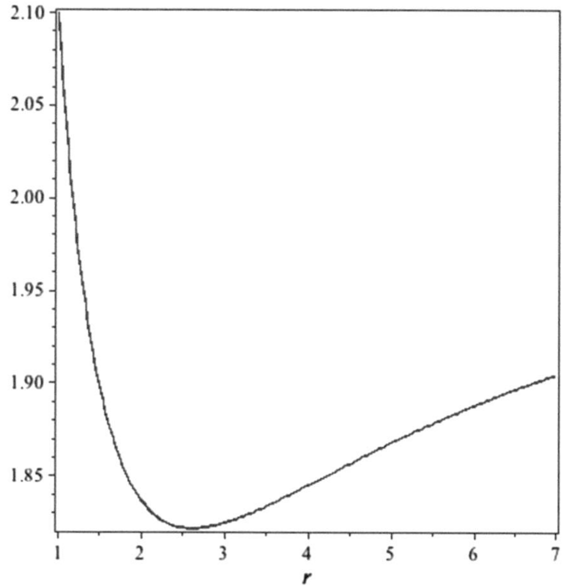

Fig. 4.2 Ratio between $N_R(r)$ and $N_S(r)$ for $r = 1–7$

Thanks to this coefficient, we can now compare the costs per point of interpolation of the two approaches.

In Fig. 4.3, we compare the cost of tetrahedra with matrix reconstruction with the cost of hexahedra with Gauss rule of rth order and Gauss–Lobatto rule of $(r + 1)$th order. Since, as we saw in Fig. 4.1, the two hexahedral approaches have almost the same cost with the same number of elements, we use n_{th} for both comparisons. One can notice the huge gain of computational time for hexahedra versus tetrahedra.

We give the same comparisons in Fig. 4.4 in the case with matrix storage.

Finally, in Fig. 4.5, we compare the cost of both approaches for tetrahedra with the cost of hexahedra with Gauss rule of rth order.

From the storage point of view, tetrahedra are of course much cheaper than hexahedra, since one must only store 9 coefficients per cell for tetrahedra versus $6(r + 1)^3$ coefficients for one hexahedron. Even by taking into account n_{th}, the ratio remains very important (Fig. 4.6). However, the storage for hexahedra is very small compared to classical finite element methods.

4.4.3.2 Numerical Comparison

We compared tetrahedra with matrix reconstruction with hexahedra using a Gauss rule by solving the Maxwell's system without right-hand side and with the initial

Fig. 4.3 $n_{th} N_R(r)/N_G(r)$ (in *red*) and $n_{th} N_R(r)/N_{GL}(r+1)$ (in *blue*) for $r = 1$–7

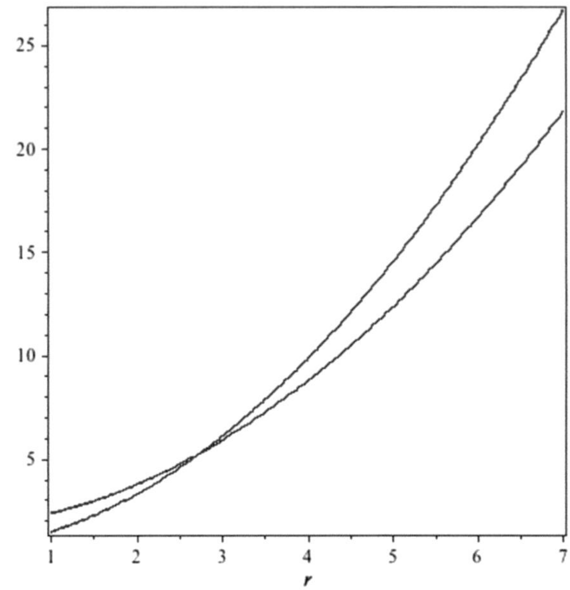

Fig. 4.4 $n_{th} N_S(r)/N_G(r)$ (in *red*) and $n_{th} N_S(r)/N_{GL}(r+1)$ (in *blue*) for $r = 1$–7

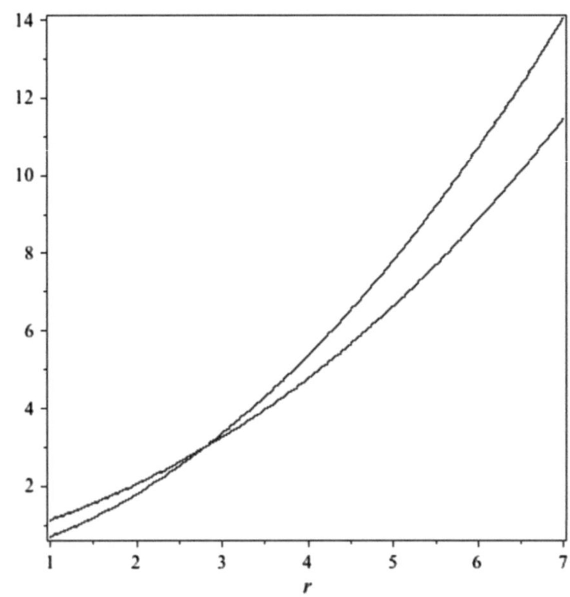

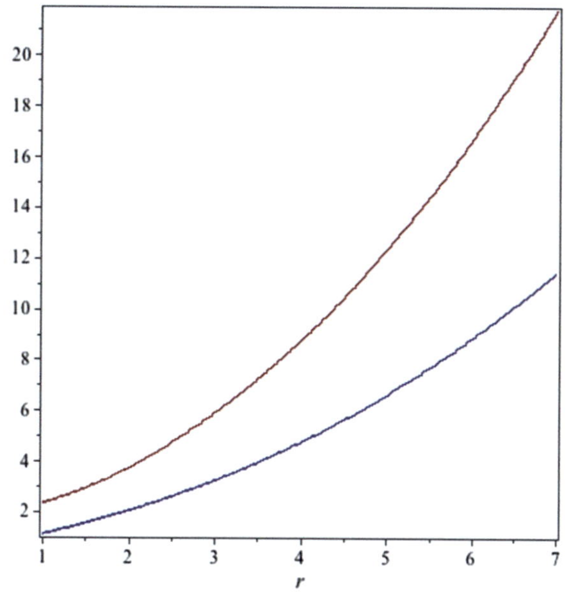

Fig. 4.5 $n_{th}N_R(r)/N_G(r)$ (in *red*) and $n_{th}N_S(r)/N_G(r)$ (in *blue*) for $r = 1$–7

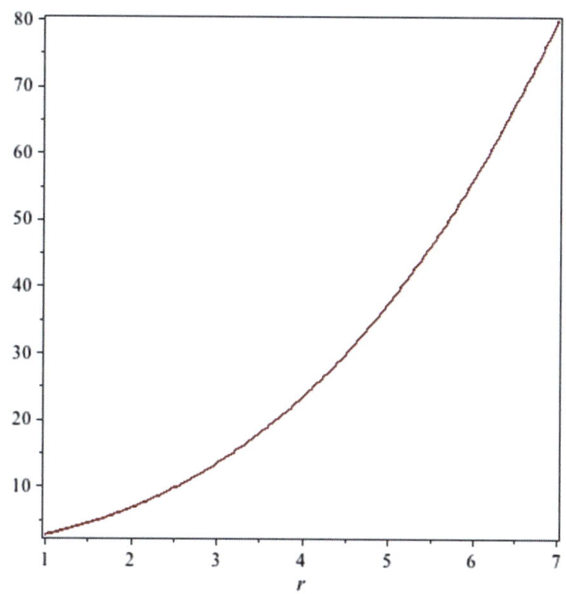

Fig. 4.6 Ratio between the storage required by an hexahedral and a tetrahedral mesh with the same number of degrees of freedom for $r = 1$–7

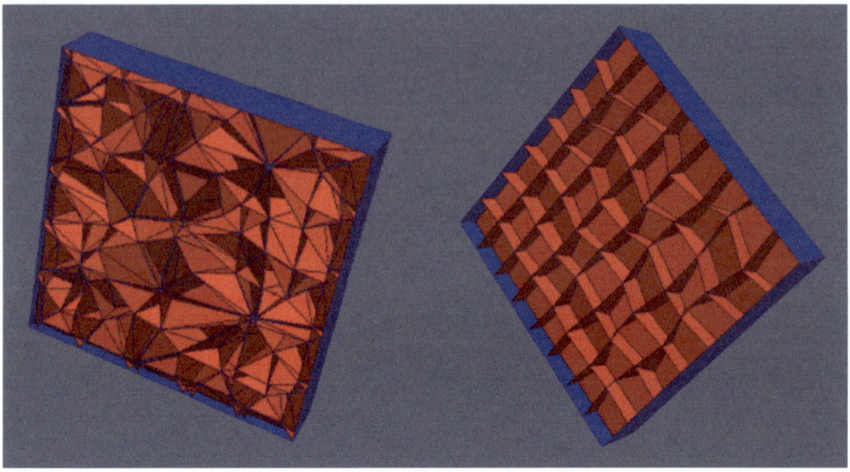

Fig. 4.7 Snapshots of the tetrahedral mesh of 1036 elements (*left*) and of the hexahedral mesh of 336 elements (*right*)

solution $E_1 = E_2 = 0,\ E_3 = \sin(m\,\pi\,\hat{x}_1)\sin(n\,\pi\,\hat{x}_2)\cos(\omega\,t),\quad \omega = c_0((m\pi)^2 + (n\pi)^2)^{1/2}$, $c_0 = 299792458$ m/s,[9] $\underline{H} = 0$ on $\Omega = [0, 1]^3$ and $\underline{E} \times \underline{n} = 0$ on $\partial\Omega$, $\underline{n}$ being the unit exterior normal to $\partial\Omega$. The total time of simulation is $T_f = 5 \times 10^{-8}$ s, which corresponds to a wave propagation across 15 m. The analytical solution of this problem reads:

$$\begin{aligned} &E_1 = 0,\ \ E_2 = 0,\ \ E_3 = \sin(m\,\pi\,\hat{x}_1)\sin(n\,\pi\,\hat{x}_2)\cos(\omega\,t),\\ &H_1 = \pi\,n\sin(m\,\pi\,\hat{x}_1)\cos(n\,\pi\,\hat{x}_2)\sin(\omega\,t)/(\omega\,\mu_0),\\ &H_2 = \pi\,n\cos(m\,\pi\,\hat{x}_1)\sin(n\,\pi\,\hat{x}_2)\sin(\omega\,t)/(\omega\,\mu_0),\ \ H_3 = 0. \end{aligned} \tag{4.122}$$

In order to solve this problem, we used one tetrahedral mesh containing 1036 tetrahedra and two structured hexahedral meshes containing 336 and 288 hexahedra whose summits were randomly moved. In Fig. 4.7, one can see snapshots of these meshes.

In Table 4.1, we give the characteristics of the meshes and of their numerical performance. One can see that all the meshes have roughly the same number of degrees of freedom, except for $r = 2$, for which the hexahedral mesh has less degrees of freedom than the tetrahedral one.

Table 4.2 provides the ratios of the costs. One can see on this table that the ratios of the costs of one iteration agree with the upper curve of Fig. 4.5.

In Table 4.3 are the frequencies and the number of wavelengths crossed by the solution versus m and n. The increase of these two parameters increases the frequency

---

[9]From a physical point of view, it seems that rounding c_0 to 10^8 m/s could make a difference in some applications.

Table 4.1 Characteristics and CPU times for the different meshes applied to the same experiment

Order	Nb. of elements	Nb. of DOF	CPU (s)	Δt (s)	Cost of one iteration (s)
2 (Tetra)	1036	31,080	22.4	4.3447e−11	1.9464e−02
2 (Hexa)	336	27,216	5.36	4.9586e−11	5.3160e−03
3 (Tetra)	1036	62,160	116	2.9149E−11	6.7626e−02
3 (Hexa)	336	64,512	27.5	2.5744e−11	1.4159e−02
4 (Tetra)	1036	108,780	440	2.0262e−11	1.7831e−01
4 (Hexa)	288	108,000	77	1.5062e−11	2.3192e−02

Table 4.2 Comparison of the performance of tetrahedra and hexahedra for different orders

	order 2	order 3	order 4
Ratio of the CPU times	4.18	4.22	5.71
Ratio of the costs of one iteration	3.66	4.77	7.70

Table 4.3 Frequencies, wavelengths λ, number N_λ of wavelengths crossed by the solution at $t = T_f$ versus m and n

(m, n)	(3, 3)	(4, 3)	(4, 4)	(5, 4)	(5, 5)	(6, 5)
Frequency (GHz)	0.636	0.7494	0.8479	0.9598	1.06	1.171
λ	0.4714	0.4	0.3536	0.3123	0.2828	0.256
N_λ	31.8	37.5	42.4	48	53	58.6

Table 4.4 Relative L^1-error ($\leq 10\,\%$) committed on the solution at $t = T_f$ versus m and n

(m, n)	(3, 3)	(4, 3)	(4, 4)	(5, 4)	(5, 5)	(6, 5)
Error order 2 (Hex)	3.26e−2	0.345	–	–	–	–
Error order 2 (Tet)	8.57e−2	0.316	–	–	–	–
Error order 3 (Hex)	1.65e−2	2.86e−2	2.24e−2	4.15e−2	8.12e−2	–
Error order 3 (Tet)	2.59e−2	3.78e−2	5.36e−2	0.15	0.27	–
Error order 4 (Hex)	1.99e−3	3.16e−3	5.33e−3	5.60e−3	1.27e−2	3.37e−2
Error order 4 (Tet)	7.63e−3	5.11e−3	1.70e−2	2.21e−2	4.41e−2	0.126

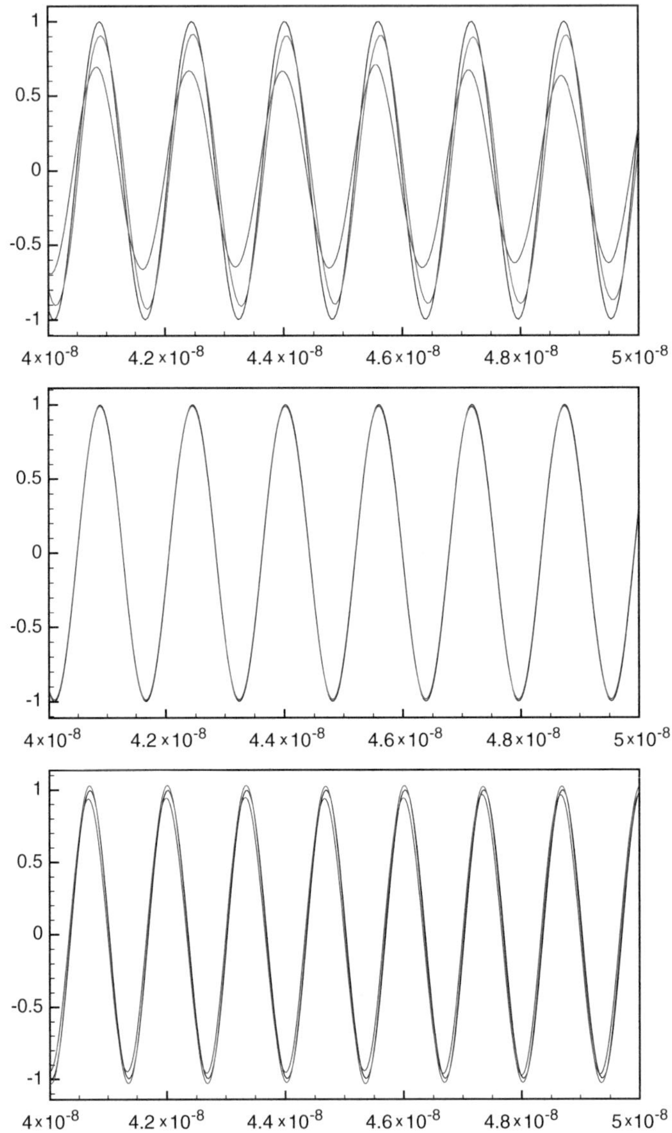

Fig. 4.8 Seismograms for $m = 3$, $n = 3$, $r = 2$ (*above*), $m = 3$, $n = 3$, $r = 3$ (*center*), $m = 4$, $n = 3$, $r = 3$ (*below*). Tetras in *red*, hexas in *blue* and exact solution in *black*

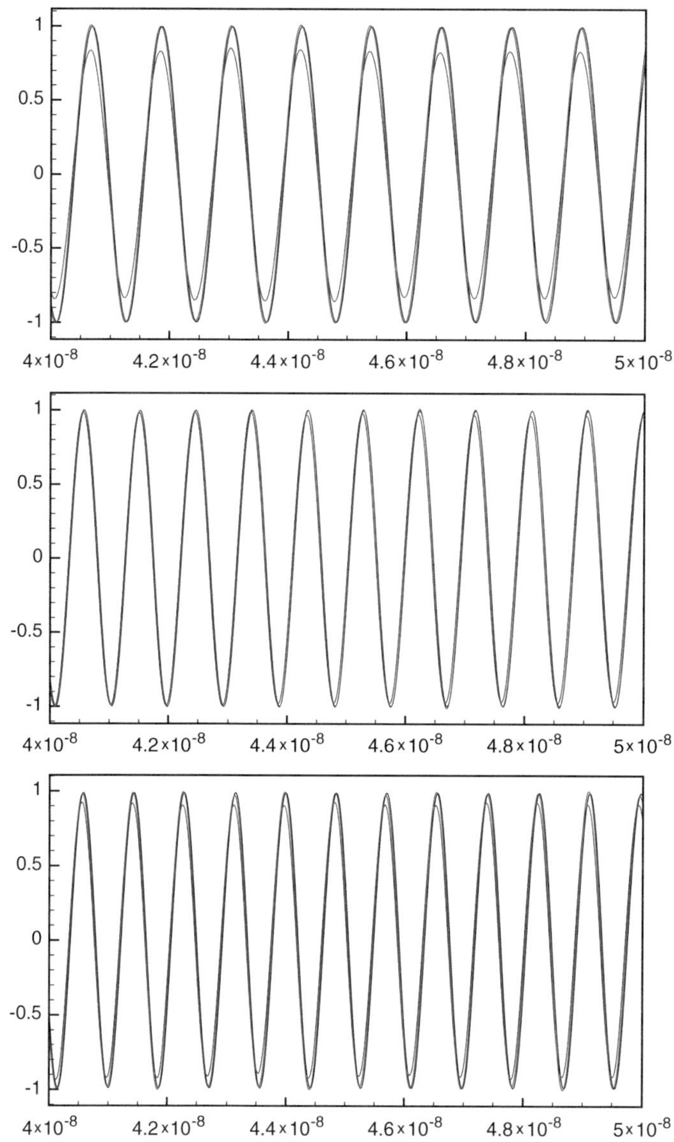

Fig. 4.9 Seismograms for $m = 4$, $n = 4$, $r = 3$ (*above*), $m = 5$, $n = 5$, $r = 4$ (*center*), $m = 6$, $n = 5$, $r = 4$ (*below*). Tetras in *red*, hexas in *blue* and exact solution in *black*

and enlarge the experiment. The number of wavelengths crossed goes from almost 32 to almost 59, which induces an additional error for the solution.

Table 4.4 gives the relative L^1-error committed on the solution at $t = T_f$. The values are computed at the same points for both methods. This table shows that, for a number of degrees of freedom less or equal to this of tetrahedra, hexahedra are more accurate for a much cheaper computation.

Figures 4.8 and 4.9 provide the seismograms at the center of the cube on the last fifth interval in time. These seismograms confirm the results of Table 4.4.

4.5 Plane Wave Analysis

4.5.1 The Eigenvalue Problem for the 1D Model

The plane wave analysis is done as in Sect. 3.4. However, the local character of discontinuous Galerkin methods enables us to derive the eigenvalue problem in a general framework, as shown in [10]. Let us first write the discontinuous Galerkin approximation of the 1D problem.

4.5.1.1 Discontinuous Galerkin Approximation of the 1D Problem

We consider the 1D wave equations:
Find $H(x,t) : \mathbb{R} \times \mathbb{R} \longrightarrow \mathbb{R}$, $E(x,t) : \mathbb{R} \times \mathbb{R} \longrightarrow \mathbb{R}$ such that

$$\frac{\partial E}{\partial t} + \frac{\partial H}{\partial x} = 0, \tag{4.123a}$$

$$\frac{\partial H}{\partial t} + \frac{\partial E}{\partial x} = 0. \tag{4.123b}$$

Let us now set:

$$\mathbb{R} = \bigcup_{p \in \mathbb{Z}} I_p, \quad I_p = [ph, (p+1)h] \tag{4.124}$$

and

$$U_h^r = \left\{ \underline{w}_h \text{ such that } \underline{w}_{h|I_p} \in P_r \right\}. \tag{4.125}$$

With these notations, the discontinuous Galerkin formulation of (4.123a)–(4.123b) reads:

Find $E_h(\cdot,t),\ H_h(\cdot,t) \in U_h^r$ such that $\forall p \in \mathbb{N}$ and $\phi_1, \phi_2 \in U_h^r$,

$$\begin{aligned}
&\frac{\mathrm{d}}{\mathrm{d}t}\int_{I_p}^{Q} E_{h,p}\phi_{1,p}\,\mathrm{d}x + \int_{I_p}^{Q} \frac{\partial H_{h,p}}{\partial x}\phi_{1,p}\,\mathrm{d}x \\
&\quad = -\frac{1}{2}\big(H_{h,p}(ph) - H_{h,p-1}(ph)\big)\phi_{1,p}(ph) \\
&\quad\quad - \frac{1}{2}\big(H_{h,p+1}((p+1)h) - H_{h,p}((p+1)h)\big)\phi_{1,p}((p+1)h), \\
&\quad \alpha\big(E_{h,p-1}(ph) - E_{h,p}(ph)\big)\phi_{1,p}(ph) \\
&\quad\quad + \alpha\big(E_{h,p+1}((p+1)h) - E_{h,p}((p+1)h)\big)\phi_{1,p}((p+1)h),
\end{aligned} \tag{4.126a}$$

$$\begin{aligned}
&\frac{\mathrm{d}}{\mathrm{d}t}\int_{I_p}^{Q} H_{h,p}\phi_{2,p}\,\mathrm{d}x + \int_{I_p}^{Q} \frac{\partial E_{h,p}}{\partial x}\phi_{2,p}\,\mathrm{d}x \\
&\quad = -\frac{1}{2}\big(E_{h,p}(ph) - E_{h,p-1}(ph)\big)\phi_{2,p}(ph) \\
&\quad\quad - \frac{1}{2}\big(E_{h,p+1}((p+1)h) - E_{h,p}((p+1)h)\big)\phi_{2,p}((p+1)h) \\
&\quad\quad + \delta\big(H_{h,p-1}(ph) - H_{h,p}(ph)\big)\phi_{2,p}(ph) \\
&\quad\quad + \delta\big(H_{h,p+1}((p+1)h) - H_{h,p}((p+1)h)\big)\phi_{2,p}((p+1)h),
\end{aligned} \tag{4.126b}$$

where $E_{h,p} = E_{h|I_p}$, $\phi_{i,p} = \phi_{i|I_p}$, $i = 1, 2$ and

$$\int_{I_p}^{Q} f(x)\,\mathrm{d}x = h\sum_{i=1}^{r+1} \hat{\omega}_i^Q f \circ F_p(\hat{\xi}_i^Q) \simeq \int_{I_p} f(x)\,\mathrm{d}x,$$

with $F_p(\hat{x}) = ph + h\hat{x}$.

If we decompose $E_{h,p}$ and $H_{h,p}$ on a basis of U_h^r as follows:

$$E_{h,p} \circ F_p = \sum_{l=1}^{r+1} E_l^p \hat{\varphi}_l^Q, \quad \forall p \in \mathbb{N}, \tag{4.127a}$$

$$H_{h,p} \circ F_p = \sum_{l=1}^{r+1} H_l^p \hat{\varphi}_l^Q, \quad \forall p \in \mathbb{N}, \tag{4.127b}$$

and we denote ϕ_1, ϕ_2 be two basis functions such that $\mathrm{supp}(\phi_1) = \mathrm{supp}(\phi_2) = I_p$, $p \in \mathbb{N}$; $\phi_{1,p} \circ F_p = \phi_{2,p} \circ F_p = \hat{\varphi}_{l_0}^Q$ and $\hat{\varphi}_{l_0}^Q(\hat{\xi}_l^Q) = \delta_{l_0 l}$, we get

$$\int_{I_p}^{Q} E_{h,p}\phi_{1,p}\,\mathrm{d}x = h\sum_{l=1}^{r+1} E_l^p \int_{[0,1]}^{Q} \hat{\varphi}_l^Q \hat{\varphi}_{l_0}^Q\,\mathrm{d}\hat{x} = h\sum_{l=1}^{r+1} E_l^p \hat{\omega}_l^Q \delta_{l_0 l} = \hat{\omega}_{l_0}^Q h E_{l_0}^p, \tag{4.128a}$$

$$\begin{aligned}\int_{I_p}^{Q} \frac{\partial H_{h,p}}{\partial x}\phi_{1,p}\,\mathrm{d}x &= \sum_{l=1}^{r+1} H_l^p \left(\sum_{i=1}^{r+1} \hat{\omega}_i^Q \frac{\partial \hat{\varphi}_l^Q}{\partial \hat{x}}(\hat{\xi}_i^Q)\,\hat{\varphi}_{l_0}^Q(\hat{\xi}_i^Q)\right)\\ &= \hat{\omega}_{l_0}^Q \sum_{l=1}^{r+1} H_l^p \frac{\partial \hat{\varphi}_l^Q}{\partial \hat{x}}(\hat{\xi}_{l_0}^Q),\end{aligned} \tag{4.128b}$$

$$\begin{aligned}&\big(H_{h,p}(ph) - H_{h,p-1}(ph)\big)\phi_{1,p}(ph)\\ &\quad= \left(\sum_{l=1}^{r+1}(H_l^p \hat{\varphi}_l^Q(0) - H_l^{p-1}\hat{\varphi}_l^Q(1))\right)\hat{\varphi}_{l_0}^Q(0).\end{aligned} \tag{4.128c}$$

Equation (4.128c) holds for Gauss rule. For Gauss–Lobatto rule, we have:

$$\big(H_{h,p}(ph) - H_{h,p-1}(ph)\big)\phi_{1,p}(ph) = H_1^p - H_{r+1}^{p-1}. \tag{4.129}$$

4.5.1.2 The Eigenvalue Problem

We search for a plane wave solution of (4.123a)–(4.123b) which reads $E = E_0\,\mathrm{e}^{\mathrm{i}(\omega t - kx)}$, $H = H_0\,\mathrm{e}^{\mathrm{i}(\omega t - kx)}$ which provides:

$$E_l^p = E_l\,\mathrm{e}^{\mathrm{i}\left(\omega_h t - khp\right)}\,\mathrm{e}^{-\mathrm{i}\hat{\xi}_l^Q kh} \tag{4.130a}$$

$$H_l^p = H_l\,\mathrm{e}^{\mathrm{i}\left(\omega_h t - khp\right)}\,\mathrm{e}^{-\mathrm{i}\hat{\xi}_l^Q kh} \tag{4.130b}$$

By inserting (4.130a)–(4.130b) in (4.126a)–(4.126b) and by taking into account (4.128a)–(4.128c), we get:

$$\begin{aligned}&i\omega_h h\hat{\omega}_{l_0}^Q E_{l_0}\\ &= \alpha\sum_{l=1}^{r+1} E_l \mathrm{e}^{\mathrm{i}\left(\hat{\xi}_{l_0}^Q - \hat{\xi}_l^Q\right)kh}\big(\hat{\varphi}_l^Q(1)\hat{\varphi}_{l_0}^Q(0)\mathrm{e}^{\mathrm{i}kh} - \hat{\varphi}_l^Q(0)\hat{\varphi}_{l_0}^Q(0) - \hat{\varphi}_l^Q(1)\hat{\varphi}_{l_0}^Q(1)\\ &\quad + \hat{\varphi}_l^Q(0)\hat{\varphi}_{l_0}^Q(1)\mathrm{e}^{-\mathrm{i}kh}\big) - \frac{1}{2}\sum_{l=1}^{r+1} H_l \mathrm{e}^{\mathrm{i}\left(\hat{\xi}_{l_0}^Q - \hat{\xi}_l^Q\right)kh}\big(-\hat{\varphi}_l^Q(1)\hat{\varphi}_{l_0}^Q(0)\mathrm{e}^{\mathrm{i}kh}\\ &\quad + \hat{\varphi}_l^Q(0)\hat{\varphi}_{l_0}^Q(0) - \hat{\varphi}_l^Q(1)\hat{\varphi}_{l_0}^Q(1) + 2\hat{\omega}_{l_0}^Q \frac{\partial \hat{\varphi}_l^Q}{\partial \hat{x}}\left(\hat{\xi}_{l_0}^Q\right) + \hat{\varphi}_l^Q(0)\hat{\varphi}_{l_0}^Q(1)\mathrm{e}^{-\mathrm{i}kh}\big)\end{aligned} \tag{4.131a}$$

$$
\begin{aligned}
& i\omega_h h\hat{\omega}_{l_0}^Q H_{l_0} \\
& = \beta\sum_{l=1}^{r+1} H_l \mathrm{e}^{\mathrm{i}\left(\hat{\xi}_{l_0}^Q - \hat{\xi}_l^Q\right)kh}\big(\hat{\varphi}_l^Q(1)\hat{\varphi}_{l_0}^Q(0)\mathrm{e}^{\mathrm{i}kh} - \hat{\varphi}_l^Q(0)\hat{\varphi}_{l_0}^Q(0) - \hat{\varphi}_l^Q(1)\hat{\varphi}_{l_0}^Q(1) \\
& +\hat{\varphi}_l^Q(0)\hat{\varphi}_{l_0}^Q(1)\mathrm{e}^{-\mathrm{i}kh}\big) - \frac{1}{2}\sum_{l=1}^{r+1} E_l \mathrm{e}^{\mathrm{i}\left(\hat{\xi}_{l_0}^Q - \hat{\xi}_l^Q\right)kh}\big(- \hat{\varphi}_l^Q(1)\hat{\varphi}_{l_0}^Q(0)\mathrm{e}^{\mathrm{i}kh} \\
& +\hat{\varphi}_l^Q(0)\hat{\varphi}_{l_0}^Q(0) - \hat{\varphi}_l^Q(1)\hat{\varphi}_{l_0}^Q(1) + 2\hat{\omega}_{l_0}^Q \frac{\partial\hat{\varphi}_l^Q}{\partial\hat{x}}\big(\hat{\xi}_{l_0}^Q\big) + \hat{\varphi}_l^Q(0)\hat{\varphi}_{l_0}^Q(1)\mathrm{e}^{-\mathrm{i}kh}\big)
\end{aligned}
\tag{4.131b}
$$

Equations (4.131a)–(4.131b) leads to the following eigenvalue problem:

$$
h\omega_{h,r}^{\alpha,\delta,Q}\begin{pmatrix}\underline{E}\\ \underline{H}\end{pmatrix} = \begin{pmatrix}\alpha A_{h,r}^Q[k] & B_{h,r}^Q[k] \\ B_{h,r}^Q[k] & \delta A_{h,r}^Q[k]\end{pmatrix}\begin{pmatrix}\underline{E}\\ \underline{H}\end{pmatrix} = C_{h,r}^Q[k]\begin{pmatrix}\underline{E}\\ \underline{H}\end{pmatrix} \tag{4.132}
$$

where $\underline{E} = (E_l)_{l=1}^{r+1}$, $\underline{H} = (H_l)_{l=1}^{r+1}$, $A_{h,r}^Q[k] = \big(A_{h,r}^Q[k](l_0,l)\big)$, $B_{h,r}^Q[k] = \big(B_{h,r}^Q[k](l_0,l)\big)$, $l_0 = 1\ldots r+1, l = 1\ldots r+1$ are $(r+1)\times(r+1)$ matrices such that

$$
\begin{aligned}
A_{h,r}^Q[k](l_0,l) = & \frac{-\mathrm{i}}{\hat{\omega}_{l_0}^Q}\mathrm{e}^{\mathrm{i}\left(\hat{\xi}_{l_0}^Q - \hat{\xi}_l^Q\right)kh}\big(\hat{\varphi}_l^Q(1)\hat{\varphi}_{l_0}^Q(0)e^{\mathrm{i}kh} \\
& - \hat{\varphi}_l^Q(0)\hat{\varphi}_{l_0}^Q(0) - \hat{\varphi}_l^Q(1)\hat{\varphi}_{l_0}^Q(1) + \hat{\varphi}_l^Q(0)\hat{\varphi}_{l_0}^Q(1)e^{-\mathrm{i}kh}\big),
\end{aligned}
\tag{4.133a}
$$

$$
\begin{aligned}
B_{h,r}^Q[k](l_0,l) = & \frac{\mathrm{i}}{2\hat{\omega}_{l_0}^Q}\mathrm{e}^{\mathrm{i}\left(\hat{\xi}_{l_0}^Q - \hat{\xi}_l^Q\right)kh}\big(- \hat{\varphi}_l^Q(1)\hat{\varphi}_{l_0}^Q(0)e^{\mathrm{i}kh} \\
& + \hat{\varphi}_l^Q(0)\hat{\varphi}_{l_0}^Q(0) - \hat{\varphi}_l^Q(1)\hat{\varphi}_{l_0}^Q(1) + 2\hat{\omega}_{l_0}^Q\frac{\partial\hat{\varphi}_l^Q}{\partial\hat{x}}\big(\hat{\xi}_{l_0}^Q\big) + \hat{\varphi}_l^Q(0)\hat{\varphi}_{l_0}^Q(1)e^{-\mathrm{i}kh}\big).
\end{aligned}
\tag{4.133b}
$$

4.5.2 Numerical Dispersion and Dissipation

By denoting $\omega_{h,r}^{\alpha,\delta,Q} = \Re\big(\omega_{h,r}^{\alpha,\delta,Q}\big) + \mathrm{i}\Im\big(\omega_{h,r}^{\alpha,\delta,Q}\big)$ the eigenvalues of (4.132), we get:

$$
E_l^p = E_l e^{-\Im\left(\omega_{h,r}^{\alpha,\delta,Q}\right)t} e^{\mathrm{i}\left(\Re\left(\omega_{h,r}^{\alpha,\delta,Q}\right)t - khp\right)} e^{-\mathrm{i}\hat{\xi}_l^Q kh} \tag{4.134a}
$$

$$
H_l^p = H_l e^{-\Im\left(\omega_{h,r}^{\alpha,\delta,Q}\right)t} e^{\mathrm{i}\left(\Re\left(\omega_{h,r}^{\alpha,\delta,Q}\right)t - khp\right)} e^{-\mathrm{i}\hat{\xi}_l^Q kh} \tag{4.134b}
$$

From $\omega_{h,r}^{\alpha,\delta,Q}$ we can deduce the following informations:

- The error committed on the velocity

$$Q_{h,r}^{\alpha,\delta,Q} = \frac{c_{h,r}^{\alpha,\delta,Q}}{c} = \frac{\Re\left(\omega_{h,r}^{\alpha,\delta,Q}\right)}{\omega}, \tag{4.135}$$

where $c = \omega/|k|$ is the exact velocity and $c_{h,r}^{\alpha,\delta,Q} = Re\left(\omega_{h,r}^{\alpha,\delta,Q}\right)/|k|$ ($\rightarrow c$) is the approximate velocity.

- The dissipation induced by the scheme

$$D_{h,r}^{\alpha,\delta,Q} = \frac{\Im\left(\omega_{h,r}^{\alpha,\delta,Q}\right)}{\omega}. \tag{4.136}$$

We now introduce the following notations:

$$K = \frac{|k|h}{2\pi(r+1)} = \frac{1}{N_\lambda} = \frac{1}{\text{Number of points per wavelength}} \tag{4.137}$$

$$Q_{h,r}^{\alpha,\delta,Q}(K) = \frac{\Re\left(\lambda_{h,r}^{\alpha,\delta,Q}\right)(K)}{2\pi(r+1)K} = 1 + C_r^{\alpha,\beta,Q} K^{p_r^{\alpha,\beta,Q}} + O\left(K^{p_r^{\alpha,\beta,Q}+1}\right) \tag{4.138a}$$

$$D_{h,r}^{\alpha,\delta,Q}(K) = \frac{\Im\left(\lambda_{h,r}^{\alpha,\delta,Q}\right)(K)}{2\pi(r+1)K} = \tilde{C}_r^{\alpha,\beta,Q} K^{q_r^{\alpha,\beta,Q}} + O\left(K^{q_r^{\alpha,\beta,Q}+1}\right) \tag{4.138b}$$

Actually, we have two forms of $Q_{h,r}^{\alpha,\delta,Q}$. If $\lim\limits_{K\to 0} Q_{h,r}^{\alpha,\beta,Q}(K) = 1$, it corresponds to a physical wave. If $\lim\limits_{K\to 0} Q_{h,r}^{\alpha,\beta,Q}(K) \neq 1$ (generally $= \infty$), we have a parasitic wave (see Sects. 3.4.4 and 3.4.5). Although non physical, this second kind of wave have an influence on stability, as we shall see later.

In Table 4.5, we give the exponents and the coefficients of the leading terms of $Q_{h,r}^{\alpha,\delta,Q}$ and $D_{h,r}^{\alpha,\delta,Q}$. One can first see the better accuracy of schemes obtained with Gauss rules in terms of order (we get a superconvergence for these rules) and dispersion (coefficients) with an increase for schemes with dissipation. On the other hand, Gauss–Lobatto rules provide equivalent performance for both approaches with even more dispersion for some dissipative schemes. The variation of the dispersive coefficient can be pointed out for $C_4^{0,0,G}$, $C_5^{0,0,G}$, $C_4^{\frac{1}{2},\frac{1}{2},G}$ and $C_5^{\frac{1}{2},\frac{1}{2},GL}$. Finally, one can notice that the order of the dissipation increases with the order of the scheme, which legitimates its use for high-order schemes.

In Figs. 4.10, 4.11, 4.12, 4.13 and 4.14, we give the dispersion curves for $r =$ 1–5. One can notice that, although more accurate, the fourth-order curve without dissipation loses its precision for $K < 0.1$ whereas the third-order remains accurate until $K \simeq 0.15$. This phenomenon is due to the important dispersion of the fourth-order scheme, as shown in Table 4.5.

Table 4.5 Dispersion and dissipation orders and coefficients for schemes with and without dissipation with Gauss and Gauss–Lobatto quadratures

	$r = 1$	$r = 2$	$r = 3$	$r = 4$	$r = 5$
$p_r^{0,0,G}$, $C_r^{0,0,G}$	2, 3.2898	6, −2670.1817	6, 312.48	10, −931215.53	10, 47988.2293
$p_r^{0,0,GL}$, $C_r^{0,0,GL}$	2, −6.5797	4, −263.05	6, −1875.209	8, −46703.7584	10, −480341.2712
$p_r^{\frac{1}{2},\frac{1}{2},G}$, $C_r^{\frac{1}{2},\frac{1}{2},G}$	4, 92.3482	6, 1067.9225	8, 14323.7153	10, 206878.0298	12, 0.31246419×10^7
$p_r^{\frac{1}{2},\frac{1}{2},GL}$, $C_r^{\frac{1}{2},\frac{1}{2},GL}$	2, −26.3097	4, −262.9699	6, −3333.1769	8, −46685.837	10, −691621.97
$q_r^{\frac{1}{2},\frac{1}{2},G}$, $\tilde{C}_r^{\frac{1}{2},\frac{1}{2},G}$	3, 27.5597	5, 330.48	7, 4488.0686	9, 64189.67	11, 987690.8818
$q_r^{\frac{1}{2},\frac{1}{2},GL}$, $\tilde{C}_r^{\frac{1}{2},\frac{1}{2},GL}$	3, 247.923	5, 2065.506	7, 24434.8855	9, 330020.24	11, 0.4780388765×10^7

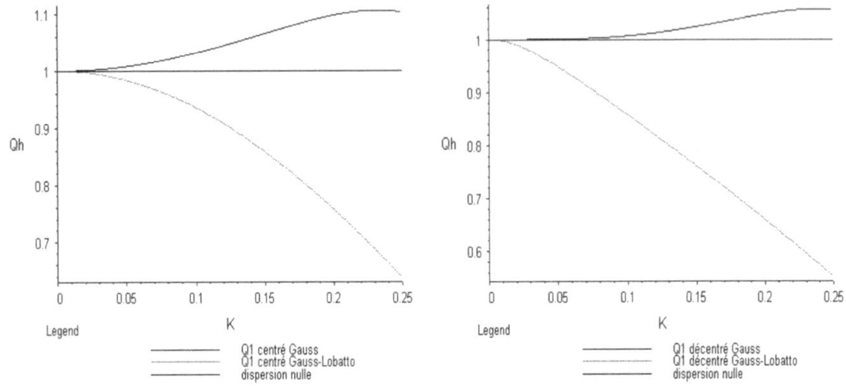

Fig. 4.10 Dispersion curves for Q_1-schemes with (*right*) and without (*left*) dissipation with Gauss (*up*) and Gauss–Lobatto (*down*) quadratures

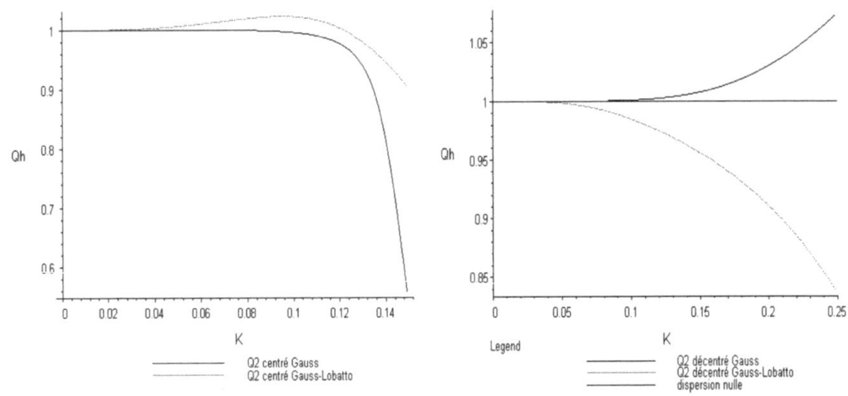

Fig. 4.11 Dispersion curves for Q_2-schemes with (*right*) and without (*left*) dissipation with Gauss (*left-down* and *right-up*) and Gauss-Lobatto (*left-up* and *right-down*) quadratures

4.5.3 Extension to Higher Dimensions

The extension of this analysis to higher dimensions can be done by tensorial products, as shown in Sect. 3.4.6. For the Maxwell's system, we get:

$$\omega_h h\varepsilon E_1 = \left(I_{r+1} \otimes I_{r+1} \otimes B^G_{h,r}[k_3]\right) H_2 - \left(I_{r+1} \otimes B^G_{h,r}[k_2] \otimes I_{r+1}\right) H_3, \quad (4.139a)$$

$$\omega_h h\varepsilon E_2 = \left(B^G_{h,r}[k_1] \otimes I_{r+1} \otimes I_{r+1}\right) H_3 - \left(I_{r+1} \otimes I_{r+1} \otimes B^G_{h,r}[k_3]\right) H_1, \quad (4.139b)$$

$$\omega_h h\varepsilon E_3 = \left(I_{r+1} \otimes B^G_{h,r}[k_2] \otimes I_{r+1}\right) H_1 - \left(B^G_{h,r}[k_1] \otimes I_{r+1} \otimes I_{r+1}\right) H_2, \quad (4.139c)$$

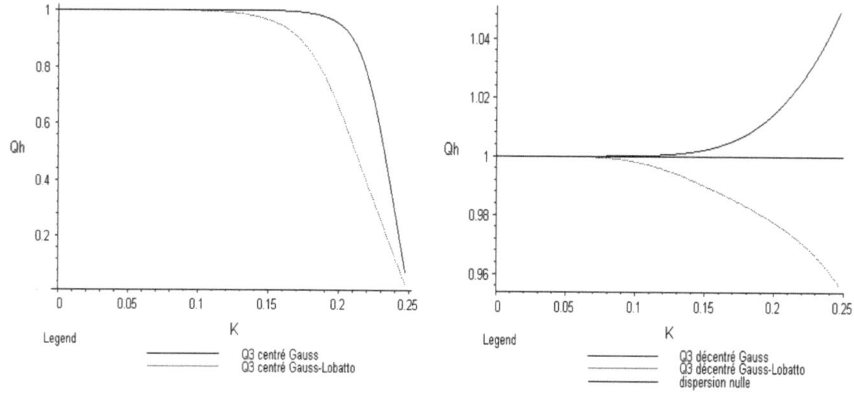

Fig. 4.12 Dispersion curves for Q_3-schemes with (*right*) and without (*left*) dissipation with Gauss (*up*) and Gauss–Lobatto (*down*) quadratures

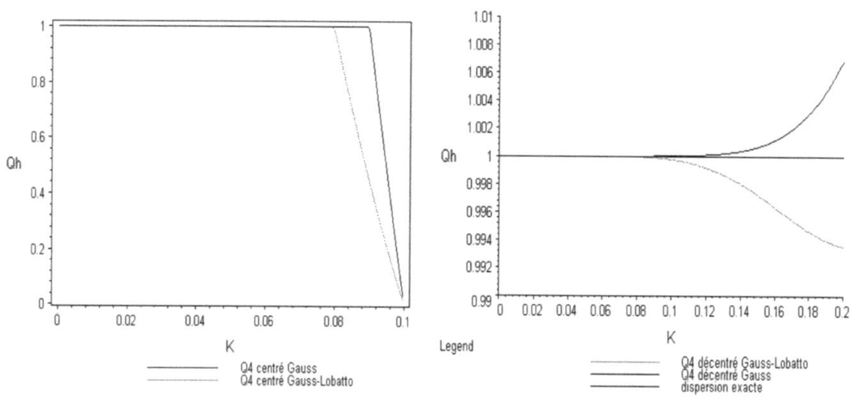

Fig. 4.13 Dispersion curves for Q_4-schemes with (*right*) and without (*left*) dissipation with Gauss (*up*) and Gauss–Lobatto (*down*) quadratures

$$\omega_h h \mu H_1 = -\left(I_{r+1} \otimes I_{r+1} \otimes B^G_{h,r}[k_3]\right) E_2 + \left(I_{r+1} \otimes B^G_{h,r}[k_2] \otimes I_{r+1}\right) E_3, \tag{4.139d}$$

$$\omega_h h \mu H_2 = -\left(B^G_{h,r}[k_1] \otimes I_{r+1} \otimes I_{r+1}\right) E_3 + \left(I_{r+1} \otimes I_{r+1} \otimes B^G_{h,r}[k_3]\right) E_1, \tag{4.139e}$$

$$\omega_h h \mu H_3 = -\left(I_{r+1} \otimes B^G_{h,r}[k_2] \otimes I_{r+1}\right) E_1 + \left(B^G_{h,r}[k_1] \otimes I_{r+1} \otimes I_{r+1}\right) E_2, \tag{4.139f}$$

where I_{r+1} is the identity matrix of order $r + 1$.

This eigenvalue problem finally provides:

$$\omega_h^2 = \frac{c^2}{h^2}\left(\omega_h[k_1]^2 + \omega_h[k_2]^2 + \omega_h[k_3]^2\right). \tag{4.140}$$

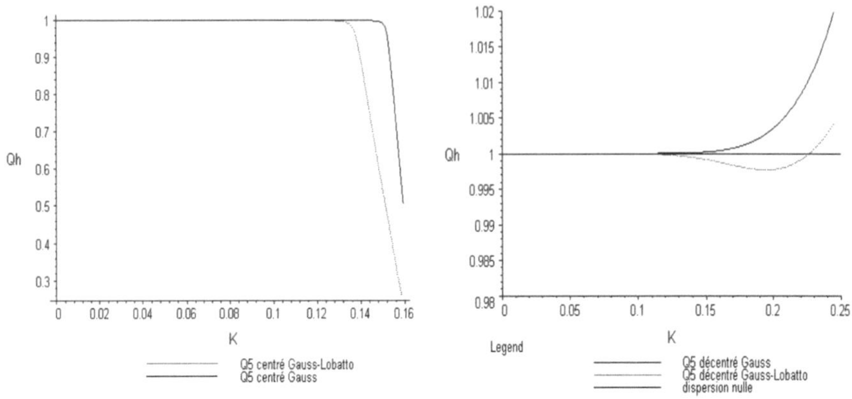

Fig. 4.14 Dispersion curves for Q_5-schemes with (*right*) and without (*left*) dissipation with Gauss (*up*) and Gauss–Lobatto (*down*) quadratures

4.6 Interior Penalty Discontinuous Galerkin Methods

Interior penalty discontinuous Galerkin methods (IPDG) were introduced in order to apply discontinuous Galerkin methods to second-order operators [11] and were developed in different forms for several equations [12]. For wave equations, one can see for instance [13, 14].

4.6.1 General Formulation

We want to solve the second-order hyperbolic problem:
Find $\underline{u} : \Omega \times [0, T] \rightarrow \mathbb{R}^p$

$$\underline{\underline{\lambda}} \frac{\partial^2 \underline{u}}{\partial t^2} + A \underline{\underline{\mu}}^{-1} A^* \underline{u} = \underline{F}, \tag{4.141a}$$

$$A(n)^* \underline{u}(x, t) = 0, \ \forall x \in \Gamma \text{ and } \forall t \in [0, T], \tag{4.141b}$$

$$\underline{u}(x, 0) = \underline{u}_0(x), \ \forall x \in \Omega, \tag{4.141c}$$

$$\frac{\partial \underline{u}}{\partial t}(x, 0) = \underline{v}_0(x), \ \forall x \in \Omega \tag{4.141d}$$

where $A = \sum_{i=1}^{d} A_i \frac{\partial}{\partial x_i}$ and $A^* = -\sum_{i=1}^{d} A_i^T \frac{\partial}{\partial x_i}$ with $A_i \in \mathbb{R}^{p \times q}$, by a discontinuous Galerkin method.

Let K be an element of a mesh $\mathscr{T}_h$ of Ω, and $\underline{v} : \Omega \to \mathbb{R}^p$ a test-function, we have, in a formal way[10]:

$$\begin{aligned}
\int_K A\underline{\underline{\mu}}^{-1}A^*\underline{u}\cdot\underline{v}\,\mathrm{d}\underline{x} &= \int_K \sum_{i=1}^d A_i\frac{\partial}{\partial x_i}\left(\underline{\underline{\mu}}^{-1}A^*\underline{u}\right)\cdot\underline{v}\,\mathrm{d}\underline{x}\\
&= \sum_{i=1}^d\left[-\int_K\left(\underline{\underline{\mu}}^{-1}A^*\underline{u}\right)\cdot\frac{\partial}{\partial x_i}\left(A_i^T\underline{v}\right)\mathrm{d}\underline{x}+\int_{\partial K}\left(\underline{\underline{\mu}}^{-1}A^*\underline{u}\right)\cdot\left(A_i^T n_i\underline{v}\right)\mathrm{d}\sigma\right]\\
&= \int_K\left(\underline{\underline{\mu}}^{-1}A^*\underline{u}\right)\cdot\left(A^*\underline{v}\right)\mathrm{d}\underline{x} - \int_{\partial K}\left(A(n)\underline{\underline{\mu}}^{-1}A^*\underline{u}\right)\cdot\underline{v}\,\mathrm{d}\sigma \qquad (4.142)
\end{aligned}$$

Equation (4.142) can be reorganized as follows:

$$\begin{aligned}
&\sum_K\int_{\partial K}\left(A(n)\underline{\underline{\mu}}^{-1}A^*\underline{u}\right)\cdot\underline{v}\,\mathrm{d}\sigma\\
&\quad=\sum_F\int_F\left[\left(A(n^+)\underline{\underline{\mu}}^{-1}A^*\underline{u}^+\right)\cdot\underline{v}^+ + \left(A(n^-)\underline{\underline{\mu}}^{-1}A^*\underline{u}^-\right)\cdot\underline{v}^-\right]\mathrm{d}\sigma\\
&\quad=\sum_{F\in\mathscr{F}_h}\int_F\left[\frac{1}{2}\left(A(n)\underline{\underline{\mu}}^{-1}A^*\underline{u}^+ + A(n)\underline{\underline{\mu}}^{-1}A^*\underline{u}^-\right)\cdot\left(\underline{v}^+-\underline{v}^-\right)\right.\\
&\quad\quad\left.+\frac{1}{2}\left(A(n)\underline{\underline{\mu}}^{-1}A^*\underline{u}^+ - A(n)\underline{\underline{\mu}}^{-1}A^*\underline{u}^-\right)\cdot\left(\underline{v}^++\underline{v}^-\right)\right]\mathrm{d}\sigma, \qquad (4.143)
\end{aligned}$$

$F \in \mathscr{F}_h$ being a face (or an edge) of ∂K.

Since (at least) $A\underline{\underline{\mu}}^{-1}A^*\underline{u} \in L^2(\Omega)^p$, we have:

$$[\![A(n)\underline{\underline{\mu}}^{-1}A^*\underline{u}]\!] = A(n)\underline{\underline{\mu}}^{-1}A^*\underline{u}^+ - A(n)\underline{\underline{\mu}}^{-1}A^*\underline{u}^- = 0 \text{ for all } F, \qquad (4.144)$$

so that

$$\begin{aligned}
\sum_{K\in\mathscr{T}_h}\int_{\partial K}\left(A(n)\underline{\underline{\mu}}^{-1}A^*\underline{u}\right)\cdot\underline{v}\,\mathrm{d}\sigma &= \sum_{F\in\mathscr{F}_h}\int_F\{\!\{A(n)\underline{\underline{\mu}}^{-1}A^*\underline{u}\}\!\}\cdot[\![\underline{v}]\!]\,\mathrm{d}\sigma\\
&= \sum_{F\in\mathscr{F}_h}\int_F\{\!\{\underline{\underline{\mu}}^{-1}A^*\underline{u}\}\!\}\cdot[\![A(n)^*\underline{v}]\!]\,\mathrm{d}\sigma \qquad (4.145)
\end{aligned}$$

where $\{\!\{\underline{w}\}\!\}$ is defined as in (4.35) and $\{\!\{\underline{\underline{\mu}}^{-1}A^*\underline{u}\}\!\} = \underline{\underline{\mu}}^{-1}A^*\underline{u}$ and $[\![A(n)^*\underline{v}]\!] = A(n)^*\underline{v}$ if $F \subset \partial\Omega$.

We can now define a discontinuous Galerkin approximation of (4.141a)–(4.141d) as follows:

---

[10] We could introduce a rigourous functional framework to derive this identity, but it is useless since we are only interested in the final formulation of the problem.

Find $u_h(\cdot, t) \in V_h$ such that:

$$\int_\Omega \underline{\underline{\lambda}} \frac{\partial^2 \underline{u}_h}{\partial t^2} \cdot \underline{v}_h \, \mathrm{d}\sigma + a_h(\underline{u}_h, \underline{v}_h) = \int_\Omega \underline{F} \cdot \underline{v}_h \, \mathrm{d}\underline{x}, \forall \underline{v}_h \in V_h, \tag{4.146}$$

where

$$a_h(\underline{u}_h, \underline{v}_h) = \sum_{K \in \mathscr{T}_h} \int_K \left(\underline{\underline{\mu}}^{-1} A^* \underline{u}_h\right) \cdot \left(A^* \underline{v}_h\right) \mathrm{d}\underline{x} - \sum_{F \in \mathscr{F}_h} \int_F \{\!\{\underline{\underline{\mu}}^{-1} A^* \underline{u}_h\}\!\} \cdot [\![A(n)^* \underline{v}_h]\!] \, \mathrm{d}\sigma$$

and $V_h \subset L^2(\Omega)^p$ is a space of approximation.

Unfortunately, this DG approximation of the operator $A \underline{\underline{\mu}}^{-1} A^*$ is not inf-sup stable because the sign of

$$\int_F \{\!\{\underline{\underline{\mu}}^{-1} A^* \underline{u}_h\}\!\} \cdot [\![A(n)^* \underline{v}_h]\!] \, \mathrm{d}\sigma$$

is not constant. So, the coercivity cannot be ensured!

4.6.2 *Coercivity of the Discrete Operator*

4.6.2.1 A New Formulation

In order to stabilize our formulation, we use another *a priori* information on the exact solution:

$$\begin{aligned} A^* \underline{u} \in \left[L^2(\Omega)\right]^q \; + \; \text{boundary condition} \; &\Rightarrow [\![A(n)^* \underline{u}]\!] = 0, \;\; \forall F \in \mathscr{F}_h \\ &\Rightarrow \int_F [\![A(n)^* \underline{u}]\!] \cdot [\![A(n)^* \underline{v}]\!] \, \mathrm{d}\sigma = 0, \;\; \forall F \in \mathscr{F}_h \text{ and } \forall \underline{v} \in V_h. \end{aligned}$$

This information enables us to introduce a new discontinuous Galerkin approximation of $A \underline{\underline{\mu}}^{-1} A^*$:$\underline{u}_h, \underline{v}_h \in V_h$ which reads:

$$\begin{aligned} a_h(\underline{u}_h, \underline{v}_h) = & \sum_{K \in \mathscr{T}_h} \int_K \left(\underline{\underline{\mu}}^{-1} A^* \underline{u}_h\right) \cdot \left(A^* \underline{v}_h\right) \mathrm{d}\underline{x} \\ & - \sum_{F \in \mathscr{F}_h} \int_F \{\!\{\underline{\underline{\mu}}^{-1} A^* \underline{u}_h\}\!\} \cdot [\![A(n)^* \underline{v}_h]\!] \, \mathrm{d}\sigma \\ & + \sum_{F \in \mathscr{F}_h} \int_F \alpha [\![A(n)^* \underline{u}_h]\!] \cdot [\![A(n)^* \underline{v}_h]\!] \, \mathrm{d}\sigma. \end{aligned} \tag{4.147}$$

Choosing $\alpha \geq 0$ sufficiently large, we expect that $a_h(\underline{v}_h, \underline{v}_h) > 0$ forall $\underline{v}_h \in V_h$. Now, since $\int_F \{\{\underline{\underline{\mu}}^{-1} A^* \underline{v}\}\} \cdot [\![A(n)^* \underline{u}]\!] \, \mathrm{d}\sigma = 0$, we can introduce a last new DG approximation:

$$\begin{aligned} a_h(\underline{u}_h, \underline{v}_h) = & \sum_{K \in \mathscr{T}_h} \int_K \left(\underline{\underline{\mu}}^{-1} A^* \underline{u}_h\right) \cdot \left(A^* \underline{v}_h\right) \mathrm{d}\underline{x} \\ & - \sum_{F \in \mathscr{F}_h} \int_F \{\{\underline{\underline{\mu}}^{-1} A^* \underline{u}_h\}\} \cdot [\![A(n)^* \underline{v}_h]\!] \, \mathrm{d}\sigma \\ & - \varepsilon \sum_{F \in \mathscr{F}_h} \int_F [\![A(n)^* \underline{u}_h]\!] \cdot \{\{\underline{\underline{\mu}}^{-1} A^* \underline{v}_h\}\} \, \mathrm{d}\sigma \\ & + \sum_{F \in \mathscr{F}_h} \int_F \alpha [\![A(n)^* \underline{u}_h]\!] \cdot [\![A(n)^* \underline{v}_h]\!] \, \mathrm{d}\sigma. \end{aligned} \tag{4.148}$$

The advantage of this last formulation is that

- if $\varepsilon = 1$, the bilinear form a_h is symmetric,
- if $\varepsilon = -1$, we immediately obtain, thanks to (4.142) and (4.144), $a_h(\underline{v}_h, \underline{v}_h) > 0$ forall $\alpha \geq 0$.

4.6.2.2 Conditions of Coercivity

In order to obtain the conditions of coercivity of a_h, we use the following inverse estimate (see Lemma 5.3 for the proof of this kind of result):

$\forall \underline{v}_h \in V_h$ and $\forall K \in \mathscr{T}_h$, $\exists\, C_{inv}(p_K) > 0$ (p_K being the order of the polynomial approximation on K) such that, $\forall F \subset \partial K$:

$$\|\underline{v}_h\|_{0,F} \leq \frac{C_{inv}(p_K)}{h_F^{1/2}} \|\underline{v}_h\|_{0,K}, \tag{4.149}$$

where h_F is the diameter of F.

This result can be extended as follows:

$\forall \underline{v}_h \in V_h$ and $\forall K \in \mathscr{T}_h$, $\exists\, C_{inv}(p_K) > 0$ such that, $\forall F \subset \partial K$:

$$\|A^* \underline{v}_h\|_{0,F} \leq \frac{C_{inv}(p_K)}{h_F^{1/2}} \|A^* \underline{v}_h\|_{0,K}. \tag{4.150}$$

We can deduce from (4.149) and (4.150) the inequality

$$\begin{aligned}\int_{F=K\cap T} \{\{\underline{\underline{\mu}}^{-1} A^* \underline{v}_h\}\} \cdot [\![A(n)^* \underline{v}_h]\!] \, d\sigma &\leq \frac{1}{2} \left\| \underline{\underline{\mu}}_K^{-1} A^* \underline{v}_K \right\|_{0,F} \| [\![A(n)^* \underline{v}]\!] \|_{0,F} \\ &+ \frac{1}{2} \left\| \underline{\underline{\mu}}_T^{-1} A^* \underline{v}_T \right\|_{0,F} \| [\![A(n)^* \underline{v}]\!] \|_{0,F} \\ &\leq \frac{1}{2} \left[\frac{C_{inv}(p_K)}{h_F^{1/2}} \mu_{K,min}^{-1/2} \left\| \underline{\underline{\mu}}_K^{-1/2} A^* \underline{v}_K \right\|_{0,K} \right. \\ &\left. + \frac{C_{inv}(p_T)}{h_F^{1/2}} \mu_{T,min}^{-1/2} \left\| \underline{\underline{\mu}}_T^{-1/2} A^* \underline{v}_T \right\|_{0,T} \right] \| [\![A(n)^* \underline{v}]\!] \|_{0,F} ,\end{aligned} \tag{4.151}$$

which can be rewritten as (K could be replaced by T)

$$\begin{aligned}&\int_{F=K\cap T} \{\{\underline{\underline{\mu}}^{-1} A^* \underline{v}_h\}\} \cdot [\![A(n)^* \underline{v}_h]\!] \, d\sigma \\ &\leq \frac{C_{inv}(p_K)}{h_F^{1/2}} \mu_{K,min}^{-1/2} \left\| \underline{\underline{\mu}}_K^{-1/2} A^* \underline{v}_K \right\|_{0,K} \left\| [\![A(n)^* \underline{v}_h]\!] \right\|_{0,F} .\end{aligned} \tag{4.152}$$

Now, by using the Young inequality $ab \leq \xi^2 a^2 + \frac{1}{4\xi^2} b^2$, we obtain:

$$\begin{aligned}\sum_{F\in\mathscr{F}_h} \int_F \{\{\underline{\underline{\mu}}^{-1} A^* \underline{v}_h\}\} \cdot [\![A(n)^* \underline{v}_h]\!] \, d\sigma &\leq \xi^2 \sum_{K\in\mathscr{T}_h} C(K) \left\| \underline{\underline{\mu}}_K^{-1/2} A^* \underline{v}_K \right\|_{0,K}^2 \\ &+ \frac{1}{4\xi^2} \sum_{F\in\mathscr{F}_h} \frac{\{\{C_{inv}(p)^2 \mu_{min}^{-1/2}\}\}}{h_F} \left\| [\![A(n)^* \underline{v}_h]\!] \right\|_{0,F}^2\end{aligned} \tag{4.153}$$

where $C(K)$ is the number of faces of the cell K.

$\diamond$ For $\varepsilon = -1$, we immediately have:

$$a_h(\underline{v}_h, \underline{v}_h) \geq \|\underline{v}_h\|_{h,*}^2, \tag{4.154}$$

where $\|\underline{v}_h\|_{h,*}^2 = \sum_{K\in\mathscr{T}_h} \left\| \underline{\underline{\mu}}_K^{-1/2} A^* \underline{v}_K \right\|_{0,K}^2 + \sum_{F\in\mathscr{F}_h} \alpha \frac{\{\{C_{inv}(p)^2 \mu_{min}^{-1/2}\}\}}{h_F}$ $\left\| [\![A(n)^* \underline{v}_h]\!] \right\|_{0,F}^2$, which of course implies the coercivity of a_h $\forall \alpha > 0$.

$\diamond$ For $\varepsilon = 1$, we get, by using (4.153):

$$a_h(\underline{v}_h, \underline{v}_h) \geq \sum_{K \in \mathcal{T}_h} \left(1 - 2C(K)\xi^2\right) \left\| \underline{\underline{\mu}}_K^{-1/2} A^* \underline{v}_K \right\|_{0,K}^2 + \sum_{F \in \mathcal{F}_h} \left(\alpha_F - \frac{1}{2\xi^2} \frac{\{\{C_{inv}(p)^2 \mu_{min}^{-1/2}\}\}}{h_F} \right) \left\| [\![A(n)^* \underline{v}_h]\!] \right\|_{0,F}^2. \quad (4.155)$$

Let $C_{ell} \in (0, 1]$ be the coercivity constant, i.e.

$$a_h(\underline{v}_h, \underline{v}_h) \geq C_{ell} \|\underline{v}_h\|_{h,*}^2$$

We can ensure the coercivity by taking the following parameters:

- $\xi^2 = \dfrac{1 - C_{ell}}{2C_{max}}$ where $C_{max} = \max\limits_{K \in \mathcal{T}_h} C(K)$,
- $\alpha_F = \alpha \dfrac{\{\{C_{inv}(p)^2 \mu_{min}^{-1/2}\}\}}{h_F}$ and $\alpha = \dfrac{C_{max}}{(1 - C_{ell})^2}$.

In particular, $C_{ell} \to 1^- \implies \alpha \to +\infty$.
$\diamond$ For $\varepsilon = 0$, we obtain:

$$a_h(\underline{v}_h, \underline{v}_h) \geq \sum_{K \in \mathcal{T}_h} \left(1 - C(K)\xi^2\right) \left\| \underline{\underline{\mu}}_K^{-1/2} A^* \underline{v}_K \right\|_{0,K}^2 + \sum_{F \in \mathcal{F}_h} \left(\alpha_F - \frac{1}{4\xi^2} \frac{\{\{C_{inv}(p)^2 \mu_{min}^{-1/2}\}\}}{h_F} \right) \left\| [\![A(n)^* \underline{v}_h]\!] \right\|_{0,F}^2 \quad (4.156)$$

To ensure the C_{ell} stability, we take the following parameters:

- $\xi^2 = \dfrac{1 - C_{ell}}{C_{max}}$ where $C_{max} = \max\limits_{K \in \mathcal{T}_h} C(K)$,
- $\alpha_F = \alpha \dfrac{\{\{C_{inv}(p)^2 \mu_{min}^{-1/2}\}\}}{h_F}$ and $\alpha = \dfrac{C_{max}}{4(1 - C_{ell})^2}$.

Here also, $C_{ell} \to 1^- \implies \alpha \to +\infty$.
We summarize the properties[11] of the different formulations versus ε in Table 4.6.

[11] We shall not develop error estimates for IPDG in this book.

Table 4.6 Properties of IPDG versus ε

ε	Symmetry	Positivity	Convergence
0	No	$\alpha > 0$	Not optimal
-1	No	$\alpha > \alpha_0 > 0$	Not optimal
1	Yes	$\alpha > \alpha_0 > 0$	Optimal

References

1. Reed, W.H., Hill, T.R.: Triangular mesh methods for the neutron transport equation, Technical Report LA-UR-73-479, Los Alamos Scientific Laboratory (1973)
2. LeSaint, P., Raviart, P.A.: I On a finite element method for solving the neutron transport equation. In: de Boor, C. (ed.) Mathematical aspects of finite elements in partial differential equations, pp. 89–145. Academic Press, New York (1974)
3. Cockburn, B., Karniadakis, G.E., Shu, C.-W.: Discontinuous Galerkin methods. Theory, Computation and Applications. Springer, Berlin (1999)
4. Hesthaven, J.S., Warburton, T.: Nodal discontinuous Galerkin methods. Texts in Applied Mathematics, vol. 54. Springer, Berlin (2008)
5. Jensen, M.: Remarks on duality in graph spaces of first-order linear operators. PAMM-Proc. Appl. Math. Mech. **6**(1), 31–34 (2006)
6. Buffa, A., Perugia, I., Warburton, T.: The mortar-discontinuous Galerkin method for the 2D Maxwell eigenproblem. J. Sci. Comput. **40**(1), 86–114 (2009)
7. Cohen, G., Ferrieres, X., Pernet, S.: A spatial high-order hexahedral discontinuous Galerkin method to solve Maxwell's equations in time domain. J. Comp. Phys. **217**(2), 340–363 (2006)
8. Maday, Y., Mavriplis, C., Patera, A.T.: Nonconforming mortar element methods: application to spectral discretizations. In: Domain Decomposition Methods, (Los Angeles, CA, 1988), pp. 392–418. SIAM, Philadelphia (1989)
9. Bernardi, C., Maday, Y., Rapetti, F.: Basics and some applications of the mortar element method. GAMM-Mitt. **28**(2), 97–123 (2005)
10. Pernet, S.: Etude de méthodes d'ordre élevé pour résoudre les équations de Maxwell dans le domaine temporel : Application à la détection et à la compatibilité électromagnétique, thèse de doctorat, U. de Paris-Dauphine (2004)
11. Wheeler, M.F.: An elliptic collocation-finite element method with interior penalties. SIAM J. Numer. Anal. **15**(1), 152–161 (1978)
12. Riviere, B.: Discontinuous Galerkin Methods for Solving Elliptic and Parabolic Equations: Theory and Implementation. SIAM, Philadelphia (2008)
13. Grote, M., Schötzau, D.: Optimal error estimates for the fully discrete interiorpenalty DG method for the wave equation. J. Sci. Comput. **40**, 257–272 (2009)
14. Grote, M., Schneebeli, A., Schötzau, D.: Interior penalty discontinuous Galerkin method for Maxwell's equations: Energy norm error estimates. IMA J. Numer. Anal. **28**, 440–468 (2008)

Chapter 5
The Maxwell's System and Spurious Modes

Abstract This chapter begins with the construction of different approximations of Maxwell's equations by finite element methods and the comparison of their performance versus discontinuous Galekin methods. Then, we treat the important and not obvious problem of spurious modes which appear in most approximations and we indicate how to suppress these modes. The last section contains error estimates of discontinuous Galekin methods with mass-lumping on hexahedra.

This chapter is devoted to the 3D Maxwell's system (2D is of less interest and can easily be derived from 3D) which presents important properties in terms of definition of its unknowns and of the behaviour of its solution which is related to the non empty character of the kernel of the curl operator. In the previous chapter, we provided a discontinuous Galerkin approximation of this system including a dissipative term whose presence, which is related to the property of the curl operator, will be justified in this chapter and its use extended to other approximations.

5.1 A First Model and Its Approximation

5.1.1 The Continuous Model

Let Ω be an open subdomain of $\mathbb{R}^3$. Equations (1.6a)–(1.6d) can be rewritten as

$$\varepsilon(\underline{x})\frac{\partial \underline{E}}{\partial t}(\underline{x},t) - \nabla \times (\mu(\underline{x})^{-1}\underline{B}(\underline{x},t)) = -\underline{J}(\underline{x},t) \ \text{ in } \Omega, \tag{5.1a}$$

$$\frac{\partial \underline{B}}{\partial t}(\underline{x},t) + \nabla \times \underline{E}(\underline{x},t) = 0 \ \text{ in } \Omega, \tag{5.1b}$$

to which we add the following boundary and initial conditions:

$$\underline{E}(\underline{x},t) \times \underline{\nu}(\underline{x}) = 0 \ \text{ on } \partial\Omega, \tag{5.2a}$$

G. Cohen and S. Pernet, *Finite Element and Discontinuous Galerkin Methods for Transient Wave Equations*, Scientific Computation,
DOI 10.1007/978-94-017-7761-2_5

$$\underline{E}(\underline{x},0)=\underline{E}_0(\underline{x}),\ \ \underline{H}(\underline{x},0)=\underline{H}_0(\underline{x})\ \text{ in } \Omega, \tag{5.2b}$$

The variational formulation of (5.1a)–(5.1b) and (5.2a) is[1]:
Find $\underline{E}$ and $\underline{B}$ such that $\underline{E}(.,t)\in H_0(\mathbf{curl},\Omega)$ and $\underline{B}(.,t)\in V$ and

$$\frac{\mathrm{d}}{\mathrm{d}t}\int_\Omega \varepsilon\underline{E}\cdot\underline{\varphi}\,\mathrm{d}\underline{x}-\int_\Omega \mu^{-1}\underline{B}\cdot\nabla\times\underline{\varphi}\,\mathrm{d}\underline{x}=-\int_\Omega \underline{J}\cdot\underline{\varphi}\,\mathrm{d}\underline{x},\ \forall\underline{\varphi}\in H_0(\mathbf{curl},\Omega), \tag{5.3a}$$

$$\frac{\mathrm{d}}{\mathrm{d}t}\int_\Omega \underline{B}\cdot\underline{\psi}\,\mathrm{d}\underline{x}+\int_\Omega \nabla\times\underline{E}\cdot\underline{\psi}\,\mathrm{d}\underline{x}=0,\ \forall\underline{\psi}\in V, \tag{5.3b}$$

where V is a functional space to be defined.

Following (1.84a)–(1.84b), we should take $V=\left[L^2(\Omega)\right]^3$. However, we cannot write the condition $\nabla\cdot\underline{B}=0$ for this choice. So, the appropriate definition is $V=H(\mathrm{div},\Omega)$, for which $\nabla\cdot\underline{B}=0$ makes sense [1]. Let us now define the approximate formulation of (5.3a)–(5.3b).

5.1.2 The Approximate Model

The appropriate approximation of this problem uses the first family of edge elements for $\underline{E}$ (defined in Sects. 2.5.2, 2.6.1 and 2.8.1) and the first family of $H(div)$ elements for $\underline{B}$ (defined in Sects. 2.7.1.1, 2.7.2.1 and 2.8.2). Let $\mathscr{T}_h$ be a mesh of Ω $\widehat{K}$ the unit element, $K\in\mathscr{T}_h$ such that $K=\underline{F}_K(\widehat{K})$ and DF_K the Jacobian matrix of $\underline{F}_K$ and J_K its determinant. With these notations, the approximate problem reads:
Find $\underline{E}_h$ and $\underline{B}_h$ such that $\underline{E}_h(.,t)\in U_h^r$ and $\underline{B}_h(.,t)\in V_h^r$ and

$$\frac{\mathrm{d}}{\mathrm{d}t}\int_\Omega \varepsilon\underline{E}_h\cdot\underline{\varphi}_h\,\mathrm{d}\underline{x}-\int_\Omega \mu^{-1}\underline{B}_h\cdot\nabla\times\underline{\varphi}_h\,\mathrm{d}\underline{x}=-\int_\Omega \underline{J}\cdot\underline{\varphi}_h\,\mathrm{d}\underline{x},\ \forall\underline{\varphi}_h\in U_h^r, \tag{5.4a}$$

$$\frac{\mathrm{d}}{\mathrm{d}t}\int_\Omega \underline{B}_h\cdot\underline{\psi}_h\,\mathrm{d}\underline{x}+\int_\Omega \nabla\times\underline{E}_h\cdot\underline{\psi}_h\,\mathrm{d}\underline{x}=0,\ \forall\underline{\psi}_h\in V_h^r, \tag{5.4b}$$

where

$$U_h^r=\left\{\underline{u}\in H_0(\mathbf{curl},\Omega)\text{ such that } DF_K^T\underline{u}_{|K}\in\mathscr{A}_r\right\} \tag{5.5}$$

with $\mathscr{A}_r=\mathscr{R}_r$ (defined in (2.56)) for tetrahedra, $\mathscr{A}_r=\mathscr{Q}_r$ (defined in (2.93)) for hexahedra, $\mathscr{A}_r=\mathscr{P}_r^e$ (defined in (2.108)) for pyramids, $\mathscr{A}_r=\mathscr{P}_{e,r}^w$ (defined in (2.110)) for wedges and

$$V_h^r=\left\{\underline{v}\in H(\mathrm{div},\Omega)\text{ such that } J_K\,DF_K^{-1}\underline{v}_{|K}\in\mathscr{B}_r\right\} \tag{5.6}$$

[1]One can multiply the second equation by μ^{-1} in order to obtain transposed stiffness matrices in the two equations.

with $\mathscr{B}_r = D_r$ (defined in (2.101)) for tetrahedra, $\mathscr{B}_r = \mathscr{Q}'_r$ (defined in (2.103)) for hexahedra, $\mathscr{B}_r = \mathscr{P}_r^{div}$ (defined in (2.112)) for pyramids, $\mathscr{B}_r = \mathscr{P}^w_{div,r}$ (defined in (2.113)) for wedges.

5.1.3 The Discrete Mass Integral

5.1.3.1 Functions in Two Different Directions

Obviously, neither tetrahedra nor pyramids lead to block-diagonal mass matrices. Following the nodal definitions of the basis functions of $\mathscr{Q}_r$ (in (2.97a)–(2.97c)) and $\mathscr{Q}'_r$ (in (2.104a)–(2.104c)), we could expect a mass-lumping if we use a quadrature rule.

So, let

$$\underline{\hat{\varphi}}^1_{i,j,k}(\hat{x}_1, \hat{x}_2, \hat{x}_3) = \hat{\varphi}^G_i(\hat{x}_1)\,\hat{\varphi}^{GL}_j(\hat{x}_2)\,\hat{\varphi}^{GL}_k(\hat{x}_3)\,\underline{e}_1$$

and

$$\underline{\hat{\varphi}}^2_{i',j',k'}(\hat{x}_1, \hat{x}_2, \hat{x}_3) = \hat{\varphi}^{GL}_{i'}(\hat{x}_1)\,\hat{\varphi}^{G}_{j'}(\hat{x}_2)\,\hat{\varphi}^{GL}_{k'}(\hat{x}_3)\,\underline{e}_2$$

be two basis functions of U_h^r on the unit cube $\widehat{K}$ and $\underline{\varphi}^1_{K,i,j,k} = DF_K^{-T}\underline{\hat{\varphi}}^1_{i,j,k}$, $\underline{\varphi}^2_{K,i',j',k'} = DF_K^{-T}\underline{\hat{\varphi}}^2_{i',j',k'}$. By using the definition of U_h^r, the elementary mass integral on K derived from (5.4a) reads

$$I_M = \int_K \varepsilon\,\underline{\varphi}^1_{K,i,j,k}\cdot\underline{\varphi}^2_{K,i',j',k'}\,\mathrm{d}\underline{x} = \int_{\widehat{K}} \hat{\varepsilon}\,M_K\,\underline{\hat{\varphi}}^1_{i,j,k}\cdot\underline{\hat{\varphi}}^2_{i',j',k'}\,\mathrm{d}\underline{\hat{x}}, \tag{5.7}$$

where $\hat{\varepsilon} = \varepsilon\circ\underline{F}_K$ and $M_K = DF_K^{-1}DF_K^{-T} = (m^K_{l,m})$, $l = 1,\dots,3$, $m = 1,\dots,3$.

Now, let $\left(\hat{\omega}^Q_p\right)_{p=1}^N$ and $\left(\hat{\xi}^Q_p\right)_{p=1}^N$ ($Q = G$ for Gauss and $Q = GL$ for Gauss–Lobatto) the weights and the points of a one-dimensional quadrature rule. Since $M_K\underline{e}_1\cdot\underline{e}_2 = m^K_{1,2}$, we can write, after setting $\lambda^K_{1,2} = \hat{\varepsilon}\,m^K_{1,2}$,

$$\begin{aligned} I_M &\simeq \sum_{n=1}^{N_1}\sum_{p=1}^{N_2}\sum_{q=1}^{r+1}\hat{\omega}^{Q_1}_n\,\hat{\omega}^{Q_2}_p\,\hat{\omega}^{GL}_q\,\lambda^K_{1,2}(\hat{\xi}^{Q_1}_n,\hat{\xi}^{Q_2}_p,\hat{\xi}^{GL}_q)\,\hat{\varphi}^G_i(\hat{\xi}^{Q_1}_n)\,\hat{\varphi}^{GL}_{i'}(\hat{\xi}^{Q_1}_n)\\ &\quad\times\hat{\varphi}^{GL}_j(\hat{\xi}^{Q_2}_p)\,\hat{\varphi}^G_{j'}(\hat{\xi}^{Q_2}_p)\,\hat{\varphi}^{GL}_k(\hat{\xi}^{GL}_q)\,\hat{\varphi}^{GL}_{k'}(\hat{\xi}^{GL}_q)\\ &\simeq \hat{\omega}^{GL}_{k'}\sum_{n=1}^{N_1}\sum_{p=1}^{N_2}\hat{\omega}^{Q_1}_n\,\hat{\omega}^{Q_2}_p\,\lambda^K_{1,2}(\hat{\xi}^{Q_1}_n,\hat{\xi}^{Q_2}_p,\hat{\xi}^{GL}_{k'})\,\hat{\varphi}^G_i(\hat{\xi}^{Q_1}_n)\,\hat{\varphi}^{GL}_{i'}(\hat{\xi}^{Q_1}_n)\\ &\quad\times\hat{\varphi}^{GL}_j(\hat{\xi}^{Q_2}_p)\,\hat{\varphi}^G_{j'}(\hat{\xi}^{Q_2}_p)\delta_{kk'} \end{aligned} \tag{5.8}$$

The presence of $\delta_{kk'}$ shows that only functions located on coplanar points of $\widehat{K}$ interact. Let us set $k = k'$, so that $\delta_{kk'} = 1$. We get:

- $I_M \simeq \hat{\omega}_i^G\,\hat{\omega}_{j'}^G\,\hat{\omega}_{k'}^{GL}\,\lambda_{1,2}^K(\hat{\xi}_i^G,\hat{\xi}_{j'}^G,\hat{\xi}_{k'}^{GL})\,\hat{\varphi}_{i'}^{GL}(\hat{\xi}_i^G)\,\hat{\varphi}_j^{GL}(\hat{\xi}_{j'}^G)$ for $Q_1 = Q_2 = G$.
- $I_M \simeq \hat{\omega}_{i'}^{GL}\,\hat{\omega}_j^{GL}\,\hat{\omega}_{k'}^{GL}\,\lambda_{1,2}^K(\hat{\xi}_{i'}^{GL},\hat{\xi}_j^{GL},\hat{\xi}_{k'}^{GL})\,\hat{\varphi}_i^G(\hat{\xi}_{i'}^{GL})\,\hat{\varphi}_{j'}^G(\hat{\xi}_j^{GL})$ for $Q_1 = Q_2 = GL$.
- $I_M \simeq \hat{\omega}_i^G\,\hat{\omega}_j^G\,\hat{\omega}_{k'}^{GL}\,\lambda_{1,2}^K(\hat{\xi}_i^G,\hat{\xi}_j^G,\hat{\xi}_{k'}^{GL})\,\hat{\varphi}_{i'}^{GL}(\hat{\xi}_i^G)\,\hat{\varphi}_{j'}^G(\hat{\xi}_j^{GL})$ for $Q_1 = G$, $Q_2 = GL$.
- $I_M \simeq \hat{\omega}_{i'}^{GL}\,\hat{\omega}_{j'}^G\,\hat{\omega}_{k'}^{GL}\,\lambda_{1,2}^K(\hat{\xi}_{i'}^{GL},\hat{\xi}_{j'}^{GL},\hat{\xi}_{k'}^{GL})\,\hat{\varphi}_i^G(\hat{\xi}_{i'}^{GL})\,\hat{\varphi}_j^{GL}(\hat{\xi}_{j'}^G)$ for $Q_1 = GL$, $Q_2 = G$.

In conclusion, two basis functions in two different directions have a non-zero interaction, which shows that, taking into account the tangential continuity, we do not obtain a block-diagonal matrix.

Remark: Since the mass matrix is not block-diagonal, one can also use either Gauss or Gauss–Lobatto rule for the three variables in space in order to compute the integrals. This will be discussed below.

5.1.3.2 Functions in the Same Direction

Let us now consider two functions in the same direction, for instance, $\underline{e}_1$. We have

$$\underline{\hat{\varphi}}^1_{i,j,k}(\hat{x}_1,\hat{x}_2,\hat{x}_3) = \hat{\varphi}_i^G(\hat{x}_1)\,\hat{\varphi}_j^{GL}(\hat{x}_2)\,\hat{\varphi}_k^{GL}(\hat{x}_3)\,\underline{e}_1,$$

$$\underline{\hat{\varphi}}^1_{i',j',k'}(\hat{x}_1,\hat{x}_2,\hat{x}_3) = \hat{\varphi}_{i'}^G(\hat{x}_1)\,\hat{\varphi}_{j'}^{GL}(\hat{x}_2)\,\hat{\varphi}_{k'}^{GL}(\hat{x}_3)\,\underline{e}_1.$$

In this case, we denote $\hat{\omega}_{n,p,q} = \hat{\omega}_n^G\,\hat{\omega}_p^{GL}\,\hat{\omega}_q^{GL}$ and $\underline{\hat{\xi}}_{n,p,q} = (\hat{\xi}_n^G,\hat{\xi}_p^{GL},\hat{\xi}_q^{GL})$. Then we get:

$$\begin{aligned} I_M &\simeq \sum_{n=1}^{r}\sum_{p=1}^{r+1}\sum_{q=1}^{r+1} \hat{\omega}_{n,p,q}\,\lambda_{1,1}^K(\underline{\hat{\xi}}_{n,p,q})\,\underline{\hat{\varphi}}^1_{i,j,k}(\underline{\hat{\xi}}_{n,p,q})\,\underline{\hat{\varphi}}^1_{i',j',k'}(\underline{\hat{\xi}}_{n,p,q}) \\ &\simeq \hat{\omega}_{i',j',k'}\,\lambda_{1,1}^K(\underline{\hat{\xi}}_{i',j',k'})\,\delta_{ii'}\,\delta_{jj'}\,\delta_{kk'}. \end{aligned} \tag{5.9}$$

We see that, in this case, two functions located at two different points have no interaction.

5.1.3.3 Orthogonal Meshes

Let $\mathcal{T}_h$ be an orthogonal mesh, i.e. a mesh whose elements have their edges parallel to the canonical basis of $\mathbb{R}^3$. In this case, one can easily see that M_K is diagonal. So, I_M is non zero if and only if the two basis functions have the same direction. On the other hand, (5.9) shows that two collinear basis functions located at two different points have no interactions. So, on an orthogonal mesh, we get a diagonal mass matrix for hexahedra. Actually, this approximation is equivalent to the Yee scheme [2] when ε and μ are continuous for $r = 1$ and can be seen as a variational generalization (different from the finite difference point of view) of this scheme to higher orders of approximation. More details on this approach can be found in [3].

Finally, although fitting to the Maxwell's system, this approximation leads in general to a mass matrix without mass-lumping. Moreover, the structure of the stiffness integral shows that we do not have the local property of the stiffness matrix obtained for discontinuous Galerkin methods, which induces an important storage. Of course, these two points are major drawbacks to efficiency.

Remark: As we shall see later, taking $V = \left[L^2(\Omega)\right]^3$ with a judicious quadrature rule leads to an approximation on a general mesh which can be useful for time-harmonic problems.

5.2 A Second Model and Its Approximations

In this section, we get rid of the constraint $\nabla \cdot \underline{B} = 0$, which is somehow a functional crime but leads to much more efficient numerical methods with some trouble which will be addressed in the next section.

5.2.1 *The Continuous Model*

Since $\underline{B} = \underline{\underline{\mu}}\,\underline{H}$, we rewrite the Maxwell's system in an anisotropic and inhomogeneous medium as in (1.8a)–(1.8b).

$$\underline{\underline{\varepsilon}}\frac{\partial \underline{E}}{\partial t}(\underline{x}, t) - \nabla \times \underline{H}(\underline{x}, t) = -\underline{J}(\underline{x}, t), \tag{5.10a}$$

$$\underline{\underline{\mu}}\frac{\partial \underline{H}}{\partial t}(\underline{x}, t) + \nabla \times \underline{E}(\underline{x}, t) = 0, \tag{5.10b}$$

$$\underline{E}(\underline{x}, t) \times \underline{\nu}(\underline{x}) = 0 \;\text{ on } \partial\Omega, \tag{5.10c}$$

$$\underline{E}(\underline{x}, 0) = \underline{E}_0(\underline{x}),\; \underline{B}(\underline{x}, 0) = \underline{B}_0(\underline{x}) \;\text{ in } \Omega. \tag{5.10d}$$

The variational formulation of (5.10a)–(5.10c) is then written as in (1.84a)–(1.84b).

Find $\underline{E}$ and $\underline{H}$ such that $\underline{E}(., t) \in U$ and $\underline{H}(., t) \in \left[L^2(\Omega)\right]^3$ and

$$\frac{\mathrm{d}}{\mathrm{d}t}\int_\Omega \underline{\underline{\varepsilon}}\,\underline{E} \cdot \underline{\varphi}\mathrm{d}\underline{x} - \int_\Omega \underline{H} \cdot (\nabla \times \underline{\varphi})\,\mathrm{d}\underline{x} = -\int_\Omega \underline{J}\underline{\varphi}\,\mathrm{d}\underline{x},\; \forall \underline{\varphi} \in U \tag{5.11a}$$

$$\frac{\mathrm{d}}{\mathrm{d}t}\int_{\Omega} \underline{\underline{\mu}}\,\underline{H} \cdot \underline{\psi} + \int_{\Omega} (\nabla \times \underline{E}) \cdot \underline{\psi}\, \mathrm{d}\underline{x} = 0, \ \forall \underline{\psi} \in [L^2(\Omega)]^3. \tag{5.11b}$$

5.2.2 General Formulations of the Approximations

We have two choices for U. The most natural one is of course $U = H_0(\mathbf{curl}, \Omega)$ which ensures a good approximation of discontinuous media since it keeps tangential continuity. However, $U = \left[H^1(\Omega)\right]^3$[2] can be another possibility for smooth coefficients [4]. So, if $\mathscr{T}_h$ is a mesh composed of hexahedra of any shape, the approximate formulation of (5.11a)–(5.11b) can be written as

Find $\underline{E}_h$ and $\underline{H}_h$ such that $\underline{E}_h(., t) \in W_h^r$ and $\underline{H}_h(., t) \in X_h^r$ and

$$\frac{\mathrm{d}}{\mathrm{d}t}\int_{\Omega} \underline{\underline{\varepsilon}}\,\underline{E}_h \cdot \underline{\varphi}_h \mathrm{d}\underline{x} - \int_{\Omega} \underline{H}_h \cdot (\nabla \times \underline{\varphi}_h)\, \mathrm{d}\underline{x} = -\int_{\Omega} \underline{J} \cdot \underline{\varphi}_h\, \mathrm{d}\underline{x}, \ \forall \underline{\varphi}_h \in W_h^r \tag{5.12a}$$

$$\frac{\mathrm{d}}{\mathrm{d}t}\int_{\Omega} \underline{\underline{\mu}}\,\underline{H}_h \cdot \underline{\psi}_h + \int_{\Omega} (\nabla \times \underline{E}_h) \cdot \underline{\psi}_h\, \mathrm{d}\underline{x} = 0, \ \forall \underline{\psi}_h \in X_h^r. \tag{5.12b}$$

where

$$W_h^r = \left\{\underline{u} \in U \text{ such that } \forall K \in \mathscr{T}_h,\ A_K \underline{u}_{|K} \in \mathscr{A}_r\right\} \tag{5.13}$$

with $A_K = DF_K^T$ when $U = H_0(\mathbf{curl}, \Omega)$ and $A_K = I_3$ when $U = \left[H^1(\Omega)\right]^3$, I_3 being the identity matrix of $\mathbb{R}^3$ and

$$X_h^r = \left\{\underline{u} \in \left[L^2(\Omega)\right]^3 \text{ such that } \forall K \in \mathscr{T}_h,\ A_K \underline{u}_{|K} \in \mathscr{A}_r\right\} \tag{5.14}$$

with $\mathscr{A}_r$ defined below.

We could of course take for $\mathscr{A}_r$ the same definitions as in the previous section. These choices have no interest for transient equation since the condition $\nabla \cdot \underline{B} = 0$ is now violated and, on the other hand, these choices lead to mass matrices with no mass-lumping. Moreover, as we saw, no approximation by other elements than hexahedra have a chance to provide block-diagonal mass matrices. Therefore, we shall only set $\mathscr{A}_r = [Q_r]^3$, Q_r being defined in (2.17). The continuous character of these approximations obviously imply to locate the interpolation points at the Gauss–Lobatto points. However, the use of the first family can be of some interest for the time-harmonic equation, for which no inversion of the mass-matrix is required [5, 6].

[2] Which is possible since $\left[H^1(\Omega)\right]^3 \subset H(\mathbf{curl}, \Omega)$. The boundary condition will be treated below.

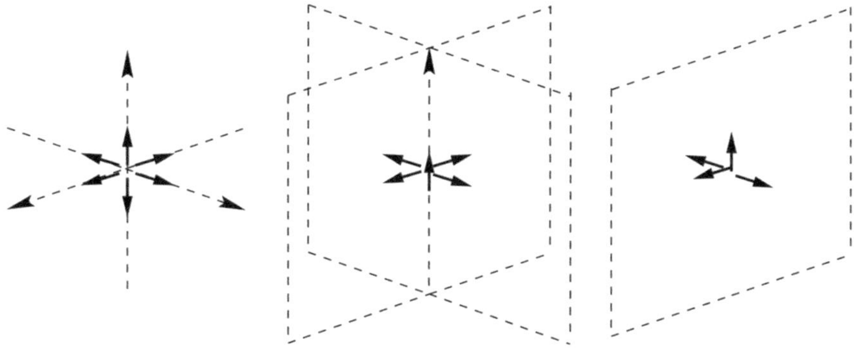

Fig. 5.1 The degrees of freedom for W_h^r around a vertex (*left*), a point on an edge (*center*) and a point on a face (*right*) when the mesh is orthogonal for an $H(curl)$ approximation

5.2.3 *Approximation in* $H(\mathbf{Curl}, \Omega)$

In this case, the mass and stiffness matrices are computed as for discontinuous Galerkin methods in Sect. 4.3.4.2 when $U = H_0(\mathbf{curl}, \Omega)$. However, for Eq. (5.12a), the continuous character of the test functions induces a coupling of the integrals on the elements surrounding the point at which the test function is defined, so that the mass matrices are no longer 3×3, but $N \times N$, N denoting the number of degrees of freedom around one point. The stiffness matrix is defined in the same way. However, in order to keep its local definition, we assembly this matrix at each time-step, as we did for the acoustics equation in Sect. 3.2.3. In particular, for a structured mesh (i.e. on a mesh which can be derived from an orthogonal mesh by a conform mapping), we obtain (Fig. 5.1)[3]

- six degrees of freedom for W_h^r around a vertex,
- five degrees of freedom for W_h^r around a point on an edge,
- four degrees of freedom for W_h^r around a point on a face,
- three degrees of freedom for both spaces around an interior point.

Now, if we denote

$$\frac{d}{dt} \widetilde{B}_h\, E_h + R_h\, H_h = J_h, \tag{5.15a}$$

$$\frac{d}{dt} B_h\, H_h - R_h^T\, E_h = 0, \tag{5.15b}$$

where $\widetilde{B}_h$ and B_h are the mass matrices and R_h the stiffness matrix, the discrete form of (5.12a)–(5.12b) and

[3]For non-structured meshes, one can obtain less or more degrees of freedom for a vertex and a point on an edge.

$$\frac{\mathrm{d}}{\mathrm{d}t}\widetilde{B}_h\,\widetilde{E}_h + K_h\,\widetilde{E}_h = J'_h, \tag{5.16}$$

where K_h is the stiffness matrix, the discrete form of the approximation in W_h^r of

$$\underline{\underline{\varepsilon}}\frac{\partial^2 \underline{E}}{\partial t^2}(\underline{x}, t) - \nabla \times (\underline{\underline{\mu}}\nabla \times \underline{E}(\underline{x}, t)) = -\underline{J}'(\underline{x}, t), \tag{5.17}$$

we have the following theorem of equivalence

Theorem 12 *Problems (5.16) and (5.15a)–(5.15b) are equivalent. We actually have*

$$K_h = R_h B_h^{-1} R_h^T \tag{5.18}$$

which ensures that

$$\widetilde{E}_h = E_h. \tag{5.19}$$

The proof of Theorem (5.152) is similar to this of Theorem 5 in Sect. 3.3.2.

5.2.4 Approximation in $\left[H^1(\Omega)\right]^3$

When $U = \left[H^1(\Omega)\right]^3$, the approximation is slightly different because of $A_K = I_3$. First, it is easy to check that, by this mapping, the mass matrix is diagonal and not block-diagonal. Let us now compute the stiffness term. If $\underline{E}_h \in W_h^r$ and $\underline{\psi}_h \in X_h^r$ is a function such that $K = \mathrm{supp}\underline{\psi}_h$, where $K = \underline{F}_K(\widehat{K})$ is an hexahedron of $\mathscr{T}_h$, we have:

$$\begin{aligned} I_S = \int_K (\nabla \times \underline{E}_h) \cdot \underline{\psi}_h \,\mathrm{d}\underline{x} &= \int_{\widehat{K}} (\nabla \times \underline{E}_h) \circ \underline{F}_K \cdot \underline{\psi}_h \circ \underline{F}_K \,\mathrm{d}\underline{\hat{x}} \\ &= \int_{\widehat{K}} |J_K| (DF_K^{-T}\hat{\nabla} \times \underline{\hat{E}}_h) \cdot \underline{\hat{\psi}}_h \,\mathrm{d}\underline{\hat{x}} \\ &= \int_{\widehat{K}} |J_K| (DF_K^{-T}\hat{\nabla} \times DF_K^{-T}(DF_K^T \underline{\hat{E}}_h)) \cdot \underline{\hat{\psi}}_h \,\mathrm{d}\underline{\hat{x}}, \end{aligned} \tag{5.20}$$

where $J_K = \det DF_K$.

By setting $\underline{\hat{\theta}} = DF_K^T \underline{\hat{E}}_h$ and using (4.105), we get:

$$I_S = s_K \int_{\widehat{K}} \hat{\nabla} \times \underline{\hat{\theta}} \cdot \underline{\hat{\psi}}_h \,\mathrm{d}\underline{\hat{x}}, \tag{5.21}$$

where $s_K = \mathrm{sgn}(J_K)$.

Now, let us denote $\left(\hat{\underline{\xi}}_\ell\right)_{\ell=1}^{(r+1)^3}$ and $\left(\hat{\omega}_\ell\right)_{\ell=1}^{(r+1)^3}$ the Gauss–Lobatto points and weights, $(DF_K^{-T})_{m,p} = \left(a_{m,p}\right)$ and let us set $\hat{\underline{\phi}}_h = \hat{\varphi}_i\, \underline{e}_p$ and $\hat{\underline{\theta}}_h = \sum_{m=1}^{3} \sum_{j=1}^{(r+1)^3} \theta_j^m\, \hat{\varphi}_j\, \underline{e}_m$, so that $\hat{\varphi}_i(\hat{\underline{\xi}}_\ell) = \delta_{i\ell}$ and $\hat{\varphi}_j(\hat{\underline{\xi}}_\ell) = \delta_{j\ell}$. $\hat{\underline{\theta}}_h$ is given by

$$\int_{\widehat{K}} DF_K^{-T} \hat{\underline{\theta}}_h \cdot \hat{\underline{\phi}}_h \, \mathrm{d}\hat{\underline{x}} = \int_{\widehat{K}} \hat{\underline{E}}_h \cdot \hat{\underline{\phi}}_h \, \mathrm{d}\hat{\underline{x}}. \tag{5.22}$$

On the other hand, we have:

$$\begin{aligned}
\int_{\widehat{K}} DF_K^{-T} \hat{\underline{\theta}}_h \cdot \hat{\underline{\phi}}_h \, \mathrm{d}\hat{\underline{x}} &= \sum_{m=1}^{3} \sum_{j=1}^{(r+1)^3} \int_{\widehat{K}} \theta_j^m\, \hat{\varphi}_j\, \hat{\varphi}_i\, DF_K^{-T} \underline{e}_m \cdot \underline{e}_p \, \mathrm{d}\hat{\underline{x}}. \\
&\simeq \sum_{m=1}^{3} \sum_{j=1}^{(r+1)^3} \sum_{\ell=1}^{(r+1)^3} \hat{\omega}_\ell\, \theta_j^m\, \hat{\varphi}_j(\hat{\underline{\xi}}_\ell)\, \hat{\varphi}_i(\hat{\underline{\xi}}_\ell)\, a_{m,p} \\
&\simeq \sum_{m=1}^{3} \hat{\omega}_i\, \theta_i^m\, a_{m,p}
\end{aligned} \tag{5.23}$$

Now, if we denote $\hat{\underline{E}}_h = \sum_{m=1}^{3} \sum_{j=1}^{(r+1)^3} E_j^m \hat{\varphi}_j\, \underline{e}_m$,

$$\begin{aligned}
\int_{\widehat{K}} \hat{\underline{E}}_h \cdot \hat{\underline{\phi}}_h \, \mathrm{d}\hat{\underline{x}} &= \sum_{m=1}^{3} \sum_{j=1}^{(r+1)^3} \int_{\widehat{K}} E_j^m \hat{\varphi}_j\, \hat{\varphi}_i\, \underline{e}_m \cdot \underline{e}_p \, \mathrm{d}\hat{\underline{x}}. \\
&\simeq \sum_{m=1}^{3} \sum_{j=1}^{(r+1)^3} \sum_{\ell=1}^{(r+1)^3} \hat{\omega}_\ell\, E_j^m \hat{\varphi}_j(\hat{\underline{\xi}}_\ell)\, \hat{\varphi}_i(\hat{\underline{\xi}}_\ell)\, \delta_{mp} = \hat{\omega}_i\, E_i^p
\end{aligned} \tag{5.24}$$

Equations (5.23) and (5.24) show that, if we denote $\underline{\Theta}_i = (\theta_i^1, \theta_i^2, \theta_i^3)^T$ and $\underline{U} = (E_i^1, E_i^2, E_i^3)^T$, we get

$$\underline{\Theta}_i = DF_K^T\, \underline{U}. \tag{5.25}$$

By using the above notations and by setting $\hat{\underline{\psi}}_h = \hat{\varphi}_i\, \underline{e}_p$, (5.21) provides:

$$I_S = s_K \sum_{m=1}^{3} \sum_{j=1}^{(r+1)^3} \theta_j^m \int_{\widehat{K}} \hat{\nabla} \times \hat{\varphi}_j\, \underline{e}_m \cdot \hat{\varphi}_i\, \underline{e}_p \, \mathrm{d}\hat{\underline{x}}$$

$$=s_K \sum_{m=1}^{3} \sum_{j=1}^{(r+1)^3} \theta_j^m S_{i,j}^{m,p}. \tag{5.26}$$

By setting $j = (j_1, j_2, j_3)$, $i = (i_1, i_2, i_3)$, $\hat{\underline{\xi}}_\ell = (\hat{\xi}_{i_1}, \hat{\xi}_{i_2}, \hat{\xi}_{i_3})$, $\hat{\varphi}_j(\hat{x}_1, \hat{x}_2, \hat{x}_3) = \hat{\varphi}_{j_1}^{GL}(\hat{x}_1)\,\hat{\varphi}_{j_2}^{GL}(\hat{x}_2)\,\hat{\varphi}_{j_3}^{GL}(\hat{x}_3)$ and taking into account (4.108), we get:

$$I_S \simeq \hat{\omega}_i \sum_{j_\kappa=1}^{r+1} \theta_{j_\kappa}^{\kappa} \frac{d\hat{\varphi}_{j_\kappa}^{GL}}{d\hat{x}_\kappa}(\hat{\xi}_{i_\kappa}), \tag{5.27}$$

with $m \neq p$ and $\kappa = 6 - m - p$.

Remarks:

1. One can also replace $\hat{\underline{E}}_h$ by $\hat{\varphi}_j\, \underline{e}_m$ and set $\hat{\underline{\theta}}_h = DF_K^T \hat{\varphi}_j\, \underline{e}_m$ and $\hat{\underline{\psi}}_h = \hat{\varphi}_i\, \underline{e}_p$ in (5.21) (which corresponds to a change of variables after decomposition of $\underline{E}_h$ on the basis functions). The computation of I_S is then more complex. For instance, we obtain, for $m = 1$ and $p = 2$

$$I_S \simeq \hat{\omega}_i \left(\frac{\partial(b_{1,1}\hat{\varphi}_{j_3}^{GL})}{\partial \hat{x}_3}(\hat{\xi}_{i_3}) - \frac{\partial(b_{3,1}\hat{\varphi}_{j_1}^{GL})}{\partial \hat{x}_1}(\hat{\xi}_{i_1}) \right), \tag{5.28}$$

   with $(DF_K^T)_{m,p} = \left(b_{m,p}\right)$.
2. This equivalence theorem is formulated as Corollary 2 in [3, 7] and is extended to the linear elastodynamics system in [8] but its proof, based on matrix analysis, is much longer and less elegant than the proof given here which is due to S. Pernet.
3. The treatment of the boundary condition $\underline{E} \times \underline{n} = 0$ is not obvious on a boundary not parallel to the axes (see for instance [9]).
4. Both approximations behave like the approximation of the acoustics equation defined in Sect. 3.2 in terms of dispersion and stability (see for instance [10] for $U = H(\mathbf{curl}, \Omega)$).

5.2.5 *Comparison of the Approximations*

5.2.5.1 Comparison of the Performance

Obviously, both approximations need less storage than discontinuous Galerkin methods (between 20 and 30 %). In terms of computational time, they are also more performant than discontinuous Galerkin methods. However, approximation in $\left[H^1(\Omega)\right]^3$ is less performant than approximation in $H(\mathbf{curl}, \Omega)$. These results are illustrated by the following experiment: we propagate a spherical source whose component in time is a Ricker (second derivative of a Gaussian function in time) in

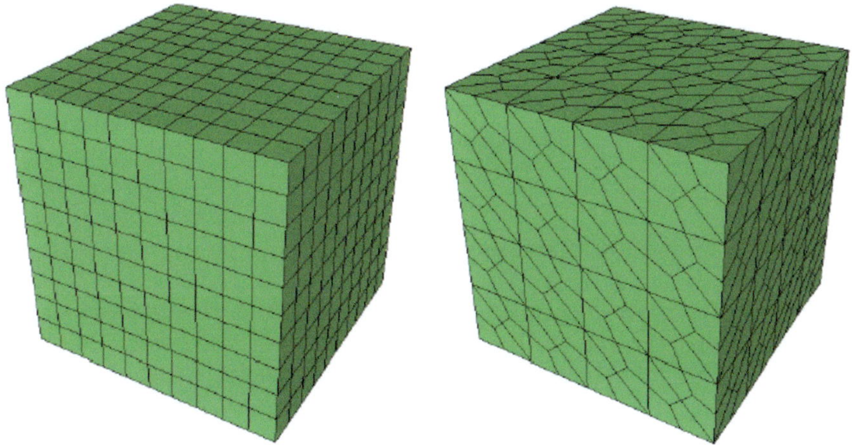

Fig. 5.2 Regular and non regular meshes

Table 5.1 Performance for the regular mesh for a Q_5 approximation

	Δt	N_{DOFs}	$CPU(s)/step$	N_{steps}	$CPU(s)$	$L^2 - error\ (\%)$
H^1	2.3e-2	1045953	0.69	173	119	2.58
$H(curl)$	2.3e-2	1116180	0.46	173	80	2.58
DGM	2.28e-2	1296000	0.75	174	130	2.58

Table 5.2 Performance for the non regular mesh for a Q_5 approximation

	Δt	N_{DOFs}	$CPU(s)/step$	N_{steps}	$CPU(s)$	$L^2 - error\ (\%)$
H^1	8.83e-3	1595211	1.04	453	471	3.41
$H(curl)$	8.83e-3	1704720	0.92	453	417	3.18
DGM	8.83e-3	1990656	1.40	453	634	3.73

a cube with an orthogonal mesh and a mesh composed of tetrahedra split into four hexahedra given in Fig. 5.2 for $0 \leq t \leq 4$ (we set $\varepsilon = \mu = 1$). In Tables 5.1, 5.2 and 5.3, we provide the performance of both meshes for a Q_5 approximation (DGM with Gauss–Lobatto). These tables show, first, that the split tetrahedra are much less performant than a regular mesh and cannot be an efficient alternative to obtain purely hexahedral mesh. On the other hand, the $H(curl)$ approximation is the best one in terms of CPU time and storage. In the next section, we show another fundamental advantage of this approximation.

Remark: Table 5.3 shows a substantial increase of the performance of hexahedral meshes when we use an $H(curl)$-L^2 instead a DGM approximation.

Table 5.3 Comparison of the CPU times of the different methods

	$H^1/H(curl)$	H^1/DGM	$H(curl)/DGM$
Regular	1.48	0.92	0.61
Non-regular	1.13	0.74	0.66

5.2.5.2 Treatment of a Discontinuity

An important property of the $H(curl)$ approximation is the continuity of the tangential component and the discontinuity of the normal component of the electric field. For the H^1 approximation, both component are continuous. This additional continuity (which is not physical) induces a substantial error on the solution in presence of a discontinuity of coefficients, as shown in the following experiment.

We consider a 2D squared domain (Fig. 5.3) composed of a left part in which $\varepsilon = \mu = 1$ and a right part in which $\varepsilon = 0.01$ and $\mu = 100$ (which provides the same velocity in the whole domain) in which we propagate a wave generated by a spherical source. In Table 5.4, we give the error committed on both components for different approximations (the reference solution was computed by a $H(curl)$ approximation with a very fine mesh). The better approximation provided by $H(curl)$ is then obvious.

Fig. 5.3 The discontinuous domain

Table 5.4 L^2-error committed on the tangential and normal components of the electric field

		Order 3 (%)	Order 5 (%)	Order 7 (%)
Tangential component	H^1	24.16	5.96	1.47
	$H(curl)$	24.76	5.80	1.40
Normal component	H^1	23.16	12.59	11.27
	$H(curl)$	24.44	7.64	2.72

5.3 Suppressing Spurious Modes

5.3.1 *Some Background About the Spurious Modes*

Localizing eigenmodes of a complex matrix is a difficult problem [11], in particular, when, as in our case, a matrix depends on parameters. In this part, we give a brief overview about the problem of spurious modes which may pollute the numerical solutions obtained from the Maxwell's equations. This topic was widely studied and we only give the main results.

We begin by recalling the link between the time-dependent solution and the solutions of the associated spectral problem. For that, we consider the following model problem: find $\underline{E}$, $\underline{H}$: $\Omega \to \mathbb{R}^3$ such that

$$\varepsilon \frac{\partial \underline{E}}{\partial t} - \nabla \times \underline{H} = \underline{J} \text{ on } \Omega, \tag{5.29a}$$

$$\mu \frac{\partial \underline{H}}{\partial t} + \nabla \times \underline{E} = 0 \text{ on } \Omega, \tag{5.29b}$$

$$\underline{E} \times \underline{n} = 0 \text{ on } \Gamma, \tag{5.29c}$$

$$\underline{E}(0, x) = \underline{E}_0(x) \text{ on } \Omega, \tag{5.29d}$$

$$\underline{H}(0, x) = \underline{H}_0(x) \text{ on } \Omega, \tag{5.29e}$$

where Ω is a simply connected bounded Lipschtz domain and $\nabla \cdot \underline{J}(t, \cdot) = 0\ \forall t$.

Equation (5.29a)–(5.29e) can be rewritten in the following equivalent form:

$$\varepsilon \frac{\partial^2 \underline{E}}{\partial t^2} + \nabla \frac{1}{\mu} \nabla \times \underline{E} = \partial_t \underline{J} =: \underline{\tilde{J}} \text{ on } \Omega, \tag{5.30a}$$

$$\underline{E} \times \underline{n} = 0 \text{ on } \Gamma, \tag{5.30b}$$

$$\underline{E}(0, x) = \underline{E}_0(x) \text{ on } \Omega, \tag{5.30c}$$

$$\frac{\partial \underline{E}}{\partial t}(0, x) = \underline{E}_1(x) \text{ on } \Omega. \tag{5.30d}$$

Let us define the following functional spaces:

$$V = H_0(curl, \Omega), \tag{5.31a}$$

$$V^0 = \{\underline{v} \in V \; : \; \nabla \times \underline{v} = 0\}, \tag{5.31b}$$

$$W = \{\underline{v} \in V \; : \; \nabla \cdot (\varepsilon \underline{v}) = 0\} \tag{5.31c}$$

and the bilinear form

$$a(\underline{u}, \underline{v}) = \int_{\Omega} \frac{1}{\mu} \nabla \times \underline{u} \cdot \nabla \times \underline{v} \, \mathrm{d}\underline{x}, \;\; \forall (\underline{u}, \;\underline{v}) \in V^2. \tag{5.32}$$

Equations (5.30a)–(5.30d) can be then written in the weak form:
Find $\underline{E}(t, \cdot) \in V$ solution of

$$\left(\varepsilon \frac{\partial^2 \underline{E}}{\partial t^2}, \underline{v}\right)_0 + a(\underline{u}, \underline{v}) = \left(\tilde{\underline{J}}, \underline{v}\right)_0, \;\; \forall \underline{v} \in V, \tag{5.33a}$$

$$\underline{E}(0, \cdot) = \underline{E}_0, \tag{5.33b}$$

$$\frac{\partial \underline{E}}{\partial t}(0, \cdot) = \underline{E}_1, \tag{5.33c}$$

where $(,)_0$ denotes the L^2 scalar product.

By using $\nabla \left(\varepsilon \underline{E}\right)(t\,,\;\cdot) = 0 \; \forall t$, we have $\underline{E} \in W$ and the previous weak formulation (5.33a)–(5.33c) can be directly defined on the subspace W of V as follows:
Find $\underline{E}(t, \cdot) \in W$ such that

$$\left(\frac{\partial^2 \underline{E}}{\partial t^2}, \underline{v}\right)_{0,\varepsilon} + a(\underline{E}, \underline{v}) = \left(\tilde{\underline{J}}, \underline{v}\right)_0, \;\; \forall \underline{v} \in W, \tag{5.34a}$$

$$\underline{E}(0, \cdot) = \underline{E}_0 \in W, \tag{5.34b}$$

$$\frac{\partial \underline{E}}{\partial t}(0, \cdot) = \underline{E}_1 \in W, \tag{5.34c}$$

where $(\underline{u}\,,\;\underline{v})_{0,\varepsilon} = (\varepsilon \underline{u}\,,\;\underline{v})_0, \forall \underline{u},\;\underline{v} \in L^2(\Omega)^d$.

Now, we introduce the following eigenvalue problem:
Find $(\underline{u}\,,\;\omega) \in W \times \mathbb{C}$ such that

$$a(\underline{u}, \underline{v}) = \omega^2 \left(\underline{u}\,,\;\underline{v}\right)_{0,\varepsilon}, \;\; \forall \underline{v} \in W. \tag{5.35}$$

It is well-known that the space W is compactly embedded in $L^2(\Omega)$ [12] (It is not the case for the space V). This property leads to the existence of discrete family $\left(\omega_i\,,\;\underline{u}_i\right)_{i=1,\cdots,+\infty}$ such that

1. $a(\underline{u}_i, \underline{v}) = \omega_i^2 \left(\underline{u}_i \,,\, \underline{v}\right)_{0,\varepsilon}$, $\forall \underline{v} \in W$,
2. $0 < \omega_1 \leq \cdots \leq \omega_i \leq \cdots$ and $\lim\limits_{i \to +\infty} \omega_i = +\infty$ (the only accumulation point),
3. $(\underline{u}_i \,,\, \underline{u}_j)_{0,\varepsilon} = \delta_{i,j}$ and consequently $a(\underline{u}_i, \underline{u}_j) = \omega_i^2 \delta_{i,j}$, $\forall i, j = 1, \ldots, +\infty$,
4. $\left(\underline{u}_i\right)_i$ is a basis of both W and $\{\underline{v} \in L^2(\Omega)^3 \,:\, \nabla \cdot \left(\varepsilon \underline{v}\right) = 0\}$.

So, one can decompose the electric field $\underline{E}$ in this way: $\underline{E}(t, x) = \sum\limits_{j \geq 1} E_j(t) \underline{u}_j(x)$.

By injecting this decomposition in (5.34a)–(5.34c) and by choosing as a test-function $\underline{v} = \underline{u}_i$, we get the following ordinary differential equations: $\forall i$, find $E_i(t) \in \mathbb{R}$ such that

$$E_i^{''} + \omega_i^2 E_i = \tilde{J}_i, \tag{5.36a}$$

$$E_i(0) = \left(\underline{E}_0(x) \,,\, \underline{u}_i\right)_{0,\varepsilon}, \tag{5.36b}$$

$$E_i^{'}(0) = \left(\underline{E}_1(x) \,,\, \underline{u}_i\right)_{0,\varepsilon}. \tag{5.36c}$$

where $\underline{\widetilde{J}}(t, x) = \sum\limits_{j \geq 1} \tilde{J}_j(t) \underline{u}_j(x)$ is obtained by developing $\dfrac{1}{\varepsilon} \underline{\widetilde{J}} \in W$ because $\nabla \cdot \underline{\widetilde{J}} = 0$.

We have now to solve a non-homogeneous second-order ordinary differential equation with constant coefficients of the form:

$$v'' + \omega^2 v = f, \tag{5.37a}$$

$$v(0) = v_0, \tag{5.37b}$$

$$v^{'}(0) = v_1. \tag{5.37c}$$

The computation of the solution of (5.37a)–(5.37c) is classical and is given by

$$v(t) = v_0 \cos(\omega t) + \frac{v_1}{\omega} \sin(\omega t) + \int_0^t \frac{\sin(\omega(t - s))}{\omega} f(s) \, \mathrm{d}s. \tag{5.38}$$

Finally, the solution of (5.34a)–(5.34c) can be decomposed as follows:

$$\underline{E}(t, x) = \sum_{i \geq 1} \left(\left(\underline{E}_0(x) \,,\, \underline{u}_i\right)_{0,\varepsilon} \cos(\omega_i t) + \frac{1}{\omega_i} \left(\underline{E}_1(x) \,,\, \underline{u}_i\right)_{0,\varepsilon} \sin(\omega_i t) \right) \underline{u}_i$$
$$+ \sum_{i \geq 1} \int_0^t \frac{\sin(\omega_i (t - s))}{\omega_i} \left(\underline{\tilde{J}} \,,\, \underline{u}_i\right)_0 (s) \, \underline{u}_i \, \mathrm{d}s. \tag{5.39}$$

The approximation of (5.30a)–(5.30d) is constructed from the weak formulation (5.33a)–(5.33c) and not from the formulation (5.34a)–(5.34c). Actually, it is not possible (except in simple particular situations) to construct a conformal approximation of W while the finite element spaces proposed by J.C. Nédélec [1, 13] allow to easily define a discrete version of problem (5.33a)–(5.33c):

Find $\underline{E}_h(t,\cdot) \in V_h \subset V$ such that

$$\left(\varepsilon\partial_{tt}^2\underline{E}_h\,,\,\underline{v}_h\right)_0 + a(\underline{u}_h,\underline{v}_h) = \left(\underline{\tilde{J}}\,,\,\underline{v}_h\right)_0,\quad \forall \underline{v}_h \in V_h, \tag{5.40a}$$

$$\underline{E}_h(0,\cdot) = \pi_h\underline{E}_0, \tag{5.40b}$$

$$\partial_t\underline{E}_h(0,\cdot) = \pi_h\underline{E}_1, \tag{5.40c}$$

where π_h is a projection on the space V_h.

We can now construct the discrete version of (5.39). For that, we introduce the following discrete eigenvalues problem
Find $(\underline{u}_h\,,\,\omega_h) \in V_h \times \mathbb{C}$ such that

$$a(\underline{u}_h,\underline{v}_h) = \omega_h^2\left(\underline{u}_h\,,\,\underline{v}_h\right)_{0,\varepsilon},\quad \forall \underline{v}_h \in V_h. \tag{5.41}$$

One can notice that (5.41) is not an approximation of (5.35) but rather of the following problem:
Find $(\underline{u}\,,\,\omega) \in V \times \mathbb{C}$ such that

$$a(\underline{u},\underline{v}) = \omega^2\left(\underline{u}\,,\,\underline{v}\right)_{0,\varepsilon},\quad \forall \underline{v} \in V. \tag{5.42}$$

Remark: Unlike (5.35), (5.42) has the essential eigenvalue $\{0\}$ which is associated to the infinite dimensional eigenspace V^0.

It is easy to see that the eigenvalues of (5.41) are all real positive numbers and $\omega_h = 0$ is an eigenvalue associated to the eigenspace $V_h^0 = V_h \cap V^0$. From now, the strictly positive eigenvalues and their associated eigenfunctions are represented by $(\omega_h^i\,,\,\underline{w}_h^i)_{i=1,\ldots,N_p}$ and the eigenfunctions of the eigenvalue $\omega_h = 0$ are $(\underline{w}_{0,h}^i)_{i=1,\ldots,N_0}$. Recall that $(\underline{w}_h^i)_{i=1,\ldots,N_p}$ and $(\underline{w}_{0,h}^i)_{i=1,\ldots,N_0}$ form orthonormal basis of V_h with respect to the scalar product $(\cdot\,,\,\cdot)_{0,\varepsilon}$.

So, we can decompose the discrete solution $\underline{E}_h$ in this way

$$\underline{E}_h(t,x) = \sum_{j=1}^{N_p} E_j^p(t)\underline{w}_h^j(x) + \sum_{j=1}^{N_0} E_j^0\underline{w}_{0,h}^j(x).$$

By injecting this decomposition in (5.31a)–(5.40c) and by choosing as test-functions $\underline{v}_h = \underline{w}_h^i$ or $\underline{w}_{0,h}^i$, we obtain the following system of ordinary differential equations:

$$E_i^{p''} + \omega_h^{i2}E_i^p = \tilde{J}_i^p,\ \forall i = 1,\ldots,N_p, \tag{5.43a}$$

$$E_i^{0''} = 0, \ \forall i = 1, \ldots, N_0, \tag{5.43b}$$

$$E_i^p(0) = \left(\pi_h \underline{E}_0(x)\,,\ \underline{w}_h^i\right)_{0,\varepsilon}, \ \forall i = 1, \ldots, N_p, \tag{5.43c}$$

$$E_i^{p'}(0) = \left(\pi_h \underline{E}_1(x)\,,\ \underline{w}_h^i\right)_{0,\varepsilon}, \ \forall i = 1, \ldots, N_p, \tag{5.43d}$$

$$E_i^0(0) = \left(\pi_h \underline{E}_0(x)\,,\ \underline{w}_{0,h}^i\right)_{0,\varepsilon}, \ \forall i = 1, \ldots, N_0, \tag{5.43e}$$

$$E_i^{0'}(0) = \left(\pi_h \underline{E}_1(x)\,,\ \underline{w}_{0,h}^i\right)_{0,\varepsilon}, \ \forall i = 1, \ldots, N_0. \tag{5.43f}$$

Finally, we obtain the discrete version of (5.39):

$$\begin{aligned}\underline{E}_h(t,x) = \sum_{i=1}^{N_p} \Bigg(&\left(\pi_h \underline{E}_0(x)\,,\ \underline{w}_h^i\right)_{0,\varepsilon} \cos(\omega_h^i t) \\ &+ \frac{1}{w_h^i} \left(\pi_h \underline{E}_1(x)\,,\ \underline{w}_h^i\right)_{0,\varepsilon} \sin(\omega_h^i t) \\ &+ \int_0^t \frac{\sin(\omega_h^i (t-s))}{w_h^i} \left(\underline{\tilde{J}}\,,\ \underline{w}_h^i\right)_0 (s) ds \Bigg) \underline{w}_h^i \\ + \sum_{i=1}^{N_0} &\left(\left(\pi_h \underline{E}_0(x)\,,\ \underline{w}_{0,h}^i\right)_{0,\varepsilon} + \left(\pi_h \underline{E}_1(x)\,,\ \underline{w}_{0,h}^i\right)_{0,\varepsilon} t \right) \underline{w}_{0,h}^i.\end{aligned} \tag{5.44}$$

The decomposition (5.44) shows that the discrete solution of the time-dependent problem is closely linked to the eigenvalue problem (5.41). In particular, if the latter is "correctly" solved, then the time-dependent solution is not affected by spurious modes. In this case, we say that we have a spectrally correct approximation. We are now going to rigorously define this notion (see for more details [14–16] and the works of Boffi et al. [17–21]).

We denote A and A_h the underlying operators to formulations (5.42) and (5.41) respectively. Recall that the spectrum of the operator A is $\sigma(A) = \{0\} \cup \sigma_p(A)$ with $\sigma_p(A)$ is a denumerable set of real isolated eigenvalues with finite multiplicities which form a sequence accumulating only at $\{+\infty\}$ and $\{0\}$ is an isolated eigenvalue with infinite multiplicity. Moreover, the eigenspace associated to this essential spectrum is $E_0(A) = V_0$.

For any $h \in I = (0, 1)$, $\sigma(A_h)$ is a finite set of real isolated eigenvalues with finite multiplicities. In particular, if $V_h^0 = V_h \cap V_0 \neq \emptyset$, then $\omega_h = 0$ is an eigenvalue and $E_0(A_h) = V_h^0$. We now define a neighborhood of the spectrum of A: Let $\varepsilon > 0$. $\forall \omega \in \sigma(A)$, we define $V(\omega) \subset (\omega - \varepsilon, \omega + \varepsilon)$ such that $V(\omega) \cap V(\mu) = \emptyset$, $\forall \omega \neq \mu \in \sigma(A)$. We define $\Omega_s = \cup_{\omega \in \sigma(A)} V(\omega)$.

Finally, we say that we have a spectrally correct approximation if the following conditions are fulfilled

1. Completeness of the spectrum (CS):

$$\forall \omega \in \sigma(A), \lim_{h \to 0^+} \left(\inf_{\omega_h \in \sigma(A_h)} |\omega - \omega_h| \right) = 0, \tag{5.45}$$

   i.e. all continuous eigenvalues smaller than an arbitrarily large fixed value are approximated when the mesh is sufficiently fine.
2. Non-pollution of the spectrum (NPS):

$$\exists h_0,\ \sigma(A_h) \subset \Omega,\ \forall h < h_0,\ h \in I, \tag{5.46}$$

   i.e. there are no discrete spurious eigenvalues.
3. Completeness of the eigenspaces (CE):

$$\forall \omega \in \sigma_p(A), \forall \underline{u} \in E_\omega(A), \lim_{h \to 0^+} \left(\inf_{\underline{u}_h \in E_{V(\omega)}(A_h)} \|\underline{u} - \underline{u}_h\|_V \right) = 0 \tag{5.47}$$

   with $E_{V(\omega)}(A_h) = \bigoplus_{\omega_h \in \sigma(A_h) \cap V(\omega)} E_{\omega_h}(A_h)$

   Condition (CE) says that, for any element of any eigenspace of (5.42) we can find a sequence of elements of the corresponding approximate eigenspace converging to it as $h \to 0$. In other words, no eigenspace of (5.42) is missed by (5.41), not even partially.
4. Non-pollution of the eigenspace (NPE):

$$\forall \omega \in \sigma_p(A), \lim_{h \to 0^+} \left(\sup_{\substack{\underline{u}_h \in E_{V(\omega)}(A_h) \\ \|\underline{u}_h\|_V \le 1}} \inf_{\underline{u} \in E_\omega(A)} \|\underline{u} - \underline{u}_h\|_V \right) = 0 \tag{5.48}$$

   Condition (NPE) says that, for any eigenspace of (5.42), the greatest distance that a normalized element of the corresponding approximate eigenspace can have from the eigenspace of (5.42) itself vanishes as $h \to 0$. This condition means that no sequence of normalized eigenvectors of (5.41) corresponding to a bounded sequence of eigenvalues can have a nonvanishing distance from the union of all the eigenspaces of (5.42) as $h \to 0$. In other words, (5.41), as an approximation of (5.42), does not introduce any eigenvector extraneous to the original problem.
5. Isolation of the discrete essential kernel: All the discrete eigenvalues approaching $\omega = 0$ are separated from the other ones.

The main difficulty to ensure these five conditions when we focus on the approximation of Maxwell's eigenvalues problem (5.42) is the presence of an eigenvalue with an infinite multiplicity (non-compact inverse operator). However, many works deal with this issue in the literature and we refer the reader to the papers mentioned in this section in order to obtain more informations. We only give here the most standard necessary and sufficient conditions to check that we have a spectrally correct approximation.

We begin by the conformal approximation (5.41) of (5.42). In this case, the approximate space V_h must verify the following conditions:

1. Completeness of the approximation space V_h: $\forall \underline{v} \in V$,

$$\lim_{h\to 0} \inf_{\underline{v}_h \in V_h} \|\underline{v} - \underline{v}_h\|_V = 0 \tag{5.49}$$

   This condition is generally satisfied by most approximation spaces.
2. Discrete Friedrichs inequality: let

$$W_h = (V_h^0)^\perp = \{\underline{v}_h \in V_h \;:\; (\underline{v}_h\,,\, \underline{w}_h)_V = 0,\; \forall \underline{w}_h \in V_h^0\},$$

   $\exists \alpha > 0$ such that

$$\left\|\mu^{-1/2}\nabla \times \underline{v}_h\right\|_0^2 = a(\underline{v}_h\,,\, \underline{v}_h) \geq \alpha \left\|\underline{v}_h\right\|_0^2, \;\forall \underline{v}_h \in W_h. \tag{5.50}$$

   This condition is equivalent to the isolation of the discrete essential spectrum.
3. Discrete compactness property: Any $\{\underline{v}_h\}_h \subset W_h$ $(h \in \mathbb{R})$ such that $\|\underline{v}_h\|_V \leq 1$ for all h contains a subsequence $\{\underline{v}_{h_j}\}_j$ such that

$$\exists \underline{v} \in L^2(\Omega)^d \;:\; \lim_{i\to+\infty} \|\underline{v} - \underline{v}_{h_i}\|_0 = 0. \tag{5.51}$$

   Coupled with the first condition, this condition implies the convergence in a mesh dependent norm of $\{A_h\}_h$ to A and consequently, the non-pollution of the spectrum.

Remark: The discrete compactness property can be viewed as the discrete version of the compact embedding of W in $L^2(\Omega)^d$.

Now, we recall some results about the classical finite element spaces used to approximate Maxwell's equations. The most famous ones are the first and the second family of Nédélec's finite elements [1, 13]. The two families are defined on triangles or tetrahedra and all the approximation orders have a correct spectral behaviour for generic non-constant dielectric parameters ε and μ (see for example [19]). Nevertheless, on meshes based on quadrilaterals or hexahedra (even regular), the second family may suffer of the presence of spurious modes.

On the other hand, it is well-known that standard continuous finite elements methods usually produce spurious modes. In particular, the computed eigenvalues, or the multiplicity of the eigenvalues, is incorrect regardless of the size of the mesh parameter h. In this case, it is the discrete compactness property and the discrete Friedrichs inequality which are not satisfied on general meshes.

We finish this part by some results about the discontinuous Galerkin approximation of the Maxwell equations (see [22–26] for more information).

In the case of a discontinuous approximation of the problem (5.42), the previous necessary and sufficient conditions must be changed in this way:

1. Completeness of the approximation space V_h: $\forall \underline{v} \in V$,

$$\lim_{h \to 0} \inf_{\underline{v}_h \in V_h} \|\underline{v} - \underline{v}_h\|_{V(h)} = 0, \tag{5.52}$$

where $V(h) = V + V_h$.
2. Discrete Friedrichs Inequality: let $W_h = (V_h^0)^{\perp} = \{\underline{v}_h \in V_h : (\underline{v}_h, \underline{w}_h)_V = 0, \forall \underline{w}_h \in V_h^0\}$, $\exists \alpha > 0$ such that

$$\left\| \mu^{-1/2} \nabla \times \underline{v}_h \right\|_0^2 = a(\underline{v}_h, \underline{v}_h) \geq \alpha \left\| \underline{v}_h \right\|_0^2, \ \forall \underline{v}_h \in W_h. \tag{5.53}$$

3. Gap property: For h small enough, for any $\underline{w}_h \in W_h$, $\exists \underline{w} \in W$ such that

$$\|\underline{w} - \underline{w}_h\|_0 \leq \eta_h \|\underline{w}_h\|_{V(h)} \tag{5.54}$$

with $\eta_h \to 0$ as $h \to 0$.

Remark: The gap property is an equivalent counterpart of the discrete compactness property.

To conclude, we give some results about the standard discontinuous Galerkin approximation of the Maxwell equations. The Interior Penalty (SIP, NIP, IIP) and the local discontinuous Galerkin formulations provide spectrally correct approximations on simplicial meshes with no hanging node. However, spurious modes may be occur on quadrilaterals and hexahedra and general non-regular meshes.

All the properties given above hold for the approximation of the second-order formulation of Maxwell's equations. To our knowledge, no theoretical studies were devoted to the first-order formulation. In particular, it is not obvious that an approximation using the second family of Nédélec edge elements for $\underline{E}$ and the corresponding family in $H(div)$ (or L^2) for $\underline{H}$ on tetrahedra would provide a spurious free approximation. We provide a partial numerical answer in the next sections.

5.3.2 *Computation of the Eigenvalues of* $\nabla \times \nabla \times$ *on a Cube*

We want to find $\omega \neq 0$ such that

$$\nabla \times (\nabla \times \underline{E}) = \omega^2 \underline{E} \text{ in } C, \tag{5.55a}$$

$$\nabla \cdot \underline{E} = 0 \text{ in } C, \tag{5.55b}$$

$$\underline{n} \times \underline{E} = 0 \text{ on } \partial C, \tag{5.55c}$$

where $C = [0, 1]^3$ and $\underline{n}$ is the outward unit normal to C.

Since $\nabla \times (\nabla \times \underline{E}) = \nabla(\nabla \cdot \underline{E}) - \Delta \underline{E}$, by denoting $\underline{E} = (E_1, E_2, E_3)^T$ and taking into account (5.55b), (5.55a)–(5.55c) reads:

$$\Delta E_i + \omega^2 E_i = 0 \text{ in } C, \ \forall i = 1, \ldots, 3, \tag{5.56a}$$

$$\underline{n} \times \underline{E} = 0 \text{ on } \partial C. \tag{5.56b}$$

By setting $\tilde{E}_i = \alpha_i \, f_{1i}(x_1) \, f_{2i}(x_2) \, f_{3i}(x_3)$, (5.56a) provides:

$$\begin{aligned} &f''_{1i}(x_1) \, f_{2i}(x_2) \, f_{3i}(x_3) + f_{1i}(x_1) \, f''_{2i}(x_2) \, f_{3i}(x_3) + f_{1i}(x_1) \, f_{2i}(x_2) \, f''_{3i}(x_3) \\ &+\omega^2 \, f_{1i}(x_1) \, f_{2i}(x_2) \, f_{3i}(x_3) = 0, \ \forall i = 1..3. \end{aligned} \tag{5.57}$$

By dividing by $\tilde{E}_i$, we get:

$$\frac{f''_{1i}(x_1)}{f_{1i}(x_1)} + \frac{f''_{2i}(x_2)}{f_{2i}(x_2)} + \frac{f''_{3i}(x_3)}{f_{3i}(x_3)} = -\omega^2, \ \forall i = 1..3. \tag{5.58}$$

Since we have three functions of different variables whose sum is constant for any value their variables, each function is constant. In particular:

$$f''_{ji} + \gamma_{ji} \, f_{ji} = 0, \ \forall i = 1, \ldots, 3, \ \forall j = 1, \ldots, 3, \ i \neq j, \ \gamma_{ji} \in \mathbb{R}. \tag{5.59}$$

On the other hand, (5.55c) implies that

$$f_{ji}(0) = f_{ji}(1) = 0, \ \forall i = 1..3, \ \forall j = 1, \ldots, 3, \ i \neq j. \tag{5.60}$$

Equations (5.59) and (5.60) and $\omega \neq 0$ lead to

$$f_{ji}(x_j) = a_{ji} \, \sin(\ell_j \, \pi \, x_j), \ a_{ji} \in \mathbb{R}^*, \ \ell_j \in \mathbb{N}^*. \tag{5.61}$$

So, we can write:

$$\tilde{E}_i = \lambda_i \, f_{ii}(x_i) \, \sin(\ell_j \, \pi \, x_j) \, \sin(\ell_k \, \pi \, x_k), \ j \neq i, \ k \neq i, \ k \neq j, \ \lambda_i \in \mathbb{R}^*. \tag{5.62}$$

We have now to take into account (5.55b). By plugging our solution in (5.55b), we obtain:

$$\begin{aligned} \nabla \cdot \underline{\tilde{E}} =& \lambda_1 \, f'_{11}(x_1) \, \sin(\ell_2 \, \pi \, x_2) \, \sin(\ell_3 \, \pi \, x_3) \\ &+ \lambda_2 \, \sin(\ell_1 \, \pi \, x_1) \, f'_{22}(x_2) \, \sin(\ell_3 \, \pi \, x_3) \\ &+ \lambda_3 \, \sin(\ell_1 \, \pi \, x_1) \, \sin(\ell_2 \, \pi \, x_2) \, f'_{33}(x_3) = 0. \end{aligned} \tag{5.63}$$

By dividing (5.63) by $\sin(\ell_1 \, \pi \, x_1) \, \sin(\ell_3 \, \pi \, x_2) \, \sin(\ell_3 \, \pi \, x_3)$, we get:

$$\lambda_1 \frac{f'_{11}(x_1)}{\sin(\ell_1 \pi x_1)} + \lambda_2 \frac{f'_{22}(x_2)}{\sin(\ell_2 \pi x_2)} + \lambda_3 \frac{f'_{33}(x_3)}{\sin(\ell_3 \pi x_3)} = 0. \tag{5.64}$$

Here also, (5.64) implies:

$$f'_{ii} = b_i \sin(\ell_i \pi x_i), \ \forall i = 1..3, \ b_i \in \mathbb{R}, \tag{5.65}$$

which provides

$$f_{ii} = c_i \cos(\ell_i \pi x_i) + d_i, \ \forall i = 1, \dots, 3, \ c_i \in \mathbb{R}, \ d_i \in \mathbb{R}, \ \ell_i \in \mathbb{N}. \tag{5.66}$$

Since (5.58) also implies that there exists $\gamma_i \in \mathbb{R}$ such that $f''_{ii} + \gamma_i f_{ii} = 0$ and, on the other hand, (5.63) implies that $\gamma_i \neq 0$, we have $d_i = 0$. We finally get:

$$\begin{aligned} &\tilde{E}_i = \lambda'_i \cos(\ell_i \pi x_i) \sin(\ell_j \pi x_j) \sin(\ell_k \pi x_k), \ (i, j, k) \in \{1, 2, 3\}^3, \\ &j \neq i, \ k \neq i, \ k \neq j, \ \ell_i \in \mathbb{N}, (\ell_j, \ell_k) \in (\mathbb{N}^*)^2, \ \lambda'_i \in \mathbb{R}^*. \end{aligned} \tag{5.67}$$

The corresponding eigenvalues are:

$$\omega^2 = \pi^2(\ell_i^2 + \ell_j^2 + \ell_k^2), \tag{5.68}$$

with (i, j, k) defined as in (5.67).

Let us now plug the eigenvectors in (5.55b). We obtain:

$$\sum_{i=1}^{3} \ell_i \pi \lambda'_i \sin(\ell_1 \pi x_1) \sin(\ell_2 \pi x_2) \sin(\ell_3 \pi x_3) = 0, \tag{5.69}$$

which implies, if $\ell_1 \ell_2 \ell_3 \neq 0$,

$$\sum_{k=1}^{3} \ell_k \lambda'_k = 0. \tag{5.70}$$

Equation (5.70) shows that $\underline{\tilde{E}}$ can be written as

$$\underline{\tilde{E}} = \theta_1 \underline{V}_1 + \theta_2 \underline{V}_2, \tag{5.71}$$

where $(\theta_1, \theta_2) \in \mathbb{R}^2$, $\theta_1 + \theta_2 \neq 0$ if $\theta_1 \theta_2 = 0$ and

$$\underline{V}_1 = \begin{pmatrix} \ell_2 \cos(\ell_1 \pi x_1) \sin(\ell_2 \pi x_2) \sin(\ell_3 \pi x_3) \\ -\ell_1 \sin(\ell_1 \pi x_1) \cos(\ell_2 \pi x_2) \sin(\ell_3 \pi x_3) \\ 0 \end{pmatrix},$$

$$\underline{V}_2 = \begin{pmatrix} 0 \\ \ell_3 \sin(\ell_1 \pi x_1) \cos(\ell_2 \pi x_2) \sin(\ell_3 \pi x_3) \\ -\ell_2 \sin(\ell_1 \pi x_1) \sin(\ell_2 \pi x_2) \cos(\ell_3 \pi x_3) \end{pmatrix}.$$

Now, if $\ell_1\,\ell_2\,\ell_3 = 0$, we obviously have:

$$\tilde{E}_i = \lambda_i' \,\sin(\ell_j\,\pi\,x_j)\,\sin(\ell_k\,\pi\,x_k)\delta_{in}, \tag{5.72}$$

where n is such that $\ell_n = 0$.

So, $\underline{\tilde{E}} = \theta_3\,\underline{V}_3$, $\theta_3 \in \mathbb{R}^*$, where

- if $\ell_1 = 0$, $\underline{V}_3 = (\sin(\ell_2\,\pi\,x_2)\,\sin(\ell_3\,\pi\,x_3), 0, 0)^T$,
- if $\ell_2 = 0$, $\underline{V}_3 = (0, \sin(\ell_1\,\pi\,x_1)\,\sin(\ell_3\,\pi\,x_3), 0)^T$,
- if $\ell_3 = 0$, $\underline{V}_3 = (0, 0, \sin(\ell_1\,\pi\,x_1)\,\sin(\ell_2\,\pi\,x_2))^T$.

From the above computations, we can deduce the multiplicity m_ω of the eigenvalues given in (5.68). If we define (i, j, k) as in (5.67), we have:

- For $\ell_i = 0$ and $\ell_j = \ell_k$, $m_\omega = 3$,
- For $\ell_i = 0$ and $\ell_j \neq \ell_k$, $m_\omega = 6$,
- For $\ell_i = \ell_j = \ell_k$, $m_\omega = 2$,
- For $\ell_j = \ell_k$ and $\ell_i \neq \ell_j$, $m_\omega = 6$,
- For $\ell_i \neq \ell_j$, $\ell_i \neq \ell_k$ and $\ell_j \neq \ell_k$, $m_\omega = 12$.

Remark: For sake of simplicity, we carried our computations on the unit cube. On a cube $C_L = [0, L]^3$, we obtain the same results with π/L instead of π in the solution.

5.3.3 Discontinuous Galerkin Methods

5.3.3.1 The Numerical Framework

For all the methods studied below, we use two kinds of mesh for the unit cube C: one orthogonal regular mesh (denoted $\mathscr{M}_R$) and one mesh composed of tetrahedra split into four hexahedra (denoted $\mathscr{M}_{NR}$) (Fig. 5.4). Our study is mainly focused on the second mesh.

5.3.3.2 The Spurious Modes

If we solve (5.55a) and (5.55c) by a discontinuous Galerkin method (which a priori does not take into account (5.55b)) without dissipative term (centered DGM) on $\mathscr{M}_R$ and $\mathscr{M}_{NR}$ for a Q_3 approximation, we obtain the eigenmodes given in Fig. 5.5.[4] The physical eigenmodes involved in these figures are given in Table 5.5.

[4]Since they were made by different students, the values of the eigenmodes vary in the following figures, but they represent the same eigenmodes with a multiplying coefficient.

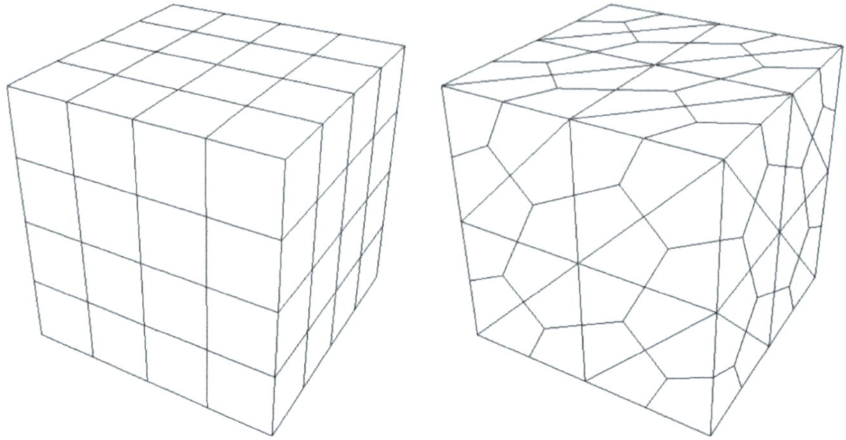

Fig. 5.4 Regular (*left*) and non regular (*right*) meshes for eigenmodes computation

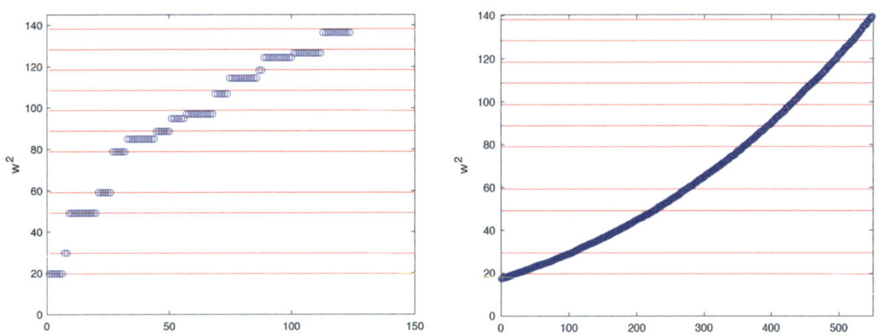

Fig. 5.5 Eigenmodes on $\mathscr{M}_R$ (*left*) and $\mathscr{M}_{NR}$ (*right*) for Q_3 centered DGM. In all the figures, the lines indicate the physical eigenmodes and the number of eigenmodes are in abscissa and their values in ordinate

Table 5.5 The values of the physical eigenmodes and their multiplicity

(ℓ_1, ℓ_2, ℓ_3)	(0, 1, 1)	(1, 1, 1)	(0, 1, 2)	(1, 1, 2)	(0, 2, 2)	(1, 2, 2)
m_ω, ω^2	3, 19.73921	2, 29.60881	6, 49.34802	6, 59.21763	3, 78.95684	6, 88.82644
(ℓ_1, ℓ_2, ℓ_3)	(0, 1, 3)	(1, 1, 3)	(2, 2, 2)	(0, 2, 3)	(1, 2, 3)	(2, 2, 3)
m_ω, ω^2	6, 98.69604	6, 108.5656	2, 118.4353	6, 128.3049	12,138.1745	6, 167.7833

Figure 5.5 obviously shows, as expected, the presence of parasitic modes for this approximate operator. For $\mathscr{M}_R$, we get about 120 modes instead of 58 and for $\mathscr{M}_{NR}$, more than 500 instead of 58! One can notice, by the way, that for $\mathscr{M}_R$, most of the spurious modes are identical to physical and we mainly have a default of multiplicity, which is not the case for $\mathscr{M}_{NR}$. This difference has a substantial impact on the solution for a default of multiplicity only induces an error on the amplitudes which is

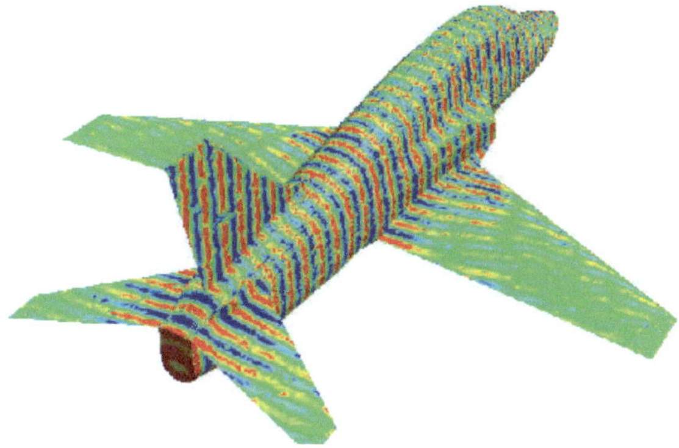

Fig. 5.6 Solution of a scattering experiment by an airplane using centered DGM on a mesh composed of split tetrahedra. One can notice the important pollution induced by the spurious modes

not significant for high-order approximations whereas non physical modes produce parasitic waves which seriously pollute the physical solution. as shown in Fig. 5.6

5.3.3.3 The Dissipative Term Effect

All the above results were obtained by computing the eigenmodes without dissipative term in the discontinuous Galerkin method. Now, if we compute these eigenmodes by adding this term (uncentered DGM), we get, for both meshes, the eigenmodes given in Fig. 5.7. One can notice that we only get the physical eigenmodes with their exact multiplicity. So, what happened to the spurious modes? Actually, without the dissipative term, the discrete operator is symmetric, positive, which implies that all its eigenvalues are real. By adding the dissipative term, we get an operator whose eigenvalues are no longer all real. Actually, only the real eigenvalues are represented in Fig. 5.7. The spurious eigenvalues become all complex with the same sign of the imaginary part (Fig. 5.8), which implies that they generate evanescent waves. This phenomenon "wipes out" the solution, as shown in Fig. 5.9.

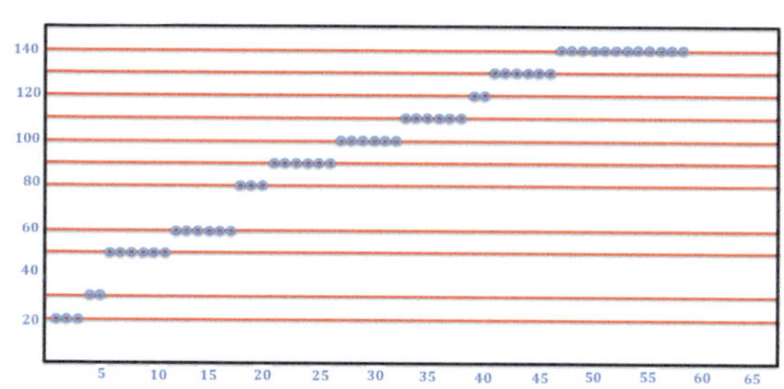

Fig. 5.7 Eigenmodes on $\mathscr{M}_R$ and $\mathscr{M}_{NR}$ for Q_3 uncentered DGM

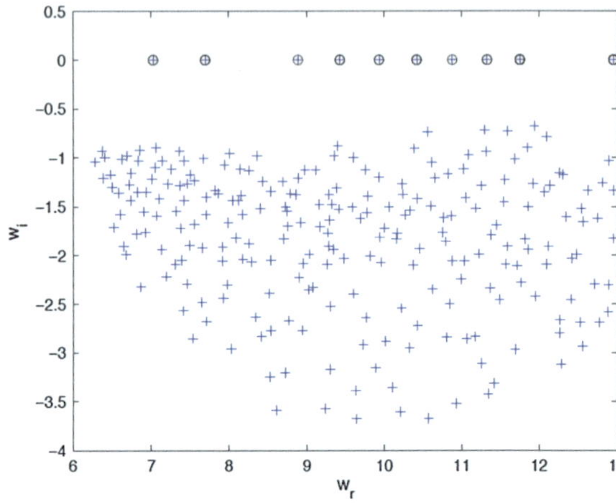

Fig. 5.8 Eigenmodes in the complex plane on $\mathcal{M}_{NR}$ for Q_3 uncentered DGM. The real eigenmodes fit to to the physical modes

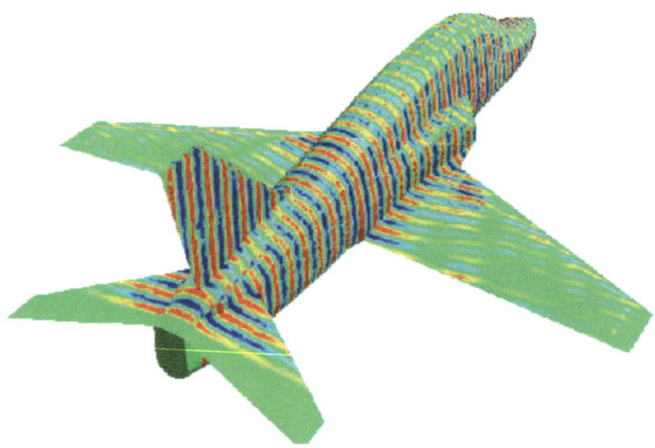

Fig. 5.9 Solution of a scattering experiment by an airplane using uncentered DGM on a mesh composed of split tetrahedra. One can notice the dramatical attenuation the spurious modes effect

5.3.4 The Second Family of Edge Elements

5.3.4.1 Spurious Modes

The approximation in $H(curl)$ of $\underline{E}$ defined in (5.12a) and (5.12b) should a priori produce spurious modes since it does not take into account (5.55b). However, the result seems much better than for centered DGM. As shown in Fig. 5.10 (right), the number of spurious modes is much smaller than for centered DGM.

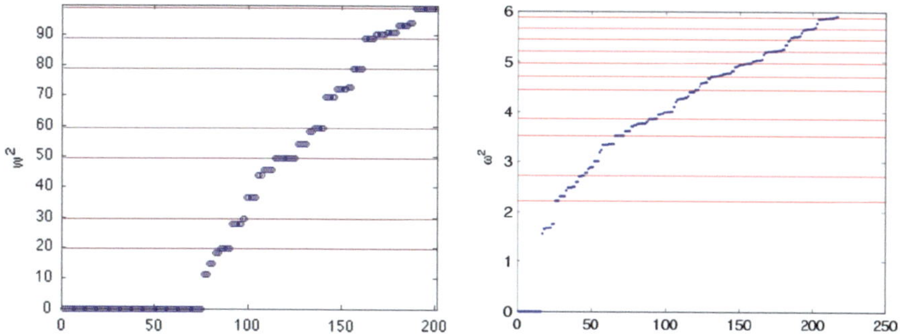

Fig. 5.10 Eigenmodes on $\mathscr{M}_R$ (*right*) and $\mathscr{M}_{NR}$ (*left*) for Q_3 approximation using the second family of edge elements

5.3.4.2 Penalty Terms and Their Effect

We now face a problem: How can we get rid of the spurious modes? One could use a divergence penalty term as in [4]. However, this approach has two drawbacks: first, it requires the addition of the integral of a divergence which is as costly as the curl term. On the other hand, it does not properly take into account the diffraction by corners, drawback which is not easy to correct [9, 27]. We can also use an edge penalty term, as for DGM. In this case, the discontinuous character of the normal component of $\underline{E}$ enables us to introduce two penalty terms: one using the jump of the normal component of $\underline{E}$ in (5.12a) and the other one identical to the dissipative term of uncentered DGM. So, (5.12a)–(5.12b) can be rewritten as

Find $\underline{E}_h$ and $\underline{H}_h$ such that $\underline{E}_h(.,t) \in W_h^r$ and $\underline{H}_h(.,t) \in X_h^r$ and

$$\sum_{K\in\mathscr{T}_h} \left\{ \frac{\mathrm{d}}{\mathrm{d}t}\int_K \underline{\underline{\varepsilon}}\underline{E}_h \cdot \underline{\varphi}_h \,\mathrm{d}\underline{x} - \int_K \underline{H}_h \cdot \nabla\times\underline{\varphi}_h \,\mathrm{d}\underline{x} \right.$$
$$\left. -\alpha'\int_{\partial K} [\underline{E}\cdot\underline{n}]_{\partial K}^K (\underline{\varphi}_h\cdot\underline{n})\,\mathrm{d}\sigma + \int_K \underline{J}\cdot\underline{\varphi}_h \,\mathrm{d}\underline{x} \right\} = 0,\ \forall \underline{\varphi}_h \in W_h^r \quad (5.73\text{a})$$

$$\sum_{K\in\mathscr{T}_h} \left\{ \frac{\mathrm{d}}{\mathrm{d}t}\int_K \underline{\underline{\mu}}\underline{H}_h \cdot \underline{\psi}_h \,\mathrm{d}\underline{x} + \int_K \nabla\times\underline{E}_h \cdot \underline{\psi}_h \,\mathrm{d}\underline{x} \right.$$
$$\left. -\ \delta'\int_{\partial K} [\underline{n}\times\underline{H}_h]_{\partial K}^K \cdot (\underline{n}\times\underline{\psi}_h)\,\mathrm{d}\sigma. \right\} = 0,\ \forall \underline{\psi}_h \in X_h^r. \quad (5.73\text{b})$$

where W_h^r and X_h^r are defined as in (5.13) and (5.14) and with $U = H_0(\mathbf{curl}, \Omega)$ and $A_K = DF_K^T$.

Let us now test the effect of these two penalty terms. The first one seems to have the expected effect, as shown in Fig. 5.11. However, the normal component on ∂K

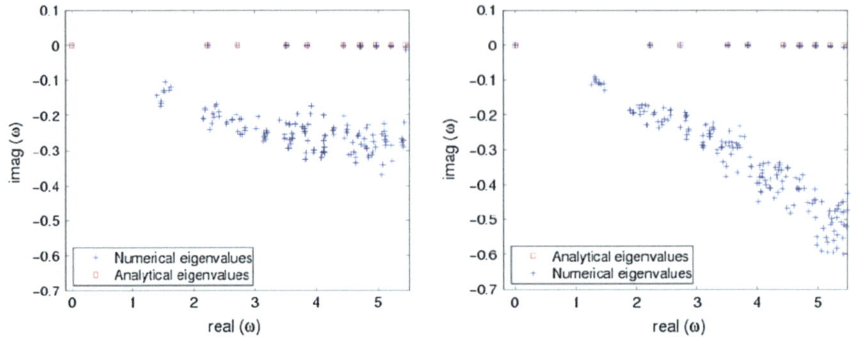

Fig. 5.11 Eigenmodes in the complex plane for $\mathscr{M}_{NR}$ for a Q_3 approximation with $\alpha' = 0.1$ (*left*) and $\alpha' = 0.5$ (*right* and $\delta' = 0$) for the second family of edge elements. The real eigenmodes fit to to the physical modes. First published in Journal of Computational Mathematics in volume 25 (3), 2007, published by Global Science Press

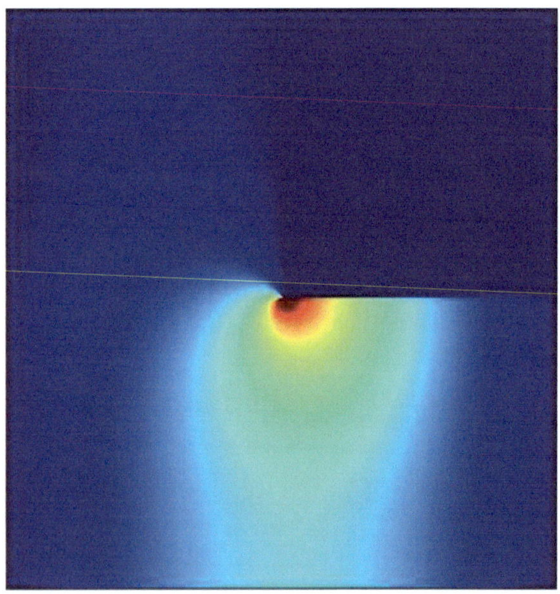

Fig. 5.12 Singularity of the solution generated by a corner

appears in the boundary term of the integration by parts of a divergence.[5] So, it seems interesting to test our penalty term on a corner. Actually, a corner creates a singularity in the solution represented in Fig. 5.12. Figure 5.13 shows that the term in δ' does much better than the term in α'.

[5]This judicious remark was made by I. Perugia during a workshop.

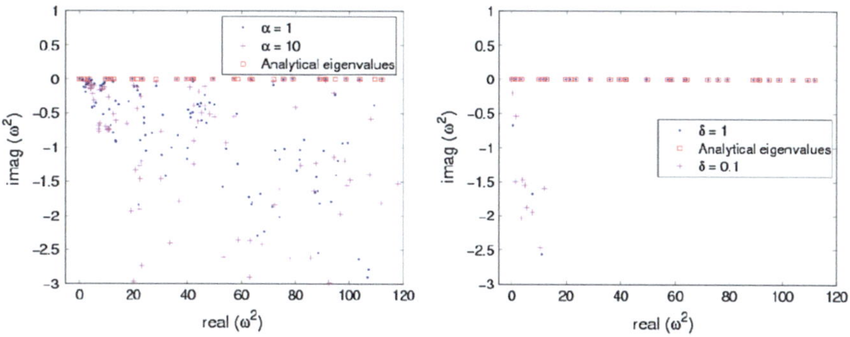

Fig. 5.13 Eigenmodes in the complex plane for a corner with $\delta' = 0$ and $\alpha' = 1, 10$ (*left*) and $\delta' = 0.1, 1$ and $\alpha' = 0$ (*right*) for the second family of edge elements

5.3.5 Continuous Elements

Continuous elements seem to provide less spurious modes than edge elements, as shown in Fig. 5.14.

By setting $U = H_0^1(\Omega)$ and $A_K = I_3$ and $\alpha' = 0$ in (5.73a)–(5.73b), we obtain a penalized formulation which gets rid of these spurious modes, as shown in Fig. 5.15.

5.3.6 The Case of the First Family of Edge Elements

As we saw in Sect. 5.1.3, we can either combine Gauss and Gauss–Lobatto quadrature rules or use one of them to compute the mass (as well as stiffness) matrix in (5.4a)–(5.4b) on a hexahedral mesh without obtaining mass-lumping. On the other hand, this formulation needs no additional term to be spurious free. For storage

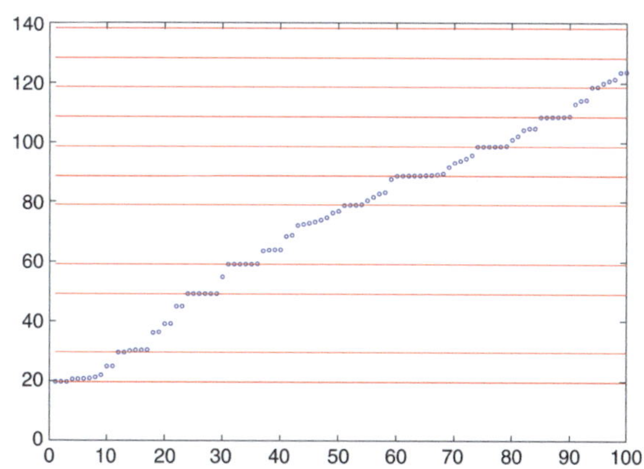

Fig. 5.14 Eigenmodes on $\mathscr{M}_{NR}$ for Q_3 approximation using continuous elements

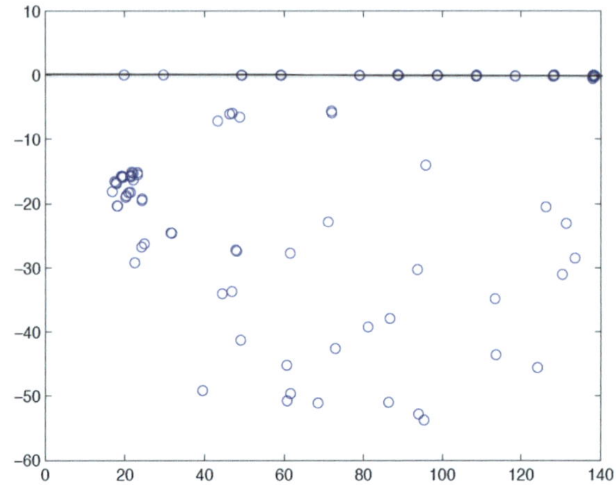

Fig. 5.15 Eigenmodes in the complex plane for $\mathscr{M}_{NR}$ for Q_3 approximation with $\delta' = 0.5$ and $\alpha' = 0$ for continuous elements

and computational time considerations, these elements could be efficient for the resolution of time-harmonic problems by iterative methods (based on conjugate gradient) which do not need to invert the mass matrix. However, its actual form is not easy to use because of the definition of $\underline{B}$ in $H(div)$ for a) it does not provide the local definition of the stiffness matrix obtained when $\underline{B}$ is sought in L^2, b) it requires the construction of a mass matrix in $H(div)$ which is not easy because of the normal continuity. For all these reasons, we tried to replace in (5.4a)–(5.4b) $\underline{B} \in H(div)$ by $\underline{B} \in L^2$ which, of course, does no longer ensure the spurious free property of the method. We used all Gauss or all Gauss–Lobatto quadrature rules to compute the integrals of the system and we computed the eigenmodes for these two approaches with $r = 3$. Surprisingly, the Gauss quadrature rule only provides physical eigenmodes, as shown in Fig. 5.16. Numerical experiments confirm the spurious free character of this method [5, 6].

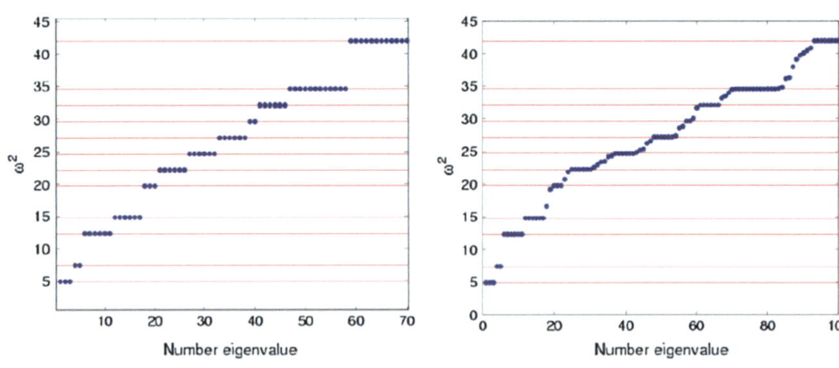

Fig. 5.16 Eigenmodes on $\mathscr{M}_{NR}$ for an approximation using the first family of edge elements with Gauss (*left*) and Gauss–Lobatto (*right*) quadrature rules. First published in Journal of Computational Mathematics in volume 25 (3), 2007, published by Global Science Press

Remarks:

1. At this stage, we see that three kinds of parasitic waves can pollute a numerical solution of Maxwell's equations based on finite element methods:
   a. Dispersive parasitic waves coming from too large space-steps.
   b. Parasitic waves coming from numerical velocities induced by the different types of degrees of freedom in one element. In order to avoid these waves, one must take a smaller space-step. For this reason, high-order approximations are efficient up tp 7th or 8th order. For higher orders, the number of parasitic modes implies a too small space step to get an interesting performance.
   c. Parasitic waves coming from the violation of $\nabla \cdot \underline{B} = 0$.
2. All the above computations show that the second family of edge elements on an hexahedral mesh is the most efficient for transient Maxwell's equations in terms of storage, computational time and accuracy.
3. The penalty term also works for DGM based on triangle and tetrahedra [28].

5.4 Error Estimates for DGM

In this section, we present the hp-convergence analysis of a non-dissipative high-order discontinuous Galerkin method on unstructured hexahedral meshes to solve the time-dependent Maxwell's equations. In particular, we underline the spectral convergence of the method (in the sense that when the solutions and the data are very smooth, the discretization is of unlimited order). Moreover, we see that the choice of a non-standard approximate space (for a discontinuous formulation) with the absence of dissipation can imply a loss of spatial convergence.

5.4.1 The Discontinuous Galerkin Formulation

We assume that the computational domain Ω is split into a set of cells $\mathcal{T}_h$ such that $\Omega = \cup_{i=1}^{N_e} K_i$, where $K_i \in \mathcal{T}_h$, $\dot{K}_i \cap \dot{K}_j = \emptyset$, $\forall i \neq j$ and K_i is a hexahedron. We denote the set of faces of $\mathcal{T}_h$ by $\mathcal{F}_h = \mathcal{F}_h^i \cup \mathcal{F}_h^b$ with $\Gamma \in \mathcal{F}_h^i \Leftrightarrow \Gamma = K' \cap K$ and $\Gamma \in \mathcal{F}_h^b \Leftrightarrow \Gamma = K \cap \partial\Omega$, so that $\mathcal{F}_h^i$ and $\mathcal{F}_h^b$ are the sets of the interior and boundary faces. To each element $K \in \mathcal{T}_h$, we associate the outward unit normal $\underline{n}_K$.

Given a nonnegative integer r and $E \subset \mathbb{R}^d$, $Q_r(E)$ is the space of polynomials of degree at most equal to r in each variable on E. Let us introduce the standard unit cube $\hat{K} = [0, 1]^3$. $\forall K \in \mathcal{T}_h$, $F_K : \hat{K} \rightarrow K$ denotes the trilinear mapping which associates the vertices of each element. $(\hat{x}_1, \hat{x}_2, \hat{x}_3)$ are the coordinates on the reference element and (x_1, x_2, x_3) the coordinates on the elements of the mesh. DF_K and J_K denote the Jacobian matrix and its determinant associated with the mapping F_K.

We use the discontinuous finite element space:

$$U_h = \{\underline{v}_h \in L^2(\Omega)^d \ : \ \forall K \in \mathscr{T}_h,\ DF_K^* \underline{v}_{h|K} o F_K \in [Q_r(\hat{K})]^d\} \tag{5.74}$$

where $r \in \mathbb{N}$.

In (5.74), the Jacobian matrix is the essential ingredient to build a conform H-curl approximation [1, 13]. As shown above, in our case, it allows to reduce the storage of the stiffness and the jump matrices.

We consider the following non-dissipative discontinuous Galerkin formulation of the first order Maxwell's system:

Find $(\underline{E}_h(\cdot,t), \underline{H}_h(\cdot,t)) \in U_h \times U_h$ such that $\forall K \in \mathscr{T}_h$ and $(\underline{u}_h, \underline{v}_h) \in U_h \times U_h$,

$$\begin{aligned}\frac{\mathrm{d}}{\mathrm{d}t}\int_K \varepsilon \underline{E}_{hK} \cdot \underline{u}_{hK} \mathrm{d}\underline{x} - \int_K \nabla \times \underline{H}_{hK} \cdot \underline{u}_{hK} \mathrm{d}\underline{x} + \int_K \underline{J} \cdot \underline{u}_{hK} \mathrm{d}\underline{x} \\ = \int_{\partial K} \beta [\![\underline{H}_h \times \underline{n}_K]\!]_{\partial K}^K \cdot \underline{u}_{hK} d\sigma,\end{aligned} \tag{5.75a}$$

$$\frac{\mathrm{d}}{\mathrm{d}t}\int_K \mu \underline{H}_{hK} \cdot \underline{v}_{hK} \mathrm{d}\underline{x} + \int_K \nabla \times \underline{E}_{hK} \cdot \underline{v}_{hK} \mathrm{d}\underline{x} = \int_{\partial K} \gamma [\![\underline{E}_h \times \underline{n}_K]\!]_{\partial K}^K \cdot \underline{v}_{hK} d\sigma. \tag{5.75b}$$

In (5.75a)–(5.75b), the parameters are chosen as follows:

1. $\forall \Gamma \in \mathscr{F}_h^i$, $\beta = -\frac{1}{2}$ and $\gamma = \frac{1}{2}$,
2. $\forall \Gamma \in \mathscr{F}_h^b$, $\beta = 0$ and $\gamma = 1$.

Remark: The choice of the parameter $\gamma = 1$ on the boundary faces allows to take into account the boundary condition $\underline{E} \times \underline{n} = 0$.

5.4.2 *Choice of a Projector*

When one derives error estimates, an important point is the choice of a "good" projector on the approximate space used for discretization. Actually, the use of an inappropriate projector can lead to sub-optimal estimates which do not provide any interesting information about the numerical scheme. This part aims at justifying our choice.

For our DG scheme (5.75a)–(5.75b), the first idea is to use a L^2 projector. In particular, one can use the projector defined in the following way.

First, we can split the approximate space U_h as follows:

$$U_h = \oplus_{K \in \mathscr{T}_h} U_K \tag{5.76}$$

where $U_K = \{\underline{v} \in L^2(K)^3 \ : \ DF_K^* \underline{v} o F_K \in [Q_r(\hat{K})]^3\}$.

Then, in a first step, we define the L^2 projector $\hat{\pi}_r^0$ on $[Q_r(\hat{K})]^3$.

Definition 5.1 (*L^2 Projector*) Let $\hat{\underline{v}} \in [L^2(\hat{K})]^3$ and $r \geq 0$. We define the projector L^2 $\hat{\pi}_r^0\hat{\underline{v}}$ of $\hat{\underline{v}}$ on $[Q_r(\hat{K})]^3$ by: $\forall \hat{\underline{\varphi}} \in [Q_r(\hat{K})]^3$,

$$\int_{\hat{K}} \hat{\pi}_r^0 \hat{\underline{v}} \cdot \hat{\underline{\varphi}} \mathrm{d}\hat{\underline{x}} = \int_{\hat{K}} \hat{\underline{v}} \cdot \hat{\underline{\varphi}} \mathrm{d}\hat{\underline{x}}. \tag{5.77}$$

In the second step, we come back to U_K by defining the projector π_K^0.

Definition 5.2 (*Projector on U_K*) Let $\underline{v} \in [L^2(K)]^3$. We define the projection $\pi_K^0 \underline{v}$ of $\underline{v}$ on U_K by:

$$(\pi_K^0 \underline{v}) o F_K = DF_K^{-T} \hat{\pi}_r^0 \hat{\underline{v}} \tag{5.78}$$

where $\hat{\underline{v}} = DF_K^T \underline{v} \circ F_K$.

Finally, we define the projection operator on U_h.

Definition 5.3 (*Projector on U_h*) Let $\underline{v} \in L^2(\Omega)^3$. We define the projection $\pi_h^0 \underline{v}$ of $\underline{v}$ on U_h by: for $K \in \mathscr{T}_h$

$$(\pi_h^0 \underline{v})_{|K} = \pi_K^0 \underline{v}_{|K}. \tag{5.79}$$

When examining in more details the DG scheme, one sees that it is necessary to know error estimates of the first-order derivatives of the projector used (because of the presence of rational terms). So, a H^1 type projector on U_h can be a possibility for this study. In particular, we have considered the projector defined as follows:

We first define the H^1 projector $\hat{\pi}_r^1$ on $[Q_r(\hat{K})]^3$

Definition 5.4 (*H^1 Projector*) Let $\hat{\underline{v}} \in H^1(\hat{K})^3$ and $r \geq 0$. We define the H^1 projection $\hat{\pi}_r^1 \hat{\underline{v}}$ of $\hat{\underline{v}}$ on $[Q_r(\hat{K})]^3$ by: $\forall \hat{\underline{\varphi}} \in [Q_r(\hat{K})]^3$, we have

$$\int_{\hat{K}} (\hat{\pi}_r^1 \hat{\underline{v}} - \hat{\underline{v}}) \cdot \hat{\underline{\varphi}} \mathrm{d}\hat{\underline{x}} + \sum_{k=1}^{3} \int_{\hat{K}} \frac{\partial}{\partial \hat{x}_k} (\hat{\pi}_r^1 \hat{\underline{v}} - \hat{\underline{v}}) \cdot \frac{\partial}{\partial \hat{x}_k} \hat{\underline{\varphi}} \mathrm{d}\hat{\underline{x}} = 0. \tag{5.80}$$

Remark: In (5.80), $\dfrac{\partial \underline{w}}{\partial \hat{x}_k}$ means $\left(\dfrac{\partial w_1}{\partial \hat{x}_k}, \dfrac{\partial w_2}{\partial \hat{x}_k}, \dfrac{\partial w_3}{\partial \hat{x}_k} \right)^T$.

Then, we come back to U_K. Let $K \in \mathscr{T}_h$ and $\underline{v} \in H^s(K)^3$ with $s \geq 1$. We define the projector π_K^1 on U_K by

$$(\pi_K^1 \underline{v}) o F_K = DF_K^{-T} (\hat{\pi}_r^1 \hat{\underline{v}}) \tag{5.81}$$

where $\hat{\underline{v}} = DF_K^T (\underline{v} o F_K)$.

Finally we define the projection operator on U_h.

Definition 5.5 (*Projector on U_h*) Let $\underline{v} \in L^2(\Omega)^3$. We define the projection $\pi_h^1 \underline{v}$ of $\underline{v}$ on U_h by: for $K \in \mathscr{T}_h$

$$(\pi_h^1 \underline{v})_{|K} = \pi_K^1 \underline{v}_{|K}. \tag{5.82}$$

In the following, we choose the best projector for the hp convergence analysis by comparing their spectral properties. Actually, only these properties reveal a difference between the projectors since, as shown in Sect. 5.4.3, they have the same h convergence rates.

Using theorem 57 of [29] as well as a tensorisation argument (i.e. $\hat{\pi}_r^0 = \hat{\pi}^0_{r,\hat{x}_3} o \hat{\pi}^0_{r,\hat{x}_2} o \hat{\pi}^0_{r,\hat{x}_1}$), we obtain the projection errors for $\hat{\pi}_r^0$:

Theorem 5.1 *$\forall \underline{\hat{u}} \in H^p(\hat{K})^3$, it exists a constant C such that*

$$\|\underline{\hat{u}} - \hat{\pi}_r^0 \underline{\hat{u}}\|_{q,\hat{K}} \leq C r^{\sigma(p,q)} \|\underline{\hat{u}}\|_{p,\hat{K}} \tag{5.83}$$

where

$$\sigma(p,q) = \begin{cases} \frac{3}{2}q - p, \; 0 \leq q \leq 1 \\ 2q - p - \frac{1}{2}q, \; q \geq 1 \end{cases} \tag{5.84}$$

and $0 \leq q \leq p$.

As already mentioned, we need the H^1 projection error to estimate the error of the GD scheme. Theorem 5.1 leads to

$$\|\underline{\hat{u}} - \hat{\pi}_r^0 \underline{\hat{u}}\|_{1,\hat{K}} \leq C r^{\frac{3}{2}-p} \|\underline{\hat{u}}\|_{p,\hat{K}}. \tag{5.85}$$

Estimation (5.85) shows that we do not have the optimality for the H^1 norm.

However for $\hat{\pi}_r^1$, we can find in [30] the following estimate: $\forall t, s \in \mathbb{R}$ verifying $0 \leq t \leq 1 \leq s$, then for $\underline{\hat{v}} \in H^s(\hat{K})^3$, it exists a constant $C > 0$ independent of r such that

$$\|\underline{\hat{v}} - \hat{\pi}_r^1 \underline{\hat{v}}\|_{t,\hat{K}} \leq C r^{t-s} \|\underline{\hat{v}}\|_{s,\hat{K}} \tag{5.86}$$

In particular, we will use the two estimates $t = 0, 1$ in (5.86).

Proposition 5.1 *For $\underline{\hat{v}} \in H^s(\hat{K})^3$, $s \geq 1$,*

$$\|\underline{\hat{v}} - \hat{\pi}_r^1 \underline{\hat{v}}\|_{0,\hat{K}} \leq \frac{C}{r^s} \|\underline{\hat{v}}\|_{s,\hat{K}}, \tag{5.87a}$$

$$\|\underline{\hat{v}} - \hat{\pi}_r^1 \underline{\hat{v}}\|_{1,\hat{K}} \leq \frac{C}{r^{s-1}} \|\underline{\hat{v}}\|_{s,\hat{K}} \tag{5.87b}$$

where $C > 0$ is a constant independent of r.

In this case, we obtain the optimal projection errors ($1/r^s$ and $1/r^{s-1}$ for the L^2 and the H^1 norms respectively). In conclusion, we decide to use the projector π_h^1 to analyze the convergence properties of the DG scheme in the hp-version.

5.4.3 hp-Projection Errors

To study the projection error introduced by $\hat{\pi}_r^1$, we use the Bramble-Hilbert Lemma 3.1 introduced in Chap. 3 and we immediately obtain

Proposition 5.2 *For $r \geq 0$ and $m \leq r+1$, there exists C dependent on $\hat{K}$ and r such that*

$$\forall \underline{\hat{v}} \in H^{r+1}(\hat{K})^3,\ |\underline{\hat{v}} - \hat{\pi}_r^1 \underline{\hat{v}}|_{m,\hat{K}} \leq C[\underline{v}]_{r+1,\hat{K}}. \tag{5.88}$$

In order to derive the hp-projection error estimates for π_h^1, we must specify the exact r-dependence of constant C in (5.88). To do so, we come back to the proof of the Bramble-Hilbert lemma but directly considering π_h^1. The first step, to prove this type of result, is to write [31]: $\forall \underline{\hat{v}} \in H^{r+1}(\hat{K})^3$,

$$\begin{aligned} |\underline{\hat{v}} - \hat{\pi}_r^1 \underline{\hat{v}}|_{m,\hat{K}} &\leq \|I - \hat{\pi}_r^1\|_{\mathscr{L}(H^{r+1}(\hat{K})^3, H^m(\hat{K})^3)} \inf_{\hat{p} \in [Q_r(\hat{K})]^3} \|\underline{\hat{v}} + p\|_{r+1,\hat{K}} \\ &\leq C_1 \|I - \hat{\pi}_r^1\|_{\mathscr{L}(H^{r+1}(\hat{K})^3, H^m(\hat{K})^3)} [\underline{v}]_{r+1,\hat{K}}, \end{aligned} \tag{5.89}$$

where C_1 is independent of r.

Equations (5.86) and (5.89) immediately provide:

$$|\underline{\hat{v}} - \hat{\pi}_r^1 \underline{\hat{v}}|_{m,\hat{K}} \leq \frac{C_2(\hat{K})}{r^{r+1-m}} [\underline{v}]_{r+1,\hat{K}},\ 0 \leq m \leq r+1. \tag{5.90}$$

Lemma 5.1 *Let $\underline{v} \in H^{r+1}(K)^3$, $r \geq 0$. It exists C independent of K and r such that*

$$\|\underline{v} - \pi_K^1 \underline{v}\|_{0,K} \leq C \frac{h_K^{\frac{1}{2}}}{r^{r+1}} [\underline{\hat{v}}]_{r+1,\hat{K}}, \tag{5.91a}$$

$$|\underline{v} - \pi_K^1 \underline{v}|_{1,K} \leq \frac{C}{h_K^{\frac{1}{2}} r^r} [\underline{\hat{v}}]_{r+1,\hat{K}}. \tag{5.91b}$$

Proof We only prove the second inequality. The same process leads to the first one.

Write $\underline{w} = \underline{v} - \pi_K^1 \underline{v} = (w_1, w_2, w_3)^T$. We have:

$$|\underline{w}|_{1,K}^2 = \sum_{i=1}^3 \sum_{l=1}^3 \int_{\hat{K}} |J_K| |(\partial_{x_l} w_i) \circ F_K|^2 d\hat{\hat{x}}, \tag{5.92}$$

where ∂_{x_l} means $\dfrac{\partial}{\partial x_l}$.

By definition, we have $\underline{w} = (DF_K^{-T}\, \underline{\hat{w}}) \circ F_K^{-1}$, where $\underline{\hat{w}} = \underline{\hat{v}} - \hat{\pi}_r^1 \underline{\hat{v}}$ and DF_K^T reads:

$$DF_K^T = \begin{pmatrix} \partial_{\hat{x}_1} x_1 & \partial_{\hat{x}_1} x_2 & \partial_{\hat{x}_1} x_3 \\ \partial_{\hat{x}_2} x_1 & \partial_{\hat{x}_2} x_2 & \partial_{\hat{x}_2} x_3 \\ \partial_{\hat{x}_3} x_1 & \partial_{\hat{x}_3} x_2 & \partial_{\hat{x}_3} x_3 \end{pmatrix} \tag{5.93}$$

with $x_i = F_K^i(\hat{x})$ for $i = 1, 2, 3$.

Inverting this matrix with the help of the cofactors formula, we obtain:

$$DF_K^{-T} = \frac{1}{J_K}$$
$$\begin{pmatrix} \partial_{\hat{x}_2} x_2 \partial_{\hat{x}_3} x_3 - \partial_{\hat{x}_2} x_3 \partial_{\hat{x}_3} x_2 & -\partial_{\hat{x}_2} x_1 \partial_{\hat{x}_3} x_3 + \partial_{\hat{x}_2} x_3 \partial_{\hat{x}_3} x_1 & \partial_{\hat{x}_2} x_1 \partial_{\hat{x}_3} x_2 - \partial_{\hat{x}_2} x_2 \partial_{\hat{x}_3} x_1 \\ -\partial_{\hat{x}_1} x_2 \partial_{\hat{x}_3} x_3 + \partial_{\hat{x}_1} x_3 \partial_{\hat{x}_3} x_2 & \partial_{\hat{x}_1} x_1 \partial_{\hat{x}_3} x_3 - \partial_{\hat{x}_1} x_3 \partial_{\hat{x}_3} x_1 & -\partial_{\hat{x}_1} x_1 \partial_{\hat{x}_3} x_2 + \partial_{\hat{x}_1} x_2 \partial_{\hat{x}_3} x_1 \\ \partial_{\hat{x}_1} x_2 \partial_{\hat{x}_2} x_3 - \partial_{\hat{x}_1} x_3 \partial_{\hat{x}_2} x_2 & -\partial_{\hat{x}_1} x_1 \partial_{\hat{x}_2} x_3 + \partial_{\hat{x}_1} x_3 \partial_{\hat{x}_2} x_1 & \partial_{\hat{x}_1} x_1 \partial_{\hat{x}_2} x_2 - \partial_{\hat{x}_1} x_2 \partial_{\hat{x}_2} x_1 \end{pmatrix}$$

We denote $DF_K^{-T} = \dfrac{1}{J_K}(m_{i,j})_{i,j=1,\dots,3}$. so we have: $w_i = \displaystyle\sum_{j=1}^{3} \frac{m_{i,j} \circ F_K^{-1}}{J_K \circ F_K^{-1}} \hat{w}_j \circ F_K^{-1}$.

Now, we derive the last expression with respect to x_l.

$$\begin{aligned} \partial_{x_l} w_i &= \sum_{j=1}^{3} \Bigg[\frac{\partial_{x_l}(m_{i,j} \circ F_K^{-1}) J_K \circ F_K^{-1} - m_{i,j} \circ F_K^{-1} \partial_{x_l}(J_K \circ F_K^{-1})}{(J_K \circ F_K^{-1})^2} \hat{w}_j \circ F_K^{-1} \\ &\quad + \frac{m_{i,j} \circ F_K^{-1}}{J_K \circ F_K^{-1}} \partial_{x_l}(\hat{w}_j \circ F_K^{-1}) \Bigg] \\ &= \sum_{j=1}^{3} \Bigg[\sum_{k=1}^{3} \frac{(\partial_{\hat{x}_k} m_{i,j}) \circ F_K^{-1} \partial_{x_l} \hat{x}_k J_K \circ F_K^{-1} - m_{i,j} \circ F_K^{-1} (\partial_{\hat{x}_k} J_K) \circ F_K^{-1} \partial_{x_l} \hat{x}_k}{(J_K \circ F_K^{-1})^2} \hat{w}_j \circ F_K^{-1} \\ &\quad + \frac{m_{i,j} \circ F_K^{-1}}{J_K \circ F_K^{-1}} (\partial_{\hat{x}_k} \hat{w}_j) \circ F_K^{-1} \partial_{x_l} \hat{x}_k \Bigg]. \end{aligned} \tag{5.94}$$

By denoting:

$$T_{i,j}^{k,l} = \frac{(\partial_{\hat{x}_k} m_{i,j}) \partial_{x_l} \hat{x}_k \circ F_K J_K - m_{i,j} (\partial_{\hat{x}_k} J_K) \partial_{x_l} \hat{x}_k \circ F_K}{(J_K)^2}, \tag{5.95a}$$

$$\tilde{T}_{i,j}^{k,l} = \frac{m_{i,j}}{J_K} \partial_{x_l} \hat{x}_k \circ F_K, \tag{5.95b}$$

we can write

$$(\partial_{x_l} w_i) \circ F_K = \sum_{j,k=1}^{3} \left[T_{i,j}^{k,l} \hat{w}^j + \tilde{T}_{i,j}^{k,l} \partial_{\hat{x}_k} \hat{w}^j \right]. \tag{5.96}$$

The regularity of the mesh leads to

$$|T_{i,j}^{k,l}| \le \frac{C}{h_K^2}, \tag{5.97a}$$

$$|\tilde{T}_{i,j}^{k,l}| \leq \frac{C}{h_K^2}. \tag{5.97b}$$

where $C > 0$ independent of K and r.

Actually, the definition of $m_{i,j}$ leads to $|m_{i,j}| \leq Ch_K^2$ and $|\partial_{\hat{x}_k} m_{i,j}| \leq Ch_K^2$ (keep in mind that $x_i = F_K^i(\hat{x})$ for $i \in [\![1, 3]\!]$). Moreover, estimates (3.164a)–(3.164c) imply $|\partial_{x_l} \hat{x}_k \circ F_K| \leq C/h_K$, $|J_K| \leq Ch_K^3$, $|\partial_{\hat{x}_k} J_K| \leq Ch_K^3$ and $|J_K| \geq C'h_K^3$.

That enables us to obtain:

$$|(\partial_{x_l} w_i) \circ F_K|^2 \leq \frac{C}{h_K^4} \sum_{j,k=1}^{3} \Big[|\hat{w}_j|^2 + |\partial_{\hat{x}_k} \hat{w}_j|^2\Big]. \tag{5.98}$$

Let us come back to our semi-norm: by using (5.98), (5.92) leads to

$$\begin{aligned} |\underline{w}|_{1,K}^2 &\leq C \frac{\|J_K\|_{\infty,\hat{K}}}{h_K^4} \sum_{i=1}^{3} \sum_{l=1}^{3} \sum_{j,k=1}^{3} \int_{\hat{K}} \Big[|\hat{w}_j|^2 + |\partial_{\hat{x}_k} \hat{w}_j|^2\Big] d\hat{x} \\ &\leq \frac{C}{h_K} \|\underline{\hat{w}}\|_{1,\hat{K}}^2, \end{aligned} \tag{5.99}$$

which concludes our proof.

The following step is to estimate $|\underline{w}|_{1,K}$ by using a power of h_K and $\|\underline{v}\|_{m,K}$.

Lemma 5.2 *Let* $\underline{v} \in H^m(K)^3$. *We have the estimate*

$$[\underline{\hat{v}}]_{m,\hat{K}} \leq C \sum_{l=0}^{1} |F_K|_{l+1,\infty,\hat{K}} [\underline{v} \circ F_K]_{m-l,\hat{K}} \tag{5.100}$$

where $C > 0$ *is independent of* K *and* r.

Proof We have $\underline{\hat{v}} = DF_K^T \, \underline{v} \circ F_K$ and $[\underline{\hat{v}}]_{m,\hat{K}}^2 = \sum_{i=1}^{3} \sum_{j=1}^{3} \int_{\hat{K}} |\frac{\partial^m \hat{v}_j}{\partial \hat{x}_i^m}|^2 d\hat{x}$.

We can write $\hat{v}_j = \sum_{k=1}^{3} J_{j,k} v_k \circ F$, where $DF_K^T = (J_{j,k})_{j,k=1,\ldots,3}$ and the Leibniz formula leads to

$$\frac{\partial^m \hat{v}_j}{\partial \hat{x}_i^m} = \sum_{k=1}^{3} \sum_{l=0}^{m} \binom{l}{m} \frac{\partial^l (J_{j,k})}{\partial \hat{x}_i^l} \frac{\partial^{m-l} (v_k \circ F)}{\partial \hat{x}_i^{m-l}}. \tag{5.101}$$

For $l \geq 2$, we have: $\frac{\partial^l (J_{j,k})}{\partial \hat{x}_i^l} = 0$ (actually $F_K \in [Q_1(\hat{K})]^3$). That implies

$$\int_{\hat{K}} |\frac{\partial^m \hat{v}_j}{\partial \hat{x}_i^m}|^2 d\hat{x} \leq C \sum_{k=1}^{3} \sum_{l=0}^{1} |F_K|_{l+1,\infty,\hat{K}}^2 \int_{\hat{K}} |\frac{\partial^{m-l}(v_k \circ F)}{\partial \hat{x}_i^{m-l}}|^2 d\hat{x}$$
$$\leq C \sum_{k=1}^{3} \sum_{l=0}^{1} |F_K|_{l+1,\infty,\hat{K}}^2 [v_k \circ F_K]_{m-l,\hat{K}}^2. \tag{5.102}$$

So, we obtain:

$$[\underline{\hat{v}}]_{m,\hat{K}}^2 \leq C \sum_{i=1}^{3} \sum_{j=1}^{3} \sum_{k=1}^{3} \sum_{l=0}^{1} |F_K|_{l+1,\infty,\hat{K}}^2 [v_k \circ F_K]_{m-l,\hat{K}}^2$$
$$\leq C \sum_{l=0}^{1} |F_K|_{l+1,\infty,\hat{K}}^2 [\underline{v} \circ F_K]_{m-l,\hat{K}}^2. \tag{5.103}$$

Finally, by grouping (3.183a), (3.181) and (5.100) together, we obtain the following error estimates.

Proposition 5.3 *Let $\underline{v} \in H^{r+1}(K)^3$, then it exists C independent of the cell K and r such that*

$$\|\underline{v} - \pi_K^1 \underline{v}\|_{0,K} \leq C \frac{h_K^r}{r^{r+1}} \left(|\underline{v}|_{r,K} + h_K |\underline{v}|_{r+1,K}\right), \tag{5.104a}$$

$$|\underline{v} - \pi_K^1 \underline{v}|_{1,K} \leq C \frac{h_K^{r-1}}{r^r} \left(|\underline{v}|_{r,K} + h_K |\underline{v}|_{r+1,K}\right). \tag{5.104b}$$

Now, by using the interpolation theorem 1.4 of [32], we extent the result to real exponents.

Proposition 5.4 *Let $\underline{v} \in H^{s+1}(K)^3$, for $0 \leq s \leq r$ real and assuming that $0 < h_K \leq 1$, it exists C independent of K and r and such that*

$$\|\underline{v} - \pi_K^1 \underline{v}\|_{0,K} \leq C \frac{h_K^s}{r^{s+1}} \|\underline{v}\|_{s+1,K}, \tag{5.105a}$$

$$|\underline{v} - \pi_K^1 \underline{v}|_{1,K} \leq C \frac{h_K^{s-1}}{r^s} \|\underline{v}\|_{s+1,K}. \tag{5.105b}$$

Proof Let $r_1 < r_2$ be two positive integers and $\theta \in [0, 1]$.

Assume that $\pi_K^0 \in \mathscr{L}(\underline{H}^{r1+1}(K), H^m(K)) \cap \mathscr{L}(H^{r2+1}(K)^3, H^m(K)^3)$ for $m = 0, 1$.

Then, we have:

$$\|I - \pi_K^1\|_{\mathscr{L}(H^{\theta r1+(1-\theta)r2+1}(K)^3, H^m(K)^3)} \leq$$
$$C \|I - \pi_K^1\|_{\mathscr{L}(H^{r1+1}(K)^3, H^m(K)^3)}^{\theta} \|I - \pi_K^1\|_{\mathscr{L}(H^{r2+1}(K)^3, H^m(K)^3)}^{1-\theta}. \tag{5.106}$$

Inequalities (5.104a)–(5.104b) lead to

$$\|I-\pi_K^1\|_{\mathscr{L}(H^{r1+1}(K)^3,H^m(K)^3)} \leq C\frac{h_K^{r1-m}}{r^{r1+1-m}}, \tag{5.107a}$$

$$\|I-\pi_K^1\|_{\mathscr{L}(H^{r2+1}(K)^3,H^m(K)^3)} \leq C\frac{h_K^{r2-m}}{r^{r2+1-m}}. \tag{5.107b}$$

So, we obtain:

$$\|I-\pi_K^1\|_{\mathscr{L}(H^{\theta r1+(1-\theta)r2+1}(K)^3,H^m(K)^3)} \leq C\frac{h_K^{\theta r1+(1-\theta)r2-m}}{r^{\theta r1+(1-\theta)r2+1-m}}. \tag{5.108}$$

Finally, by taking $r_1 = 0$, $r_2 = r$ and $s = (1-\theta)$. We can write the inequality:

$$\begin{aligned}\|\underline{v}-\pi_K^1\underline{v}\|_{m,K} &\leq \|I-\pi_K^1\|_{\mathscr{L}(H^{s+1}(K)^3,H^m(K)^3)}\|\underline{v}\|_{s+1,K}\\ &\leq C\frac{h_K^{s-m}}{r^{s+1-m}}\|\underline{v}\|_{s+1,K}.\end{aligned} \tag{5.109}$$

Now, if we take $\underline{v} \in H^s(K)^3$ with $s \geq r+1$, we easily prove the error estimates:

$$\|\underline{v}-\pi_K^1\underline{v}\|_{0,K} \leq C\frac{h_K^r}{r^s}\|\underline{v}\|_{s,K}, \tag{5.110a}$$

$$|\underline{v}-\pi_K^1\underline{v}|_{1,K} \leq C\frac{h_K^{r-1}}{r^{s-1}}\|\underline{v}\|_{s,K}. \tag{5.110b}$$

Finally, (5.104a)–(5.104b) and (5.110) lead to the global result: let $\underline{v} \in H^{s+1}(K)^3$ with $s \geq 0$,

$$\|\underline{v}-\pi_K^1\underline{v}\|_{0,K} \leq C\frac{h_K^{\min(s,r)}}{r^{s+1}}\|\underline{v}\|_{s+1,K}, \tag{5.111a}$$

$$|\underline{v}-\pi_K^1\underline{v}|_{1,K} \leq C\frac{h_K^{\min(s-1,r-1)}}{r^s}\|\underline{v}\|_{s+1,K} \tag{5.111b}$$

where C independent of K and r.

5.4.4 *Trace Lemmas*

To estimate the surface terms appearing in the jumps, we need several intermediate results.

Lemma 5.3 *Let $\underline{u}_h \in U_h$ and $K \in \mathscr{T}_h$. It exists a constant $C > 0$ independent of K and r such that*

$$\|, \underline{u}_{h|K}\|_{0,\partial K}^2 \leq C\sigma_K^{11}\frac{r^2}{\rho_K}\|\underline{u}_h\|_{0,K}^2. \tag{5.112}$$

Moreover, if $\mathscr{T}_h$ is a regular family of meshes, we have:

$$\|\underline{u}_{h|K}\|_{0,\partial K}^2 \leq C\frac{r^2}{h_K}\|\underline{u}_h\|_{0,K}^2. \tag{5.113}$$

Proof We have:

$$\frac{\displaystyle\int_{\partial K}|\underline{u}_{hK}|^2 d\sigma}{\displaystyle\int_K |\underline{u}_{hK}|^2 \mathrm{d}\underline{x}} = \frac{\displaystyle\int_{\partial\hat{K}}|J_K|\|DF_K^{-T}\hat{n}\|\left(DF_K^{-1}DF_K^{-T}\underline{\hat{u}}_K\right)\cdot\underline{\hat{u}}_K d\hat{\sigma}}{\displaystyle\int_{\hat{K}}|J_K|\left(DF_K^{-1}DF_K^{-T}\underline{\hat{u}}_K\right)\cdot\underline{\hat{u}}_K d\hat{x}}. \tag{5.114}$$

Estimations (3.164a)–(3.164c) lead to

$$\frac{\displaystyle\int_{\partial K}|\underline{u}_{hK}|^2 d\sigma}{\displaystyle\int_K |\underline{u}_{hK}|^2 \mathrm{d}\underline{x}} \leq C\frac{\sigma_K^{11}}{\rho_K}\frac{\displaystyle\int_{\partial\hat{K}}\underline{\hat{u}}_K\cdot\underline{\hat{u}}_K d\hat{\sigma}}{\displaystyle\int_{\hat{K}}\underline{\hat{u}}_K\cdot\underline{\hat{u}}_K d\hat{x}}. \tag{5.115}$$

In [33], we find the estimation:

$$\frac{\displaystyle\int_{\partial\hat{K}}\underline{\hat{u}}_K\cdot\underline{\hat{u}}_K d\hat{\sigma}}{\displaystyle\int_{\hat{K}}\underline{\hat{u}}_K\cdot\underline{\hat{u}}_K d\hat{x}} \leq Cr^2. \tag{5.116}$$

So, we obtain the wanted result.

We also need the trace inequality:

Lemma 5.4 *Let $K \in \mathscr{T}_h$. It exists $C > 0$ independent of K and r such that $\forall v \in H^1(K)$*

$$\|v\|_{0,\partial K}^2 \leq C(\|v\|_{0,K}\|\nabla v\|_{0,K} + \rho_K^{-1}\sigma_K^{-1}\|v\|_{0,K}^2). \tag{5.117}$$

Moreover, if $\mathscr{T}_h$ is a regular family of meshes, we have:

$$\|v\|_{0,\partial K}^2 \leq C(\|v\|_{0,K}\|\nabla v\|_{0,K} + h_K^{-1}\|v\|_{0,K}^2). \tag{5.118}$$

Proof Let $K \in \mathscr{T}_h$ and $\underline{v} \in H^1(K)$. We define $\hat{v} = v \circ F_K$. So, we have the trace inequality:

$$\|\hat{v}\|_{0,\partial\hat{K}}^2 \leq C(\|\hat{v}\|_{0,\hat{K}}\|\hat{\nabla}\hat{v}\|_{0,\hat{K}} + \|\hat{v}\|_{0,\hat{K}}^2). \tag{5.119}$$

See for example the annexes of [34] to obtain a proof of this result.

Now, we come back to the cell K. We have the estimations:

1.

$$\|\hat{v}\|_{0,\partial\hat{K}}^2 = \int_{\partial\hat{K}} \hat{v}^2 d\hat{\sigma} = \int_{\partial K} \frac{1}{|J_K| \|DF_K^{-T}\underline{\hat{n}}\|} v^2 d\sigma$$
$$\geq \frac{1}{\|J_K\|_{\infty,\hat{K}} |F_K^{-1}|_{1,\infty,K}} \|v\|_{0,\partial K}^2$$
$$\geq C\frac{\sigma_K^3}{h_K^2}\|v\|_{0,\partial K}^2 \text{ by using (3.164a)–(3.164c).} \quad (5.120)$$

2.

$$\|\hat{v}\|_{0,\hat{K}}^2 = \int_{\hat{K}} \hat{v}^2 d\hat{\hat{x}} = \int_K \frac{1}{|J_K|} v^2 d\hat{x} \leq \|J_K^{-1}\|_{\infty,K} \|v\|_{0,K}^2 = \rho_K^{-3}\|v\|_{0,K}^2. \quad (5.121)$$

3.

$$\|\hat{\nabla}\hat{v}\|_{0,\hat{K}}^2 = \int_{\hat{K}} \hat{\nabla}\hat{v}\cdot\hat{\nabla}\hat{v} d\hat{\hat{x}} = \int_K \frac{1}{|J_K|} DF_K^*\nabla v\cdot DF_K^*\nabla v d\hat{x}$$
$$\leq C\frac{\sigma_K^2}{\rho_K}\|\nabla v\|_{0,K}^2 \text{ (see Sect. 3.6.1).} \quad (5.122)$$

(5.119) becomes:

$$C_1\frac{\sigma_K^3}{h_K^2}\|v\|_{0,\partial K}^2 \leq C(C_2\frac{\sigma_K}{\rho_K^2}\|v\|_{0,K}\|\nabla v\|_{0,K} + C_3\frac{1}{\rho_K^3}\|v\|_{0,K}^2) \quad (5.123)$$

We then obtain the wanted result.

5.4.5 A Priori Error Estimates in Energy Norm

Let $(\underline{E}, \underline{H})$ and $(\underline{E}_h, \underline{H}_h)$ be respectively the exact solution and the DG solution (5.75a)–(5.75b). Our goal is to estimate $\|\underline{E} - \underline{E}_h\|_{0,\Omega}$ and $\|\underline{H} - \underline{H}_h\|_{0,\Omega}$.

For that, we introduce the energy norm:

$$\|(\underline{E}, \underline{H})\|_*^2 = \|\underline{E}\|_{0,\Omega,\varepsilon}^2 + \|\underline{H}\|_{0,\Omega,\mu}^2. \quad (5.124)$$

The norm (5.124) is more adapted for our estimations because it appears naturally in Maxwell's equations. So, we prefer to estimate:

$$\|(\underline{E} - \underline{E}_h, \underline{H} - \underline{H}_h)\|_* = \sqrt{\|\underline{E} - \underline{E}_h\|_{0,\Omega,\varepsilon}^2 + \|\underline{H} - \underline{H}_h\|_{0,\Omega,\mu}^2}. \quad (5.125)$$

We introduce the projection of the exact solution $(\underline{E}, \underline{H})$ i.e. $(\pi_h^1 \underline{E}, \pi_h^1 \underline{H})$ (we assume that $\underline{E}$ and $\underline{H}$ have the necessary regularity for the definition of projections in (5.125))

$$\begin{aligned}
&\|(\underline{E} - \underline{E}_h, \underline{H} - \underline{H}_h)\|_*^2 \\
&= \|\underline{E} - \pi_h^1 \underline{E} + \pi_h^1 \underline{E} - \underline{E}_h\|_{0,\Omega,\varepsilon}^2 + \|\underline{H} - \pi_h^1 \underline{H} + \pi_h^1 \underline{H} - \underline{H}_h\|_{0,\Omega,\mu}^2 \\
&\leq \|\Delta_{\underline{E}}^P\|_{0,\Omega,\varepsilon}^2 + \|\Delta_{\underline{E}}^I\|_{0,\Omega,\varepsilon}^2 + 2\|\Delta_{\underline{E}}^P\|_{0,\Omega,\varepsilon}\|\Delta_{\underline{E}}^I\|_{0,\Omega,\varepsilon} \\
&\quad + \|\Delta_{\underline{H}}^P\|_{0,\Omega,\mu}^2 + \|\Delta_{\underline{H}}^I\|_{0,\Omega,\mu}^2 + 2\|\Delta_{\underline{H}}^P\|_{0,\Omega,\mu}\|\Delta_{\underline{H}}^I\|_{0,\Omega,\mu}
\end{aligned} \tag{5.126}$$

where $\Delta_{\underline{E}}^P = \underline{E} - \pi_h^1 \underline{E}$ (projection error) and $\Delta_{\underline{E}}^I = \underline{E}_h - \pi_h^1 \underline{E}$ (interpolation error). We have the same thing for $\underline{H}$.

Using the inequality: $2ab \leq a^2 + b^2$, (5.126) becomes:

$$\|(\underline{E} - \underline{E}_h, \underline{H} - \underline{H}_h)\|_*^2 \leq 2\left(\|(\Delta_{\underline{E}}^P, \Delta_{\underline{H}}^P)\|_*^2 + \|\left(\Delta_{\underline{E}}^I, \Delta_{\underline{H}}^I\right)\|_*^2\right). \tag{5.127}$$

To estimate the error introduced by the spatial approximation, we have to evaluate $\|(\Delta_{\underline{E}}^P, \Delta_{\underline{H}}^P)\|_*$ and $\|\left(\Delta_{\underline{E}}^I, \Delta_{\underline{H}}^I\right)\|_*$. The estimation of the first term do not rise any problem, it is sufficient to use the projection errors of the previous section, on the other hand the second term requires more work. This is done in two steps.

5.4.5.1 Orthogonal Property

Introducing $\left(\pi_h^1 \underline{E}, \pi_h^1 \underline{H}\right)$ in the semi-discrete DG system (5.75a)–(5.75b) and taking $\underline{u}_h = \Delta_{\underline{E}}^I$, we obtain:

$$\begin{aligned}
\frac{1}{2}\frac{\mathrm{d}}{\mathrm{d}t}(\Delta_{\underline{E}}^I, \Delta_{\underline{E}}^I)_{0,K,\varepsilon} = &-\left(\frac{\partial}{\partial t}\pi_h^1 \underline{E}, \Delta_{\underline{E}}^I\right)_{0,K,\varepsilon} + (\nabla \times \Delta_{\underline{H}}^I, \Delta_{\underline{E}}^I)_{0,K} \\
&+ (\nabla \times \pi_h^1 \underline{H}, \Delta_{\underline{E}}^I)_{0,K} - (\underline{J}, \Delta_{\underline{E}}^I)_{0,K} \\
&+ (\beta[\![\Delta_{\underline{H}}^I \times \underline{n}_K]\!]_{\partial K}^K, \Delta_{\underline{E}|K}^I)_{0,\partial K} + (\beta[\![\pi_h^1 \underline{H} \times \underline{n}_K]\!]_{\partial K}^K, \Delta_{\underline{E}|K}^I)_{0,\partial K}.
\end{aligned} \tag{5.128}$$

It is easy to see that the exact solution verifies:

$$\begin{aligned}
&\left(\frac{\partial}{\partial t}\Delta_{\underline{E}}^P, \Delta_{\underline{E}}^I\right)_{0,K,\varepsilon} + \left(\frac{\partial}{\partial t}\pi_h^1 \underline{E}, \Delta_{\underline{E}}^I\right)_{0,K,\varepsilon} - (\nabla \times \Delta_{\underline{H}}^P, \Delta_{\underline{E}}^I)_{0,K} \\
&- (\nabla \times \pi_h^1 \underline{H}, \Delta_{\underline{E}}^I)_{0,K} + \left(\underline{J}, \Delta_{\underline{E}}^I\right)_{0,K} = 0.
\end{aligned} \tag{5.129}$$

Combining (5.128) and (5.129), we get:

$$\frac{1}{2}\frac{\mathrm{d}}{\mathrm{d}t}(\Delta^I_{\underline{E}}, \Delta^I_{\underline{E}})_{0,K,\varepsilon} = \left(\frac{\partial}{\partial t}\Delta^P_{\underline{E}}, \Delta^I_{\underline{E}}\right)_{0,K,\varepsilon} + (\nabla \times \Delta^I_{\underline{H}}, \Delta^I_{\underline{E}})_{0,K} - (\nabla \times \Delta^P_{\underline{H}}, \Delta^I_{\underline{E}})_{0,K} + (\beta[\![\Delta^I_{\underline{H}} \times \underline{n}_K]\!]^K_{\partial K}, \Delta^I_{\underline{E}|K})_{0,\partial K} + (\beta[\![\pi^1_h \underline{H} \times \underline{n}_K]\!]^K_{\partial K}, \Delta^I_{\underline{E}|K})_{0,\partial K} \tag{5.130}$$

By applying the same process to equation in $\underline{H}$, we have:

$$\frac{1}{2}\frac{\mathrm{d}}{\mathrm{d}t}(\Delta^I_{\underline{H}}, \Delta^I_{\underline{H}})_{0,K,\mu} = \left(\frac{\partial}{\partial t}\Delta^P_{\underline{H}}, \Delta^I_{\underline{H}}\right)_{0,K,\mu} - (\nabla \times \Delta^I_{\underline{E}}, \Delta^I_{\underline{H}})_{0,K} + (\nabla \times \Delta^P_{\underline{E}}, \Delta^I_{\underline{H}})_{0,K} + \left(\gamma[\![\Delta^I_{\underline{E}} \times \underline{n}_K]\!]^K_{\partial K}, \Delta^I_{\underline{H}|K}\right)_{0,\partial K} + \left(\gamma[\![\pi^1_h \underline{E} \times \underline{n}_K]\!]^K_{\partial K}, \Delta^I_{\underline{H}|K}\right)_{0,\partial K} \tag{5.131}$$

Green's formula provides

$$(\nabla \times \Delta^I_{\underline{H}}, \Delta^I_{\underline{E}})_{0,K} = (\Delta^I_{\underline{H}}, \nabla \times \Delta^I_{\underline{E}})_{0,K} + (\Delta^I_{\underline{H}|K}, \Delta^I_{\underline{E}|K} \times \underline{n}_K)_{0,\partial K} \tag{5.132}$$

Adding (5.130) and (5.131), we obtain:

$$\begin{aligned}\frac{1}{2}\frac{\mathrm{d}}{\mathrm{d}t}\left[(\Delta^I_{\underline{E}}, \Delta^I_{\underline{E}})_{0,K,\varepsilon} + (\Delta^I_{\underline{H}}, \Delta^I_{\underline{H}})_{0,K,\mu}\right] &= \left(\frac{\partial}{\partial t}\Delta^P_{\underline{E}}, \Delta^I_{\underline{E}}\right)_{0,K,\varepsilon} \\ &+ \left(\frac{\partial}{\partial t}\Delta^P_{\underline{H}}, \Delta^I_{\underline{H}}\right)_{0,K,\mu} + (\nabla \times \Delta^P_{\underline{E}}, \Delta^I_{\underline{H}})_{0,K} \\ &- (\nabla \times \Delta^P_{\underline{H}}, \Delta^I_{\underline{E}})_{0,K} \\ &+ \left(\beta[\![\Delta^I_{\underline{H}} \times \underline{n}_K]\!]^K_{\partial K}, \Delta^I_{\underline{E}|K}\right)_{0,\partial K} \\ &+ (\beta[\![\pi^1_h \underline{H} \times \underline{n}_K]\!]^K_{\partial K}, \Delta^I_{\underline{E}|K})_{0,\partial K} \\ &+ (\Delta^I_{\underline{H}|K}, \Delta^I_{\underline{E}|K} \times \underline{n}_K)_{0,\partial K} \\ &- \left(\gamma[\![\Delta^I_{\underline{E}} \times \underline{n}_K]\!]^K_{\partial K}, \Delta^I_{\underline{H}|K}\right)_{0,\partial K} \\ &+ \left(\gamma[\![\pi^1_h \underline{E} \times \underline{n}_K]\!]^K_{\partial K}, \Delta^I_{\underline{H}|K}\right)_{0,\partial K}\end{aligned} \tag{5.133}$$

From Sect. 1.2.4, we know that

$$\forall t \in (0, T),\ (\underline{E}, \underline{H})(t) \in H_0(rot, \Omega) \times H(rot, \Omega),$$

so that we have:

$$\forall \Gamma = (K \cap K') \in \mathscr{F}_h^i,\ [\![\underline{E} \times \underline{n}_K]\!]_\Gamma^{K \text{ or } K'} = 0 \text{ and } [\![\underline{H} \times \underline{n}_K]\!]_\Gamma^{K \text{ or } K'} = 0.$$

Moreover, we must keep in mind that $\forall \Gamma \in \mathscr{F}_h^b,\ \beta = 0$.

By summing (5.133) over all the cells of the mesh and using the previous properties, we can write:

$$\begin{aligned}
\frac{1}{2}\frac{\mathrm{d}}{\mathrm{d}t}\|(\Delta_E^I, \Delta_H^I)\|_*^2 &= \frac{1}{2}\frac{\mathrm{d}}{\mathrm{d}t}\sum_{K\in\mathscr{T}_h}\left[\left(\Delta_{\underline{E}}^I, \Delta_{\underline{E}}^I\right)_{0,K,\varepsilon} + \left(\Delta_{\underline{H}}^I, \Delta_{\underline{H}}^I\right)_{0,K,\mu}\right] \\
&\leq \sum_{K\in\mathscr{T}_h}\Bigg[\left|\left(\frac{\partial}{\partial t}\Delta_{\underline{E}}^P, \Delta_{\underline{E}}^I\right)_{0,K,\varepsilon}\right| + \left|\left(\frac{\partial}{\partial t}\Delta_{\underline{H}}^P, \Delta_{\underline{H}}^I\right)_{0,K,\mu}\right| \\
&\qquad + |(\nabla \times \Delta_{\underline{E}}^P, \Delta_{\underline{H}}^I)_{0,K}| + \left|\left(\nabla \times \Delta_{\underline{H}}^P, \Delta_{\underline{E}}^I\right)_{0,K}\right| \\
&\qquad + |(\Delta_{\underline{E}}^P, \Delta_{\underline{E}}^I)_{0,K,\sigma}| + \left|\left(\beta[\![\Delta_{\underline{H}}^P \times \underline{n}_K]\!], \Delta_{\underline{E},K}^I\right)_{0,\partial K}\right| \\
&\qquad + \left|\left(\gamma[\![\Delta_{\underline{E}}^P \times \underline{n}_K]\!], \Delta_{\underline{H},K}^I\right)_{0,\partial K}\right|\Bigg].
\end{aligned} \tag{5.134}$$

To eliminate surface terms in $\Delta_{\underline{E}}^I$ and $\Delta_{\underline{H}}^I$, we use the identity:

$$\begin{aligned}
&\sum_{K\in\mathscr{T}_h}\Big(\beta[\![\Delta_{\underline{H}}^I \times \underline{n}_K]\!]_{\partial K}^K, \Delta_{\underline{E}|K}^I\Big)_{0,\partial K} - \Big(\gamma[\![\Delta_{\underline{E}}^I \times \underline{n}_K]\!]_{\partial K}^K, \Delta_{\underline{H}|K}^I\Big)_{0,\partial K} \\
&\quad + (\Delta_{\underline{H}|K}^I, \Delta_{\underline{E}|K}^I \times \underline{n}_K)_{0,\partial K}\Big) = 0.
\end{aligned}$$

5.4.5.2 *hp*-Error Estimates

Equation (5.112) allows us to establish the following estimations of the surface terms:

$$\begin{aligned}
&\sum_{K\in\mathscr{T}_h}\left[\left|\left(\beta[\![\Delta_{\underline{H}}^P \times \underline{n}_K]\!], \Delta_{\underline{E}|K}^I\right)_{0,\partial K}\right| + \left|\left(\gamma[\![\Delta_{\underline{E}}^P \times \underline{n}_K]\!], \Delta_{\underline{H}|K}^I\right)_{0,\partial K}\right|\right] \\
&\leq \sum_{K\in\mathscr{T}_h}\Big[|\beta|\|[\![\Delta_{\underline{H}}^P \times \underline{n}_K]\!]\|_{0,\partial K}\|\Delta_{\underline{E}|K}^I\|_{0,\partial K} \\
&\qquad + |\gamma|\|[\![\Delta_{\underline{E}}^P \times \underline{n}_K]\!]\|_{0,\partial K}\|\Delta_{\underline{H}|K}^I\|_{0,\partial K}\Big] \\
&\leq C\sum_{K\in\mathscr{T}_h}\sigma_K^{\frac{11}{2}}\frac{r}{\rho_K^{\frac{1}{2}}}\Big[|\beta|\|[\![\Delta_{\underline{H}}^P \times \underline{n}_K]\!]\|_{0,\partial K}\|\Delta_{\underline{E}}^I\|_{0,K,\underline{\underline{\varepsilon}}} \\
&\qquad + |\gamma|\|[\![\Delta_{\underline{E}}^P \times \underline{n}_K]\!]\|_{0,\partial K}\|\Delta_{\underline{H}}^I\|_{0,K,\underline{\underline{\mu}}}\Big],
\end{aligned} \tag{5.135}$$

where C is a constant independent of K and r.

Estimation (5.117) gives:

$$\|\Delta^P_{\underline{E}|K} \times \underline{n}_K\|^2_{0,\partial K} \leq C\left(\|\Delta^P_{\underline{E}}\|_{0,K}\|\Delta^P_{\underline{E}}\|_{1,K} + \sigma_K^{-1}\rho_K^{-1}\|\Delta^P_{\underline{E}}\|^2_{0,K}\right)$$
$$\|\Delta^P_{\underline{H}|K} \times \underline{n}_K\|^2_{0,\partial K} \leq C\left(\|\Delta^P_{\underline{H}}\|_{0,K}\|\Delta^P_{\underline{H}}\|_{1,K} + \sigma_K^{-1}\rho_K^{-1}\|\Delta^P_{\underline{H}}\|^2_{0,K}\right) \quad (5.136)$$

where C is a constant independent of K and r.

Actually, notice that $\underline{v} = \Delta^P_{\underline{E}|K} = (v_1, v_2, v_3)^T$ and $\underline{n}_K = (n_1, n_2, n_3)^T$. We can then write:

$$\Delta^P_{\underline{E}|K} \times \underline{n}_K = \underline{v} \times \underline{n}_K = (v_2 n_3 - v_3 n_2, v_3 n_1 - v_1 n_3, v_1 n_2 - v_2 n_1)^T. \quad (5.137)$$

Now, by developing $\|\Delta^P_{\underline{E}|K} \times \underline{n}_K\|^2_{0,\partial K}$, we get:

$$\begin{aligned}\|\Delta^P_{\underline{E}|K} \times \underline{n}_K\|^2_{0,\partial K} &= \int_{\partial K} \left((v_2 n_3 - v_3 n_2)^2 + (v_3 n_1 - v_1 n_3)^2 + (v_1 n_2 - v_2 n_1)^2\right) d\sigma \\ &\leq 2\int_{\partial K} \left(v_2^2 n_3^2 + v_3^2 n_2^2 + v_3^2 n_1^2 + v_1^2 n_3^2 + v_1^2 n_2^2 + v_2^2 n_1^2\right) d\sigma \\ &\leq 4\int_{\partial K} (v_1^2 + v_2^2 + v_3^2) d\sigma = 4\left(\|v_1\|^2_{0,\partial K} + \|v_2\|^2_{0,\partial K} + \|v_3\|^2_{0,\partial K}\right).\end{aligned} \quad (5.138)$$

To obtain the last inequality, we have used the fact that $n_1^2 + n_2^2 + n_3^2 = 1$.

Now, by applying (5.118) to $v_i \in H^1(K)$ (for $1 \leq i \leq 3$), one deduces the inequalities:

$$\|v_i\|^2_{0,\partial K} \leq C(\|v_i\|_{0,K}\|\nabla v_i\|_{0,K} + \sigma_K^{-1}\rho_K^{-1}\|v_i\|^2_{0,K}) \quad (5.139)$$

Finally, introducing (5.139) into (5.138), we have:

$$\begin{aligned}\|\Delta^P_{\underline{E}|K} \times \underline{n}_K\|^2_{0,\partial K} &\leq 4C\sum_{i=1}^{3}(\|v_i\|_{0,K}\|\nabla v_i\|_{0,K} + \sigma_K^{-1}\rho_K^{-1}\|v_i\|^2_{0,K}) \\ &\leq 4C\sum_{i=1}^{3}(\|\underline{v}\|_{0,K}\|\underline{v}\|_{1,K} + \sigma_K^{-1}\rho_K^{-1}\|\underline{v}\|^2_{0,K}) \\ &\leq 12C(\|\underline{v}\|_{0,K}\|\underline{v}\|_{1,K} + \sigma_K^{-1}\rho_K^{-1}\|\underline{v}\|^2_{0,K}).\end{aligned} \quad (5.140)$$

We obtain the wanted result.

For the estimation of $\|\Delta^P_{\underline{H}|K} \times \underline{n}_K\|^2_{0,\partial K}$, we proceed in the same way.

For the other terms of (5.134), we have the following estimates:

$$\sum_{K\in\mathscr{T}_h} \left[|(\nabla \times \Delta^P_{\underline{E}}, \Delta^I_{\underline{H}})_{0,K}| + |(\nabla \times \Delta^P_{\underline{H}}, \Delta^I_{\underline{E}})_{0,K}|\right]$$
$$\leq C \sum_{K\in\mathscr{T}_h} \left[\|\Delta^P_{\underline{E}}\|_{1,K}\|\Delta^I_{\underline{H}}\|_{0,K,\mu} + \|\Delta^P_{\underline{H}}\|_{1,K}\|\Delta^I_{\underline{E}}\|_{0,K,\varepsilon}\right] \tag{5.141a}$$

$$\sum_{K\in\mathscr{T}_h} |(\frac{\partial}{\partial t}\Delta^P_{\underline{E}}, \Delta^I_{\underline{E}})_{0,K,\varepsilon}| + |(\frac{\partial}{\partial t}\Delta^P_{\underline{H}}, \Delta^I_{\underline{H}})_{0,K,\mu}|$$
$$\leq C \sum_{K\in\mathscr{T}_h} \left(\|\Delta^P_{\underline{E}_t}\|_{0,K}\|\Delta^I_{\underline{E}}\|_{0,K,\varepsilon} + \|\Delta^P_{\underline{H}_t}\|_{0,K}\|\Delta^I_{\underline{H}}\|_{0,K,\mu}\right) \tag{5.141b}$$

$$\sum_{K\in\mathscr{T}_h} |(\Delta^P_{\underline{E}}, \Delta^I_{\underline{E}})_{0,K,\sigma}| \leq C \sum_{K\in\mathscr{T}_h} \|\Delta^P_{\underline{E}}\|_{0,K}\|\Delta^I_{\underline{E}}\|_{0,K,\varepsilon} \tag{5.141c}$$

where C is a constant independent of K and r and $\underline{u}_t = \dfrac{\partial \underline{u}}{\partial t}$.

Remark: To obtain the second inequality (5.141b), we used the property:

$$\frac{\partial}{\partial t}\Delta^P_{\underline{E}} = \Delta^P_{\frac{\partial}{\partial t}\underline{E}} = \Delta^P_{\underline{E}_t} \tag{5.142}$$

This property holds for $\Delta^P_{\underline{H}}$.

Now, we are going to recombine the established estimates and use the projection errors of the previous section. Keep in mind that we use a regular family, $(\mathscr{T}_h)_{h>0}$, of meshes. We assume that $\underline{E},\ \underline{H} \in [H^{s+1}(\mathscr{T}_h)]^3 \cap H(\text{rot}, \Omega)$ and $\underline{E}_t,\ \underline{H}_t \in [H^{s'+1}(\mathscr{T}_h)]^3$ with $s,\ s' \geq 0$ and $0 < h_K \leq 1,\ \forall K \in \mathscr{T}_h$.

Using (5.111), (5.136) becomes:

$$\|\Delta^P_{\underline{E}|K} \times \underline{n}_K\|^2_{0,\partial K} \leq C\frac{h_K^{\min(2s-1,2r-1)}}{r^{2s+1}}\|\underline{E}\|_{s+1,K}, \tag{5.143a}$$

$$\|\Delta^P_{\underline{H}|K} \times \underline{n}_K\|^2_{0,\partial K} \leq C\frac{h_K^{\min(2s-1,2r-1)}}{r^{2s+1}}\|\underline{H}\|_{s+1,K}. \tag{5.143b}$$

Thus, the boundary terms are bounded by

$$\sum_{K\in\mathscr{T}_h} \left[\left|(\beta[\![\Delta^P_{\underline{H}} \times \underline{n}_K]\!], \Delta^I_{\underline{E},K})_{0,\partial K}\right| + \left|(\gamma[\![\Delta^P_{\underline{E}} \times \underline{n}_K]\!], \Delta^I_{\underline{H},K})_{0,\partial K}\right|\right]$$
$$\leq C \sum_{K\in\mathscr{T}_h} \frac{h_K^{\min(s-1,r-1)}}{r^{s-\frac{1}{2}}}\left[\|\underline{E}\|_{s+1,K}\|\Delta^I_{\underline{E},K}\|_{0,K,\varepsilon} + \|\underline{H}\|_{s+1,K}\|\Delta^I_{\underline{H},K}\|_{0,K,\mu}\right]. \tag{5.144}$$

We also have the estimates of (5.141a)–(5.141c):

$$\sum_{K\in\mathscr{T}_h}\left[|(\nabla\times\Delta_{\underline{E}}^{P},\Delta_{\underline{H}}^{I})_{0,K}|+|\big(\nabla\times\Delta_{\underline{H}}^{P},\Delta_{\underline{E}}^{I}\big)_{0,K}|\right]$$
$$\le C\sum_{K\in\mathscr{T}_h}\frac{h_K^{\min(s-1,r-1)}}{r^s}\left[\|\underline{E}\|_{s+1,K}\|\Delta_{\underline{H}}^{I}\|_{0,K,\mu}+\|\underline{H}\|_{s+1,K}\|\Delta_{\underline{E}}^{I}\|_{0,K,\varepsilon}\right] \quad (5.145a)$$

$$\sum_{K\in\mathscr{T}_h}\left|\big(\frac{\partial}{\partial t}\Delta_{\underline{E}}^{P},\Delta_{\underline{E}}^{I}\big)_{0,K,\varepsilon}\right|+\left|\big(\frac{\partial}{\partial t}\Delta_{\underline{H}}^{P},\Delta_{\underline{H}}^{I}\big)_{0,K,\mu}\right|$$
$$\le C\sum_{K\in\mathscr{T}_h}\frac{h_K^{\min(s',r)}}{r^{s'+1}}\left[\|\underline{E}_t\|_{s'+1,K}\|\Delta_{\underline{E}}^{I}\|_{0,K,\varepsilon}+\|\underline{H}_t\|_{s'+1,K}\|\Delta_{\underline{H}}^{I}\|_{0,K,\mu}\right] \quad (5.145b)$$

$$\sum_{K\in\mathscr{T}_h}|(\Delta_{\underline{E}}^{P},\Delta_{\underline{E}}^{I})_{0,K,\sigma}|\le C\sum_{K\in\mathscr{T}_h}\frac{h_K^{\min(s,r)}}{r^{s+1}}\|\underline{E}\|_{s+1,K}\|\Delta_{\underline{E}}^{I}\|_{0,K,\varepsilon} \quad (5.145c)$$

Now, by using the fact that $\dfrac{\|\Delta_{\underline{E}}^{I}\|_{0,K,\varepsilon}}{\|(\Delta_{E}^{I},\Delta_{H}^{I})\|_*}\le 1$ and $\dfrac{\|\Delta_{\underline{H}}^{I}\|_{0,K,\mu}}{\|(\Delta_{E}^{I},\Delta_{H}^{I})\|_*}\le 1$, (5.134) leads to

$$\frac{\mathrm{d}}{\mathrm{d}t}\|(\Delta_{E}^{I},\Delta_{H}^{I})\|_*\le C\sum_{K\in\mathscr{T}_h}\left[\frac{h_K^{\min(s-1,r-1)}}{r^{s-\frac{1}{2}}}\big(\|\underline{E}\|_{s+1,K}+\|\underline{H}\|_{s+1,K}\big)\right.$$
$$\left.+\frac{h_K^{\min(s',r)}}{r^{s'+1}}\big(\|\underline{E}_t\|_{s'+1,K}+\|\underline{H}_t\|_{s'+1,K}\big)\right] \quad (5.146)$$

Finally, by integrating (5.146) over the interval $(0, T)$, we have the following theorem:

Theorem 5.2 *Let r be a positive integer. Assume that the exact solution verifies $(\underline{E},\underline{H})\in\underline{H}^{s+1}\big(\mathscr{T}_h\big)$ and $(\underline{E}_t,\underline{H}_t)\in\underline{H}^{s'+1}\big(\mathscr{T}_h\big)$ for $s, s'\ge 0$ real and $0<h_K\le 1$ $\forall K\in\mathscr{T}_h$. Then, we have the global estimate of the interpolation error*

$$\|(\Delta_{E}^{I},\Delta_{H}^{I})\|_*(T)\le\|(\Delta_{E}^{I},\Delta_{H}^{I})\|_*(0)+CT\frac{h^{\min(s-1,s',r-1)}}{r^{\min(s-\frac{1}{2},s'+1)}}\beta(\underline{E},\underline{H}), \quad (5.147)$$

where $C>0$ is a constant independent of K and r, $h=\max_{K\in\mathscr{T}_h}h_K$ and

$$\beta(\underline{E},\underline{H})=\max_{t\in(0,T)}\big(\|\underline{E}\|_{s+1,h}(t),\|\underline{H}\|_{s+1,h}(t),\|\underline{E}_t\|_{s'+1,h}(t),\|\underline{H}_t\|_{s'+1,h}(t)\big). \quad (5.148)$$

Let us go back to the error of the scheme: according to (5.127), we have:

$$\begin{aligned}\|(\underline{E}-\underline{E}_h, \underline{H}-\underline{H}_h)\|_*(T) &\leq \sqrt{2}(\|(\Delta_{\underline{E}}^P, \Delta_{\underline{H}}^P)\|_* + \|(\Delta_{\underline{E}}^I, \Delta_{\underline{H}}^I)\|_*)(T)\\ &\leq \sqrt{2}\|(\Delta_{\underline{E}}^I, \Delta_{\underline{H}}^I)\|_*(0)\\ &\quad + C\sqrt{2}\big[h^s \max\big(\|\underline{E}\|_{s+1,h}(T), \|\underline{H}\|_{s+1,h}(T)\big)\\ &\quad + T\frac{h^{\min(s-1,s',r-1)}}{r^{\min(s-\frac{1}{2},s'+1)}}\beta(\underline{E},\underline{H})\big]. \end{aligned} \tag{5.149}$$

Remarks:

1. We see that the error seems to be sub-optimal and it increases at most linearly in time. Moreover, for $r = 1$, the previous estimate does not prove the consistency of the scheme. In [35], we show with a simple numerical example that it is not clear that this scheme is consistent for some meshes.
2. If the mesh used is orthogonal or almost parallelepipedic, we find an exponent h^s. In this case, we are both in an affine case and with second derivatives of F_K bounded by Ch_K^2.
3. In [35], we also study the effect of the use of the Gauss quadrature rule to compute the integrals on the previous error estimates. We conclude that the use of the Gauss quadrature formula can generate a deterioration of the spatial convergence when the exact solution of the problem is not very regular inside at least one cell. Nevertheless, if the data of the treated problem are regular, the mass-lumping, thus does not generate a deterioration of the h convergence (i.e. h_K^{r-1}).

5.4.6 *Extension to the Dissipative Case*

In [36], we study the a priori error analysis in h-version of the dissipative DG scheme i.e. $\delta, \gamma > 0$ in formulation (4.37a) and (4.37b). We show a gain in the order of convergence of the scheme which implies the convergence for all spatial orders of approximation. In this part, we only give the main results and we refer the reader to the paper for the technical proofs. The main ingredient is the choice of a well-adapted projector on the approximate space (see Proposition 5.5).

Consider $(\underline{u}, \underline{w}) \in U_h \times U_h$, we write $\underline{E} - \underline{E}_h = \underline{E} - \underline{v} + \underline{v} - \underline{E}_h = \Delta_{\underline{E}}^P - \Delta_{\underline{E}}^I$ and $\underline{H} - \underline{H}_h = \underline{H} - \underline{w} + \underline{w} - \underline{H}_h = \Delta_{\underline{H}}^P - \Delta_{\underline{H}}^I$ with $\Delta_{\underline{E}}^P = \underline{E} - \underline{v}$, $\Delta_{\underline{E}}^I = \underline{E}_h - \underline{v}$, $\Delta_{\underline{H}}^P = H - w$ and $\Delta_{\underline{H}}^I = \underline{H}_h - w$.

We have the two following propositions:

Proposition 5.5 *Let* $(\underline{v}, \underline{w}) \in U_h \times U_h$ *be the solution of the problem:* $\forall(\underline{v}', \underline{w}') \in U_h \times U_h$ *and* $\forall K \in \mathscr{T}_h$,

$$\begin{aligned}
&\int_K \varepsilon \underline{v} \cdot \underline{v}' \mathrm{d}\underline{x} + \int_K \nabla \times \underline{w} \cdot \underline{v}' \mathrm{d}\underline{x} \\
&\quad - \int_{\partial K} \left(\beta [\![\underline{w} \times \underline{n}]\!]_{\partial K}^K + \alpha [\![\underline{n} \times \left(\underline{v} \times \underline{n} \right)]\!]_{\partial K}^K \right) \cdot \underline{v}' ds = l_1(\underline{v}') \\
&\int_K \mu \underline{w} \cdot \underline{w}' \mathrm{d}\underline{x} - \int_K \nabla \times \underline{v} \cdot \underline{w}' \mathrm{d}\underline{x} \\
&\quad - \int_{\partial K} \left(\gamma [\![\underline{v} \times \underline{n}]\!]_{\partial K}^K + \delta [\![\underline{n} \times \left(\underline{w} \times \underline{n} \right)]\!]_{\partial K}^K \right) \cdot \underline{w}' ds = l_2(w')
\end{aligned} \tag{5.150}$$

where l_1, l_2 are the two linear forms on U_h defined by

$$l_1(\underline{v}') = \int_K \varepsilon \underline{E} \cdot \underline{v}' \mathrm{d}\underline{x} + \int_K \nabla \times \underline{H} \cdot \underline{v}' \mathrm{d}\underline{x},$$

$$l_2(\underline{w}') = \int_K \mu \underline{H} \cdot \underline{w}' \mathrm{d}\underline{x} - \int_K \nabla \times \underline{E} \cdot \underline{w}' \mathrm{d}\underline{x}.$$

Then, we have:

$$\frac{\mathrm{d}}{\mathrm{d}t} \left\| (\Delta_E^I, \Delta_H^I) \right\|_* \leq \left\| \Delta_{\frac{\partial E}{\partial t}}^P \right\|_{0,\varepsilon,\Omega} + \left\| \Delta_{\frac{\partial H}{\partial t}}^P \right\|_{0,\mu,\Omega} + \left\| \Delta_E^P \right\|_{0,\varepsilon,\Omega} + \left\| \Delta_H^P \right\|_{0,\mu,\Omega} \tag{5.151}$$

Proposition 5.6 *If we assume that the exact solution verifies $(\underline{E}, \underline{H}) \in \left[H^{s+1}(\mathcal{T}_h)\right]^3$ for $s \geq 0$, then there exists a constant $C > 0$ such that*

$$\left\| (\Delta_{\underline{E}}^P, \Delta_{\underline{H}}^P) \right\|_* \leq C h^{\min(s-\frac{1}{2}, r-\frac{1}{2})} \max(\left\| \underline{E} \right\|_{s+1,h}, \left\| \underline{H} \right\|_{s+1,h})$$

where $H^s(\mathcal{T}_h) = \{\underline{v} \in [L^2(\Omega)]^3 \ : \ \forall K \in \mathcal{T}_h, \ \underline{v}_{|K} \in [H^s(K)]^3\}$ and $\left\| \underline{v} \right\|_{s,h}^2 = \sum_{K \in \mathcal{T}_h} \left\| \underline{v} \right\|_{s,K}^2$.

By using Propositions 5.5 and 5.6 and by integrating the result over the interval $(0, T)$, we have:

Theorem 5.3 *Let r be a positive integer. Assume that the exact solution verifies $(\underline{E}, \underline{H}) \in [H^{s+1}(\mathcal{T}_h)]^3$ and $(\underline{E}_t, \underline{H}_t) \in [H^{s'+1}(\mathcal{T}_h)]^3$ for $s, s' \geq 0$ real and $0 < h_K \leq 1$. Then, we have the global estimate of the interpolation error*

$$\left\| (\Delta_{\underline{E}}^I, \Delta_{\underline{H}}^I) \right\|_* (T) \leq \left\| (\Delta_{\underline{E}}^I, \Delta_{\underline{H}}^I) \right\|_* (0) + C T h^{\min(s-\frac{1}{2}, s'-\frac{1}{2}, r-\frac{1}{2})} A(T, \underline{E}, \underline{H}) \tag{5.152}$$

where

$$A(T, \underline{E}, \underline{H}) = \max_{t \in (0,T)} \left(\left\| \underline{E} \right\|_{s+1,h}(t), \left\| \underline{vH} \right\|_{s+1,h}(t), \left\| \underline{E}_t \right\|_{s'+1,h}(t), \left\| \underline{H}_t \right\|_{s'+1,h}(t) \right).$$

Finally, by using (5.152) and (5.6), we deduce the error of the DG-scheme by:

$$\left\|(\underline{E} - \underline{E}_h, \underline{H} - \underline{H}_h)\right\|_*^2 \leq 2(\left\|(\Delta_{\underline{E}}^P, \Delta_{\underline{H}}^P)\right\|_*^2 + \left\|(\Delta_{\underline{E}}^I, \Delta_{\underline{H}}^I)\right\|_*^2). \qquad (5.153)$$

In conclusion, if the exact solution is smooth enough, the convergence rate for the penalized scheme is $r - 1/2$ versus $r - 1$ for the non-dissipative scheme. So, the dissipative terms ensure the L^2-convergence for Q_1 approximation.

References

1. Nédélec, J.-C.: Mixed finite elements in $I\!R^3$. Numer. Math. **35**(3), 315–341 (1980)
2. Yee, K.: Numerical solutions of initial boundary value problems involving Maxwell's equations in isotropic media. IEEE Trans. Antennas Propag. **14**(3), 302–307 (1966)
3. Cohen, G.: High order numerical methods for transient wave equations. Scientific computation. Springer, Berlin (2001)
4. Assous, F., Degond, P., Heintze, E., Raviart, P.-A., Segre, J.: On a finite-element method for solving the three-dimensional Maxwell equations. J. Comput. Phys. **109**(2), 222–237 (1993)
5. Cohen, G., Duruflé, M.: Non spurious spectral-like element methods for Maxwell's equations. J. Comp. Math. **25**(3), 282–304 (2007)
6. M. Duruflé, *Intégration numérique et éléments finis d'ordre élevé appliqués aux équations de Maxwell en régime harmonique*, thèse de doctorat, U. de Paris-Dauphine, 2006
7. Cohen, G., Fauqueux, S.: Mixed finite elements with mass-lumping for the transient wave equation. J. Comput. Acoust. **8**(1), 171–188 (2000)
8. Cohen, G., Fauqueux, S.: Mixed spectral finite elements for the linear elasticity system in unbounded domains. SIAM J. Sci. Comput. **26**(3), 864–884 (2005)
9. Costabel, M., Dauge, M., Martin, D., Vial, G.: Numerical mathematics and advanced applications. In: Weighted Regularization of Maxwell Equations: Computations in Curvilinear Polygons, pp. 273-280. Springer, Milan (2003)
10. Cohen, G., Monk, P.: Mur-Nédélec finite element schemes for Maxwell's equations. Comput. Methods Appl. Mech. Engrg. **169**(3–4), 197–217 (1999)
11. Gurari, M.: The Location of Eigenvalues and Eigenvectors of Complex Matrices. J. Approx. Theory **22**(2), 119–149 (1978)
12. Monk, P.: Finite Element Methods for Maxwell's Equations. Oxford University Press, Numerical Mathematics and Scientific Computation (2003)
13. Nédélec, J.-C.: A new family of mixed finite elements in $I\!R^3$. Numer. Math. **50**(1), 57–81 (1986)
14. Descloux, J., Nassif, N., Rappaz, J.: On spectral approximation Part 1. The problem of convergence. RAIRO Numer. Anal. **12**, pp. 97–112 (1978)
15. Descloux, J., Nassif, N., Rappaz, J.: On spectral approximation Part 2. Error estimates for the Galerkin method convergence. RAIRO Numer. Anal. **12**, pp. 113–119 (1978)
16. Caorsi, S., Fernandes, P., Raffetto, M.: On the convergence of Galerkin finite element approximations of electromagnetic eigenproblems. SIAM J. Numer. Anal. **38**(2), 580–607 (2000)
17. Boffi, D.: Fortin operator and discrete compactness for edge elements. Numer. Math. **87**, 229–246 (2000)
18. Boffi, D.: A note on the de Rham complex and a discrete compactness property. Appl. Math. Lett. **14**, 33–38 (2001)
19. Boffi, D., Costabel, M., Dauge, M., Demkowicz, L., Hiptmair, R.: Discrete Compactness for the p-Version of Discrete Differential Forms. SIAM J. Numer. Anal. **49**(1), 135–158 (2009)

20. Boffi, D., Fernandez, P., Perugia, I.: Computational models of electromagnetic resonators: Analysis of edge element approximation. SIAM J. Numer. Anal. **36**(4), 1264–1290 (1999)
21. Boffi, D., Gastaldi, L.: Edge finite elements for the approximation of Maxwell resolvent operator. RAIRO - Math. Model. Numer. Anal. **36**(2), 293–305 (2002)
22. Buffa, A., Perugia, I.: Discontinuous Galerkin approximation of the Maxwell eigenproblem. SIAM J. Numer. Anal. **44**(5), 2198–2226 (2006)
23. Warburton, T., Embree, M.: The role of the penalty in the local discontinuous Galerkin method for Maxwell's eigenvalue problem. Comput. Methods Appl. Mech. Engrg. **195**(25–28), 3205–3223 (2006)
24. Creusé, E., Nicaise, S.: Discrete compactness for a discontinuous Galerkin approximation of Maxwell's system. ESAIM Math. Model. Numer. Anal. **40**(2), 413–430 (2006)
25. Buffa, A., Houston, P., Perugia, I.: Discontinuous Galerkin computation of the maxwell eigenvalues on simplicial meshes. J. Comput. Appl. Math. **204**(2), 317–333 (2007)
26. Buffa, A., Perugia, I., Warburton, T.: The mortar-discontinuous Galerkin method for the 2D Maxwell eigenproblem. J. Sci. Comput. **40**(1), 86–114 (2009)
27. Assous, F., Ciarlet, P., Raviart, P.-A., Sonnendrücker, E.: A characterization of the singular part of the solution to Maxwell's equations in a polyhedral domain. Math. Meth. Appl. Sci. **22**(6), 485–499 (1999)
28. Hesthaven, J.S., Warburton, T.: Nodal discontinuous Galerkin methods. Texts in Applied Mathematics, **54**, Springer, Berlin (2008)
29. Hesthaven, J.S., Gottlieb, S., Gottlieb, D.: Cambridge monographs on applied and computational mathematics spectral methods for time-dependent problems. Spectral Methods for Time-Dependent Problems, Cambridge University Press. Cambridge (2007)
30. Bernardi, C., Maday, Y.: Spectral methods. In: Ciarlet, P.G., Lions, J.L. (eds.) Handbook of Numerical Analysis V. Elsevier, North-Holland, The Netherlands (1999)
31. Ciarlet, P.G.: The Finite Element Method for Elliptic Problems. Elsevier, North-Holland (2002)
32. Girault, V., Raviart, P.-A.: Finite Element Methods for Navier-Stokes Equations. Sringer, New York (1986)
33. Rivière, B., Wheeler, M.F., Girault, V.: A priori error estimates for finite element methods based on discontinuous approximation spaces for elliptic problems. SIAM J. Numer. Anal. **39**(3), 902–931 (2001)
34. Prudhomme, S., Pascal, F., Oden, T., Romkes, A.: Review of a Priori Error Estimation for Discontinuous Galerkin, Orsay (2000)
35. Pernet, S., Ferrieres, X.: HP a-priori error estimates for a non-dissipative spectral discontinuous Galerkin method to solve the Maxwell equations in the time domain. Math. Comp. **76**(260), 1801–1832 (2007)
36. Montseny, E., Pernet, S., Ferrières, X., Cohen, G.: Dissipative terms and local time-stepping improvements in a spatial high order Discontinuous Galerkin scheme for the time-domain Maxwell's equations. J. Comput. Phys. **227**(14), 6795–6820 (2008)

Chapter 6
Approximating Unbounded Domains

Abstract This chapter provides the construction and approximation of two approaches for the treatment of unbounded domains: first by absorbing boundary conditions (ABC), then by perfectly matched layers (PML) for the three wave equations.

This chapter speaks about a mathematical approximation of an unreal model: the unbounded (or infinite) domains. These domains are unreal since we live in a finite universe. However, some experiments in very large domains (such as the propagation of an electromagnetic wave in free space) can be regarded as experiments in infinite domains from a physical point of view. One could use the physical (bounded) domain to compute the numerical resolution but it would be very expensive. Moreover, the interesting part of the solutions is often around a small bounded domain containing scatterers or sources. These considerations lead to investigate a way to replace a large domain by a smaller one (almost) without altering the solution.

A first direction in this way was to add a layer with a damping material at the boundary of the numerical domain in order to reduce the amplitude of the incoming wave and consequently its reflection (see for instance [1]). Unfortunately this approach presented a major drawback: the difference of media generated a reflection at the interface.

A second and important step was made by introducing a transparent condition at the boundary, which then behaved as an open boundary that waves can cross without coming back. Unfortunately, this condition is not local and would be very expensive to use for problems in time-domain. For this reason, this condition is approximated by a local condition. The efficiency of this condition, called absorbing boundary condition (ABC), depends on the order of the approximation of the transparent condition [2–4]. These conditions were derived to any order for the wave equation [5] and to some orders for Maxwell's equations [6, 7] and the linear elastodynamics system [8, 9]. The main difficulties of this approach are (a) obtaining a stable approximation of the transparent condition, which is not always obvious, (b) writing an adequate treatment of the corners or the edges of a domain [5, 10]. Some analysis of ABC can be found in [11, 12] and a review of different approaches of ABC (or NRBC) is given in [13].

G. Cohen and S. Pernet, *Finite Element and Discontinuous Galerkin Methods for Transient Wave Equations*, Scientific Computation,
DOI 10.1007/978-94-017-7761-2_6

The third approach was a true revolution. In 1994, a French engineer, J.-P. Bérenger, published a paper where he constructed damping layers which produced **no reflection at all** at their interface. He called them "perfectly matched layers" and their popular name became their initials, i.e. PML. This method, which was built for the second-order finite difference approximation of Maxwell's equations (Yee scheme [14]), first in 2D [15], then in 3D [16], was extended, by a different approach, to the continuous problem [17, 18], then to other equations [19–24]. On the other hand, these miraculous layers raised many theoretical problem such as their well posedness and numerous papers were devoted to these studies [18, 25–30]. A new class of PML, called "convolutional PML" or CPML, which solved some problems of the wave absorption by PML, such as low frequency waves was introduced in [31, 32]. Finally, different solutions [33, 34] were proposed to the not obvious problem of stability of PML for linear elastodynamics in non-isotropic media [35].

6.1 Absorbing Boundary Conditions (ABC)

6.1.1 Transparent Condition of the Wave Equation

6.1.1.1 The 1D Case

Let us write the 1D homogeneous wave equation in $\mathbb{R}$:

$$\frac{\partial^2 u}{\partial t^2} - c^2 \frac{\partial^2 u}{\partial x^2} = 0, \ x \in \mathbb{R}. \tag{6.1}$$

By taking the Fourier transform in (x, t) of (6.1) ((ω, k) being the dual variables of (t, x)), we get the dispersion relation:

$$\omega^2 = c^2 k^2. \tag{6.2}$$

Now, we want to avoid the incoming waves for $x \leq 0$. The plane wave solution implies $k \leq 0$ and then:

$$\omega + c\, k = 0. \tag{6.3}$$

By using the inverse Fourier transform, (6.3) provides:

$$\frac{\partial u}{\partial t} + c \frac{\partial u}{\partial x} = 0. \tag{6.4}$$

Equation (6.4) is the *transparent condition* at $x = 0$ such that if u_T is the solution of

$$\begin{aligned} \frac{\partial^2 u_T}{\partial t^2} - c^2 \frac{\partial^2 u_T}{\partial x^2} &= 0, \ x < 0, \\ \frac{\partial u_T}{\partial t} + c \frac{\partial u_T}{\partial x} &= 0 \text{ at } x = 0, \end{aligned} \tag{6.5}$$

we have $u_T = u$ in R^{*-}.

6.1.1.2 Higher Dimensions

The extension of this process to higher dimensions is quite easy but unfortunately does not provide simple conditions. Let us compute the transparent condition in 2D. The Fourier transform in (x_1, x_2, t) of the 2D homogeneous wave equation in $\mathbb{R}^2$:

$$\frac{\partial^2 u}{\partial t^2} - c^2 \Delta u = 0, \; x \in \mathbb{R}^2 \tag{6.6}$$

reads

$$\omega^2 \overline{\overline{u}} = c^2 \, (k_1^2 + k_2^2) \overline{\overline{u}}, \tag{6.7}$$

where (ω, k_1, k_2) are the dual variables of (t, x_1, x_2) and $\overline{\overline{u}}$ is the Fourier transform in (x_1, x_2, t) of u.

We want to kill the reflected waves in the half-plane defined by $x_1 \le 0$, so $k_1 \le 0$ and (6.7) becomes:

$$k_1 \overline{\overline{u}} + \frac{\omega}{c} \sqrt{1 - \frac{k_2^2 \, c^2}{\omega^2}} \, \overline{\overline{u}} = 0. \tag{6.8}$$

Now, by taking the inverse Fourier transform in k_1 of (6.8), we obtain:

$$\frac{\mathrm{d}\overline{u}}{\mathrm{d}x_1} + \mathrm{i} \frac{\omega}{c} \sqrt{1 - \frac{k_2^2 \, c^2}{\omega^2}} \, \overline{u} = 0, \tag{6.9}$$

where $\overline{u}$ is the Fourier transform in (x_2, t) of u.

Equation (6.9) is the transparent condition in 2D. We have a similar result in 3D.

6.1.2 Construction of ABC for the Wave Equation

6.1.2.1 The 2D Case

Obviously, we cannot derive from (6.9) a partial differential condition at $x_1 = 0$. In order to get such a condition, we must take an approximation of (6.9). Let us set $z = c\, k_2/\omega$ in (6.9). We get:

$$\frac{\mathrm{d}\overline{u}}{\mathrm{d}x_1} + \mathrm{i} \frac{\omega}{c} \sqrt{1 - z^2} \, \overline{u} = 0. \tag{6.10}$$

The principle of ABC is to get an approximation of $\sqrt{1 - z^2}$ (we assume $0 \le z \le 1$) which provides a partial differential condition after inverse Fourier transform. The most natural approximation would be the Taylor expansion which leads to a polynomial in $k_2\, c/\omega$. Unfortunately, such approximations are not stable from the third-order [3, 12] and must be replaced by Padé approximations. Let us derive the first-order condition. We have:

$$\sqrt{1-z^2} = 1 + O(z^2). \tag{6.11}$$

Taking into account (6.11) and applying the inverse Fourier transform to (6.10), we obtain the first-order ABC:

$$\frac{\partial u}{\partial t} + c\frac{\partial u}{\partial x_1} = 0, \tag{6.12}$$

which reads as the 1D transparent condition but is not exact in 2D.

The (still stable) second-order Taylor approximation of (6.9) leads to

$$\sqrt{1-z^2} = 1 - \frac{1}{2}z^2 + O(z^4) = 1 - \frac{1}{2}\frac{c^2\,k_2^2}{\omega^2} + O\left(\frac{c^4\,k_2^4}{\omega^4}\right). \tag{6.13}$$

The inverse Fourier transform of (6.13) provides to the second-order ABC:

$$\frac{\partial^2 u}{\partial t^2} + c\frac{\partial^2 u}{\partial x_1 \partial t} - \frac{c^2}{2}\frac{\partial^2 u}{\partial x_2^2} = 0. \tag{6.14}$$

The general Padé approximations read:

$$\sqrt{1-z^2} \simeq 1 - \sum_{i=1}^{N} \frac{\beta_i z^2}{1-\alpha_i z^2}, \tag{6.15}$$

where

$$\alpha_i = \cos^2\left(\frac{i\pi}{2N+1}\right), \tag{6.16a}$$

$$\beta_i = \frac{2\sin^2\left(\dfrac{i\pi}{2N+1}\right)}{2N+1}. \tag{6.16b}$$

By reporting this expansion in (6.9), we get:

$$\frac{\mathrm{d}\overline{u}}{\mathrm{d}x_1} + \mathrm{i}\frac{\omega}{c}\left(1 - \sum_{i=1}^{N}\frac{\beta_i c^2\,k_2^2}{\omega^2 - \alpha_i c^2\,k_2^2}\right)\overline{u} = 0. \tag{6.17}$$

Following [5], one can derive the higher-order ABC. We set:

$$\overline{\phi}_i = \frac{c^2 k_2^2}{\omega^2 - \alpha_i c^2 k_2^2} \overline{u}, \tag{6.18}$$

so that (6.17) becomes

$$\frac{\mathrm{d}\overline{u}}{\mathrm{d}x_1} + \mathrm{i}\frac{\omega}{c}\overline{u} - \mathrm{i}\frac{\omega}{c}\sum_{i=1}^{N} \beta_i \overline{\phi}_i = 0. \tag{6.19}$$

After inverse Fourier transform, (6.18) and (6.19) finally provide:

$$\frac{\partial u}{\partial t} + c\frac{\partial u}{\partial x_1} - \sum_{i=1}^{N} \beta_i \frac{\partial \phi_i}{\partial t} = 0, \tag{6.20a}$$

$$\frac{\partial^2 \phi_i}{\partial t^2} - c^2 \alpha_i \frac{\partial^2 \phi_i}{\partial x_2^2} - c^2 \frac{\partial^2 u}{\partial x_2^2} = 0, \ i = 1 \dots N. \tag{6.20b}$$

Remarks:

1. In (6.20a)–(6.20b), ϕ_i is a function of t and x_2 only.
2. Similar conditions can be obtained for domains defined by $x_1 \geq a$ or $x_1 \leq a$, $a \in \mathbb{R}$ by changing the directions of the waves.
3. ABC are generally written on the boundaries of squared domains, so that we must write absorbing conditions at the corners. This difficult problem founds a complete solution for finite difference methods [5, 10] but is not obvious for finite element methods.

6.1.2.2 The 3D Case

Little literature has been devoted to the 3D case. Actually, one can derive the boundary conditions in the same way. Let us write the transparent condition which avoids incoming waves in the domain defined by $x_1 \leq 0$. As in (6.9), we have:

$$\frac{\mathrm{d}\overline{u}}{x_1} + \mathrm{i}\frac{\omega}{c}\sqrt{1 - \frac{k_2^2 c^2}{\omega^2} - \frac{k_3^2 c^2}{\omega^2}}\, \overline{u} = 0. \tag{6.21}$$

By setting $y = c\,k_2/\omega$ and $z = c\,k_3/\omega$, (6.17) reads:

$$\frac{\mathrm{d}\overline{u}}{\mathrm{d}x_1} + \mathrm{i}\frac{\omega}{c}\sqrt{1 - y^2 - z^2}\, \overline{u} = 0. \tag{6.22}$$

One can easily check that the first-order ABC is given by (6.12). Let us write the second-order derived from a Taylor expansion of (6.18). We have:

$$\begin{aligned}\sqrt{1-y^2-z^2} &= 1-\frac{1}{2}y^2-\frac{1}{2}z^2-\frac{1}{4}y^2z^2+O(y^4)+O(z^4)\\ &\simeq 1-\frac{1}{2}\frac{c^2\,k_2^2}{\omega^2}-\frac{1}{2}\frac{c^2\,k_3^2}{\omega^2}-\frac{1}{4}\frac{c^4\,k_2^2\,k_3^2}{\omega^4}.\end{aligned} \tag{6.23}$$

which provides, after inverse Fourier transform:

$$\frac{\partial^4 u}{\partial t^4}+c\frac{\partial^4 u}{\partial x_1\partial t^3}+\frac{c^2}{2}\frac{\partial^2}{\partial t^2}(\Delta_T u)+\frac{c^4}{4}\frac{\partial^4 u}{\partial x_2^2\partial x_3^2}=0, \tag{6.24}$$

with $\Delta_T=\partial^2/\partial x_2^2+\partial^2/\partial x_3^2$.

As we can see, (6.24) is much more complex than (6.14). A Padé approximation of (6.17) can be written as follows. We set:

$$\sqrt{1-y^2-z^2}\simeq 1-\sum_{i=1}^{N}\frac{\beta_i(y^2+z^2)}{1-\alpha_i(y^2+z^2)}, \tag{6.25}$$

which provides

$$\frac{\mathrm{d}\overline{u}}{\mathrm{d}x_1}+\mathrm{i}\frac{\omega}{c}\left(1-\sum_{i=1}^{N}\frac{\beta_i\,c^2\,(k_2^2+k_3^2)}{\omega^2-\alpha_i c^2\,(k_2^2+k_3^2)}\right)\overline{u}=0. \tag{6.26}$$

By setting:

$$\overline{\phi}_i=\frac{c^2\,(k_2^2+k_3^2)}{\omega^2-\alpha_i c^2\,(k_2^2+k_3^2)}\overline{u}, \tag{6.27}$$

we finally get, after inverse Fourier transform:

$$\frac{\partial u}{\partial t}+c\frac{\partial u}{\partial x_1}-\sum_{i=1}^{N}\beta_i\frac{\partial\phi_i}{\partial t}=0, \tag{6.28a}$$

$$\frac{\partial^2\phi_i}{\partial t^2}-c^2\alpha_i\,\Delta_T\phi_i-c^2\,\Delta_T u=0,\ i=1\ldots N. \tag{6.28b}$$

The stability is not proven. One could use methods described in [36] to prove it.

6.1.3 *Plane Wave Analysis*

The aim of this section is to evaluate the amplitude of the wave reflected by the absorbing boundary in 2D. The solution in the domain of $\mathbb{R}^2$ such that $x_1 \leq 0$ reads:

$$u = \mathrm{e}^{\mathrm{i}(\omega t - k_1 x_1 - k_2 x_2)} + R\,\mathrm{e}^{\mathrm{i}(\omega t + k_1 x_1 - k_2 x_2)} = u_1 + R\,u_2, \tag{6.29}$$

where R is the amplitude of the reflected wave.

In order to compute R, we write the value of u at $x_1 = 0$. The first-order ABC given by (6.12) provides:

$$\mathrm{i}\,\omega\,(u_1 + R\,u_2) - \mathrm{i}\,c\,k_1(u_1 - R\,u_2) = 0. \tag{6.30}$$

Since $u_1 = u_2$ for $x_1 = 0$, we get:

$$\omega\,(1 + R) - c\,k_1(1 - R) = 0. \tag{6.31}$$

By setting $k_1 = |\underline{k}|\sin\theta$ ($\underline{k} = (k_1, k_2)$), (6.31) leads to

$$R = \frac{\sin\theta - 1}{\sin\theta + 1}. \tag{6.32}$$

The second-order ABC provides:

$$-\,2\,\omega^2\,(1 + R) + 2\,c\,\omega\,k_1(1 - R) - c^2\,k_2^2\,(1 + R) = 0. \tag{6.33}$$

By setting $k_2 = |\underline{k}|\cos\theta$, we finally get:

$$R = \frac{2\,(\cos\theta - 1) - \sin^2\theta}{2\,(\cos\theta + 1) + \sin^2\theta}. \tag{6.34}$$

Let us now compute R for the ABC given by (6.20a)–(6.20b). By reporting (6.29) into (6.20a)–(6.20b), we get, since $u_1 = u_2$ at $x_1 = 0$:

$$\mathrm{i}\,\omega\,(1 + R)\,u_1 - \mathrm{i}\,c\,k_1(1 - R)\,u_1 - \sum_{i=1}^{N}\beta_i\frac{\partial\phi_i}{\partial t} = 0, \tag{6.35a}$$

$$\frac{\partial^2\phi_i}{\partial t^2} - c^2\alpha_i\frac{\partial^2\phi_i}{\partial x_2^2} + c^2\,k_2^2\,(1 + R)\,u_1 = 0,\; i = 1\ldots N. \tag{6.35b}$$

We apply the Fourier transform in x_2 and t to (6.35a)–(6.35b). By denoting $\overline{u}_1$ and $\overline{\phi}_i$ the Fourier transforms of u_1 and ϕ_i, we have:

$$\mathrm{i}\,\omega\,(1+R)\,\overline{u}_1 - \mathrm{i}\,c\,k_1(1-R)\,\overline{u}_1 - \mathrm{i}\,\omega\sum_{i=1}^{N}\beta_i\,\overline{\phi}_i = 0, \tag{6.36a}$$

$$-\,\omega^2\,\overline{\phi}_i + \alpha_i\,c^2\,k_2^2\,\overline{\phi}_i + c^2\,k_2^2\,(1+R)\,\overline{u}_1 = 0,\ \ i = 1\ldots N. \tag{6.36b}$$

We then eliminate $\overline{\phi}_i$ from (6.36a)–(6.36b) and we obtain the equation:

$$\mathrm{i}\,\omega\,(1+R) - \mathrm{i}\,c\,k_1(1-R) - \mathrm{i}\,\omega\,c^2\,k_2^2\,(1+R)\sum_{i=1}^{N}\frac{\beta_i}{\omega^2 - \alpha_i\,c^2\,k_2^2} = 0. \tag{6.37}$$

By dividing by $\mathrm{i}\,\omega$ and by taking into account the equality $c\,|\underline{k}|/\omega = 1$, we finally get:

$$R = \frac{\cos\theta + S - 1}{\cos\theta - S + 1}, \tag{6.38}$$

where

$$S = \sum_{i=1}^{N}\frac{\beta_i\,\sin^2\theta}{1-\alpha_i\,\sin^2\theta}.$$

In Fig. 6.1, we give the absolute values of the reflection coefficient $|R|$ defined by (6.38) for $N = 0$ to 6 versus the angle of incidence (defined by the normal to the boundary). One can notice the dramatical increase of the absorption when one shifts from the first-order to the second-order. Approximations greater than 6th-order seem to have no use in terms of accuracy. On the other hand, one can notice that waves close to the x_1-direction never vanish.

In Fig. 6.2, we compare the second-order approximations obtained by Taylor expansion and Padé approximations. One can see the better accuracy provided by the Padé approximation.

Remark: Of course, a similar analysis can be developed in 3D.

6.1.4 Finite Element Implementation

6.1.4.1 Continuous Finite Elements

Let us write the H^1 approximation of the wave equation defined by (6.6) in $\Omega = \mathbb{R}^{-*}\times\mathbb{R}^d$, $d = 0\ldots 2$:

$$\frac{\mathrm{d}^2}{\mathrm{d}t^2}\int_\Omega u\,v\,\mathrm{d}\underline{x} + c^2\int_\Omega \nabla u\cdot\nabla v\,\mathrm{d}\underline{x} - c^2\int_{\partial\Omega} v\frac{\partial u}{\partial x_1}\,\mathrm{d}\sigma = \int_\Omega f v\,\mathrm{d}\underline{x},\ \forall v\in H^1(\Omega). \tag{6.39}$$

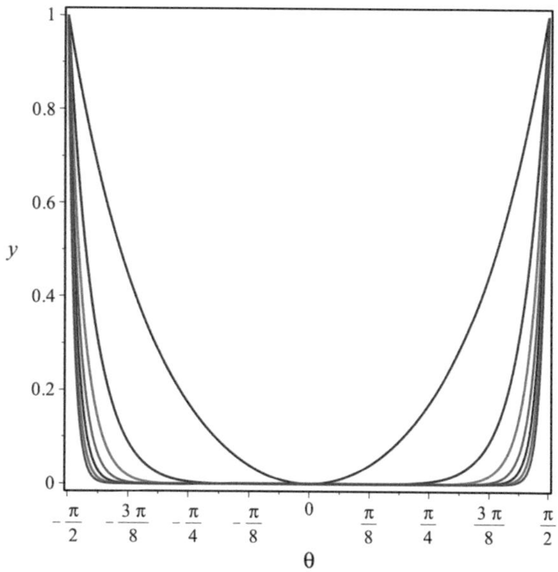

Fig. 6.1 $|R|$ defined by (6.38) for $N = 0$ (*upper curve*) to 6 (*lower curve*)

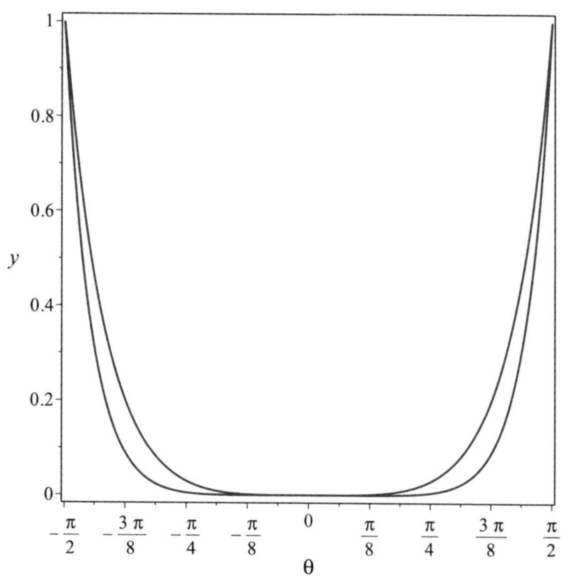

Fig. 6.2 $|R|$ defined by (6.34) (*upper curve*) and by (6.38) for $N = 1$ (*lower curve*)

By plugging the first-order ABC (for any value of d) in (6.39), we get:

$$\frac{\mathrm{d}^2}{\mathrm{d}t^2}\int_{\Omega} u\,v\,\mathrm{d}\underline{x} + c^2\int_{\Omega} \nabla u \cdot \nabla v\,\mathrm{d}\underline{x} + c\,\frac{\mathrm{d}}{\mathrm{d}t}\int_{\partial\Omega} u\,v\,\mathrm{d}\sigma = \int_{\Omega} fv\,\mathrm{d}\underline{x},\ \forall v \in H^1(\Omega). \tag{6.40}$$

So, for all v such that $\mathrm{supp}(v) \cap \partial\Omega \neq \emptyset$, we add a first-order term in time to our equation.

The second-order Taylor approximation is not obvious to implement neither in 2D nor in 3D. It requires us to derive in time the equations on the boundary of the domain. Let us implement the Padé approximations in 2D (3D being working in the same way). By inserting (6.20a) in (6.39), we obtain:

$$\begin{aligned}&\frac{\mathrm{d}^2}{\mathrm{d}t^2}\int_\Omega u\,v\,\mathrm{d}\underline{x} + c^2\int_\Omega \nabla u\cdot\nabla v\,\mathrm{d}\underline{x} + c\,\frac{\mathrm{d}}{\mathrm{d}t}\int_{\partial\Omega} u\,v\,\mathrm{d}\sigma\\ &-c\,\frac{\mathrm{d}}{\mathrm{d}t}\sum_{i=1}^{N}\int_{\partial\Omega}\beta_i\,\phi_i\,v\,\mathrm{d}\sigma = \int_\Omega fv\,\mathrm{d}\underline{x},\ \forall v\in H^1(\Omega).\end{aligned} \tag{6.41}$$

On the other hand, by multiplying (6.20b) by $w(x_2) \in H^1(\mathbb{R})$ and summing all over $\mathbb{R}$, we get the set of equations:

$$\begin{aligned}&\frac{\mathrm{d}^2}{\mathrm{d}t^2}\int_{\mathbb{R}}\phi_i\,w\,\mathrm{d}x_2 + c^2\,\alpha_i\int_{\mathbb{R}}\frac{\partial\phi_i}{\partial x_2}\,\frac{\partial w}{\partial x_2}\,\mathrm{d}x_2 + c^2\int_{\mathbb{R}}\frac{\partial u}{\partial x_2}\,\frac{\partial w}{\partial x_2}\,\mathrm{d}x_2 = 0,\\ &\forall w\in H^1(\mathbb{R}),\ i = 1\ldots N.\end{aligned} \tag{6.42}$$

The solution is then given by the system of $N+1$ equations on $\partial\Omega$ defined by (6.41) and (6.42).

6.1.4.2 Discontinuous Galerkin Methods

In order to implement ABC to DGM, we have to use a second-order formulation of the wave equation. For this purpose, one can either use the IPDG method described in Sect. 4.6 or use the second-order mixed formulation given in (3.30a)–(3.30b). We choose the second possibility. Let us rewrite the formulation in the homogeneous case. For sake of simplicity, we set $f = 0$.

$$\frac{\partial^2 u}{\partial t^2} - c^2\nabla\cdot\underline{v} = 0, \tag{6.43a}$$

$$\underline{v} - \nabla u = 0. \tag{6.43b}$$

The DG formulation of (6.43a)–(6.43b) reads:

Find $u_h(.,t) \in H^s(\mathscr{T}_h)$, $\underline{v}_h(.,t) \in \underline{H}^s_d(\mathscr{T}_h)$ such that $\forall K_\ell \in \mathscr{T}_h$ and $\forall \varphi_h \in H^s(\mathscr{T}_h)$, $\forall \underline{\psi}_h \in \underline{H}^s_d(\mathscr{T}_h)$,

$$\frac{\mathrm{d}^2}{\mathrm{d}^2 t}\int_{K_\ell} u\,\varphi_\ell\,\mathrm{d}\underline{x} = c^2\int_{K_\ell}\nabla\cdot\underline{v}_\ell\,\varphi_\ell\,\mathrm{d}\underline{x} + \frac{c^2}{2}\int_{\partial K_\ell}\underline{n}\cdot[\underline{v}_h]_{\partial K_\ell}^{K_\ell}\varphi_\ell\,\mathrm{d}\sigma, \tag{6.44a}$$

$$\int_{K_\ell} \underline{v}_\ell \cdot \underline{\psi}_\ell \,\mathrm{d}\underline{x} = \int_{K_\ell} \nabla u_\ell \cdot \underline{\psi}_\ell \,\mathrm{d}\underline{x} + \frac{1}{2}\int_{\partial K_\ell} \underline{n}[u]^{K_\ell}_{\partial K_\ell} \cdot \underline{\psi}_\ell \,\mathrm{d}\sigma, \tag{6.44b}$$

As said in Chap. 4, there is no jump on the boundary. So, for $\Gamma_\ell \subset \partial\Omega$ ($= \{x_1 = 0\}$), the right-hand side of (6.44a) reads:

$$\int_{K_\ell} \nabla \cdot \underline{v}_\ell\, \varphi_\ell \,\mathrm{d}\underline{x} = \int_{K_\ell} \underline{v}_\ell \cdot \nabla\varphi_\ell \,\mathrm{d}\underline{x} + \int_{\partial K_\ell \setminus \Gamma_\ell} \underline{n} \cdot \underline{v}_h \varphi_\ell \,\mathrm{d}\sigma + \int_{\Gamma_\ell} \underline{n} \cdot \underline{v}_h \varphi_\ell \,\mathrm{d}\sigma. \tag{6.45}$$

We directly address the Padé approximation given by (6.20a)–(6.20b), first order being easily deduced. By taking into account (6.43b) and by inserting (6.20a) in (6.45), we then obtain:

$$\begin{aligned}
&\frac{\mathrm{d}^2}{\mathrm{d}^2 t}\int_{K_\ell} u\,\varphi_\ell \,\mathrm{d}\underline{x} = -c^2 \int_{K_\ell} \underline{v}_\ell \cdot \nabla\varphi_\ell \,\mathrm{d}\underline{x} + c^2 \int_{\partial K_\ell \setminus \Gamma_\ell} \underline{n} \cdot \{\underline{v}_h\}^{K_\ell}_{\partial K_\ell} \varphi_\ell \,\mathrm{d}\sigma \\
&+c\,\frac{\mathrm{d}}{\mathrm{d}t}\int_{\Gamma_\ell} u\,\varphi_\ell \,\mathrm{d}\sigma - c\,\frac{\mathrm{d}}{\mathrm{d}t}\sum_{i=1}^{N}\int_{\Gamma_\ell} \beta_i\,\phi_i\,\varphi_\ell \,\mathrm{d}\sigma.
\end{aligned} \tag{6.46}$$

Of course, this equation is coupled with (6.44b) and the DG form of (6.20b) which reads:

$$\begin{aligned}
&\frac{\mathrm{d}^2}{\mathrm{d}t^2}\int_{\Gamma_\ell} \phi_i\, w \,\mathrm{d}x_2 + c^2\,\alpha_i \int_{\Gamma_\ell} \frac{\partial \phi_i}{\partial x_2}\frac{\partial w}{\partial x_2}\,\mathrm{d}x_2 + c^2 \int_{\Gamma_\ell} \frac{\partial u}{\partial x_2}\frac{\partial w}{\partial x_2}\,\mathrm{d}x_2 = 0, \\
&\forall w \in H^1(\Gamma_\ell),\ i = 1 \ldots N.
\end{aligned} \tag{6.47}$$

Remarks:

1. By using centered approximations in time $\left(i.e. \dfrac{\mathrm{d}^2 U}{\mathrm{d}t^2} \simeq \dfrac{U^{n+1} - 2U^n + U^{n-1}}{\Delta t^2}\right.$ and $\left.\dfrac{\mathrm{d}U}{\mathrm{d}t} \simeq \dfrac{U^{n+1} - U^{n-1}}{2\Delta t}\right)$, one can easily see that the $N+1$ equations derived from the Padé approximations are uncoupled. However, the centered approximation of the first-order derivative induces a substantial decrease of the CFL. On the other hand, an uncentered approximation $\left(i.e. \dfrac{\mathrm{d}U}{\mathrm{d}t} \simeq \dfrac{U^n - U^{n-1}}{\Delta t} 6\right)$ would produce parasitic reflections.
2. These ABC can be applied to the mixed formulation defined by (6.43a)–(6.43b) in the continuous case.
3. On a squarred domain, we have to replace $\mathbb{R}$ by a segment in (6.42). If we suppose that we have a Neumann condition at the corner, we get no boundary term. The corners then behave as punctuals scatterers which produce some reflected waves.
4. These implementations can easily be extended to the 3D case for continuous and discontinuous formulations.

6.1.5 The Maxwell's System

The derivation of the transparent condition for the Maxwell's system needs some complex computations (done here in a way different than [6]). We recall the Maxwell's system in an isotropic homogeneous medium:

$$\frac{\partial \underline{E}}{\partial t} - \nabla \times \underline{H} = 0, \tag{6.48a}$$

$$\frac{\partial \underline{H}}{\partial t} + \nabla \times \underline{E} = 0. \tag{6.48b}$$

We apply to (6.48a)–(6.48b) the Fourier transform in t, x_2, x_3 and we denote $\varepsilon = (\varepsilon_2, \varepsilon_3)$ with $\varepsilon_i = k_i/\omega$, $i = 2, 3$ and $\hat{\underline{E}}$, $\hat{\underline{H}}$ the Fourier transforms of $\underline{E}$, $\underline{H}$. By eliminating $\hat{E}_1$ and $\hat{H}_1$ by using the first components of (6.48a)–(6.48b) and by setting $\underline{v} = (\hat{E}_2, \hat{E}_3, \hat{H}_2, \hat{H}_3)^T$, we finally get the ODE system:

$$\frac{\partial \underline{v}}{\partial x_1} = \mathrm{i}\,\omega\, M(\varepsilon)\, \underline{v}, \tag{6.49}$$

where

$$M(\varepsilon) = \frac{1}{2}\begin{pmatrix} 0 & 0 & -\varepsilon_2\,\varepsilon_3 & {\varepsilon_2}^2 - 1 \\ 0 & 0 & 1 - {\varepsilon_3}^2 & \varepsilon_2\,\varepsilon_3 \\ \varepsilon_2\,\varepsilon_3 & 1 - {\varepsilon_2}^2 & 0 & 0 \\ {\varepsilon_3}^2 - 1 & -\varepsilon_2\,\varepsilon_3 & 0 & 0 \end{pmatrix}.$$

This matrix has two opposite double eigenvalues λ and $-\lambda$, where

$$\lambda = \sqrt{1 - {\varepsilon_2}^2 - {\varepsilon_3}^2} = \sqrt{1 - ||\varepsilon||^2}.$$

Now, let T denote the matrix of eigenvectors corresponding to these eigenvalues (stored in the above order). (6.49) can be then rewritten as

$$\frac{\partial \underline{w}}{\partial x_1} = \mathrm{i}\,\omega\, D\, \underline{w}, \tag{6.50}$$

where $\underline{w} = T^{-1}\underline{v}$, $D = \begin{pmatrix} \lambda & 0 & 0 & 0 \\ 0 & \lambda & 0 & 0 \\ 0 & 0 & -\lambda & 0 \\ 0 & 0 & 0 & -\lambda \end{pmatrix}$ and $T^{-1} = \dfrac{1}{2\,\lambda}\begin{pmatrix} \varepsilon_2\,\varepsilon_3 & 1 - {\varepsilon_2}^2 & 1 & 0 \\ {\varepsilon_3}^2 - 1 & -\varepsilon_2\,\varepsilon_3 & 0 & 1 \\ -\varepsilon_2\,\varepsilon_3 & {\varepsilon_2}^2 - 1 & 1 & 0 \\ 1 - {\varepsilon_3}^2 & \varepsilon_2\,\varepsilon_3 & 0 & 1 \end{pmatrix}$.

Obviously, $\underline{w} = (\exp(\mathrm{i}\,\omega\,\lambda\, x_1), \exp(\mathrm{i}\,\omega\,\lambda\, x_1), \exp(-\mathrm{i}\,\omega\,\lambda\, x_1), \exp(-\mathrm{i}\,\omega\,\lambda\, x_1))^T$. So, in order to kill reflected waves in the half-space defined by $x_1 \leq 0$, one must have $w_3 = w_4 = 0$, which provides, after taking into account the relation $\underline{w} = T^{-1}\underline{v}$, the transparent condition for the Maxwell's system:

$$\varepsilon_2\,\varepsilon_3\,\hat{E}_2 + (1-\varepsilon_2{}^2)\,\hat{E}_3 - \lambda\,\hat{H}_2, \tag{6.51a}$$

$$(1-\varepsilon_2{}^2)\,\hat{E}_2 + \varepsilon_2\,\varepsilon_3\,\hat{E}_3 - \lambda\,\hat{H}_3. \tag{6.51b}$$

By setting $\varepsilon_2 = \varepsilon_3 = 0$ in (6.51a)–(6.51b), we get the Silver–Müller condition (see Sect. 1.1.4.2) after applying the inverse Fourier transform. Higher-order ABC are not obvious to derive and, in particular, to implement in finite element or discontinuous Galerkin approximations. However, for DGM, one can use the fact that (6.48a)–(6.48b) is equivalent to a system of wave equations in a free homogenous space (as written in (1.14) and (1.15)) and apply the ABC defined by (6.28a)–(6.28b) to each component. For this purpose, on can replace (6.48a)–(6.48b) by (1.14) and (1.15) in the elements leaning on the boundary.

ABC for the linear elastodynamics system are very complex and not obvious to implement. Moreover, it is difficult to obtain stable ABC. For all these reasons, we do not describe them in this book.

6.2 Perfectly Matched Layers (PML)

6.2.1 Interpretation of the Method

In this section, we show the principle of PML and its non-reflecting property on a simple case. Let us consider the first-order system formulation of the wave equation in $\mathbb{R}^2$:

$$\frac{\partial u}{\partial t} = \nabla \cdot \underline{v}, \tag{6.52a}$$

$$\frac{\partial \underline{v}}{\partial t} = \nabla u. \tag{6.52b}$$

Bérenger's formulation of this system is based on the decomposition of u into $u_x + u_y$. With this notation, (6.52a)–(6.52b) read:

$$\frac{\partial u_x}{\partial t} = \frac{\partial v_1}{\partial x_1}, \tag{6.53a}$$

$$\frac{\partial u_y}{\partial t} = \frac{\partial v_2}{\partial x_2}, \tag{6.53b}$$

$$\frac{\partial \underline{v}}{\partial t} = \nabla u. \tag{6.53c}$$

In a second, step, one adds a damping term multiplied by a continuous function $\zeta(x_1)$ such that $\zeta(x_1) = 0$ for $x_1 \leq 0$ (the PML zone is then $\mathbb{R}^+ \times \mathbb{R}$). We obtain:

$$\frac{\partial u_x}{\partial t} + \zeta u_x = \frac{\partial v_1}{\partial x_1}, \tag{6.54a}$$

$$\frac{\partial u_y}{\partial t} = \frac{\partial v_2}{\partial x_2}, \tag{6.54b}$$

$$\frac{\partial v_1}{\partial t} + \zeta v_1 = \frac{\partial u}{\partial x_1}, \tag{6.54c}$$

$$\frac{\partial v_2}{\partial t} = \frac{\partial u}{\partial x_2}. \tag{6.54d}$$

By applying the Fourier transform in time to (6.54a)–(6.54d), we get:

$$(-\mathrm{i}\omega + \zeta)\hat{u}_x = \frac{\partial \hat{v}_1}{\partial x_1}, \tag{6.55a}$$

$$-\mathrm{i}\omega\hat{u}_y = \frac{\partial \hat{v}_2}{\partial x_2}, \tag{6.55b}$$

$$(-\mathrm{i}\omega + \zeta)\hat{v}_1 = \frac{\partial \hat{u}}{\partial x_1}, \tag{6.55c}$$

$$-\mathrm{i}\omega\hat{v}_2 = \frac{\partial \hat{u}}{\partial x_2}. \tag{6.55d}$$

After some algebra, (6.55a)–(6.55b) provide:

$$(\mathrm{i}\omega)^2(\hat{u}_x + \hat{u}_y) + \mathrm{i}\omega\frac{\partial \hat{v}_2}{\partial x_2} - \frac{(i\omega)^2}{-\mathrm{i}\omega + \zeta}\frac{\partial \hat{v}_1}{\partial x_1} = 0 \tag{6.56}$$

By plugging (6.55c)–(6.55d) into (6.56), we finally obtain the equation in $\hat{u}$:

$$\omega^2\hat{u} + \frac{\partial^2 \hat{u}}{\partial x_2^2} + \frac{\mathrm{i}\omega}{-\mathrm{i}\omega + \zeta}\frac{\partial}{\partial x_1}\left(\frac{\mathrm{i}\omega}{-\mathrm{i}\omega + \zeta}\frac{\partial \hat{u}}{\partial x_1}\right) = 0. \tag{6.57}$$

Now, by setting:

$$\tilde{x}_1 = x_1 + \frac{\mathrm{i}}{\omega}\int_0^{x_1} \zeta(s)\,\mathrm{d}s, \tag{6.58}$$

(6.57) takes the form of the Helmoltz equation:

$$\omega^2 \hat{u} + \frac{\partial^2 \hat{u}}{\partial \tilde{x}_1^2} + \frac{\partial^2 \hat{u}}{\partial x_2^2} = 0. \tag{6.59}$$

By applying the Fourier transform in x_2 to (6.59), we obtain the following ODE:

$$\frac{\partial^2 \check{u}}{\partial \tilde{x}_1^2} + (\omega^2 - k_2^2)\check{u} = 0. \tag{6.60}$$

By setting $k_1 = \sqrt{\omega^2 - k_2^2}$, we obtain the equation:

$$\frac{\partial^2 \check{u}}{\partial \tilde{x}_1^2} + k_1^2 \check{u} = 0, \tag{6.61}$$

whose solution can be written as

$$\check{u}(\tilde{x}_1) = A(k_2, \omega)\, \mathrm{e}^{\mathrm{i}k_1\tilde{x}_1} + B(k_2, \omega)\, \mathrm{e}^{-\mathrm{i}k_1\tilde{x}_1}. \tag{6.62}$$

After applying the inverse Fourier transform in x_2 to (6.62), we obtain:

$$\begin{aligned} \hat{u}(\tilde{x}_1, x_2, \omega) = {} & \frac{1}{2\pi}\int_{-\infty}^{+\infty} A(k_2, \omega)\, \mathrm{e}^{\mathrm{i}k_1\tilde{x}_1} \mathrm{e}^{\mathrm{i}k_2x_2}\, \mathrm{d}k_2 \\ & + \frac{1}{2\pi}\int_{-\infty}^{+\infty} B(k_2, \omega)\, \mathrm{e}^{-\mathrm{i}k_1\tilde{x}_1} \mathrm{e}^{\mathrm{i}k_2x_2}\, \mathrm{d}k_2. \end{aligned} \tag{6.63}$$

By replacing $\tilde{x}_1$ by its value, one can easily check that for both $k_2^2 < \omega^2$ or $k_2^2 > \omega^2$, the second integral of (6.63) provides an exponentially increasing wave which is not an acceptable solution of the problem. For this reason, we set $B = 0$ in the following. (6.62), then reads:

$$A(k_2, \omega) = \check{u}(0) = \int_{-\infty}^{+\infty} \hat{u}(0, x_2, \omega)\mathrm{e}^{-\mathrm{i}k_2x_2}\, \mathrm{d}k_2, \tag{6.64}$$

so that, finally

$$\hat{u}(\tilde{x}_1, x_2, \omega) = \frac{1}{2\pi}\int_{-\infty}^{+\infty} \check{u}(0)\, \mathrm{e}^{\mathrm{i}k_1\tilde{x}_1} \mathrm{e}^{\mathrm{i}k_2x_2}\, \mathrm{d}k_2, \tag{6.65}$$

with $\check{u}(0)$ defined in (6.64).

Now, by using (6.58), we can rewrite (6.65) in the following form:

$$\hat{u}(\tilde{x}_1, x_2, \omega) = \frac{1}{2\pi}\int_{-\infty}^{+\infty} \check{u}(0)\, \mathrm{e}^{\mathrm{i}(k_1x_1 + k_2x_2)}\, \mathrm{e}^{-k_1\theta(x_1)/\omega}\, \mathrm{d}k_2, \tag{6.66}$$

where $\theta(x_1) = \int_0^{x_1} \zeta(s)\, \mathrm{d}s$.

By using the dispersion relation of (6.52a)–(6.52b) (which can be written $\omega^2 = c^2\,(k_1^2 + k_2^2)$) and since $k_1 = |\underline{k}|\,\cos\varphi$, (6.66) can be rewritten as

$$\hat{u}(\tilde{x}_1, x_2, \omega) = \frac{1}{2\pi}\int_{-\infty}^{+\infty} \check{u}(0)\,\mathrm{e}^{\mathrm{i}(k_1 x_1 + k_2 x_2)}\,\mathrm{e}^{-\theta(x_1)\cos\varphi}\,\mathrm{d}k_2. \tag{6.67}$$

So, for $x_1 \le 0$, the solution is exactly that of the wave equation and, for $x_1 > 0$, it is an evanescent wave exponentially decreasing and no reflection appears at the interface. This shows that the PML absorbs the incident wave without any reflection and that is true for any value of (k_1, k_2), i.e. for any angle of incidence.

Remarks:

1. The change of variable (6.58) can be interpreted as the extension of the variable x_1 to a path of $\mathbb{C}$.
2. Equation (6.67) shows that the absorption depends on the angle of incidence but is a priori independent of the frequency.

6.2.2 *The Acoustics System*

6.2.2.1 The 2D Case

The above interpretation can be in fact used for constructing PML for a continuous hyperbolic system as described in [17, 18]. In this section, we apply this approach to the 2D acoustics system [37]. This system reads, in $\Omega = (\mathbb{R}^-)^2$:

$$\rho\,\frac{\partial u}{\partial t} = \nabla\cdot\underline{v}, \tag{6.68a}$$

$$\frac{\partial \underline{v}}{\partial t} = \mu\,\nabla u. \tag{6.68b}$$

After Fourier transform in time, we get:

$$-\,\mathrm{i}\omega\rho\hat{u} - \frac{\partial \hat{v}_1}{\partial x_1} - \frac{\partial \hat{v}_2}{\partial x_2} = 0, \tag{6.69a}$$

$$-\,\mathrm{i}\omega\mu^{-1}\hat{v}_1 - \frac{\partial \hat{u}}{\partial x_1} = 0 \tag{6.69b}$$

$$-\,\mathrm{i}\omega\mu^{-1}\hat{v}_2 - \frac{\partial \hat{u}}{\partial x_2} = 0. \tag{6.69c}$$

We now introduce the change of variables for $i = 1, 2$:

$$\tilde{x}_i = \begin{cases} x_i \text{ if } x_i < 0, \\ x_i + \dfrac{\mathrm{i}}{\omega} \displaystyle\int_0^{x_i} \zeta_i(s)\, \mathrm{d}s \text{ otherwise }, \end{cases} \tag{6.70}$$

where, as in the previous section, $\zeta_i(x_i) = 0$ in Ω.

Then, we extend (6.69a)–(6.69c) to $\mathbb{C}^2$ by using this change of variables. We obtain:

$$-\mathrm{i}\omega\rho\hat{u} - \frac{\partial \hat{v}_1}{\partial \tilde{x}_1} - \frac{\partial \hat{v}_2}{\partial \tilde{x}_2} = 0, \tag{6.71a}$$

$$-\mathrm{i}\omega\mu^{-1}\hat{v}_1 - \frac{\partial \hat{u}}{\partial \tilde{x}_1} = 0, \tag{6.71b}$$

$$-\mathrm{i}\omega\mu^{-1}\hat{v}_2 - \frac{\partial \hat{u}}{\partial \tilde{x}_2} = 0. \tag{6.71c}$$

Now, we have:

$$\mathrm{d}\hat{v}_1 = \frac{\partial \hat{v}_1}{\partial x_1}\mathrm{d}x_1 + \frac{\partial \hat{v}_1}{\partial x_2}\mathrm{d}x_2 = \frac{\partial \hat{v}_1}{\partial \tilde{x}_1}\mathrm{d}\tilde{x}_1 + \frac{\partial \hat{v}_1}{\partial \tilde{x}_2}\mathrm{d}\tilde{x}_2.$$

Since

$$\mathrm{d}\tilde{x}_i = \begin{cases} \mathrm{d}x_i \text{ if } x_i < 0, \\ \mathrm{d}x_i + \dfrac{\mathrm{i}\zeta_i}{\omega}\,\mathrm{d}x_i \text{ if } x_i \geq 0, \;\; i = 1, 2 \end{cases}$$

and $\zeta_i(s) = 0$ for $s < 0$, we get the following relation

$$\frac{\partial \hat{v}_i}{\partial x_i} = \left(1 + \frac{\mathrm{i}\zeta_i}{\omega}\right) \frac{\partial \hat{v}_i}{\partial \tilde{x}_i}.$$

Therefore,

$$\frac{\partial}{\partial \tilde{x}_i} = \frac{1}{1 + \mathrm{i}\zeta_i/\omega} \frac{\partial}{\partial x_i}. \tag{6.72}$$

So, (6.71a)–(6.71c) can be then rewritten as

$$-\mathrm{i}\omega\rho\hat{u} - \frac{1}{1 + \mathrm{i}\zeta_1/\omega} \frac{\partial \hat{v}_1}{\partial x_1} - \frac{1}{1 + \mathrm{i}\zeta_2/\omega} \frac{\partial \hat{v}_2}{\partial x_2} = 0, \tag{6.73a}$$

$$-\mathrm{i}\omega\mu^{-1}\hat{v}_1 - \frac{1}{1 + \mathrm{i}\zeta_1/\omega} \frac{\partial \hat{u}}{\partial x_1} = 0, \tag{6.73b}$$

$$- \mathrm{i}\omega\mu^{-1}\hat{v}_2 - \frac{1}{1+\mathrm{i}\zeta_2/\omega}\frac{\partial \hat{u}}{\partial x_2} = 0, \tag{6.73c}$$

which can be set into the following form

$$-\left(1+\frac{\mathrm{i}\zeta_1}{\omega}\right)\left(1+\frac{\mathrm{i}\zeta_2}{\omega}\right)\mathrm{i}\omega\rho\hat{u} - \left(1+\frac{\mathrm{i}\zeta_2}{\omega}\right)\frac{\partial \hat{v}_1}{\partial x_1} - \left(1+\frac{\mathrm{i}\zeta_1}{\omega}\right)\frac{\partial \hat{v}_2}{\partial x_2} = 0, \tag{6.74a}$$

$$-\left(1+\frac{\mathrm{i}\zeta_1}{\omega}\right)\mathrm{i}\omega\mu^{-1}\hat{v}_1 - \frac{\partial \hat{u}}{\partial x_1} = 0, \tag{6.74b}$$

$$-\left(1+\frac{\mathrm{i}\zeta_2}{\omega}\right)\mathrm{i}\omega\mu^{-1}\hat{v}_2 - \frac{\partial \hat{u}}{\partial x_2} = 0. \tag{6.74c}$$

By setting:

$$\tilde{v}_1 = \left(1+\frac{\mathrm{i}\zeta_2}{\omega}\right)\hat{v}_1, \tag{6.75a}$$

$$\tilde{v}_2 = \left(1+\frac{\mathrm{i}\zeta_1}{\omega}\right)\hat{v}_2, \tag{6.75b}$$

we get:

$$-\left(1+\frac{\mathrm{i}\zeta_1}{\omega}\right)\left(1+\frac{\mathrm{i}\zeta_2}{\omega}\right)\mathrm{i}\omega\rho\hat{u} - \frac{\partial \tilde{v}_1}{\partial x_1} - \frac{\partial \tilde{v}_2}{\partial x_2} = 0, \tag{6.76a}$$

$$-\frac{1+\mathrm{i}\zeta_1/\omega}{1+\mathrm{i}\zeta_2/\omega}\mathrm{i}\omega\mu^{-1}\tilde{v}_1 - \frac{\partial \hat{u}}{\partial x_1} = 0, \tag{6.76b}$$

$$-\frac{1+\mathrm{i}\zeta_2/\omega}{1+\mathrm{i}\zeta_1/\omega}\mathrm{i}\omega\mu^{-1}\tilde{v}_2 - \frac{\partial \hat{u}}{\partial x_2} = 0. \tag{6.76c}$$

We can now define the following Bérenger-like variables:

$$\hat{v}_1^* = \frac{1+\mathrm{i}\zeta_1/\omega}{1+\mathrm{i}\zeta_2/\omega}\tilde{v}_1, \tag{6.77a}$$

$$\hat{v}_2^* = \frac{1+\mathrm{i}\zeta_2/\omega}{1+\mathrm{i}\zeta_1/\omega}\tilde{v}_2, \tag{6.77b}$$

$$\hat{u}^* = (1+\mathrm{i}\zeta_1/\omega)(1+\mathrm{i}\zeta_2/\omega)\hat{u}, \tag{6.77c}$$

which can be rewritten as

$$(-\mathrm{i}\omega+\zeta_2)\hat{v}_1^* = (-\mathrm{i}\omega+\zeta_1)\tilde{v}_1, \tag{6.78a}$$

$$(-\mathrm{i}\omega + \zeta_1)\hat{v}_2^* = (-\mathrm{i}\omega + \zeta_2)\tilde{v}_2, \tag{6.78b}$$

$$-\omega^2\hat{u}^* = (\mathrm{i}\omega - \zeta_1)(\mathrm{i}\omega - \zeta_2)\hat{u} = (-\omega^2 - \mathrm{i}\omega(\zeta_1 + \zeta_2) + \zeta_1\zeta_2)\hat{u}. \tag{6.78c}$$

By combining (6.77a)–(6.77c) with (6.76a)–(6.76c) and (6.78a)–(6.78c), we get the following system:

$$-\mathrm{i}\omega\rho\hat{u}^* - \frac{\partial\tilde{v}_1}{\partial x_1} - \frac{\partial\tilde{v}_2}{\partial x_2} = 0, \tag{6.79a}$$

$$-\mathrm{i}\omega\mu^{-1}\hat{v}_1^* - \frac{\partial\hat{u}}{\partial x_1} = 0, \tag{6.79b}$$

$$-\mathrm{i}\omega\mu^{-1}\hat{v}_2^* - \frac{\partial\hat{u}}{\partial x_2} = 0, \tag{6.79c}$$

$$(-\mathrm{i}\omega + \zeta_2)\hat{v}_1^* = (-\mathrm{i}\omega + \zeta_1)\tilde{v}_1, \tag{6.79d}$$

$$(-\mathrm{i}\omega + \zeta_1)\hat{v}_2^* = (-\mathrm{i}\omega + \zeta_2)\tilde{v}_2, \tag{6.79e}$$

$$-\omega^2\hat{u}^* = (-\omega^2 - \mathrm{i}\omega(\zeta_1 + \zeta_2) + \zeta_1\zeta_2)\hat{u}. \tag{6.79f}$$

Now, by calling v_1 and v_2 the inverse Fourier transforms of $\tilde{v}_1$ and $\tilde{v}_2$ (which is legitimate since $\tilde{v}_1$ and $\tilde{v}_2$ coincide in Ω with $\hat{v}_1$ and $\hat{v}_2$), after inverse Fourier transform in time, we obtain the system:

$$\rho\,\frac{\partial u^*}{\partial t} = \nabla\cdot\underline{v}, \tag{6.80a}$$

$$\frac{\partial\underline{v}^*}{\partial t} = \mu\,\nabla u. \tag{6.80b}$$

$$\frac{\partial v_1}{\partial t} + \zeta_1 v_1 = \frac{\partial v_1^*}{\partial t} + \zeta_2 v_1^*, \tag{6.80c}$$

$$\frac{\partial v_2}{\partial t} + \zeta_2 v_2 = \frac{\partial v_2^*}{\partial t} + \zeta_1 v_2^*, \tag{6.80d}$$

$$\frac{\partial^2 u}{\partial t^2} + (\zeta_1 + \zeta_2)\frac{\partial u}{\partial t} + \zeta_1\zeta_2 u = \frac{\partial^2 u^*}{\partial t^2}. \tag{6.80e}$$

So, applying the PML is equivalent to solve the same system (6.80a)–(6.80c) as that defined in Ω plus three ODEs in u, v_1 and v_2 defined in (6.80d)–(6.80e). The practical definition of $\zeta_j, j = 1, 2$ can be of the form:

$$\zeta_j = \begin{cases} 0 & \text{if } x \leq 0, \\ \zeta \left(\frac{x}{a}\right)^2 & \text{otherwise,} \end{cases} \tag{6.81}$$

with

$$\zeta = \frac{3c_0}{2a} \log(R),$$

where a is the thickness of the layer, $c_0 = \sqrt{\mu/\rho}$ is the velocity in this layer and $R = 1000$.

In practice, PML are computed in a strip defined by $0 \leq x_i \leq L$. In general, L equal to 1 or 2 wavelengths provides an excellent absorption, i.e. such that the reflected waves are of the order of the dispersion of the numerical method. One can close the PML domain by a (partially) reflecting boundary condition. This additional domain is a drawback of PML since in 3D, this domain can use as many degrees of freedom as the physical (interior) domain. This can induce a substantial additional cost.

6.2.2.2 The 3D Case

Let the system (6.68a)–(6.68b) be set in the open set $\Omega = (\mathbb{R}^-)^3$. After applying the Fourier transform in time to this system, we transform it, as in 2D, by extending the change of variables defined in (6.70) to $i = 1 \ldots 3$. We get:

$$-\mathrm{i}\omega\rho\hat{u} - \frac{1}{1+\mathrm{i}\zeta_1/\omega}\frac{\partial \hat{v}_1}{\partial x_1} - \frac{1}{1+\mathrm{i}\zeta_2/\omega}\frac{\partial \hat{v}_2}{\partial x_2} - \frac{1}{1+\mathrm{i}\zeta_3/\omega}\frac{\partial \hat{v}_3}{\partial x_2} = 0, \tag{6.82a}$$

$$-\mathrm{i}\omega\mu^{-1}\hat{v}_i - \frac{1}{1+\mathrm{i}\zeta_i/\omega}\frac{\partial \hat{u}}{\partial x_i} = 0, \quad i = 1..3, \tag{6.82b}$$

which can be rewritten as

$$\begin{aligned} &-\left(1+\frac{\mathrm{i}\zeta_1}{\omega}\right)\left(1+\frac{\mathrm{i}\zeta_2}{\omega}\right)\left(1+\frac{\mathrm{i}\zeta_3}{\omega}\right)\mathrm{i}\omega\rho\hat{u} - \left(1+\frac{\mathrm{i}\zeta_2}{\omega}\right)\left(1+\frac{\mathrm{i}\zeta_3}{\omega}\right)\frac{\partial \hat{v}_1}{\partial x_1} \\ &-\left(1+\frac{\mathrm{i}\zeta_1}{\omega}\right)\left(1+\frac{\mathrm{i}\zeta_3}{\omega}\right)\frac{\partial \hat{v}_2}{\partial x_2} - \left(1+\frac{\mathrm{i}\zeta_1}{\omega}\right)\left(1+\frac{\mathrm{i}\zeta_2}{\omega}\right)\frac{\partial \hat{v}_3}{\partial x_3} = 0, \end{aligned} \tag{6.83a}$$

$$-\left(1+\frac{\mathrm{i}\zeta_i}{\omega}\right)\mathrm{i}\omega\mu^{-1}\hat{v}_i - \frac{\partial \hat{u}}{\partial x_i} = 0, \quad i = 1\ldots 3. \tag{6.83b}$$

By setting:

$$\tilde{v}_1 = \left(1+\frac{\mathrm{i}\zeta_2}{\omega}\right)\left(1+\frac{\mathrm{i}\zeta_3}{\omega}\right)\hat{v}_1, \tag{6.84a}$$

$$\tilde{v}_2 = \left(1+\frac{\mathrm{i}\zeta_1}{\omega}\right)\left(1+\frac{\mathrm{i}\zeta_3}{\omega}\right)\hat{v}_2, \tag{6.84b}$$

$$\tilde{v}_3 = \left(1 + \frac{\mathrm{i}\zeta_1}{\omega}\right)\left(1 + \frac{\mathrm{i}\zeta_2}{\omega}\right)\hat{v}_3, \tag{6.84c}$$

we obtain:

$$-\left(1 + \frac{\mathrm{i}\zeta_1}{\omega}\right)\left(1 + \frac{\mathrm{i}\zeta_2}{\omega}\right)\left(1 + \frac{\mathrm{i}\zeta_3}{\omega}\right)\mathrm{i}\omega\rho\hat{u} - \frac{\partial\tilde{v}_1}{\partial x_1} - \frac{\partial\tilde{v}_2}{\partial x_2} - \frac{\partial\tilde{v}_3}{\partial x_2} = 0, \tag{6.85a}$$

$$-\frac{1 + \mathrm{i}\zeta_1/\omega}{(1 + \mathrm{i}\zeta_2/\omega)(1 + \mathrm{i}\zeta_3/\omega)}\mathrm{i}\omega\mu^{-1}\tilde{v}_1 - \frac{\partial\hat{u}}{\partial x_1} = 0, \tag{6.85b}$$

$$-\frac{1 + \mathrm{i}\zeta_2/\omega}{(1 + \mathrm{i}\zeta_1/\omega)(1 + \mathrm{i}\zeta_3/\omega)}\mathrm{i}\omega\mu^{-1}\tilde{v}_2 - \frac{\partial\hat{u}}{\partial x_2} = 0. \tag{6.85c}$$

$$-\frac{1 + \mathrm{i}\zeta_3/\omega}{(1 + \mathrm{i}\zeta_1/\omega)(1 + \mathrm{i}\zeta_2/\omega)}\mathrm{i}\omega\mu^{-1}\tilde{v}_3 - \frac{\partial\hat{u}}{\partial x_3} = 0. \tag{6.85d}$$

We can now define the following Bérenger-like variables:

$$\hat{v}_1^* = \frac{1 + \mathrm{i}\zeta_1/\omega}{(1 + \mathrm{i}\zeta_2/\omega)(1 + \mathrm{i}\zeta_3/\omega)}\tilde{v}_1, \tag{6.86a}$$

$$\hat{v}_2^* = \frac{1 + \mathrm{i}\zeta_2/\omega}{(1 + \mathrm{i}\zeta_1/\omega)(1 + \mathrm{i}\zeta_3/\omega)}\tilde{v}_2, \tag{6.86b}$$

$$\hat{v}_3^* = \frac{1 + \mathrm{i}\zeta_3/\omega}{(1 + \mathrm{i}\zeta_1/\omega)(1 + \mathrm{i}\zeta_2/\omega)}\tilde{v}_3, \tag{6.86c}$$

$$\hat{u}^* = (1 + \mathrm{i}\zeta_1/\omega)(1 + \mathrm{i}\zeta_2/\omega)(1 + \mathrm{i}\zeta_3/\omega)\hat{u}, \tag{6.86d}$$

which can be rewritten as

$$(-\mathrm{i}\omega + \zeta_2)(-\mathrm{i}\omega + \zeta_3)\hat{v}_1^* = (-\mathrm{i}\omega + \zeta_1)\tilde{v}_1, \tag{6.87a}$$

$$(-\mathrm{i}\omega + \zeta_1)(-\mathrm{i}\omega + \zeta_3)\hat{v}_2^* = (-\mathrm{i}\omega + \zeta_2)\tilde{v}_2, \tag{6.87b}$$

$$(-\mathrm{i}\omega + \zeta_1)(-\mathrm{i}\omega + \zeta_2)\hat{v}_3^* = (-\mathrm{i}\omega + \zeta_3)\tilde{v}_3, \tag{6.87c}$$

$$\begin{aligned} &-\mathrm{i}\omega^3\hat{u}^* = (\mathrm{i}\omega - \zeta_1)(\mathrm{i}\omega - \zeta_2)(\mathrm{i}\omega - \zeta_3)\hat{u} \\ &= (-\mathrm{i}\omega^3 + \omega^2(\zeta_1 + \zeta_2 + \zeta_3) + \mathrm{i}\omega(\zeta_1\zeta_2 + \zeta_2\zeta_3 + \zeta_1\zeta_3) - \zeta_1\zeta_2\zeta_3)\hat{u}. \end{aligned} \tag{6.87d}$$

By combining (6.86a)–(6.86d) with (6.85a)–(6.85d) and (6.87a)–(6.87d) and by applying the inverse Fourier transform in time, we obtain the following system in the time domain:

$$\rho\frac{\partial u^*}{\partial t} = \nabla\cdot\underline{v}, \tag{6.88a}$$

$$\frac{\partial \underline{v}^*}{\partial t} = \mu \nabla u. \tag{6.88b}$$

$$\frac{\partial v_1}{\partial t} + \zeta_1 v_1 = \frac{\partial^2 v_1^*}{\partial t^2} + (\zeta_2 + \zeta_3)\frac{\partial v_1^*}{\partial t} + \zeta_2\zeta_3 v_1^*, \tag{6.88c}$$

$$\frac{\partial v_2}{\partial t} + \zeta_2 v_2 = \frac{\partial^2 v_2^*}{\partial t^2} + (\zeta_1 + \zeta_3)\frac{\partial v_2^*}{\partial t} + \zeta_1\zeta_3 v_1^*, \tag{6.88d}$$

$$\frac{\partial v_3}{\partial t} + \zeta_3 v_3 = \frac{\partial^2 v_3^*}{\partial t^2} + (\zeta_1 + \zeta_2)\frac{\partial v_3^*}{\partial t} + \zeta_1\zeta_2 v_3^*, \tag{6.88e}$$

$$\frac{\partial^3 u}{\partial t^3} - (\zeta_1 + \zeta_2 + \zeta_3)\frac{\partial^2 u}{\partial t^2} - (\zeta_1\zeta_2 + \zeta_2\zeta_3 + \zeta_1\zeta_3)\frac{\partial u}{\partial t} - \zeta_1\zeta_2\zeta_3 u = \frac{\partial^3 u^*}{\partial t^3}. \tag{6.88f}$$

So, the ODEs in v_1, v_2, v_3 remain of first-order with a second-order right-hand side whereas the ODE in u becomes of third-order instead of second-order in 2D. However, when one of the ζ_i is equal to 0, one can obtain a second-order system. Actually all the ζ_i are different from 0 only in the corners of a cubic domain. In the other cases, the equation is the same as in the 2D case. However, one can replace (6.88f) by a first-order system by adding intermediate unknowns.

6.2.3 The Maxwell's System

Let us consider the Maxwell's equations in an isotropic homogeneous medium:

$$\varepsilon \frac{\partial \underline{E}}{\partial t} - \nabla \times \underline{H} = 0, \tag{6.89a}$$

$$\mu \frac{\partial \underline{H}}{\partial t} + \nabla \times \underline{E} = 0. \tag{6.89b}$$

After Fourier transform and by using the change of variables defined in Sect. 6.2.2.2, (6.89a)–(6.89b) provide:

$$\mathrm{i}\omega\varepsilon\hat{E}_1 + \frac{1}{1 + \mathrm{i}\zeta_2/\omega}\frac{\partial \hat{H}_3}{\partial x_2} - \frac{1}{1 + \mathrm{i}\zeta_3/\omega}\frac{\partial \hat{H}_2}{\partial x_3} = 0, \tag{6.90a}$$

$$\mathrm{i}\omega\varepsilon\hat{E}_2 + \frac{1}{1 + \mathrm{i}\zeta_3/\omega}\frac{\partial \hat{H}_1}{\partial x_3} - \frac{1}{1 + \mathrm{i}\zeta_1/\omega}\frac{\partial \hat{H}_3}{\partial x_1} = 0, \tag{6.90b}$$

$$\mathrm{i}\omega\varepsilon\hat{E}_3 + \frac{1}{1 + \mathrm{i}\zeta_1/\omega}\frac{\partial \hat{H}_2}{\partial x_1} - \frac{1}{1 + \mathrm{i}\zeta_2/\omega}\frac{\partial \hat{H}_1}{\partial x_2} = 0, \tag{6.90c}$$

$$\mathrm{i}\omega\mu\hat{H}_1 - \frac{1}{1+\mathrm{i}\zeta_2/\omega}\frac{\partial\hat{E}_3}{\partial x_2} + \frac{1}{1+\mathrm{i}\zeta_3/\omega}\frac{\partial\hat{E}_2}{\partial x_3} = 0, \tag{6.90d}$$

$$\mathrm{i}\omega\mu\hat{H}_2 - \frac{1}{1+\mathrm{i}\zeta_3/\omega}\frac{\partial\hat{E}_1}{\partial x_3} + \frac{1}{1+\mathrm{i}\zeta_1/\omega}\frac{\partial\hat{E}_3}{\partial x_1} = 0, \tag{6.90e}$$

$$\mathrm{i}\omega\mu\hat{H}_3 - \frac{1}{1+\mathrm{i}\zeta_1/\omega}\frac{\partial\hat{E}_2}{\partial x_1} + \frac{1}{1+\mathrm{i}\zeta_2/\omega}\frac{\partial\hat{E}_1}{\partial x_2} = 0. \tag{6.90f}$$

Equation (6.90a)–(6.90f) is rewritten as

$$\left(1+\frac{\mathrm{i}\zeta_2}{\omega}\right)\left(1+\frac{\mathrm{i}\zeta_3}{\omega}\right)\mathrm{i}\omega\varepsilon\hat{E}_1 - \left(1+\frac{\mathrm{i}\zeta_3}{\omega}\right)\frac{\partial\hat{H}_3}{\partial x_2} + \left(1+\frac{\mathrm{i}\zeta_2}{\omega}\right)\frac{\partial\hat{H}_2}{\partial x_3} = 0, \tag{6.91a}$$

$$\left(1+\frac{\mathrm{i}\zeta_1}{\omega}\right)\left(1+\frac{\mathrm{i}\zeta_3}{\omega}\right)\mathrm{i}\omega\varepsilon\hat{E}_2 - \left(1+\frac{\mathrm{i}\zeta_1}{\omega}\right)\frac{\partial\hat{H}_1}{\partial x_3} + \left(1+\frac{\mathrm{i}\zeta_3}{\omega}\right)\frac{\partial\hat{H}_3}{\partial x_1} = 0, \tag{6.91b}$$

$$\left(1+\frac{\mathrm{i}\zeta_1}{\omega}\right)\left(1+\frac{\mathrm{i}\zeta_2}{\omega}\right)\mathrm{i}\omega\varepsilon\hat{E}_3 - \left(1+\frac{\mathrm{i}\zeta_2}{\omega}\right)\frac{\partial\hat{H}_2}{\partial x_1} + \left(1+\frac{\mathrm{i}\zeta_1}{\omega}\right)\frac{\partial\hat{H}_1}{\partial x_2} = 0, \tag{6.91c}$$

$$\left(1+\frac{\mathrm{i}\zeta_2}{\omega}\right)\left(1+\frac{\mathrm{i}\zeta_3}{\omega}\right)\mathrm{i}\omega\varepsilon\hat{H}_1 + \left(1+\frac{\mathrm{i}\zeta_3}{\omega}\right)\frac{\partial\hat{E}_3}{\partial x_2} - \left(1+\frac{\mathrm{i}\zeta_2}{\omega}\right)\frac{\partial\hat{E}_2}{\partial x_3} = 0, \tag{6.91d}$$

$$\left(1+\frac{\mathrm{i}\zeta_1}{\omega}\right)\left(1+\frac{\mathrm{i}\zeta_3}{\omega}\right)\mathrm{i}\omega\varepsilon\hat{H}_2 + \left(1+\frac{\mathrm{i}\zeta_1}{\omega}\right)\frac{\partial\hat{E}_1}{\partial x_3} - \left(1+\frac{\mathrm{i}\zeta_3}{\omega}\right)\frac{\partial\hat{E}_3}{\partial x_1} = 0, \tag{6.91e}$$

$$\left(1+\frac{\mathrm{i}\zeta_1}{\omega}\right)\left(1+\frac{\mathrm{i}\zeta_2}{\omega}\right)\mathrm{i}\omega\varepsilon\hat{H}_3 + \left(1+\frac{\mathrm{i}\zeta_2}{\omega}\right)\frac{\partial\hat{E}_2}{\partial x_1} - \left(1+\frac{\mathrm{i}\zeta_1}{\omega}\right)\frac{\partial\hat{E}_1}{\partial x_2} = 0. \tag{6.91f}$$

Now, by setting:

$$\tilde{E}_1 = \left(1+\frac{\mathrm{i}\zeta_2}{\omega}\right)\left(1+\frac{\mathrm{i}\zeta_3}{\omega}\right)\hat{E}_1, \quad \tilde{H}_1 = \left(1+\frac{\mathrm{i}\zeta_2}{\omega}\right)\left(1+\frac{\mathrm{i}\zeta_3}{\omega}\right)\hat{H}_1, \tag{6.92a}$$

$$\tilde{E}_2 = \left(1+\frac{\mathrm{i}\zeta_1}{\omega}\right)\left(1+\frac{\mathrm{i}\zeta_3}{\omega}\right)\hat{E}_2, \quad \tilde{H}_2 = \left(1+\frac{\mathrm{i}\zeta_1}{\omega}\right)\left(1+\frac{\mathrm{i}\zeta_3}{\omega}\right)\hat{H}_2, \tag{6.92b}$$

$$\tilde{E}_3 = \left(1+\frac{\mathrm{i}\zeta_1}{\omega}\right)\left(1+\frac{\mathrm{i}\zeta_2}{\omega}\right)\hat{E}_3, \quad \tilde{H}_3 = \left(1+\frac{\mathrm{i}\zeta_1}{\omega}\right)\left(1+\frac{\mathrm{i}\zeta_2}{\omega}\right)\hat{H}_3, \tag{6.92c}$$

Equation (6.91a)–(6.91f) reads:

$$-\left(1+\frac{\mathrm{i}\zeta_1}{\omega}\right)\left(1+\frac{\mathrm{i}\zeta_2}{\omega}\right)\left(1+\frac{\mathrm{i}\zeta_3}{\omega}\right)\mathrm{i}\omega\varepsilon\underline{\hat{E}} - \nabla\times\underline{\tilde{H}} = 0, \tag{6.93a}$$

$$-\left(1+\frac{\mathrm{i}\zeta_1}{\omega}\right)\left(1+\frac{\mathrm{i}\zeta_2}{\omega}\right)\left(1+\frac{\mathrm{i}\zeta_3}{\omega}\right)\mathrm{i}\omega\varepsilon\underline{\hat{H}}+\nabla\times\underline{\tilde{E}}=0, \tag{6.93b}$$

with $\underline{\tilde{H}}=(\tilde{H}_1,\tilde{H}_2,\tilde{H}_3)^T$ and $\underline{\tilde{E}}=(\tilde{E}_1,\tilde{E}_2,\tilde{E}_3)^T$.

Here, the vector-valued Bérenger-like variables read:

$$\underline{E}^*=\left(1+\frac{\mathrm{i}\zeta_1}{\omega}\right)\left(1+\frac{\mathrm{i}\zeta_2}{\omega}\right)\left(1+\frac{\mathrm{i}\zeta_3}{\omega}\right)\underline{\hat{E}}, \tag{6.94a}$$

$$\underline{H}^*=\left(1+\frac{\mathrm{i}\zeta_1}{\omega}\right)\left(1+\frac{\mathrm{i}\zeta_2}{\omega}\right)\left(1+\frac{\mathrm{i}\zeta_3}{\omega}\right)\underline{\hat{H}}. \tag{6.94b}$$

As in Sect. 6.2.2.2, by combining (6.93a)–(6.93b) and (6.94a)–(6.94b) and applying the inverse Fourier transform in time, we finally have:

$$\varepsilon\frac{\partial\underline{E}^*}{\partial t}-\nabla\times\underline{H}=0, \tag{6.95a}$$

$$\mu\frac{\partial\underline{H}^*}{\partial t}+\nabla\times\underline{E}=0, \tag{6.95b}$$

$$\begin{aligned}\frac{\partial^3\underline{E}}{\partial t^3}&-(\zeta_1+\zeta_2+\zeta_3)\frac{\partial^2\underline{E}}{\partial t^2}\\&-(\zeta_1\zeta_2+\zeta_2\zeta_3+\zeta_1\zeta_3)\frac{\partial\underline{E}}{\partial t}-\zeta_1\zeta_2\zeta_3\underline{E}-\frac{\partial^3\underline{E}^*}{\partial t^3}=0,\end{aligned} \tag{6.95c}$$

$$\begin{aligned}\frac{\partial^3\underline{H}}{\partial t^3}&-(\zeta_1+\zeta_2+\zeta_3)\frac{\partial^2\underline{H}}{\partial t^2}\\&-(\zeta_1\zeta_2+\zeta_2\zeta_3+\zeta_1\zeta_3)\frac{\partial\underline{E}}{\partial t}-\zeta_1\zeta_2\zeta_3\underline{H}-\frac{\partial^3\underline{H}^*}{\partial t^3}=0,\end{aligned} \tag{6.95d}$$

6.2.4 The Linear Elastodynamics System

By applying the Fourier transform in time and the extension of variables defined in (6.70) to the elastics system defined in (3.260a)–(3.260c) in 2D (the 3D case can be deduced from the acoustics system), with $F_i=0$, $i=1,2$ we obtain [19]:

$$\begin{aligned}-\rho\mathrm{i}\omega\hat{v}_1&-\frac{1}{1+\mathrm{i}\zeta_1/\omega}\left(\frac{\partial\hat{\gamma}_{11,1}}{\partial x_1}+\frac{\partial\hat{\gamma}_{12,1}}{\partial x_1}\right)\\&-\frac{1}{1+\mathrm{i}\zeta_2/\omega}\left(\frac{\partial\hat{\gamma}_{11,2}}{\partial x_2}+\frac{\partial\hat{\gamma}_{22,2}}{\partial x_2}\right)=0,\end{aligned} \tag{6.96a}$$

$$-\rho i\omega\hat{v}_2 - \frac{1}{1+i\zeta_1/\omega}\left(\frac{\partial\hat{\gamma}_{21,1}}{\partial x_1}+\frac{\partial\hat{\gamma}_{22,1}}{\partial x_1}\right) - \frac{1}{1+i\zeta_2/\omega}\left(\frac{\partial\hat{\gamma}_{21,2}}{\partial x_2}+\frac{\partial\hat{\gamma}_{22,2}}{\partial x_2}\right)=0, \tag{6.96b}$$

$$-i\omega\underline{\hat{\gamma}}_1 = \left(\frac{1}{1+i\zeta_1/\omega}\frac{\partial\hat{v}_1}{\partial x_1}, \frac{1}{1+i\zeta_2/\omega}\frac{\partial\hat{v}_1}{\partial x_2}\right)^T, \tag{6.96c}$$

$$-i\omega\underline{\hat{\gamma}}_2 = \left(\frac{1}{1+i\zeta_1/\omega}\frac{\partial\hat{v}_2}{\partial x_1}, \frac{1}{1+i\zeta_2/\omega}\frac{\partial\hat{v}_2}{\partial x_2}\right)^T, \tag{6.96d}$$

$$\underline{\hat{\gamma}}_{ij}(\underline{x},\omega) = A_{ij}(\underline{x})\,\underline{\hat{\gamma}}_i(\underline{x},\omega). \tag{6.96e}$$

where $\hat{\gamma}_{ij,k}$ is the kth component of $\underline{\hat{\gamma}}_{ij}$.

Equation (6.96a)–(6.97d) can be rewritten as

$$-\left(1+\frac{i\zeta_1}{\omega}\right)\left(1+\frac{i\zeta_2}{\omega}\right)\rho i\omega\hat{v}_1 - \left(1+\frac{i\zeta_2}{\omega}\right)\left(\frac{\partial\hat{\gamma}_{11,1}}{\partial x_1}+\frac{\partial\hat{\gamma}_{12,1}}{\partial x_1}\right) - \left(1+\frac{i\zeta_1}{\omega}\right)\left(\frac{\partial\hat{\gamma}_{11,2}}{\partial x_2}+\frac{\partial\hat{\gamma}_{22,2}}{\partial x_2}\right)=0, \tag{6.97a}$$

$$-\left(1+\frac{i\zeta_1}{\omega}\right)\left(1+\frac{i\zeta_2}{\omega}\right)\rho i\omega\hat{v}_2 - \left(1+\frac{i\zeta_2}{\omega}\right)\left(\frac{\partial\hat{\gamma}_{21,1}}{\partial x_1}+\frac{\partial\hat{\gamma}_{22,1}}{\partial x_1}\right) - \left(1+\frac{i\zeta_1}{\omega}\right)\left(\frac{\partial\hat{\gamma}_{21,2}}{\partial x_2}+\frac{\partial\hat{\gamma}_{22,2}}{\partial x_2}\right)=0, \tag{6.97b}$$

$$-M\,i\omega\underline{\hat{\gamma}}_1 = \left(\frac{\partial\hat{v}_1}{\partial x_1}, \frac{\partial\hat{v}_1}{\partial x_2}\right)^T, \tag{6.97c}$$

$$-M\,i\omega\underline{\hat{\gamma}}_2 = \left(\frac{\partial\hat{v}_2}{\partial x_1}, \frac{\partial\hat{v}_2}{\partial x_2}\right)^T, \tag{6.97d}$$

where $M = \begin{pmatrix} 1+\frac{i\zeta_1}{\omega} & 0 \\ 0 & 1+\frac{i\zeta_2}{\omega} \end{pmatrix}$.

Now, by setting:

$$\gamma^*_{ij,1} = \left(1+\frac{i\zeta_2}{\omega}\right)\hat{\gamma}_{ij,1},\ \gamma^*_{ij,2} = \left(1+\frac{i\zeta_1}{\omega}\right)\hat{\gamma}_{ij,2},\ \forall i=1,2,\ \forall j=1,2,$$

$$\underline{\hat{v}}^* = \left(1+\frac{i\zeta_1}{\omega}\right)\left(1+\frac{i\zeta_2}{\omega}\right)\underline{\hat{v}},\ \underline{\gamma}^*_i = M\,\underline{\hat{\gamma}}_i,\ \forall i=1,2,$$

and by applying the inverse Fourier transform in time, we finally get:

$$\rho(\underline{x})\frac{\partial v_i^*}{\partial t} - \sum_{j=1}^{d} \nabla \cdot \underline{\gamma}_{ij}^* = 0, \quad \forall i = 1\ldots 2, \tag{6.98a}$$

$$\frac{\partial \underline{\gamma}_i^*}{\partial t} = \nabla v_i(\underline{x}, t), \quad \forall i = 1\ldots 2, \tag{6.98b}$$

$$\frac{\partial^2 \underline{v}}{\partial t^2} + (\zeta_1 + \zeta_2)\frac{\partial \underline{v}}{\partial t} + \zeta_1\,\zeta_2 \underline{v} = \frac{\partial^2 \underline{v}^*}{\partial t^2}, \tag{6.98c}$$

$$\frac{\partial \underline{\gamma}_i}{\partial t} + Z_1 \underline{\gamma}_i = \frac{\partial \underline{\gamma}_i^*}{\partial t}, \quad \forall i = 1\ldots 2, \tag{6.98d}$$

$$\underline{\gamma}_{ij}(\underline{x}, t) = A_{ij}(\underline{x})\underline{\gamma}_i(\underline{x}, t), \quad \forall i, j = 1\ldots d, \tag{6.98e}$$

$$\frac{\partial \underline{\gamma}_{ij}^*}{\partial t} = \frac{\partial \underline{\gamma}_{ij}}{\partial t} + Z_2 \underline{\gamma}_{ij}, \quad \forall i, j = 1\ldots 2, \tag{6.98f}$$

where $Z_1 = \begin{pmatrix} \zeta_1 & 0 \\ 0 & \zeta_2 \end{pmatrix}$, $Z_2 = \begin{pmatrix} \zeta_2 & 0 \\ 0 & \zeta_1 \end{pmatrix}$.

6.2.5 Modified PML

6.2.5.1 M-PML

The multiaxial PML (M-PML), introduced in [33], solves the important problem of instability of the PML in anisotropic media pointed out in [35], produced by the non uniform concavity of the slowness curves in non-isotropic media. In the following, we just indicate the principle of the method. More details can be found in [33]. Let us consider d-dimensional PML in the x_1-direction. Classical PML, defined by the change of variable:

$$\tilde{x}_1 = x_1 + \frac{\mathrm{i}}{\omega}\int_0^{x_1} \zeta(s)\,\mathrm{d}s = x_1 + \frac{\mathrm{i}\,\theta(x_1)}{\omega}, \tag{6.99}$$

are then replaced by

$$\tilde{x}_i = x_i + \frac{\mathrm{i}\,\alpha_i(x_1)}{\omega} x_i, \quad i = 1\ldots d, \tag{6.100}$$

where $\alpha_1 = \theta/x_1$, $\alpha_i = c_i\theta$ for $2 < i \le d$, c_i being real constants.

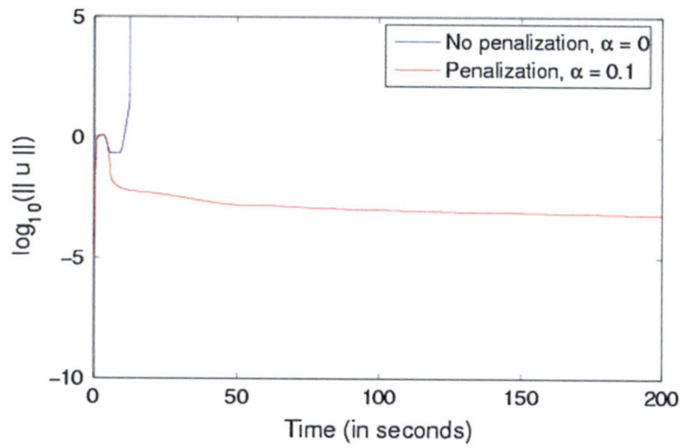

Fig. 6.3 Effect of a dissipative term on PML for a DGM approximation

This change of variables provide the following relations for the derivatives:

$$\frac{\partial}{\partial \tilde{x}_i} = \frac{1}{1 + \mathrm{i}\, c_i\, \zeta(x_1)/\omega} \frac{\partial}{\partial x_i}, \quad i = 1 \ldots d, \tag{6.101}$$

with $c_1 = 1$.

When M-PML are in all directions, (6.101) is replaced by

$$\frac{\partial}{\partial \tilde{x}_i} = \frac{1}{1 + \mathrm{i}\, \zeta^{(i)}(x_1, x_2, x_3)/\omega} \frac{\partial}{\partial x_i}, \quad i = 1 \ldots d, \tag{6.102}$$

where $\zeta^{(i)}(x_1, x_2, x_3) = \sum_{j=1}^{d} c_{ij}\, \zeta_i(x_j)$, c_{ij} being real constants and $c_{ii} = 1$.

Remarks:

1. In [33], stability is proven for isotropic media but numerical experiments show stable results in heterogeneous media.
2. M-PML were not tested for anisotropic Maxwell's equations which suffer of the same kind of instabilities, but should be also efficient in this case.

6.2.5.2 C-PML

Another variation of PML is given by C-PML [31, 32] in which (6.99) is replaced by

$$\tilde{x}_1 = x_1 + \frac{\mathrm{i}\,\theta(x_1)}{\omega - \mathrm{i}\,\lambda}, \tag{6.103}$$

λ being a real positive constant. This change of variables seems to provide a better absorption of low frequency components.

Remark: Stable continuous PML sometimes become unstable after approximation. Adding dissipative jumps can stabilize such an approximation (Fig. 6.3).

References

1. Cerjan, C., Kosloff, D., Kosloff, R., Reshef, M.: A nonreflecting boundary condition for discrete acoustic and elastic wave equations. Geophysics **50**(4), 705–708 (1985)
2. Bayliss, A., Turkel, E.: Radiation boundary conditions for wave-like equations. Commun. Pure Appl. Math. **33**, 707–725 (1980)
3. Engquist, B., Majda, A.: Absorbing boundary conditions for the numerical simulation of waves. Math. Comput. **31**(139), 629–651 (1977)
4. Engquist, B., Majda, A.: Radiation boundary conditions for acoustic and elastic wave calculations. Commun. Pure Appl. Anal. **32**(3), 313–357 (1979)
5. Collino, F.: High order absorbing boundary conditions for wave propagation models: straight line boundary and corner cases. Second International Conference on Mathematical and Numerical Aspects of Wave Propagation (Newark, DE, 1993), pp. 161–171, SIAM, Philadelphia, PA (1993)
6. Bendali, A., Halpern, L.: Conditions aux limites absorbantes pour le système de Maxwell dans le vide en dimension trois d'espace. C.R. Acad. Sci. Paris Ser. I Math. **307**(20):1011-1013 (1988)
7. Hall, W.F., Kabakian, A.V.: A sequence of absorbing boundary conditions for Maxwell's equations. J. Comput. Phys. **194**(1), 140–155 (2004)
8. Higdon, R.L.: Absorbing boundary conditions for acoustic and elastic waves in stratified media. J. Comput. Phys **101**(2), 386–418 (1992)
9. Sochacki, J.: Absorbing boundary conditions for the elastic wave equations. Appl. Math. Comput. **28**(1), 1–14 (1988)
10. Bamberger, A., Joly, P., Roberts, J.E.: Second-order absorbing boundary conditions for the wave equation: a solution for the corner problem. SIAM J. Numer. Anal. **27**(2), 323–352 (1990)
11. Halpern, L., Rauch, J.: Error analysis for absorbing boundary conditions. Numer. Math. **51**(4), 459–467 (1987)
12. Trefethen, L.N., Halpern, L.: Well-posedness of one-way wave equations and absorbing boundary conditions. Math. Comp. **47**(176), 421–435 (1986)
13. Givoli, D.: High-order local non-reflecting boundary conditions: a review. Wave Motion **39**(4), 319–326 (2004)
14. Yee, K.: Numerical solutions of initial boundary value problems involving Maxwell's equations in isotropic media. IEEE Trans. Antennas Propag. **14**(3), 302–307 (1966)
15. Bérenger, J.-P.: A perfectly matched layer for the absorption of electromagnetic waves. J. Comput. Phys. **114**(2), 185–200 (1994)
16. Bérenger, J.-P.: Three-dimensional perfectly matched layer for the absorption of electromagnetic waves. J. Comput. Phys. **127**(2), 363–379 (1996)
17. Rappaport, C.: Perfectly matched absorbing conditions based on anisotropic lossy mapping of space. IEEE Microw. Guided Wave Lett. **5**(3), 90–92 (1995)
18. Zhao, L., Cangellaris, A.C.: GT-PML: Generalized theory of perfectly matched layers and its application to reflectionless truncation of finite-difference time-domain grids. IEEE Trans. Microw. Theory Techn. **44**(12), 2555–2563 (1996)
19. Cohen, G., Fauqueux, S.: Mixed spectral finite elements for the linear elasticity system in unbounded domains. SIAM J. Sci. Comput. **26**(3), 864–884 (2005)
20. Cohen, G., Imperiale, S.: Perfectly matched layer with mixed spectral elements for the propagation of linearized water waves. Commun. Comput. Phys. **11**(2), 285–302 (2012)
21. Collino, F., Tsogka, C.: Application of the perfectly matched absorbing layer model to the linear elastodynamic problem in anisotropic heterogeneous media. Geophysics **66**(1), 294–307 (2001)

22. Hesthaven, J.S.: On the analysis and construction of perfectly matched layers for the linearized Euler equations. J. Comput. Phys. **142**(1), 129–147 (1998)
23. Nataf, F.: A new construction of perfectly matched layers for the linearized Euler equations. J. Comput. Phys. **214**(2), 757–772 (2006)
24. Tam, C.K.W., Auriault, L., Cambuli, F.: Perfectly matched layer as an absorbing boundary condition for the linearized Euler equations in open and ducted domains. J. Comput. Phys. **144**(1), 213–234 (1998)
25. Abarbanel, S., Gottlieb, D.: A mathematical analysis of the PML method. J. Comput. Phys. **134**(2), 357–363 (1997)
26. Abarbanel, S., Gottlieb, D., Hesthaven, J.S.: Well-posed perfectly matched layers for advective acoustics. J. Comput. Phys. **154**(2), 266–283 (1999)
27. Collino, F., Monk, P.: The perfectly matched layer in curvilinear coordinates. SIAM J. Sci. Comput. **19**(6), 2061–2090 (1998)
28. Halpern, L., Petit-Bergez, S., Rauch, J.: The analysis of matched layers. Conflu. Math. **3**(2), 159–236 (2011)
29. Lassas, M., Somersalo, E.: On the existence and convergence of the solution of PML equations. Computing **60**(3), 229–241 (1998)
30. Mittra, R., Pekel, U., Veihl, J.: A theoretical and numerical study of Berenger's perfectly matched layer (PML) concept for mesh truncation in time and frequency domains. Approximations and numerical methods for the solution of Maxwell's equations, pp. 1–19. Oxford University Press, Oxford (1995)
31. Kuzuoglu, M., Mittra, R.: Frequency dependence of the constitutive parameters of causal perfectly matched anisotropic absorbers. IEEE Microw. Guided Wave Lett. **6**(12), 447–449 (1996)
32. Roden, J.A., Gedney, S.D.: Convolution PML (CPML): An efficient FDTD implementation of the CFS-PML for arbitrary media. Microw. Opt. Techn. Let. **27**(5), 334–339 (2000)
33. Meza-Fajardo, K., Papageorgiou, A.: A nonconvolutional, split-field, perfectly matched layer for wave propagation in isotropic and anisotropic elastic media: stability analysis. Bull. seism. Soc. Am. **98**(4), 1811–1836 (2008)
34. Tago, J., Métivier, L., Virieux, J.: SMART layers: a simple and robust alternative to PML approaches for elastodynamics. Geophys. J. Int. **199**(2), 700–706 (2014)
35. Bécache, E., Fauqueux, S., Joly, P.: Stability of perfectly matched layers, group velocities and anisotropic waves. J. Comput. Phys. **188**(2), 399–433 (2003)
36. Kreiss, H.-O.: Initial boundary value problems for hyperbolic systems. Commun. Pure Appl. Anal. **23**, 277–298 (1970)
37. Cohen, G., Fauqueux, S.: Mixed finite elements with mass-lumping for the transient wave equation. J. Comput. Acoust. **8**(1), 171–188 (2000)

Chapter 7
Time Approximation

Abstract This chapter deals with time approximation for the methods described in this book, including local time-stepping. The first part is devoted to the schemes with a constant time-step adapted to wave phenomena. After describing their construction, stability analysis is studied by using both plane wave analysis and energy techniques. Moreover, a link with the approximation of unbounded domains is studied. The second part introduces three efficient local time-stepping schemes which can be used in order to speed up the methods.

7.1 Schemes with a Constant Time-Step

As seen in the previous chapters, numerous accurate high order methods which provide accurate solution for a reasonable computational time can be constructed in space. The challenge is now to construct time approximations which preserve this performance. Unlike parabolic equations, centered and explicit time approximations are more fitted to wave equations which are hyperbolic and not dissipative. A natural idea to construct time approximations is to approximate the time derivatives by centered finite difference schemes. The lowest order approximation is the well-known Leapfrog scheme which provides a simple second-order algorithm in time. Unfortunately, no higher-order centered (or uncentered) finite difference scheme is stable [1]. So, numerous palliative schemes were proposed to get high order centered and explicit time approximations such as the modified equation approach introduced in [2] and extended in [3, 4], pseudo-spectral approximations [5], symmetric schemes [6], etc. All these schemes are able to provide high order approximations in a more or less efficient way but have the same main drawback: they are difficult to handle with ABC or PML or any complex boundary condition, as we shall show later. Moreover, their stability condition do not generally balance their complexity, which induces a substantial additional cost. For these reasons, we shall just present in this section the Leapfrog scheme and modified equation approach which seems to be the most efficient fourth order approximation in time.

G. Cohen and S. Pernet, *Finite Element and Discontinuous Galerkin Methods for Transient Wave Equations*, Scientific Computation,
DOI 10.1007/978-94-017-7761-2_7

7.1.1 Construction of the Schemes

7.1.1.1 The Leapfrog Scheme

In this section, we present second-order centered explicit schemes for the second and first order formulations of wave equations. Originally, the Leapfrog scheme was the scheme applied to the first order system, but it is now usual to call the time approximation of the second-order equation by the same name since both schemes are equivalent.

Let us consider the matrix semi-discrete in space approximation of a wave equation (which can be the second order form of the acoustics, linear elastodynamics or Maxwell's equations) obtained by any finite element or discontinuous Galerkin method in $\mathbb{R}^d$, $d = 1, \ldots, 3$, introduced in the previous chapters:

$$M_h \frac{\mathrm{d}^2 U(t)}{\mathrm{d}t^2} - K_h\, U(t) = F(t), \tag{7.1}$$

where M_h is the mass matrix (hopefully block-diagonal) and K_h the stiffness matrix (which includes the jumps or mean values in discontinuous formulations) and $U \in \mathbb{R}^p$.

By denoting $t_n = n\,\Delta t$, Taylor expansions of $U(t)$ provide:

$$\begin{aligned} U(t_{n+1}) = U(t_n) + \Delta t\, \frac{\mathrm{d}U(t_n)}{\mathrm{d}t} + \frac{\Delta t^2}{2}\frac{\mathrm{d}^2 U(t_n)}{\mathrm{d}t^2} + \frac{\Delta t^3}{6}\frac{\mathrm{d}^3 U(t_n)}{\mathrm{d}t^3} \\ + \frac{\Delta t^4}{24}\frac{\mathrm{d}^4 U(t_n)}{\mathrm{d}t^4} + O(\Delta t^5), \end{aligned} \tag{7.2a}$$

$$\begin{aligned} U(t_{n-1}) = U(t_n) - \Delta t\, \frac{\mathrm{d}U(t_n)}{\mathrm{d}t} + \frac{\Delta t^2}{2}\frac{\mathrm{d}^2 U(t_n)}{\mathrm{d}t^2} - \frac{\Delta t^3}{6}\frac{\mathrm{d}^3 U(t_n)}{\mathrm{d}t^3} \\ + \frac{\Delta t^4}{24}\frac{\mathrm{d}^4 U(t_n)}{\mathrm{d}t^4} + O(\Delta t^5). \end{aligned} \tag{7.2b}$$

After summation of (7.2a) and (7.2b) and division by Δt^2, we get:

$$\frac{\mathrm{d}^2 U(t_n)}{\mathrm{d}t^2} = \frac{U(t_{n+1}) - 2\,U(t_n) + U(t_{n-1})}{\Delta t^2} + O(\Delta t^2). \tag{7.3}$$

So, (7.3) provides the Leapfrog approximation of (7.1), which reads

$$M_h \frac{U_h^{n+1} - 2\,U_h^n + U_h^{n-1}}{\Delta t^2} - K_h\, U_h^n = F^n, \tag{7.4}$$

where U_h^n is the time approximation of $U(t_n)$ and $F^n = F(t_n)$.

Remark: Taking $K_h\, U_h^n$ leads to a centered explicit approximation. If we use $K_h\, U_h^{n-1}$, the system is still explicit but no longer centered. $K_h\, U_h^{n+1}$ would provide a implicit uncentered approximation.

Let us now introduce the matrix space approximation of a first-order linear system (mixed formulation):

$$B_h^1 \frac{\mathrm{d}U(t)}{\mathrm{d}t} + R_h^1 V(t) = F(t), \tag{7.5a}$$

$$B_h^2 \frac{\mathrm{d}V(t)}{\mathrm{d}t} + R_h^2 U(t) = 0, \tag{7.5b}$$

where B_h^1 and B_h^2 are the mass matrices and R_h^1 and R_h^2 (possibly transposed) stiffness matrices (which include the jumps or mean values in discontinuous formulations) and $U \in \mathbb{R}^p$, $V \in \mathbb{R}^q$.

By denoting $t_n = n\,\Delta t$ and $t_{n+\frac{1}{2}} = (n+1/2)\,\Delta t$, Taylor expansions of $U(t)$ provide:

$$\begin{aligned} U(t_{n+1}) = U(t_{n+\frac{1}{2}}) &+ \frac{\Delta t}{2}\frac{\mathrm{d}U(t_{n+\frac{1}{2}})}{\mathrm{d}t} + \frac{\Delta t^2}{8}\frac{\mathrm{d}^2U(t_{n+\frac{1}{2}})}{\mathrm{d}t^2} \\ &+ \frac{\Delta t^3}{48}\frac{\mathrm{d}^3U(t_{n+\frac{1}{2}})}{\mathrm{d}t^3} + \frac{\Delta t^4}{384}\frac{\mathrm{d}^4U(t_{n+\frac{1}{2}})}{\mathrm{d}t^4} + O(\Delta t^5), \end{aligned} \tag{7.6a}$$

$$\begin{aligned} U(t_n) = U(t_{n+\frac{1}{2}}) &- \frac{\Delta t}{2}\frac{\mathrm{d}U(t_{n+\frac{1}{2}})}{\mathrm{d}t} + \frac{\Delta t^2}{8}\frac{\mathrm{d}^2U(t_{n+\frac{1}{2}})}{\mathrm{d}t^2} \\ &- \frac{\Delta t^3}{48}\frac{\mathrm{d}^3U(t_{n+\frac{1}{2}})}{\mathrm{d}t^3} + \frac{\Delta t^4}{384}\frac{\mathrm{d}^4U(t_{n+\frac{1}{2}})}{\mathrm{d}t^4} + O(\Delta t^5), \end{aligned} \tag{7.6b}$$

After subtraction and division by Δt, (7.6a) and (7.6b) give:

$$\frac{\mathrm{d}U(t_{n+\frac{1}{2}})}{\mathrm{d}t} = \frac{U(t_{n+1}) - U(t_n)}{\Delta t} + O(\Delta t^2). \tag{7.7}$$

In a same way, one can show that

$$\frac{\mathrm{d}V(t_n)}{\mathrm{d}t} = \frac{V(t_{n+\frac{1}{2}}) - V(t_{n-\frac{1}{2}})}{\Delta t} + O(\Delta t^2). \tag{7.8}$$

Now, by combining (7.5a) and (7.5b) with (7.7) and (7.8), we get the second-order (Leapfrog) approximation of (7.5a) and (7.5b):

$$B_h^1 \frac{U_h^{n+1} - U_h^n}{\Delta t} + R_h^1 V_h^{n+\frac{1}{2}} = F^{n+\frac{1}{2}}, \tag{7.9a}$$

$$B_h^2 \frac{V_h^{n+\frac{1}{2}} - V_h^{n-\frac{1}{2}}}{\Delta t} + R_h^2 U_h^n = 0, \tag{7.9b}$$

where U_h^n is the time approximation of $U(t_n)$, $V_h^{n+\frac{1}{2}}$, the time approximation of $V(t_{n+\frac{1}{2}})$ and $F^{n+\frac{1}{2}} = F(t_{n+\frac{1}{2}})$.

Remark: By setting $n = n - 1$, (7.9a) reads:

$$B_h^1 \frac{U_h^n - U_h^{n-1}}{\Delta t} + R_h^1 V_h^{n-\frac{1}{2}} = F^{n-\frac{1}{2}}, \tag{7.10}$$

By subtracting (7.10) to (7.9a), we get:

$$B_h^1 \frac{U_h^{n+1} - 2\,U_h^n + U_h^{n-1}}{\Delta t} + R_h^1 (V_h^{n+\frac{1}{2}} - V_h^{n-\frac{1}{2}}) = (F^{n+\frac{1}{2}} - F^{n-\frac{1}{2}}). \tag{7.11}$$

Now, we replace $V_h^{n+\frac{1}{2}} - V_h^{n-\frac{1}{2}}$ by its value derived from (7.9b) and we finally obtain:

$$B_h^1 \frac{U_h^{n+1} - 2\,U_h^n + U_h^{n-1}}{\Delta t^2} + R_h^1 (B_h^2)^{-1} R_h^2 U_h^n = \frac{F^{n+\frac{1}{2}} - F^{n-\frac{1}{2}}}{\Delta t}. \tag{7.12}$$

Equation (7.12) shows the equivalence of the first-order and second-order Leapfrog schemes.

7.1.1.2 The Modified Equation Approach

By expliciting the second-order term in (7.3), we get:

$$\frac{\mathrm{d}^2 U(t_n)}{\mathrm{d}t^2} = \frac{U(t_{n+1}) - 2\,U(t_n) + U(t_{n-1})}{\Delta t^2} - \frac{\Delta t^2}{12}\frac{\mathrm{d}^4 U(t_n)}{\mathrm{d}t^4} + O(\Delta t^4). \tag{7.13}$$

Now, from (7.1) (in which we set $F = 0$), we have:

$$\frac{\mathrm{d}^2}{\mathrm{d}t^2}\left(\frac{\mathrm{d}^2 U(t)}{\mathrm{d}t^2}\right) = \frac{\mathrm{d}^2}{\mathrm{d}t^2}(M_h^{-1} K_h U(t)) = M_h^{-1} K_h \frac{\mathrm{d}^2 U(t)}{\mathrm{d}t^2} = (M_h^{-1} K_h)^2 U(t). \tag{7.14}$$

Equation (7.14) enables us to replace the fourth-order derivative of (7.13) by $(M_h^{-1} K_h)^2 U(t)$, so that

$$\frac{\mathrm{d}^2 U(t_n)}{\mathrm{d}t^2} = \frac{U(t_{n+1}) - 2\,U(t_n) + U(t_{n-1})}{\Delta t^2} - \frac{\Delta t^2}{12}(M_h^{-1} K_h)^2 U(t) + O(\Delta t^4). \tag{7.15}$$

From (7.15), we can deduce the following fourth-order approximation in time of (7.1):

$$M_h \frac{U_h^{n+1} - 2\,U_h^n + U_h^{n-1}}{\Delta t^2} - \frac{\Delta t^2}{12}(K_h M_h^{-1} K_h)\,U_h^n - K_h U_h^n = F^n, \tag{7.16}$$

which also reads

$$\frac{U_h^{n+1} - 2\,U_h^n + U_h^{n-1}}{\Delta t^2} - M_h^{-1} K_h \left(\frac{\Delta t^2}{12} M_h^{-1} K_h + I_p\right) U_h^n = F^n, \tag{7.17}$$

where I_p is the identity matrix in $\mathbb{R}^p$.

Equation (7.17) is called the modified equation approach.

Remarks:

1. We should take into account F in (7.14) but this involves complex additional terms which have no significant influence on the solution.
2. The modified equation approach seems to be a priori twice more expensive than the Leapfrog scheme since it uses the square of the stiffness matrix. However, as we shall see in the next section, its stability condition is almost twice larger than this of the Leapfrog scheme. This means that the modified equation approach provides a fourth-order approximation in time for almost the same cost as the Leapfrog scheme. Unfortunately, this is not true for higher order modified equations.

Let us now apply this approach to (7.5a) and (7.5b). Equation (7.6a) and (7.6b) provide:

$$\frac{\mathrm{d}U(t_{n+\frac{1}{2}})}{\mathrm{d}t} = \frac{U(t_{n+1}) - U(t_n)}{\Delta t} - \frac{\Delta t^3}{24}\frac{\mathrm{d}^3 U(t_{n+\frac{1}{2}})}{\mathrm{d}t^3} + O(\Delta t^4), \tag{7.18a}$$

$$\frac{\mathrm{d}V(t_n)}{\mathrm{d}t} = \frac{V(t_{n+\frac{1}{2}}) - V(t_{n-\frac{1}{2}})}{\Delta t} - \frac{\Delta t^3}{24}\frac{\mathrm{d}^3 V(t_n)}{\mathrm{d}t^3} + O(\Delta t^4). \tag{7.18b}$$

By using (7.5a) and (7.5b), we have:

$$\begin{aligned}\frac{\mathrm{d}^3 U(t)}{\mathrm{d}t^3} &= \frac{\mathrm{d}^2}{\mathrm{d}t^2}\left(\frac{\mathrm{d}U(t)}{\mathrm{d}t}\right) = -\frac{\mathrm{d}^2}{\mathrm{d}t^2}(N_h^1\, V(t)) = -\frac{\mathrm{d}}{\mathrm{d}t}\left(N_h^1\, \frac{\mathrm{d}V(t)}{\mathrm{d}t}\right)\\ &= N_h^1\, N_h^2\, \frac{\mathrm{d}U(t)}{\mathrm{d}t} = -N_h^1\, N_h^2\, N_h^1\, V(t).\end{aligned} \tag{7.19}$$

where $N_h^1 = (B_h^1)^{-1}\, R_h^1$ and $N_h^2 = (B_h^2)^{-1}\, R_h^2$.

Similarly, we get:

$$\frac{\mathrm{d}^3 V(t)}{\mathrm{d}t^3} = -N_h^2\, N_h^1\, N_h^2\, U(t). \tag{7.20}$$

We then replace the third-order derivatives in time in (7.18a) and (7.18b) by using (7.19) and (7.20), which provides:

$$\frac{\mathrm{d}U(t_{n+\frac{1}{2}})}{\mathrm{d}t} = \frac{U(t_{n+1}) - U(t_n)}{\Delta t} + \frac{\Delta t^3}{24} N_h^1\, N_h^2\, N_h^1\, V(t_{n+\frac{1}{2}}) + O(\Delta t^4), \tag{7.21a}$$

$$\frac{\mathrm{d}V(t_n)}{\mathrm{d}t} = \frac{V(t_{n+\frac{1}{2}}) - V(t_{n-\frac{1}{2}})}{\Delta t} + \frac{\Delta t^3}{24} N_h^2\, N_h^1\, N_h^2\, U(t) + O(\Delta t^4). \tag{7.21b}$$

So (7.21a) and (7.21b) enable us to write the modified equation approximation of (7.5a) and (7.5b):

$$\frac{U_h^{n+1} - U_h^n}{\Delta t} + N_h^1 \left(\frac{\Delta t^3}{24} N_h^2 N_h^1 + I_q \right) V_h^{n+\frac{1}{2}} = F^{n+\frac{1}{2}}, \tag{7.22a}$$

$$\frac{V_h^{n+\frac{1}{2}} - V_h^{n-\frac{1}{2}}}{\Delta t} + N_h^2 \left(\frac{\Delta t^3}{24} N_h^1 N_h^2 + I_p \right) U_h^n = 0, \tag{7.22b}$$

where I_q is the identity matrix in $\mathbb{R}^q$.

Remark: In this case, the modified equation approach is about three times more expensive than a Leapfrog scheme. As we shall see below, this additional cost is partially balanced by the increase of stability.

7.1.2 Stability of the Schemes by Plane Wave Analysis

7.1.2.1 The Leapfrog Scheme

In order to develop a plane wave analysis study of the schemes in time, we start by the 1D case. We consider the partition of $\mathbb{R}$ defined by (3.65) and a set of n_r ($= r$ for FEM and $= r + 1$ for DGM) points $(\nu_i\, h)_{i=1}^{n_r}$ $(0 \leq \nu_i \leq 1)$ in each interval $[x_\ell, x_{\ell+1}]$. On this interval, we define a plane wave solution of (7.4) (in which we set $F = 0$) given by

$$(U_h^n)^T = \left((U_\ell)^T\right)_{\ell \in \mathbb{Z}} \, \mathrm{e}^{\mathrm{i}(\omega_h\, n\, \Delta t)} = \widetilde{U}_h \, \mathrm{e}^{\mathrm{i}(\omega_h\, n\, \Delta t)}, \tag{7.23}$$

where

$$(U_\ell)^T = \left(\alpha_i \, \mathrm{e}^{\mathrm{i}((\ell+\nu_i)\, k\, h)}\right)_{i=1}^{n_r}. \tag{7.24}$$

By plugging (7.23) in (7.4), we get, for the left-hand side:

$$\begin{aligned} M_h \frac{U_h^{n+1} - 2\, U_h^n + U_h^{n-1}}{\Delta t^2} &= M_h \, \widetilde{U}_h \, \mathrm{e}^{\mathrm{i}(\omega_h\, n\, \Delta t)} \, \frac{\mathrm{e}^{\mathrm{i}(\omega_h\, \Delta t)} - 2 + \mathrm{e}^{-\mathrm{i}(\omega_h\, \Delta t)}}{\Delta t^2} \\ &= \frac{1}{\Delta t^2} \left(2 - 2\, \cos(\omega_h\, \Delta t)\right) M_h \, \widetilde{U}_h \, \mathrm{e}^{\mathrm{i}(\omega_h\, n\, \Delta t)} \\ &= -\frac{4}{\Delta t^2} \sin^2 \frac{\omega_h\, \Delta t}{2} \, M_h \, \widetilde{U}_h \, \mathrm{e}^{\mathrm{i}(\omega_h\, n\, \Delta t)}. \end{aligned} \tag{7.25}$$

So, (7.4) finally reads:

$$\frac{4}{\Delta t^2} \sin^2 \frac{\omega_h\, \Delta t}{2} \, \widetilde{U}_h = M_h^{-1} \, K_h \, \widetilde{U}_h = N_h \, \widetilde{U}_h. \tag{7.26}$$

As shown in Sects. 3.4 and 4.5, (7.26) leads to the eigenvalue problem:

$$\lambda \, U_\alpha = \widehat{N}_h(k\, h) \, U_\alpha, \tag{7.27}$$

where $\lambda = (4\,h^2/\Delta t^2)\sin^2(\omega_h\,\Delta t/2)$, $\widehat{N}_h$ is $n_r \times n_r$ matrix with $k\,h \in [0, 2\,\pi]$ and $(U_\alpha)^T = (\alpha_i)_{i=1}^{n_r}$. $\widehat{N}_h$ is called the symbol of the discrete operator N_h.

The eigenvalues of this problem are denoted $\lambda_i = (4\,h^2/\Delta t^2)\sin^2(\omega_{h,i}\,\Delta t/2), i = 1, \ldots, n_r$. Now, (7.23) shows that, if $\omega_{h,i}$ in not real, the solution is evanescent or exponentially growing[1] (unstable). So, in order to have a stable solution, we must have:

$$0 \le \lambda_i(k\,h)) \le \frac{4\,h^2}{\Delta t^2}. \tag{7.28}$$

The left-hand side of (7.29) provides the positivity of the discrete operator in space (which should be naturally derived from its construction) and the right-hand side is the stability (or CFL[2]) condition. So, the stability condition reads:

$$\alpha = \frac{\Delta t}{h} \le \frac{2}{\sup\limits_{1\le i\le n_r,\, k\,h\in[0,2\,\pi]} \sqrt{\lambda_i(k\,h))}} = \alpha_M. \tag{7.29}$$

Of course, α_M can be computed numerically or by using a symbolic computation software. However, we provide below a general method based on the characteristic polynomial of the eigenvalue problem. Let us consider a finite element approximation which leads to the eigenvalue problems described in Sect. 3.4. For $r = 2$ and $r = 3$, the eigenvalues can be explicitly computed but their expression for $r = 3$ are not easy to exploit. By using (3.92), $(\widehat{M}_h)^{-1}\,\widehat{K}_h$ reads in this case:

$$\begin{bmatrix} 52 - 2\cos k\,h & \dfrac{-5\,a\mathrm{e}^{\mathrm{i}k\,h\mu} + 5\,b\mathrm{e}^{-\mathrm{i}k\,h\nu}}{2} & \dfrac{-5\,a\mathrm{e}^{-\mathrm{i}k\,h\mu} + 5\,b\mathrm{e}^{\mathrm{i}k\,h\nu}}{2} \\ -a\mathrm{e}^{-\mathrm{i}k\,h\mu} + b\mathrm{e}^{\mathrm{i}k\,h\nu} & 20 & -10\,\mathrm{e}^{\mathrm{i}\frac{\sqrt{5}}{5}k\,h} \\ \dfrac{-5a\mathrm{e}^{\mathrm{i}k\,h\mu} + 5b\mathrm{e}^{-\mathrm{i}k\,h\nu}}{4} & \dfrac{-25\mathrm{e}^{-\mathrm{i}\frac{\sqrt{5}}{5}k\,h}}{2} & 25 \end{bmatrix}. \tag{7.30}$$

By setting $k\,h = 2\,\pi\,K$, the characteristic polynomial of (7.30) can be written as

$$\begin{aligned} P(\lambda_j) &= \lambda_j^3 + 2(\cos 2\pi K - 46)\lambda_j^2 + 120(\cos 2\pi K + 14)\lambda_j \\ &\quad + 3600(\cos 2\pi K - 1) = 0, \quad \forall j = 1 \ldots 3. \end{aligned} \tag{7.31}$$

Since the maximum of λ_j is reached at a value of K such that

$$\frac{\mathrm{d}\lambda_j(K)}{\mathrm{d}K} = 0. \tag{7.32}$$

---

[1] Actually, for a second order operator, for any complex ω_h involved in (7.23), $\overline{\omega_h}$ is also involved. This implies that a complex value of ω_h always produces exponentially growing solutions.

[2] Abbreviation of Courant–Friedrichs–Lewy condition.

By differentiating P versus K and taking (7.32) into account, we get:

$$\frac{\mathrm{d}P(\lambda_j)}{\mathrm{d}K} = \frac{\mathrm{d}P(\lambda_j)}{\mathrm{d}\lambda_j}\frac{\mathrm{d}\lambda_j}{\mathrm{d}K} = 4\sin 2\pi K Q(\lambda_j) = 0, \tag{7.33}$$

where

$$Q(\lambda_j) = \lambda_j^2 + 60\lambda_j + 1800. \tag{7.34}$$

Since Q has no real root, this equation is satisfied for $\sin 2\pi K = 0$, which provides two classes of solutions $K = \ell$ or $K = 1/2 + \ell$, $\ell \in \mathbb{Z}$, for which λ_j can reach its maximum. The values of λ_j for these two classes can be obtained by inserting these solutions into P. So,

- For $K = \ell$, we have

$$\lambda_j^3(0) - 90\lambda_j^2(0) + 1800\lambda_j(0) = 0, \tag{7.35}$$

  whose solutions are 0, 30 and 60.

- For $K = 1/2 + \ell$, we get

$$\lambda_j^3(1/2) - 94\lambda_j^2(1/2) + 1560\lambda_j(1/2) - 7200 = 0, \tag{7.36}$$

  whose solutions are $6(7 - \sqrt{29})$, 10 and $6(7 + \sqrt{29})$.

Obviously, $6(7 + \sqrt{29})$ is the maximum of all these solutions and the stability condition is

$$\alpha \leq \alpha_M = \frac{2}{\sqrt{6(7 + \sqrt{29})}} = \frac{\sqrt{30}}{30}\sqrt{7 - \sqrt{29}} \simeq 0.232\,008\,27. \tag{7.37}$$

Such a process can also be applied, with more difficulty, to higher-order approximations.

In Table 7.1, we give the stability conditions for the Leapfrog scheme used with a Q_r approximation, $r = 1$ to 5 in 1D for discontinuous elements with Gauss–Lobatto (denoted α_M^{GL}) and Gauss points (denoted α_M^{G}) respectively. The values with Gauss–Lobatto points hold for continuous elements with Gauss–Lobatto points. These CFL conditions must be divided by $\sqrt{d}$ in dimension d. Moreover, one can show that these CFL conditions in 2D and 3D hold for the approximations of Maxwell's equations by DGM or edge elements.

Remarks:

1. The stability condition for P_3 seems to be about twice as restrictive as that of P_2 and much more restrictive than the stability condition obtained for FDM [6]. In

Table 7.1 The stability conditions for Q_r approximations in 1D with a leapfrog scheme using Gauss and Gauss–Lobatto points, when $r = 1$ to 5

r	1	2	3	4	5
$\alpha_M^{GL} \simeq$	1	0.4082	0.2320	0.1476	0.1010
$\alpha_M^{G} \simeq$	0.5	0.247	0.15	0.101	0.0732

fact, this is not true because the $\alpha = c\Delta t/h$ used for FEM takes into account the length h of the element but, actually, in terms of points, a P_r mesh of N_e elements contains $rN_e + 1$ points, i.e. r points per element, and, therefore, the stability conditions versus the number of points is 0.816 496 58 for P_2 and 0.696 024 83 for P_3 for Gauss–Lobatto points.

2. Two remarks about the CFL with Gauss points: (a) it is smaller than the CFL obtained for Gauss–Lobatto points, (b) the difference between both CFLs decreases with r. These two phenomena seem to come from the fact that Gauss points do not include the ends of the intervals and require an extrapolation to compute the solution. It is well known that extrapolation induces a loss of stability (and even instability). So, since the Gauss points get closer to the ends of the interval (and so, require less extrapolation) when r increases, the stability increases too.
3. Equation (7.11) shows that we have an equivalent CFL for the first order Leapfrog scheme.

7.1.2.2 The Modified Equation Approach

The study of the stabilty for the modified equation approach is based on the two following properties:

1. The symbol of the nth power D^n of a matricial operator D is equal to the nth power of the symbol of D.
2. If λ is the eigenvalue of a matrix M then, for any polynomial P, $P(\lambda)$ is the eigenvalue of $P(M)$.

By using the first property, one can see that the dispersion relations of the modified Eq. (7.17) are the eigenvalues of the following problem:

$$4\sin^2\frac{\omega_h\Delta t}{2}\,\underline{U}_\alpha = \left(\Delta t^2\widehat{N}_h - \frac{\Delta t^4}{12}\widehat{N}_h^2\right)\underline{U}_\alpha. \tag{7.38}$$

If $\lambda_j(k\,h)$ is an eigenvalue of $\widehat{N}_h$, we set $\lambda_j' = h^2\lambda_j/c^2$ and, by using the second property, one can write the different dispersion relations of the problem in the form:

$$4\sin^2\frac{\omega_h\Delta t}{2} = \alpha^2\lambda_j' - \frac{\alpha^4}{12}\lambda_j'^2, \tag{7.39}$$

with $\alpha = c\Delta t/h$.

So, the stability condition is given by the double inequality:

$$4 \geq \alpha^2 \lambda'_j - \frac{\alpha^4}{12} \lambda'^2_j \geq 0. \tag{7.40}$$

A simple computation shows that the right-hand inequality is always satisfied and the left-hand one holds for $\alpha \leq 2\sqrt{3/\lambda'_j}$. So, the stability condition is:

$$\alpha \leq \max_{1 \leq j \leq r} \left[\frac{2\sqrt{3}}{\sup\limits_{k\,h \in [0, 2\pi]} \sqrt{\lambda'_j}} \right]. \tag{7.41}$$

In particular, we obtain for P_2:

$$\alpha_M^{GL} = \frac{2\sqrt{3}}{\sqrt{24}} = \frac{\sqrt{2}}{2} \simeq 0.70710678 \tag{7.42}$$

and, for P_3

$$\alpha \leq \alpha_M^{GL} = \frac{2\sqrt{3}}{\sqrt{6(7+\sqrt{29})}} = \frac{\sqrt{10}}{10}\sqrt{7-\sqrt{29}} \simeq 0.40185011. \tag{7.43}$$

As we said previously, the stability condition is almost twice as large as that of the leapfrog scheme, which balances the additional cost induced by the modified equation approach.

In the same way, the stability condition for a sixth-order modified equation is given by

$$4 \geq \alpha^2 \lambda'_j - \frac{\alpha^4}{12} \lambda'^2_j + \frac{\alpha^6}{360} \lambda'^3_j \geq 0. \tag{7.44}$$

The right-hand equation is always true and the left-hand one is satisfied when $\alpha \leq \sqrt{2(5^{3/2} - 25^{1/3} + 5)/\lambda'_j}$. So, the stability condition can be written as

$$\alpha \leq \max_{1 \leq j \leq r} \left[\frac{\sqrt{2(5^{3/2} - 25^{1/3} + 5)}}{\sup\limits_{k\,h \in [0, 2\pi]} \sqrt{\lambda'_j}} \right]. \tag{7.45}$$

In particular, we obtain for P_3:

$$\alpha_M^{GL} = \frac{\sqrt{2(5^{3/2} - 25^{1/3} + 5)}}{\sqrt{6(7 + \sqrt{29})}} \simeq 0.319\,209\,92. \qquad (7.46)$$

For this approximation in time, one must compute three discrete operators instead of one for the leapfrog scheme and, obviously, the additional cost is not balanced by the stability condition which is less than 1.4 times greater than that of the Leapfrog scheme. This is also the case of higher-order approximations in time obtained by using the modified equation.

Remarks:

1. Stability of the modified equation for first order systems can be derived in the same way. We do not develop the computations in this book. The main result is that the additional cost of the method is not balanced by the increase of the CFL, even for fourth order approximation in time.
2. Taylor expansions and dispersion curves for the continuous elements with Gauss–Lobatto points can be found in [6].

7.1.3 Stability of the Schemes by Energy Techniques

We look for a discrete equivalent of the energy identity given in (1.93). Let us consider the following discrete problem in some approximation space U_h:

$$\left(\frac{u_h^{n+1} - 2u_h^n + u_h^{n-1}}{\Delta t^2}, v_h\right)_0 + a_h(u_h^n, v_h) = 0 \quad u_h^n \in U_h, v_h \in U_h, \qquad (7.47)$$

where $a_h(u_h^n, v_h)$ is associated to an approximation A_h of a second-order differential operator in space.

By setting $v_h = \dfrac{u_h^{n+1} - u_h^{n-1}}{2\Delta t} = \dfrac{1}{2}\left(\dfrac{u_h^{n+1} - u_h^n}{\Delta t} + \dfrac{u_h^n - u_h^{n-1}}{\Delta t}\right)$ in (7.47), we get:

$$\frac{1}{2\Delta t}\left(\left\|\frac{u_h^{n+1} - u_h^n}{\Delta t}\right\|^2 - \left\|\frac{u_h^n - u_h^{n-1}}{\Delta t}\right\|^2 + a_h(u_h^{n+1}, u_h^n) - a_h(u_h^n, u_h^{n-1})\right) = 0. \qquad (7.48)$$

On the basis of the following discrete energy:

$$E_h^{n+\frac{1}{2}} = \frac{1}{2}\left\|\frac{u_h^{n+1} - u_h^n}{\Delta t}\right\|^2 + \frac{1}{2}a_h(u_h^{n+1}, u_h^n), \qquad (7.49)$$

this identity reads:

$$\frac{1}{\Delta t}(E_h^{n+\frac{1}{2}} - E_h^{n-\frac{1}{2}}) = 0. \qquad (7.50)$$

Unfortunately, this definition of the discrete energy does not ensure the positivity of $a_h(u_h^{n+1}, u_h^n)$. Therefore, we must work on this formulation to obtain an adequate form.

We first notice that

$$a_h(u_h^{n+1}, u_h^n) = a_h\left(\frac{u_h^{n+1}+u_h^n}{2}, \frac{u_h^{n+1}+u_h^n}{2}\right) - \frac{\Delta t^2}{4} a_h\left(\frac{u_h^{n+1}-u_h^n}{\Delta t}, \frac{u_h^{n+1}-u_h^n}{\Delta t}\right). \tag{7.51}$$

This equality implies:

$$E_h^{n+\frac{1}{2}} = \left\| \frac{u_h^{n+1}-u_h^n}{\Delta t} \right\|^2 + a_h\left(\frac{u_h^{n+1}+u_h^n}{2}, \frac{u_h^{n+1}+u_h^n}{2}\right) - \frac{\Delta t^2}{4} a_h\left(\frac{u_h^{n+1}-u_h^n}{\Delta t}, \frac{u_h^{n+1}-u_h^n}{\Delta t}\right). \tag{7.52}$$

Now, since we have:

$$||A_h|| = \sup_{u_h \in U_h} \frac{a_h(u_h, u_h)}{||u_h||^2}, \tag{7.53}$$

then

$$a_h\left(\frac{u_h^{n+1}-u_h^n}{\Delta t}, \frac{u_h^{n+1}-u_h^n}{\Delta t}\right) \leq ||A_h|| \left\| \frac{u_h^{n+1}-u_h^n}{\Delta t} \right\|^2. \tag{7.54}$$

Finally, the discrete energy verifies:

$$E_h^{n+\frac{1}{2}} \geq \left(1 - \frac{\Delta t^2}{4}||A_h||\right) \left\| \frac{u_h^{n+1}-u_h^n}{\Delta t} \right\|^2 + a_h(u_h^{n+\frac{1}{2}}, u_h^{n+\frac{1}{2}}). \tag{7.55}$$

A sufficient condition of positivity of $E_h^{n+\frac{1}{2}}$ is then

$$\Delta t < \frac{2}{\sqrt{||A_h||}}. \tag{7.56}$$

So, the computation of the positivity condition is equivalent to the computation of the norm of A_h. In practice, this norm can be obtained by computing the largest eigenvalue of A_h, which can be done numerically. For a regular mesh in 1D, this eigenvalue can be computed exactly, as shown in the previous section.

We shall not develop this technique for the modified equation.

7.1.4 The Modified Equation and Unbounded Domains

7.1.4.1 ABC

Treating unbounded domains with a Leapfrog schemes present no difficulty. On the contrary, both ABC and PML are almost impossible to use with the modified equation approach. Let us first illustrate this difficulty by the simplest method: the first-order ABC. The matrix form of (6.39) reads:

$$M_h \frac{\mathrm{d}^2 U(t)}{\mathrm{d}t^2} - K_h\, U(t) - B_h \frac{\mathrm{d}U(t)}{\mathrm{d}t} = F(t), \tag{7.57}$$

whose leapfrog approximation is

$$M_h \frac{U_h^{n+1} - 2\,U_h^n + U_h^{n-1}}{\Delta t^2} - K_h\, U_h^n - B_h \frac{U_h^{n+1} - U_h^{n-1}}{2\,\Delta t} = F^n. \tag{7.58}$$

In order to construct a modified equation approximation, one must use the Taylor expansion given by (7.13) but also the Taylor expansion of the approximation in time of the boundary term of (7.58), which reads:

$$\frac{\mathrm{d}U(t_n)}{\mathrm{d}t} = \frac{U(t_{n+1}) - U(t_{n-1})}{2\,\Delta t} - \frac{\Delta t^2}{6} \frac{\mathrm{d}^3 U(t_n)}{\mathrm{d}t^3} + O(\Delta t^4). \tag{7.59}$$

So, in order to construct our method, we must replace the fourth-order and the third-order derivatives in time of U by operators in space. Using (7.57), we get, after setting $F = 0$:

$$\frac{\mathrm{d}^2}{\mathrm{d}t^2}\left(\frac{\mathrm{d}^2 U(t)}{\mathrm{d}t^2}\right) = M_h^{-1}\, K_h \frac{\mathrm{d}^2 U(t)}{\mathrm{d}t^2} + M_h^{-1}\, B_h \frac{\mathrm{d}^3 U(t)}{\mathrm{d}t^3}. \tag{7.60}$$

Still using (7.57), we get:

$$\frac{\mathrm{d}^2 U(t)}{\mathrm{d}t^2} = M_h^{-1}\, K_h\, U(t) + M_h^{-1}\, B_h \frac{\mathrm{d}U(t)}{\mathrm{d}t}, \tag{7.61}$$

$$\frac{\mathrm{d}^3 U(t)}{\mathrm{d}t^3} = M_h^{-1}\, K_h \frac{\mathrm{d}U(t)}{\mathrm{d}t} + M_h^{-1}\, B_h\,, \frac{\mathrm{d}^2 U(t)}{\mathrm{d}t^2}. \tag{7.62}$$

Now, by combining (7.60), (7.61) and (7.62), we finally obtain:

$$\begin{aligned} \frac{\mathrm{d}^4 U(t)}{\mathrm{d}t^4} &= (M_h^{-1}\, K_h)^2\, U(t) + (M_h^{-1}\, B_h)^2 \frac{\mathrm{d}^2 U(t)}{\mathrm{d}t^2} \\ &\quad + (M_h^{-1}\, K_h\, M_h^{-1}\, B_h + M_h^{-1}\, B_h\, M_h^{-1}\, K_h) \frac{\mathrm{d}U(t)}{\mathrm{d}t}. \end{aligned} \tag{7.63}$$

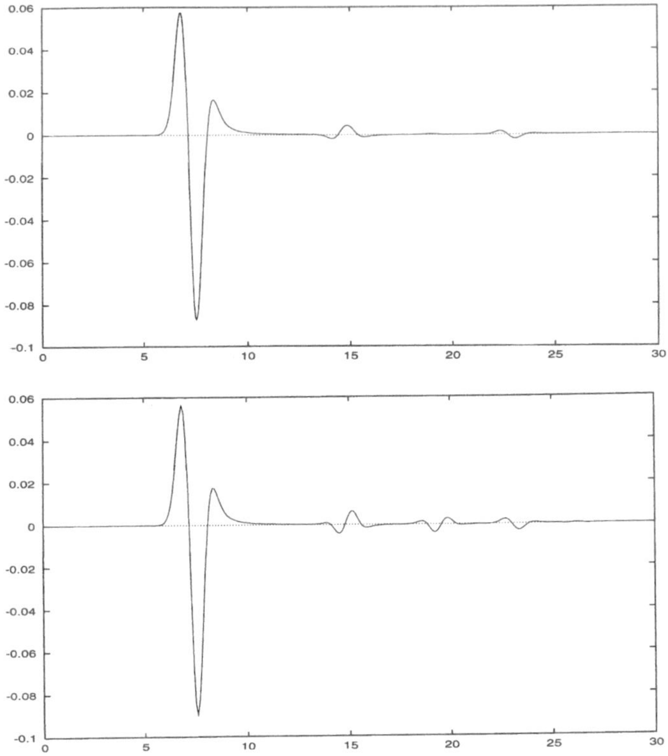

Fig. 7.1 Seismograms at a first order ABC boundary in 2D with Leapfrog (*above*) and partially modified equation (*below*) schemes

Equations (7.62) and (7.63) must be then inserted into (7.58) in order to get a fourth-order approximation in time of (7.57). All this process shows that the modified equation approach introduces heavy matrix terms at the boundary (which can be not invertible). In particular, for continuous elements with mass lumping, these matrices (which contain the stiffness matrix) are no longer (block-)diagonal. A palliative to avoid this problem is to ignore the boundary term to construct the approximation (partially modified equation). Unfortunately, this approach induces non-negligible reflections at the boundary, as shown in Figs. 7.1 and 7.2.

7.1.4.2 PML

The matrix expression of (6.80a)–(6.80e) derived from a continuous or discontinuous approximation can be written as

$$D_h^1 \frac{\mathrm{d}U^*(t)}{\mathrm{d}t} + R_h^1 V(t) = F(t), \tag{7.64a}$$

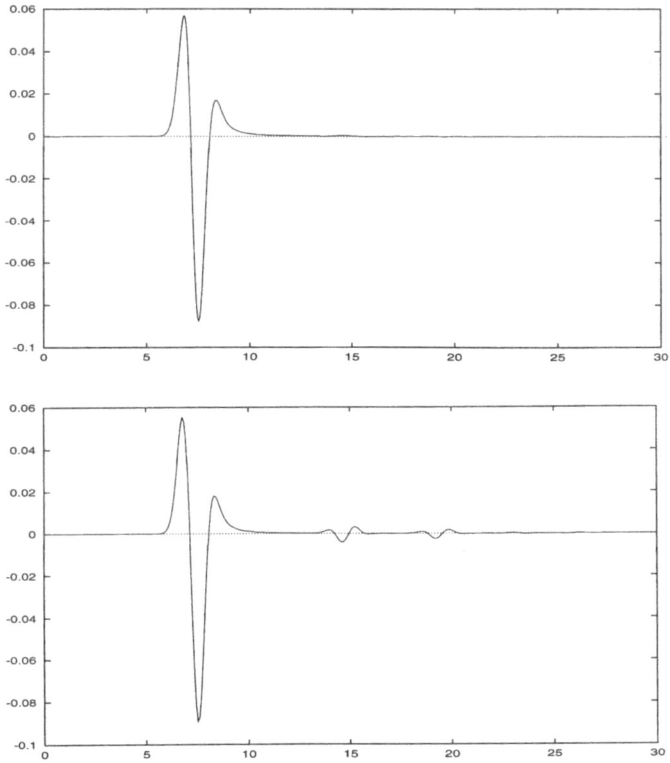

Fig. 7.2 Seismograms at a second order ABC boundary in 2D with Leapfrog (*above*) and partially modified equation (*below*) schemes

$$B_h^1 \frac{\mathrm{d}V^*(t)}{\mathrm{d}t} + R_h^2 U(t) = 0, \tag{7.64b}$$

$$B_h^2 \frac{\mathrm{d}V(t)}{\mathrm{d}t} + Z_h^1 V(t) = B_h^2 \frac{\mathrm{d}V^*(t)}{\mathrm{d}t} + Z_h^2 V^*(t), \tag{7.64c}$$

$$D_h^2 \frac{\mathrm{d}^2 U(t)}{\mathrm{d}t^2} + Z_h^3 \frac{\mathrm{d}U(t)}{\mathrm{d}t} + Z_h^4 U(t) = D_h^2 \frac{\mathrm{d}^2 U^*(t)}{\mathrm{d}t^2}. \tag{7.64d}$$

As shown in the previous section, in order to get a fourth-order approximation of the first-order derivative in time, we must replace a third-order derivative in time by an operator in space. So, let do this for $\mathrm{d}U^*(t)/\mathrm{d}t$. We have, after setting $F = 0$:

$$\frac{\mathrm{d}^3 U^*(t)}{\mathrm{d}t^3} = -R_h^1 \frac{\mathrm{d}^2 V(t)}{\mathrm{d}t^2} = -R_h^1 \frac{\mathrm{d}}{\mathrm{d}t}\left(\frac{\mathrm{d}V(t)}{\mathrm{d}t}\right)$$
$$= \frac{\mathrm{d}}{\mathrm{d}t}\left((B_h^2)^{-1}\left(-Z_h^1\, V(t) + B_h^2 \frac{\mathrm{d}V^*(t)}{\mathrm{d}t} + Z_h^2\, V^*(t)\right)\right)\ldots \tag{7.65}$$

We stop here our computation because (7.65) already shows how difficult it would be to get our approximation.

In conclusion, high order methods in time are not easy to implement and almost impossible to use with complex boundary conditions. For this reason, people generally use Leapfrog schemes (or fourth-order Runge–Kutta schemes which are dissipative and not better than Leapfrog schemes for large time-steps), which induces a theoretical loss of accuracy, Actually, FEM or DGM lean on non regular meshes with important differences between the size of the elements. Since the CFL depends on the smaller element, we rarely use the maximum time-step and the effect of the low-order approximation in time is then not significant. However, some techniques were developed to increase the time-step which are described in the next section.

7.1.5 A Remark About the Time Approximation of Dissipative DG Schemes

When we consider an uncentered DG approximation of a first-order system of wave equations, we obtain a matrix system of the form:

$$B_h^1 \frac{\mathrm{d}U}{\mathrm{d}t}(t) - R_h\, V(t) + D_h^1 U(t) = 0, \tag{7.66a}$$

$$B_h^2 \frac{\mathrm{d}V}{\mathrm{d}t}(t) + R_h^T\, U(t) + D_h^2 V(t) = 0. \tag{7.66b}$$

where D_h^1 and D_h^2 are symmetric matrices which satisfy the property $\left(D_h^1 U\,,\ U\right) \geq 0$ and $\left(D_h^2 V\,,\ V\right) \geq 0$.

The second-order Leapfrog time approximation of (7.66a) and (7.66b) reads:

$$B_h^1 \frac{U^{n+1} - U^n}{\Delta t} - R_h\, V^{n+\frac{1}{2}} + D_h^1 \frac{U^{n+1} + U^n}{2} = 0, \tag{7.67a}$$

$$B_h^2 \frac{V^{n+\frac{3}{2}} - V^{n+\frac{1}{2}}}{\Delta t} + R_h^T\, U^{n+1} + D_h^2 \frac{V^{n+\frac{3}{2}} + V^{n+\frac{1}{2}}}{2} = 0. \tag{7.67b}$$

Unfortunately, this scheme is implicit and therefore, in practice, it is more suited to use the following explicit uncentered discretization in time of (7.66a) and (7.66b):

$$B_h^1 \frac{U^{n+1} - U^n}{\Delta t} - R_h V^{n+\frac{1}{2}} + D_h^1 U^n = 0, \tag{7.68a}$$

$$B_h^2 \frac{V^{n+\frac{3}{2}} - V^{n+\frac{1}{2}}}{\Delta t} + R_h^T U^{n+1} + D_h^2 V^{n+\frac{1}{2}} = 0 \tag{7.68b}$$

where we consider the approximations $D_h^1 U(t_{n+1/2}) \simeq D_h^1 U^n$ and $D_h^2 V(t_{n+1}) \simeq D_h^2 U^{n+1/2}$ which are only first-order accurate.

Let us now study the stability of (7.68a) and (7.68b) by using an energy technique. For that, we construct a quadratic form $\mathcal{E}_h(U, V)$ satisfying under a CFL condition the property:

$$0 < \mathcal{E}_h(U^n, V^{n+\frac{1}{2}}) \leq C, \quad \forall n \tag{7.69}$$

where C is a positive constant independent of n.

We first write the scalar product of (7.68a) with U^{n+1} and U^n, we get:

$$\left(B_h^1 \frac{U^{n+1} - U^n}{\Delta t}, U^{n+1}\right) - \left(R_h V^{n+\frac{1}{2}}, U^{n+1}\right) + \left(D_h^1 U^n, U^{n+1}\right) = 0, \tag{7.70a}$$

$$\left(B_h^1 \frac{U^{n+1} - U^n}{\Delta t}, U^n\right) - \left(R_h V^{n+\frac{1}{2}}, U^n\right) + \left(D_h^1 U^n, U^n\right) = 0. \tag{7.70b}$$

In the same way, (7.68b) provides:

$$\left(B_h^2 \frac{V^{n+\frac{3}{2}} - V^{n+\frac{1}{2}}}{\Delta t}, V^{n+\frac{1}{2}}\right) + \left(R_h^T U^{n+1}, V^{n+\frac{1}{2}}\right) + \left(D_h^2 V^{n+\frac{1}{2}}, V^{n+\frac{1}{2}}\right) = 0, \tag{7.71a}$$

$$\left(B_h^2 \frac{V^{n+\frac{1}{2}} - V^{n-\frac{1}{2}}}{\Delta t}, V^{n+\frac{1}{2}}\right) + \left(R_h^T U^n, V^{n+\frac{1}{2}}\right) + \left(D_h^2 V^{n-\frac{1}{2}}, V^{n+\frac{1}{2}}\right) = 0. \tag{7.71b}$$

Let us introduce the discrete quantity:

$$E_h^n = \left(B_h^1 U^n, U^n\right) + \left(B_h^2 V^{n+\frac{1}{2}}, V^{n-\frac{1}{2}}\right). \tag{7.72}$$

After summation of (7.70a)–(7.70b) and (7.71a)–(7.71b), we obtain:

$$E_h^{n+1} - E_h^n = -\Delta t \left(D_h^1 U^n \,,\, U^{n+1}\right) - \Delta t \left(D_h^2 V^{n-\frac{1}{2}} \,,\, V^{n+\frac{1}{2}}\right)$$
$$- \Delta t \left(D_h^1 U^n \,,\, U^n\right) - \Delta t \left(D_h^2 V^{n+\frac{1}{2}} \,,\, V^{n+\frac{1}{2}}\right). \qquad (7.73)$$

Remark: If the DG scheme is non-dissipative i.e. $D_h^1 = D_h^2 = 0$, then we have the well-known result $E_h^{n+1} = E_h^n$.

Since matrices D_h^1 and D_h^2 are symmetric and positive, then we can define their square-root matrices $\left(D_h^1\right)^{\frac{1}{2}}$ and $\left(D_h^2\right)^{\frac{1}{2}}$. Now, by using the estimate $ab \leq (a^2 + b^2)/2$ and the Cauchy–Schwarz inequality, we get:

$$E_h^{n+1} - E_h^n \leq \frac{\Delta t}{2} \left(\left\| \left(D_h^1\right)^{\frac{1}{2}} U^n \right\|^2 + \left\| \left(D_h^1\right)^{\frac{1}{2}} U^{n+1} \right\|^2 \right)$$
$$+ \frac{\Delta t}{2} \left(\left\| \left(D_h^2\right)^{\frac{1}{2}} V^{n-\frac{1}{2}} \right\|^2 + \left\| \left(D_h^2\right)^{\frac{1}{2}} V^{n+\frac{1}{2}} \right\|^2 \right)$$
$$- \Delta t \left\| \left(D_h^1\right)^{\frac{1}{2}} U^n \right\|^2 - \Delta t \left\| \left(D_h^2\right)^{\frac{1}{2}} V^{n+\frac{1}{2}} \right\|^2. \qquad (7.74)$$

Equation (7.74) suggests to define the discrete quantity:

$$\tilde{E}_h^n = E_h^n - \frac{\Delta t}{2} \left\| \left(D_h^1\right)^{\frac{1}{2}} U^n \right\|^2 + \frac{\Delta t}{2} \left\| \left(D_h^2\right)^{\frac{1}{2}} V^{n-\frac{1}{2}} \right\|^2 \qquad (7.75)$$

and we have

$$\tilde{E}_h^{n+1} \leq \tilde{E}_h^n. \qquad (7.76)$$

In particular,

$$\tilde{E}_h^n \leq \tilde{E}_h^0, \ \forall n. \qquad (7.77)$$

This new quantity $\tilde{E}_h^n$ is not yet a quadratic form in U^n and $V^{n+\frac{1}{2}}$ and it still remains one last step in order to define $\mathcal{E}_h$ as in (7.69).

First, by using (7.71a), we can write:

$$\tilde{E}_h^n = \|U^n\|_{B_h^1}^2 + \left\| V^{n+\frac{1}{2}} \right\|_{B_h^2}^2 + \Delta t \left(R_h^T U^n \,,\, V^{n+\frac{1}{2}}\right) + \Delta t \left(D_h^2 V^{n-\frac{1}{2}} \,,\, V^{n+\frac{1}{2}}\right)$$
$$- \frac{\Delta t}{2} \|U^n\|_{D_h^1}^2 + \frac{\Delta t}{2} \left\| V^{n-\frac{1}{2}} \right\|_{D_h^2}^2 \qquad (7.78)$$

where $\|W\|_A = (AW, W)$.

Then, the estimate $ab \leq (a^2 + b^2)/2$ and the Cauchy–Schwarz inequality lead to

$$\tilde{E}_h^n \geq \|U^n\|^2_{B_h^1} + \left\|V^{n+\frac{1}{2}}\right\|^2_{B_h^2} + \Delta t \left(R_h^T\, U^n\,,\; V^{n+\frac{1}{2}}\right)$$
$$- \frac{\Delta t}{2}\,\|U^n\|^2_{D_h^1} - \frac{\Delta t}{2}\left\|V^{n+\frac{1}{2}}\right\|^2_{D_h^2}. \tag{7.79}$$

Let us now define the quadratic form:

$$\mathscr{E}_h(U,V) = \|U\|^2_{B_h^1} + \|V\|^2_{B_h^2} + \Delta t \left(R_h^T\, U\,,\; V\right)$$
$$-\frac{\Delta t}{2}\,\|U\|^2_{D_h^1} - \frac{\Delta t}{2}\,\|V\|^2_{D_h^2}. \tag{7.80}$$

So, (7.77) and (7.79) leads to the right-hand side of (7.69):

$$\mathscr{E}_h(U^n, V^{n+\frac{1}{2}}) \leq \tilde{E}_h^0,\ \forall n. \tag{7.81}$$

Finally, we prove the positivity of $\mathscr{E}(U^n, V^{n+\frac{1}{2}})$ under a CFL condition. For that, we use again the Cauchy–Schwarz inequality:

$$\begin{aligned}
\mathscr{E}_h(U^n, V^{n+\frac{1}{2}}) \geq\ & \|U^n\|^2_{B_h^1} + \left\|V^{n+\frac{1}{2}}\right\|^2_{B_h^2} \\
& - \Delta t \left\|\left(B_h^2\right)^{-1/2} R_h^T \left(B_h^1\right)^{-1/2}\right\| \|U^n\|_{B_h^1} \left\|V^{n+\frac{1}{2}}\right\|_{B_h^2} \\
& - \frac{\Delta t}{2}\left\|\left(B_h^1\right)^{-1/2}\,{}^tD_h^1 \left(B_h^1\right)^{-1/2}\right\| \|U^n\|^2_{B_h^1} \\
& - \frac{\Delta t}{2}\left\|\left(B_h^2\right)^{-1/2}\,{}^tD_h^2 \left(B_h^2\right)^{-1/2}\right\| \left\|V^{n+\frac{1}{2}}\right\|^2_{B_h^2} \\
\geq & \left(1 - \frac{\Delta t}{2}\mathrm{C}_1\right)\|U^n\|^2_{B_h^1} + \left(1 - \frac{\Delta t}{2}\mathrm{C}_2\right)\left\|V^{n+\frac{1}{2}}\right\|^2_{B_h^2},
\end{aligned} \tag{7.82}$$

where

$$\mathrm{C}_1 = \left\|\left(B_h^2\right)^{-1/2} R_h^T \left(B_h^1\right)^{-1/2}\right\| + \left\|\left(B_h^1\right)^{-1/2}\,{}^tD_h^1 \left(B_h^1\right)^{-1/2}\right\| \tag{7.83a}$$

$$\mathrm{C}_2 = \left\|\left(B_h^2\right)^{-1/2} R_h^T \left(B_h^1\right)^{-1/2}\right\| + \left\|\left(B_h^2\right)^{-1/2}\,{}^tD_h^2 \left(B_h^2\right)^{-1/2}\right\| \tag{7.83b}$$

Finally, it is easy to see that the scheme (7.68a) and (7.68b) is L^2 stable if Δt satisfies the condition:

$$\Delta t < \frac{2}{\max\left(\mathrm{C}_1\,,\ \mathrm{C}_2\right)}. \tag{7.84}$$

Remark: Under the CFL condition (7.84), the quadratic form $\mathscr{E}(U^n, V^{n+\frac{1}{2}})$ defines a norm which is equivalent to the L^2 norm: $\|U^n, V^{n+\frac{1}{2}}\|^2 = \|U^n\|^2_{B_h^1} + \left\|V^{n+\frac{1}{2}}\right\|^2_{B_h^2}$.

A similar way enables us to prove that the CFL condition in the context of non-dissipative scheme is:

$$\Delta t < \frac{2}{C_0} \tag{7.85}$$

where $C_0 = \left\| \left(B_h^2\right)^{-1/2} R_h^T \left(B_h^1\right)^{-1/2} \right\|$.

Consequently, (7.84) shows that the uncentered discretization in time of the diffusive terms in a DG method leads to a decrease of the CFL in comparison with the non-dissipative formulation. Unfortunately, this is the price to pay to keep an explicit scheme.

In [7], one can find a more explicit stability analysis for the scheme defined by (7.68a) and (7.68b) in the context of the DG approximation of the Maxwell equations.

7.2 Local Time Stepping

In this section, we are interested in the use of local time-stepping techniques in order to enhance the performance of explicit schemes in time. Actually, these last ones are stable under a CFL (Courant–Friedrichs–Lewy) condition of the form $\Delta t/h_{\min} \leq C(r)$ where Δt is the global time step, $h_{\min}$ is the smallest diameter of the cells of mesh and $C(r)$ is a positive constant generally dependent of the order r of the space approximation used in the considered scheme.

In some parts of the physical domain, if we want to take into account the geometrical details or capture a singularity of the solution, it is natural to use techniques of spatial local mesh refinement. So, the use of this kind of heterogeneous mesh leads to a significant decrease of the time step in order to guarantee the stability of the scheme.

Moreover if these zones are reduced to small parts of the computational domain, they induce in the complementary part a needless computational effort and are very penalizing from a cost point of view. To overcome this problem, the idea is to consider a local time-stepping method which uses time steps adapted to the different local spatial constraints to achieve the advance in time. This objective is non-trivial to achieve in the context of the wave propagation. In particular, some crucial issues for the development of a local time-stepping technique must be considered:

- Ensure the stability with the best time steps in each sub-domain,
- Keep the accuracy of the time discretization,
- The scheme must be at most "quasi" explicit and cheap.

Numerous works have been devoted to local time-stepping methods. In this section, we present three efficient approaches which can be used in the context of the approximations in space developed in this book. These methods are based on the conservation of a discrete energy which is a relevant property to ensure the stability of the scheme.

More precisely, in a first part, we present schemes based on a symplectic approach which are easy to implement and generally efficient in practice but whose stability issue is not totally solved. Then, we present a scheme which is optimal in terms of stability property but whose implementation is more complex and requires the solution of a small linear system. Finally, the last part concerns an efficient explicit local time-stepping scheme adapted to the second order wave equations.

Remark: There are also some schemes based on interpolation techniques [8–11] but they raise stability problems. We can also advise the reader to look at the interesting so-called ADER local time-stepping method [12].

7.2.1 *Symplectic Schemes for Conservative Approximations*

In this part, we present the technique proposed in [13] which is based on the symplectic schemes used for the Hamiltonian systems. The idea is to use a time approximation preserving the symplectic structure[3] to solve the ordinary differential equation in time (obtained after a non-dissipative DG discretization in space) in order to ensure the conservation of a discrete energy and thus to obtain the stability of the local time-stepping schemes.

This scheme uses a multi-class approach in which the cells (or the degrees of freedom in the context of a finite element method) are bulked into N sets or classes $1, 2, \ldots, N-1, N$ which are associated to the respective time-steps $\Delta t/2^{N-1}$, $\Delta t/2^{N-2}$, ..., $\Delta t/2$, Δt. So, the smallest cells are in class 1 and the largest cells in class N.

The method is based on the Verlet scheme, which is a reorganization of the classical Leapfrog scheme into three steps: let U^n and V^n be the fields at the time step n. The values of the fields U^{n+1} and V^{n+1} at the time $n+1$ by using the Verlet scheme are given by

$$V^{n+\frac{1}{2}} = V^n - \frac{\Delta t}{2} \left(B_h^2\right)^{-1} R_h^T\, U^n, \tag{7.86a}$$

$$U^{n+1} = U^n + \Delta t \, \left(B_h^1\right)^{-1} R_h\, V^{n+\frac{1}{2}}, \tag{7.86b}$$

$$V^{n+1} = V^{n+\frac{1}{2}} - \frac{\Delta t}{2} \left(B_h^2\right)^{-1} R_h^T\, U^{n+1}. \tag{7.86c}$$

[3] It is a structure associated to a manifold (whose the cotangent bundle is the phase space of the Hamiltonian system) which is defined by a skew-symmetric closed non-degenerate differential 2-form.

Remark: The scheme (7.86a)–(7.86c) is derived from the time-discretization of the generic system (7.5a) and (7.5b) where we have taken $R_h^1 = R_h$ and $R_h^2 = -R_h^T$ since we only want to consider the conservative discretization in space. For example, this system may be obtained after a DG approximation with centered flux.

The algorithm associated to these N-classes is defined by a recursive process labeled $R^N(\Delta t)$ and given by

1. Evaluate $R^{N-1}\left(\dfrac{\Delta t}{2}\right)$,
2. Evaluate the cells into class N by using the Verlet scheme (7.86a)–(7.86c),
3. Evaluate $R^{N-1}\left(\frac{\Delta t}{2}\right)$.

with the convention that R^0 performs no operation.

Remark: The scheme based on the algorithm $R^1(\Delta t)$ is exactly the Verlet scheme (7.86a–7.86c).

We now explicitly describe algorithm $R^2(\Delta t)$ (see [13]). For that, system (7.5a) and (7.5b) is reordered in the following obvious way:

$$\left\{\begin{aligned} B_{h1}^1 \frac{\mathrm{d}U_1}{\mathrm{d}t}(t) &= R_{h1}\, V_1(t) - A_{12}\, V_2(t) \\ B_{h1}^2 \frac{\mathrm{d}V_1}{\mathrm{d}t}(t) &= -R_{h1}^T\, U_1(t) + A_{21}^T\, U_2(t) \end{aligned}\right. \tag{7.87a}$$

$$\left\{\begin{aligned} B_{h2}^1 \frac{\mathrm{d}U_2}{\mathrm{d}t}(t) &= R_{h2}\, V_2(t) - A_{21}\, V_1(t) \\ B_{h2}^2 \frac{\mathrm{d}V_2}{\mathrm{d}t}(t) &= -R_{h2}^T\, U_2(t) + A_{12}^T\, U_1(t) \end{aligned}\right. \tag{7.87b}$$

where U_i and V_i represent the degrees of freedom of U and V in the class i and we have split the matrices as follows

$$B_h^k\, W = \begin{bmatrix} B_{h1}^k & 0 \\ 0 & B_{h1}^k \end{bmatrix} \begin{bmatrix} W_1 \\ W_2 \end{bmatrix}, \tag{7.88a}$$

$$R_h\, W = \begin{bmatrix} R_{h1} & -A_{12} \\ -A_{21} & R_{h2} \end{bmatrix} \begin{bmatrix} W_1 \\ W_2 \end{bmatrix}. \tag{7.88b}$$

Remark: The splitting in (7.88a) comes from the block diagonal structure of the mass matrices in the context of DG and mass-lumped FE approximations.

Now the steps of the algorithm $R^2(\Delta t)$ are:

$$B_{h1}^2 \frac{V_1^{n+\frac{1}{4}} - V_1^n}{\Delta t/4} = -R_{h1}^T U_1^n + A_{21}^T U_2^n \tag{7.89a}$$

$$B_{h1}^1 \frac{U_1^{n+\frac{1}{2}} - U_1^n}{\Delta t/2} = R_{h1} V_1^{n+\frac{1}{4}} - A_{12} V_2^n \tag{7.89b}$$

$$B_{h1}^2 \frac{V_1^{n+\frac{1}{2}} - V_1^{n+\frac{1}{4}}}{\Delta t/4} = -R_{h1}^T U_1^{n+\frac{1}{2}} + A_{21}^T U_2^n \tag{7.89c}$$

$$B_{h2}^2 \frac{V_2^{n+\frac{1}{2}} - V_2^n}{\Delta t/2} = -R_{h2}^T U_2^n + A_{12}^T U_1^{n+\frac{1}{2}} \tag{7.89d}$$

$$B_{h2}^1 \frac{U_2^{n+1} - U_2^n}{\Delta t} = R_{h2} V_2^{n+\frac{1}{2}} - A_{21} V_1^{n+\frac{1}{2}} \tag{7.89e}$$

$$B_{h2}^2 \frac{V_2^{n+1} - V_2^{n+\frac{1}{2}}}{\Delta t/2} = -R_{h2}^T U_2^{n+1} + A_{12}^T U_1^{n+\frac{1}{2}} \tag{7.89f}$$

$$B_{h1}^2 \frac{V_1^{n+\frac{3}{4}} - V_1^{n+\frac{1}{2}}}{\Delta t/4} = -R_{h1}^T U_1^{n+\frac{1}{2}} + A_{21}^T U_2^{n+1} \tag{7.89g}$$

$$B_{h1}^1 \frac{U_1^{n+1} - U_1^{n+\frac{1}{2}}}{\Delta t/2} = R_{h1} V_1^{n+\frac{3}{4}} - A_{12} V_2^{n+1} \tag{7.89h}$$

$$B_{h1}^2 \frac{V_1^{n+1} - V_1^{n+\frac{3}{4}}}{\Delta t/4} = -R_{h1}^T U_1^{n+1} + A_{21}^T U_2^{n+1} \tag{7.89i}$$

The main drawbacks of this approach are:

- the discrete energy is generally non-explicit and, consequently, it is not possible to derive an estimation of the CFL condition,
- these schemes are based on loop time (the Verlet method) more expensive than the classical Leapfrog scheme,
- the resulting scheme is only first-order accurate.

Remark: Although the Verlet scheme is symplectic, we can not *a priori* affirm that this recursive scheme is symplectic too. Actually, using the Verlet scheme for different classes requires values of the fields at some unknown times. To overcome this difficulty, the last known fields available are used. Consequently, the scheme which is recursively called in [13] is not exactly the Verlet scheme. Moreover, the symplectic and stability properties of this multi-classes method are still an open question, despite the fact that an energy conservation is proved in [13]. In particular, for long time simulations, it can be necessary to reduce the time steps in order to avoid instabilities. Nevertheless, this recursive scheme remains attractive. Actually, regardless of the number of classes N, this scheme is easy to implement, fully explicit, it does not need any additional storage and gives good results: the numerical solution is comparable to standard ones and the computational time is significantly reduced. Thanks to the time-coincidence of fields computed with the Verlet scheme, writing of a Verlet-based recursive scheme is natural.

From the same idea, in [7] we propose a Leapfrog based recursive method, better adapted to our scheme and more efficient in terms of computational time.

For instance, for $N = 2$, we propose the following multi-class Leapfrog method at cells located at the interface between class 1 and 2 that can be written as

$$B_{h2}^2 \frac{V_2^{n+\frac{1}{2}} - V_2^{n-\frac{1}{2}}}{\Delta t} = -R_{h2}^T U_2^n + A_{12}^T U_1^n \tag{7.90a}$$

$$B_{h1}^2 \frac{V_1^{n+\frac{1}{6}} - V_1^{n-\frac{1}{6}}}{\Delta t/3} = -R_{h1}^T U_1^n + A_{21}^T U_2^n \tag{7.90b}$$

$$B_{h1}^1 \frac{U_1^{n+\frac{2}{6}} - U_1^n}{\Delta t/3} = R_{h1} V_1^{n+\frac{1}{6}} - A_{12} \left(V_2^{n+\frac{1}{2}}\right)^\star \tag{7.90c}$$

$$B_{h1}^2 \frac{V_1^{n+\frac{1}{2}} - V_1^{n+\frac{1}{6}}}{\Delta t/3} = -R_{h1}^T U_1^{n+\frac{2}{6}} + A_{21}^T \left(U_2^n\right)^\star \tag{7.90d}$$

$$B_{h2}^1 \frac{U_2^{n+1} - U_2^n}{\Delta t} = R_{h2} V_2^{n+\frac{1}{2}} - A_{21} V_1^{n+\frac{1}{2}} \tag{7.90e}$$

$$B_{h1}^1 \frac{U_1^{n+\frac{4}{6}} - U_1^{n+\frac{2}{6}}}{\Delta t/3} = R_{h1} V_1^{n+\frac{1}{2}} - A_{12} V_2^{n+\frac{1}{2}} \tag{7.90f}$$

$$B_{h1}^2 \frac{V_1^{n+\frac{5}{6}} - V_1^{n+\frac{1}{2}}}{\Delta t/3} = -R_{h1}^T U_1^{n+\frac{4}{6}} + A_{21}^T \left(U_2^{n+1}\right)^\star \tag{7.90g}$$

$$B_{h1}^1 \frac{U_1^{n+1} - U_1^{n+\frac{4}{6}}}{\Delta t/3} = R_{h1} V_1^{n+\frac{5}{6}} - A_{12} \left(V_2^{n+\frac{1}{2}}\right)^\star \tag{7.90h}$$

As for the previous scheme (7.89a)–(7.89i), we replace fields at unknown times by the last known values, denoted * in (7.90a)–(7.90h).

Remark: For cells which are not located at the interface between cells 1 and 2, the classical Leapfrog is applied (with the time-step of the class).

More generally, if we label $LeapFrogV(n, \Delta t)$ (resp. $LeapFrogU(n, \Delta t)$) the equation (7.9a) (resp. (7.9b)) applied to the cells belonging at the class N with a time-step Δt, we can define the multi-class Leapfrog method as a recursive process. A step of integration of the recursive Leapfrog method is defined by

$$\begin{cases} 1.\ ComputeV(N, \Delta t) \\ 2.\ ComputeU(N, \Delta t) \end{cases} \tag{7.91}$$

where the recursive functions $ComputeV(N, \Delta t)$ and $ComputeU(N, \Delta t)$ are respectively defined by

$$\begin{array}{ll} \underline{ComputeV(N, \Delta t)}: & \underline{ComputeU(N, \Delta t)}: \\ \begin{cases} -\ LeapFrogV\ (N, \Delta t) \\ -\ ComputeV\left(N-1, \frac{\Delta t}{3}\right) \\ -\ ComputeU\left(N-1, \frac{\Delta t}{3}\right) \\ -\ ComputeV\left(N-1, \frac{\Delta t}{3}\right) \end{cases} & \begin{cases} -\ LeapFrogU\ (N, \Delta t) \\ -\ ComputeU\left(N-1, \frac{\Delta t}{3}\right) \\ -\ ComputeV\left(N-1, \frac{\Delta t}{3}\right) \\ -\ ComputeU\left(N-1, \frac{\Delta t}{3}\right) \end{cases} \end{array}$$

with $ComputeV\ (1, \delta t)$ defined by $LeapFrogV\ (1, \delta t)$ and $ComputeU\ (1, \delta t)$ defined by $LeapFrogU\ (1, \delta t)$, where δt denotes the time-step.

Remark: Since the Leapfrog is composed of only two steps (instead of three for the Verlet scheme), this method requires 33 % less computation than the Verlet-based recursive scheme, with the same advantages: the scheme is fully explicit, easy to write, does not require additional storage and gives good results with a CPU time significantly reduced. However, we have also the same problems for the stability study as for the Verlet-based method: the CFL must sometimes be strengthened for long-time simulations to ensure stability.

To finish this part, we give a numerical comparison (see [7] for more details) between the Recursive Verlet (R-V) and the Leapfrog (R-LF) algorithms. We consider a generic missile (see [7]) illuminated by a plane wave.

We give in Table 7.2 the repartition of cells by classes respectively obtained with the two methods. We can notice, for all the methods, the low percentage of small cells, but also important cells-size disparities. For example, there is at least a factor 2^{10} between small and largest cells of the missile mesh (because the R-V scheme uses 10 classes of cells). This explains the real efficiency of multi-class recursive methods on such meshes.

In [7], we show the gain of computational time obtained versus the standard Leapfrog scheme (without local time-stepping) and the two local time-stepping methods. For this example, the improvement obtained with recursive methods is very significant: the R-V and R-LF methods lead to schemes 11 and 15 times faster

Table 7.2 Cell repartition by classes for the missile mesh (~91000 cells)

Scheme/Class	1	2	3	4	5	6	7	8	9	10
R-LF	10	200	1400	14300	71600	3500	×	×	×	×
R-V	8	16	160	550	1500	5800	46000	33500	3300	200

than without local time-stepping, respectively. In such a case, the requirement of multi-class strategies is clearly highlighted.

7.2.2 *Scheme Based on a Lagrange Multiplier*

In this part, we briefly present the space time mesh refinement method proposed and studied in [14–17]. This method is based on domain decomposition techniques and is constructed in order to ensure an *a priori* conservation of a classical discrete energy to obtain the numerical stability of the scheme.

We now re-use the abstract formalism proposed in Chap. 4 to present the method.

$$\underline{\underline{\lambda}}(\underline{x})\frac{\partial \underline{u}}{\partial t}(\underline{x},t) + A\underline{v}(\underline{x},t) = \underline{f}(\underline{x},t) \text{ in } \Omega, \tag{7.92a}$$

$$\underline{\underline{\mu}}(\underline{x})\frac{\partial \underline{v}}{\partial t}(\underline{x},t) - A^*\underline{u}(\underline{x},t) = 0 \text{ in } \Omega, \tag{7.92b}$$

$$A^*(\underline{n})\,\underline{u}(\underline{x},t) = 0 \text{ on } \partial\Omega, \tag{7.92c}$$

where $\underline{u}(\underline{x},t) \in \mathbb{R}^p$, $\underline{v}(\underline{x},t) \in \mathbb{R}^q$, $\underline{f}(\underline{x},t) \in \mathbb{R}^p$, $\underline{\underline{\lambda}}(\underline{x}) \in \mathbb{R}^p \times \mathbb{R}^p$, $\underline{\underline{\mu}}(\underline{x}) \in \mathbb{R}^q \times \mathbb{R}^q$ are symmetric, definite, positive matrices, $\underline{n} = (n_i)_{i=1}^d$ is the exterior normal to $\partial\Omega$, $A(\underline{n}) = \sum_{i=1}^d A_i n_i$, $A^*(\underline{n}) = -\sum_{i=1}^d A_i^T n_i$ and A does not depend on space.

The domain Ω is such that $\Omega = \Omega_c \cup \Omega_f$ with $\Omega_c \cap \Omega_f = \emptyset$ and $\overline{\Omega}_c \cap \overline{\Omega}_f = \Gamma$. The unit outgoing normal vectors to Ω_c and Ω_f are $\underline{n}_c$ and $\underline{n}_f$ respectively.

The starting point of this method consists in reformulating the problem (7.92a–7.92c) as a transmission problem between Ω_c and Ω_f

$$\underline{\underline{\lambda_c}}(\underline{x})\frac{\partial \underline{u}_c}{\partial t}(\underline{x},t) + A\underline{v}_c(\underline{x},t) = \underline{f}_c(\underline{x},t) \text{ in } \Omega_c, \tag{7.93a}$$

$$\underline{\underline{\mu_c}}(\underline{x})\frac{\partial \underline{v}_c}{\partial t}(\underline{x},t) - A^*\underline{u}_c(\underline{x},t) = 0 \text{ in } \Omega_c, \tag{7.93b}$$

$$A^*(\underline{n}_c)\,\underline{u}_c(\underline{x},t) = 0 \text{ on } \partial\Omega \cap \partial\Omega_c, \tag{7.93c}$$

$$\underline{\underline{\lambda}}_f(\underline{x})\frac{\partial \underline{u}_f}{\partial t}(\underline{x},t) + A\underline{v}_f(\underline{x},t) = \underline{f}_f(\underline{x},t) \text{ in } \Omega_f, \tag{7.93d}$$

$$\underline{\underline{\mu}}_f(\underline{x})\frac{\partial \underline{v}_f}{\partial t}(\underline{x},t) - A^*\underline{u}_f(\underline{x},t) = 0 \text{ in } \Omega_f, \tag{7.93e}$$

$$A^*(\underline{n}_f)\,\underline{u}_f(\underline{x},t) = 0 \text{ on } \partial\Omega \cap \partial\Omega_f, \tag{7.93f}$$

$$A^*(\underline{n}_c)\,\underline{u}_c(\underline{x},t) = -A^*(\underline{n}_f)\,\underline{u}_f(\underline{x},t) \text{ on } \Gamma, \tag{7.93g}$$

$$\underline{j}_c = \underline{j}_f \text{ on } \Gamma, \tag{7.93h}$$

where $\underline{j}_c = A(\underline{n}_c)\,\underline{v}_c(\underline{x},t), \underline{j}_f = -A(\underline{n}_f)\,\underline{v}_f(\underline{x},t)$ and $\underline{w}_s$ denotes the restriction of the function $\underline{w}$ to the subdomain Ω_s with $s = c, f$.

Remark: Problems (7.92a)–(7.92c) and (7.93a)–(7.93h) are equivalent.

We now use the DG formalism introduced in Chap. 4 in order to present the method both for the FE and DG conservative approximations described in this book.

Let $\mathcal{T}_{h_c}(\Omega_c)$, $\mathcal{T}_{h_f}(\Omega_f)$ and $\mathcal{T}_H(\Gamma)$ be the meshes of the subdomain Ω_c, Ω_f and Γ respectively. We define for $s = c$ or f the sets:

$$\mathcal{F}_h^s = \{F : \text{face of } \mathcal{T}_{h_s} \mid F \cap \Gamma = \emptyset\}. \tag{7.94}$$

The conservative DG formulation of the problem defined by (7.93a)–(7.93f) is: Find $(\underline{u}_c, \underline{v}_c, \underline{j}_c) \in X_{h,c} \times M_{h,f} \times G_H$ and $(\underline{u}_f, \underline{v}_f, \underline{j}_f) \in X_{h,f} \times M_{h,f} \times G_H$ such that for all $(\tilde{\underline{u}}_c, \tilde{\underline{v}}_c) \in X_{h,c} \times M_{h,c}$ and $(\tilde{\underline{u}}_f, \tilde{\underline{v}}_f) \in X_{h,f} \times M_{h,f}$,

$$\begin{aligned}\frac{\mathrm{d}}{\mathrm{d}t}\int_{\Omega_c} \underline{\underline{\lambda}}\,\underline{u}_c \cdot \tilde{\underline{u}}_c\,\mathrm{d}\underline{x} + \int_{\Omega_c} \underline{v}_c \cdot A_h^*\tilde{\underline{u}}_c\,\mathrm{d}\underline{x} - \int_{\mathcal{F}_h^c} \{\underline{v}_c\} \cdot [\![A^*(\underline{n})\tilde{\underline{u}}_c]\!]\mathrm{d}\sigma \\ + \int_\Gamma \underline{j}_c \cdot \tilde{\underline{u}}_c d\sigma = \int_{\Omega_c} \underline{f}_c \cdot \tilde{\underline{u}}_c\,\mathrm{d}\underline{x},\end{aligned} \tag{7.95a}$$

$$\frac{\mathrm{d}}{\mathrm{d}t}\int_{\Omega_c} \underline{\underline{\mu}}\underline{v}_c \cdot \tilde{\underline{v}}_c\,\mathrm{d}\underline{x} - \int_{\Omega_c} A_h^*\underline{u}_c \cdot \tilde{\underline{v}}_c\,\mathrm{d}\underline{x} + \int_{\mathcal{F}_h^c} [\![A^*(\underline{n})\underline{u}_c]\!] \cdot \{\tilde{\underline{v}}_c\}\mathrm{d}\sigma = 0, \tag{7.95b}$$

$$\begin{aligned}\frac{\mathrm{d}}{\mathrm{d}t}\int_{\Omega_f} \underline{\underline{\lambda}}\underline{u}_f \cdot \tilde{\underline{u}}_f\,\mathrm{d}\underline{x} + \int_{\Omega_f} \underline{v}_f \cdot A_h^*\tilde{\underline{u}}_f\,\mathrm{d}\underline{x} - \int_{\mathcal{F}_h^f} \{\underline{v}_f\} \cdot [\![A^*(\underline{n})\tilde{\underline{u}}_f]\!]\mathrm{d}\sigma \\ + \int_\Gamma \underline{j}_f \cdot \tilde{\underline{u}}_f d\sigma = \int_{\Omega_f} \underline{f}_f \cdot \tilde{\underline{u}}_f\,\mathrm{d}\underline{x},\end{aligned} \tag{7.95c}$$

$$\frac{\mathrm{d}}{\mathrm{d}t}\int_{\Omega_f} \underline{\underline{\mu}}\,\underline{v}_f \cdot \underline{\tilde{v}}_f \,\mathrm{d}\underline{x} - \int_{\Omega_f} A_h^* \underline{u}_f \cdot \underline{\tilde{v}}_f \,\mathrm{d}\underline{x} + \int_{\mathscr{F}_h^f} [\![A^*(\underline{n})\underline{u}_f]\!] \cdot \{\underline{\tilde{v}}_f\}\mathrm{d}\sigma = 0, \tag{7.95d}$$

where A_h and A_h^* correspond to the elementwise contribution of the operators A and A^* respectively, and $(X_{h,s}\,,\, M_{h,s})_{s=c,f}$ and G_H are adequate approximate spaces.

Remark: In the context of a conformal finite element approximation of problem (7.93a)–(7.93f), we have that $\forall \underline{\tilde{u}}_s \in X_{h,s}$, $[\![A(n)^*\underline{\tilde{u}}_s]\!]_{|F} = 0, \;\; \forall F \in \mathscr{F}_h^s$ with $s = c, f$ and, consequently, formulation (7.95a)–(7.95d) leads a mixed finite element formulation.

Let us look at the conservation properties of the DG scheme. For that, we introduce the semi-discrete energy:

$$\begin{aligned} E_h(t) = {} & \frac{1}{2}\left(\int_{\Omega_c} \underline{\underline{\lambda}}_c \underline{u}_c \cdot \underline{u}_c \,\mathrm{d}\underline{x} + \int_{\Omega_c} \underline{\underline{\mu}}_c \underline{v}_c \cdot \underline{v}_c \,\mathrm{d}\underline{x}\right) \\ & + \frac{1}{2}\left(\int_{\Omega_f} \underline{\underline{\lambda}}_f \underline{u}_f \cdot \underline{u}_f \,\mathrm{d}\underline{x} + \int_{\Omega_f} \underline{\underline{\mu}}_f \underline{v}_f \cdot \underline{v}_f \,\mathrm{d}\underline{x}\right). \end{aligned} \tag{7.96}$$

By choosing $\underline{\tilde{u}}_s = \underline{u}_s$ and $\underline{\tilde{v}}_s = \underline{v}_s$, for $s = c, f$ in (7.95a)–(7.95d) and by summing all the equations, we obtain the first identity (we set $\underline{f} = 0$):

$$\frac{\mathrm{d}E_h}{\mathrm{d}t}(t) = -\int_{\Gamma} \underline{j}_c \cdot \underline{u}_c \, d\sigma - \int_{\Gamma} \underline{j}_f \cdot \underline{u}_f \,\mathrm{d}\sigma \tag{7.97}$$

From the mortar finite element theory, we know that a compatibility condition between the spaces $X_{h,c}, X_{h,f}$ and G_H is needed to ensure the well-posedness character of the semi-discrete problem. More precisely, the approximate trace operator on Γ must verify a discrete inf-sup condition (see the previous quoted papers for more details on the choice of the finite dimensional space G_H). It is well-known that the natural approach is to choose a space G_H verifying the inclusion property:

$$G_H \subset \{A(n_c)\underline{\tilde{v}}_{c|\Gamma} \;:\; \underline{\tilde{v}}_{c|\Gamma} \in X_{h,c}\}. \tag{7.98}$$

For now, we assume that (7.98) holds and we take the following discretization of the transmission conditions (7.93g)–(7.93h)

$$\int_{\Gamma} \left(\underline{u}_c + \underline{u}_f\right) \cdot \underline{\tilde{j}}\,\mathrm{d}\sigma = 0, \;\; \forall \underline{\tilde{j}} \in G_H, \tag{7.99a}$$

$$\underline{j}_c = \underline{j}_f. \tag{7.99b}$$

So, (7.97) and (7.99a)–(7.99b) lead to $\dfrac{\mathrm{d}E_h}{\mathrm{d}t}(t) = 0$ and the conservation of the discrete energy E_h.

Let U_c, V_c, J_c and U_f, V_f, J_f be the coordinate vectors of the discrete unknowns $\underline{u}_c$, $\underline{v}_c$, $\underline{j}_c$ and $\underline{u}_f$, $\underline{v}_f$, $\underline{j}_f$ respectively. We now write the matrix formulation of the semi-discrete formulation (7.95a)–(7.95d) and (7.99a)–(7.99b):

$$B^1_{h,c}\frac{\mathrm{d}}{\mathrm{d}t}U_c + R_{h,c}V_c + S_{h,c}J_c = F_c, \tag{7.100a}$$

$$B^2_{h,c}\frac{\mathrm{d}}{\mathrm{d}t}V_c - R^*_{h,c}U_c = 0, \tag{7.100b}$$

$$B^1_{h,f}\frac{\mathrm{d}}{\mathrm{d}t}U_f + R_{h,f}V_f + S_{h,f}J_f = F_f, \tag{7.100c}$$

$$B^2_{h,f}\frac{\mathrm{d}}{\mathrm{d}t}V_f - R^*_{h,f}U_f = 0, \tag{7.100d}$$

$$S^*_{h,c}U_c + S^*_{h,f}U_f = 0, \tag{7.100e}$$

$$J_c = J_f \tag{7.100f}$$

where $B^1_{h,c}$, $B^2_{h,c}$, $B^1_{h,f}$, $B^2_{h,f}$ are the mass matrices, $R_{h,c}$, $R_{h,f}$ the stiffness matrices and $S_{h,c}$, $S_{h,f}$ the matrices associated to the discrete trace operators on Γ.

Remark: If we multiply (7.100a) and (7.100c) by $S^*_{h,c}\left(B^1_{h,c}\right)^{-1}$ and $S^*_{h,f}\left(B^1_{h,f}\right)^{-1}$ and we use the discretization of the transmission conditions, we can see that $J = J_c = J_f$, we show that the Lagrange multiplier J is a solution of the linear system:

$$\begin{aligned}&\left[S^*_{h,c}\left(B^1_{h,c}\right)^{-1}S_{h,c} + S^*_{h,f}\left(B^1_{h,f}\right)^{-1}S_{h,f}\right]J\\ &= S^*_{h,c}\left(B^1_{h,c}\right)^{-1}\left(F_c - R_{h,c}V_c\right) + S^*_{h,f}\left(B^1_{h,f}\right)^{-1}\left(F_f - R_{h,f}V_f\right).\end{aligned} \tag{7.101}$$

Equation (7.98) ensures the invertibility of the system (7.101).

We now describe the time discretization of the matrix formulation (7.100a)–(7.100f). We only consider the 1-2 refinement. In this case, we have $h_c = 2h_f$ and we use a time step $\Delta t_c = 2\Delta t$ in Ω_c and $\Delta t_f = \Delta t$ in Ω_f. The fully discrete matrix formulation proposed in [15] for (7.100a–7.100d) is:

$$B^2_{h,c}\frac{V_c^{2n+1} - V_c^{2n-1}}{2\Delta t} - R^*_{h,c}U_c^{2n} = 0, \tag{7.102a}$$

$$B^1_{h,c}\frac{U_c^{2n+2} - U_c^{2n}}{2\Delta t} + R_{h,c}V_c^{2n+1} + S_{h,c}J_c^{2n+1} = F_c^{2n+1}, \tag{7.102b}$$

$$B^2_{h,f}\frac{V_f^{2n+\frac{1}{2}}-V_f^{2n-\frac{1}{2}}}{\Delta t}-R^*_{h,f}U_f^{2n}=0, \tag{7.102c}$$

$$B^1_{h,f}\frac{U_f^{2n+1}-U_f^{2n}}{\Delta t}+R_{h,f}V_f^{2n+\frac{1}{2}}+S_{h,f}J_f^{2n+\frac{1}{2}}=F_f^{2n+\frac{1}{2}}, \tag{7.102d}$$

$$B^2_{h,f}\frac{V_f^{2n+\frac{3}{2}}-V_f^{2n+\frac{1}{2}}}{\Delta t}-R^*_{h,f}U_f^{2n+1}=0, \tag{7.102e}$$

$$B^1_{h,f}\frac{U_f^{2n+2}-U_f^{2n+1}}{\Delta t}+R_{h,f}V_f^{2n+\frac{3}{2}}+S_{h,f}J_f^{2n+\frac{3}{2}}=F_f^{2n+\frac{3}{2}}. \tag{7.102f}$$

The choice of time discretization of the semi-discrete transmission conditions is guided by forcing of the conservation of the total discrete energy $E^{2n}=E_c^{2n}+E_f^{2n}$ where

$$E_c^{2n}=\left(B^1_{h,c}\,U_c^{2n}\,,\,U_c^{2n}\right)+\left(B^2_{h,c}\,V_c^{2n+1}\,,\,V_c^{2n-1}\right), \tag{7.103a}$$

$$E_f^{2n}=\left(B^1_{h,f}\,U_f^{2n}\,,\,U_f^{2n}\right)+\left(B^2_{h,f}\,V_f^{2n+\frac{1}{2}}\,,\,V_f^{2n-\frac{1}{2}}\right). \tag{7.103b}$$

A classical calculus of variations of E leads to ([18])

$$\begin{aligned}\frac{E^{2n+2}-E^{2n}}{2\Delta t}=&\left(S^*_{h,f}\frac{U_f^{2n+2}+U_f^{2n+1}}{4}\,,\,J^{2n+\frac{3}{2}}\right)\\&+\left(S^*_{h,f}\frac{U_f^{2n+1}+U_f^{2n}}{4}\,,\,J^{2n+\frac{1}{2}}\right)\\&+\left(S^*_{h,c}\frac{U_c^{2n+2}+U_c^{2n}}{2}\,,\,J^{2n+1}\right)\end{aligned} \tag{7.104}$$

Finally, [18] proposes the following time approximation of the transmission conditions:

$$S^*_{h,c}\frac{U_c^{2n+2}+U_c^{2n}}{2}+S^*_{h,f}\frac{U_f^{2n+2}+2U_f^{2n+1}+U_f^{2n}}{4}=0, \tag{7.105a}$$

$$J_c^{2n+1}=J_f^{2n+\frac{1}{2}}=J_f^{2n+\frac{3}{2}} \tag{7.105b}$$

which immediately provides the conservation of energy i.e. $E^{2n+2}=E^{2n}$.

This property and the classical form of the discrete energy enable us to prove L^2-stability of the scheme. More precisely, the scheme is L^2-stable if the following two CFL conditions hold:

$$\Delta t_s < \frac{2}{|||R^*_{h,s}|||}, \; s = c, f \tag{7.106}$$

where

$$|||R^*_{h,s}||| = \sup_{U_s} \frac{\left(\left(B^2_{h,s}\right)^{-1} R^*_{h,s} U_s \,,\, U_s\right)}{\left(B^1_{h,s} U_s \,,, U_s\right)} \tag{7.107}$$

with $(\cdot\,,\,\cdot)$ denotes the Euclidean scalar product.

Condition (7.106) represents the stability conditions of the scheme applied in each subdomain and we can see that the coupling between the two subdomains does not penalize the stability. In other words, the optimal time step on each grid can be chosen.

As we already saw for the semi-discrete matrix system, the computation of vectors V_c^{2n+1}, U_c^{2n+2}, $V_f^{2n+\frac{1}{2}}$, U_f^{2n+1}, $V_f^{2n+\frac{3}{2}}$, V_f^{2n+2} and J_c^{2n+1} needs to solve a small linear system. Actually, J_c^{2n+1} is derived from

$$\mathcal{M} J_c^{2n+1} = G \tag{7.108}$$

where $\mathcal{M} = \Delta t S^*_{h,c} \left(B^1_{h,c}\right)^{-1} S_{h,c} + \Delta t S^*_{h,f} \left(B^1_{h,f}\right)^{-1} \mathcal{N}_f \left(B^1_{h,f}\right)^{-1} S_{h,f}$ with $\mathcal{N}_f = B^1_{h,f} - \frac{\Delta t^2}{4} R_{h,f} \left(B^2_{h,f}\right)^{-1} R^*_{h,f}$ and G is a given vector.

Remarks:

1. The linear system (7.108) is well-posed under the CFL condition (7.106) on the fine grid.
2. The discretization of the transmission conditions (7.105a)–(7.105b) is not centered and it is only first-order accurate. A fine a priori error analysis in L^2-norm [19] shows that the scheme is actually of order $\frac{3}{2}$.
3. The scheme defined by (7.102a)–(7.102f) and (7.105a)–(7.105b) produces parasitic high frequency waves when the waves cross the boundary Γ. These waves are generally evanescent and it is sufficient to decrease the CFL condition to obtain a good result. Nevertheless, in [16], a post-processed scheme is proposed in order to suppress these waves. This scheme is based on the following averaging approach:

$$\overline{U}_c^{2n+1} = \frac{U_c^{2n+2} + U_c^{2n}}{2}, \tag{7.109a}$$

$$\overline{U}_f^{n} = \frac{U_f^{n+1} + 2U_f^{n} + U_f^{n-1}}{4}, \tag{7.109b}$$

$$\overline{V}_c^{2n} = \frac{V_c^{2n+1} + V_c^{2n-1}}{2}, \tag{7.109c}$$

$$\overline{V}_f^{n+\frac{1}{2}} = \frac{V_f^{n+\frac{3}{2}} + 2V_f^{n+\frac{1}{2}} + V_f^{n-\frac{1}{2}}}{4}. \qquad (7.109d)$$

Since the averaging is centered, this new scheme is second-order accurate in time.

4. This scheme can be generalized to any rational space-time refinement rate between the coarse and the fine grid [18].
5. In [18], another approach is proposed. This approach is Lagrange multiplier free and avoids the non-obvious issue related to the choice of space G_H. It is based on a coupling method in which the formulation used for the fine grid is dual to this one used for the coarse grid. This scheme still requires the inversion of a small linear system. In [17], we can find an extension of this approach to the discontinuous Galerkin approximation of symmetric first order hyperbolic systems. To our knowledge, there are few numerical results in the literature using this approach. We have only found a 1D numerical validation for the linearized Euler equations in [17]. This explains our choice to present only the version with a Lagrange multiplier whose numerical robustness has been proven.

7.2.3 An Explicit Conservative Scheme for Second-Order Wave Equations

In this part, we present the conservative local time-stepping scheme introduced in [20]. This scheme is fully explicit, stable and easy to implement.

Let us consider the matrix semi-discrete in space approximation of a wave equation introduced in (7.1):

$$M_h \frac{\mathrm{d}^2 U}{\mathrm{d}t^2}(t) - K_h\, U(t) = 0. \qquad (7.110)$$

Remark: We have chosen $F = 0$ for sake of simplicity.

The mass matrix M_h is symmetric and definite positive. So, we can rewrite the system (7.110) under the form:

$$\frac{\mathrm{d}^2 V}{\mathrm{d}t^2}(t) + A\, V(t) = 0, \qquad (7.111)$$

where $V = M_h^{1/2} U$ and $A = -M_h^{-1/2} K_h M_h^{-1/2}$.

We begin with the time discretization. We use with the integral form of the remainder:

$$V(t_{n+1}) - 2V(t_n) + V(t_{n-1}) = \Delta t^2 \int_{-1}^{1} (1 - |\theta|) V''(t_n + \theta \Delta t) \mathrm{d}\theta. \qquad (7.112)$$

Using (7.111), i.e. $V''(t) + AV(t) = 0$, we get:

$$V(t_{n+1}) - 2V(t_n) + V(t_{n-1}) = -\Delta t^2 \int_{-1}^{1} (1 - |\theta|)AV(t_n + \theta\Delta t)\mathrm{d}\theta. \quad (7.113)$$

The construction of the Leapfrog scheme is done by considering the second-order derivative constant on the interval $[t_{n-1}, t_{n+1}]$, i.e. $\forall\theta \in [-1, 1]$, $AV(t_n + \theta\Delta t) \simeq AV(t_n)$. In this case, if we denote by V^n the approximation of $V(t_n)$ and we get the following standard Leapfrog scheme:

$$V^{n+1} - 2V^n + V^{n-1} = -\Delta t^2 AV^n. \quad (7.114)$$

Unfortunately, the stability condition of this scheme, $\Delta t \leq 2/\sqrt{\lambda_{max}(A)}$, is linked to the largest eigenvalue of the matrix A, $\lambda_{max}(A)$, which is proportional to $1/h_f$ where h_f is the space-step of the fine part. In other words, Δt is globally penalized by the fine part, which is very binding and unrealistic.

The principle of the construction of the local-time stepping scheme is to improve the previous scheme by considering the following approximation:

$$\begin{aligned} AV(t_n + \theta\Delta t) &= AV_c(t_n + \theta\Delta t) + AV_f(t_n + \theta\Delta t) \\ &\simeq A(I - P)V(t_n) + APV(t_n + \theta\Delta t) \end{aligned} \quad (7.115)$$

where P is the canonical restriction to the fine part. We now denote $Q = (I - P)$ and we assume that the degrees of freedom are sorted in the following form:

$$V = \begin{pmatrix} V_c \\ \\ V_f \end{pmatrix}. \quad (7.116)$$

Remark: By using (7.116), the canonical restriction reads:

$$P \begin{pmatrix} V_c \\ \\ V_f \end{pmatrix} = \begin{pmatrix} 0 \\ \\ V_f \end{pmatrix}. \quad (7.117)$$

We now have to define an approximation of $APV(t_n + \theta\Delta t)$.

A classical way to do this is to add unknowns $V^{n+m/p}$ (for $m = -p + 1, \cdots, p - 1$) approximations of $V(t_n + m/p\Delta t)$ in the fine part. We then calculate the unknowns in the coarse and fine parts using a Leapfrog scheme and a specific treatment must be done to link the two parts while ensuring the stability and order of consistency of the overall scheme. This task is difficult to achieve optimally.

To avoid this difficulty, J. Diaz and M. Grote use a different approach. Their idea is to construct a global a priori approximation of $APV(t_n + \theta\Delta t)$ and not to make a global *a posteriori* connection.

1. We first construct a scheme with only one global time-step based on the exact computation, at each time-step, of the solution in the fine part. We call this scheme the exact scheme (see Part 7.2.3.1).
2. The exact scheme is next approximated and leads to the local-time stepping method (see Part 7.2.3.2).

7.2.3.1 Exact Scheme: Solution of the Fine Part Without Local-Time Approximation

We introduce the following second-order ordinary differential equation:
Find $\tilde{z} : [-\Delta t, \Delta t] \to \mathbb{R}^n$ solution of

$$\tilde{z}''(\tau) = -A(I-P)V(t_n) - AP\tilde{z}(\tau), \tag{7.118a}$$

$$\tilde{z}(0) = V(t_n), \tag{7.118b}$$

$$\tilde{z}'(0) = \nu, \tag{7.118c}$$

where $\nu \in \mathbb{R}^n$ is a free constant vector parameter to be made precise later on.

Remark: It is important to remember for later that $\tilde{z}$ is only intended to provide $\tilde{z}''(\tau) \simeq V''(t_n+\tau)$ and not to give an approximation of V! It should be noted that $\tilde{z}(\tau)$ is generally not a "good" approximation of $V(t_n+\tau)$. This is usually an order 1 approximation because of the first initial condition. We can nevertheless improve that by taking $\nu = V'(t_n)$ but we shall see later that this choice is not the most appropriate.

We now give some properties on the differential problem (7.118a)–(7.118c).

Proposition 4 • *The expression of $\tilde{z}$ in the fine part is*

$$P\tilde{z}(\tau) = \cos((PAP)^{1/2}\tau)PV(t_n) + \sin((PAP)^{1/2}\tau)(PAP)^{-1/2}P\nu + (\cos((PAP)^{1/2}\tau) - 1)(PAP)^{-1}PA(I-P)V(t_n). \tag{7.119}$$

• *$\tilde{z}''(\tau)$ is an approximation de $V''(t_n)$ given by*

$$\tilde{z}''(\tau) = V''(t_n) - \sum_{n=1}^{+\infty} \frac{(-1)^n}{(2n)} A(PAP)^n PV(t_n)\tau^{2n} - \sum_{n=0}^{+\infty} \frac{(-1)^n}{(2n+1)!} A(PAP)^n P\nu\tau^{2n+1} + \sum_{n=1}^{+\infty} \frac{(-1)^n}{(2n)} A(PAP)^{n-1} PA(I-P)PV(t_n)\tau^{2n}. \tag{7.120}$$

- *The expressions of* $\Theta_n = \Delta t^2 \int_{-1}^{1} (1 - |\theta|)\tilde{z}''(\theta \Delta t)\mathrm{d}\theta$ *are*

$$\Theta_n = \begin{cases} 2(\cos((PAP)^{1/2}\Delta t) - I)(PAP)^{-1}PAV(t_n) \textit{ (fine zone)}, \\ 2QA(PAP)^{-1}(\cos((PAP)^{1/2}\Delta t) - I)(PAP)^{-1}PAV(t_n), \\ -\Delta t^2(QAQ)V(t_n) + \Delta t^2 QA(PAP)^{-1}PAQV(t_n) \textit{ (coarse zone)} \end{cases} \tag{7.121}$$

Remark: $(PAP)^{-1}$ is a Moore–Penrose pseudoinverse [21] since a large part of this matrix is null, but the non null block is invertible, this is why we kept this notation.

Proof To obtain the expression of $P\tilde{z}$, we have to solve the ordinary differential equation with constant coefficients (7.118a)–(7.118c) projected on the fine part by applying P. First, we define the solution of the homogeneous problem:

$$P\tilde{z}_0''(\tau) = -PAP\tilde{z}_0(\tau). \tag{7.122}$$

Hence, we immediately get:

$$\tilde{z}_0(\tau) = \cos((PAP)^{1/2}\tau)\beta + \sin((PAP)^{1/2}\tau)\alpha, \tag{7.123}$$

where $\alpha,\ \beta \in \mathbb{R}^N$.

Matrices $\cos((PAP)^{1/2}\tau)$ and $\sin((PAP)^{1/2}\tau)$ are defined as follows: We diagonalize the symmetric positive definite matrix PAP i.e. $PAP = VDV'$ where $D = diag(\lambda_i, i = 1, \cdots, N)$ and λ_i are the eigenvalues of PAP. We thus have:

$$\cos((PAP)^{1/2}\tau) = V\, diag(\cos(\sqrt{\lambda_i}))V' (\text{idem for } \sin). \tag{7.124}$$

We can then easily verify that $\tilde{z}_1 = -(PAP)^{-1}PAQV(t_n)$ is a particular solution of (7.118a)–(7.118c) (without the initial conditions). Finally, we seek for $P\tilde{z}$ in the form $\tilde{z}_0 + \tilde{z}_1$ and the initial conditions determine the two values (α, β).

To prove (7.120), we simply use $\tilde{z}''(\tau) = -A(I - P)V(t_n) - AP\tilde{z}(\tau)$ and the power series of cosine and sine functions:

$$\cos((PAP)^{1/2}\tau) = \sum_{n=0}^{+\infty}(-1)^n \frac{\tau^{2n}}{(2n)!}(PAP)^n, \tag{7.125a}$$

$$\sin((PAP)^{1/2}\tau) = \sum_{n=0}^{+\infty}(-1)^n \frac{\tau^{2n+1}}{(2n+1)!}(PAP)^{n+1/2}. \tag{7.125b}$$

Finally, to prove (7.121), we use again $\tilde{z}''(\tau) = -A(I-P)V(t_n) - AP\tilde{z}(\tau)$ and the following integrals:

$$\int_{-1}^{1}(1-|\theta|)\cos((PAP)^{1/2}\theta\Delta t)\mathrm{d}\theta = -\frac{2}{\Delta t^2}(PAP)^{-1}(\cos((PAP)^{1/2}\Delta t)-1), \tag{7.126a}$$

$$\int_{-1}^{1}(1-|\theta|)\sin((PAP)^{1/2}\theta\Delta t)\mathrm{d}\theta = 0. \tag{7.126b}$$

Let us define a first temporal scheme (Diaz-Grote's scheme will be an approximation of this one) using the previous approximation $\tilde{z}''(\theta\Delta t)$ of $V''(t_n+\theta\Delta t)$. We have:

$$V^{n+1} - 2V^n + V^{n-1} = \Delta t^2\int_{-1}^{1}(1-|\theta|)\tilde{z}''(\theta\Delta t)\mathrm{d}\theta \tag{7.127}$$

Using (7.121), (7.127) takes the form of a Leapfrog scheme which we call the exact scheme:

$$V^{n+1} - 2V^n + V^{n-1} = -\Delta t^2 BV^n \tag{7.128}$$

where the matrix B is:

$$B = \begin{bmatrix} B_{11} & B_{12} \\ B_{21} & B_{22} \end{bmatrix} \tag{7.129}$$

with

$$B_{11} = -\frac{2}{\Delta t^2}(\cos((PAP)^{1/2}\Delta t) - I)P, \tag{7.130a}$$

$$B_{12} = B_{21}^* = -\frac{2}{\Delta t^2}(\cos((PAP)^{1/2}\Delta t) - I)(PAP)^{-1}PAQ, \tag{7.130b}$$

$$B_{22} = -\frac{2}{\Delta t^2}QA(PAP)^{-1}(\cos((PAP)^{1/2}\Delta t) - I)(PAP)^{-1}PAQ + (QAQ) - QA(PAP)^{-1}PAQ. \tag{7.130c}$$

Remarks:

1. There is an abuse of notation in the matrices B_{ij}, $i,j=1,2$. They are the restriction to the corresponding non null matrix block since the size of B is the same as the size of A. Moreover, in order to write B in this form, we use the assumption that the degrees of freedom of V are sorted as defined in (7.116).
2. We notice that matrix B is symmetric and does not depend on the initial condition ν of the differential equation (7.118a)–(7.118c).
3. Scheme (7.127) is consistent of order 2 in time.

7.2.3.2 Approximation of the Exact Scheme: Local Time Stepping Algorithm

Scheme (7.128) is mathematically appealing. However in most situations, computing B would be really expensive since it is derived from the exact solution of (7.118a)–(7.118c). The strategy proposed by J. Diaz and M. Grote is to approximate B in a Leapfrog way. Solving (7.118a)–(7.118c) with a Leapfrog scheme leads to the local time stepping algorithm proposed in [20]. We emphasize that it is only when we approximate scheme (7.127) that a local time stepping appears, and until now there was no notion of local time step.

In order to get a local time-stepping algorithm, we have to use a relation between ν and $\tilde{z}$ since we will not use the exact solution described previously (7.128). By construction, $\tilde{z}$ verifies:

$$\tilde{z}(\Delta t) - 2\tilde{z}(0) + \tilde{z}(-\Delta t) = \Delta t^2 \int_{-1}^{1} (1 - |\theta|)\tilde{z}''(\theta \Delta t)\mathrm{d}\theta \tag{7.131}$$

and since V verifies (7.127) we have

$$V^{n+1} - 2V^n + V^{n-1} = \tilde{z}(\Delta t) - 2\tilde{z}(0) + \tilde{z}(-\Delta t). \tag{7.132}$$

The first step is thus to approximate $\tilde{z}(\Delta t)$ and $\tilde{z}(-\Delta t)$ where, we recall, $\tilde{z}$ solves the following differential problem:

$$\tilde{z}''(\tau) = -A(I - P)V^n - AP\tilde{z}(\tau), \tag{7.133a}$$

$$\tilde{z}(0) = V^n, \tag{7.133b}$$

$$\tilde{z}'(0) = \nu. \tag{7.133c}$$

It is important to notice that we need to solve this equation forward and backward in time to get $\tilde{z}(\Delta t)$ and $\tilde{z}(-\Delta t)$.

Unlike scheme (7.127) where the parameter ν was of absolutely no use, for the Diaz-Grote's scheme we need to choose a value for the initial condition ν since we are going to use this value to initialize the algorithm. Taking $\nu = 0$ leads to a very convenient algorithm since we then have $\tilde{z}(\tau) = -\tilde{z}(-\tau)$. Thus we do not need any more to solve (7.133a)–(7.133c) both forward and backward. Noticing that $\tilde{z}(0) = V^n$, the temporal scheme becomes:

$$V^{n+1} + V^{n-1} = 2\tilde{z}(\Delta t). \tag{7.134}$$

The previous relation leads to the local time stepping scheme:

$$V^{n+1} + V^{n-1} = 2\tilde{z}_{p/p}, \tag{7.135}$$

where $\tilde{z}_{m/p}$ is the approximation of $\tilde{z}\,(m\Delta t/p)$ achieved by solving (7.133a)–(7.133c) with a Leapfrog scheme at the fine time step $\Delta t/p$.

Hence, we get the following algorithm:

1. Set $w = A(I - P)V^n$ and $\tilde{z}_0 = V^n$.
2. $\tilde{z}_{1/p} = \tilde{z}_0 - \dfrac{1}{2}\left(\dfrac{\Delta t}{p}\right)^2 (AP\tilde{z}_0 + w)$.
3. For $m = 1, \cdots p-1$, compute

$$\tilde{z}_{(m+1)/p} = 2\tilde{z}_{m/p} - \tilde{z}_{(m-1)/p} - \left(\frac{\Delta t}{p}\right)^2 \left(AP\tilde{z}_{m/p} + w\right). \tag{7.136}$$

4. Compute $V^{n+1} = -V^{n-1} + 2\tilde{z}_{p/p}$.

Remark: Solving the differential problem (7.133a)–(7.133c) in order to get $\tilde{z}(\Delta t)$ corresponds to steps 1–3 of the algorithm. As we can see on step 3, the differential equation (7.133a)–(7.133c) is solved with a Leapfrog scheme and a time step $\Delta t/p$. Step 2 is a standard second order accurate intialisation of the Leapfrog scheme. Finally, we just have to apply (7.135) to finish the coarse time step, which corresponds to the step 4.

The properties of this local time stepping scheme are [20]:

1. The scheme can be written as a Leapfrog scheme

$$V^{n+1} = 2V^n - V^{n-1} - \Delta t^2 A_p V^n, \tag{7.137}$$

   where A_p is symmetric.
2. The scheme is second order accurate in time.
3. The discrete energy:

$$\begin{aligned} E^{n+\frac{1}{2}} &= \frac{1}{2}\left(\left(I - \frac{\Delta t^2}{4}A_p\right)\frac{V^{n+1}-V^n}{\Delta t}, \frac{V^{n+1}-V^n}{\Delta t}\right) \\ &+ \frac{1}{2}\left(A_p\frac{V^{n+1}+V^n}{2}, \frac{V^{n+1}+V^n}{2}\right) \end{aligned} \tag{7.138}$$

   is conserved and hence, if λ_{min} and λ_{max} denote the smallest and largest eigenvalues of A_p, the numerical scheme is stable if and only if

$$0 \leq \frac{\Delta t^2}{4}\lambda_{min} \leq \frac{\Delta t^2}{4}\lambda_{max} \leq 1. \tag{7.139}$$

4. Numerical analysis of the stability shows that it is necessary to include the first coarse elements close to the fine elements in the fine time step in order to obtain the optimal coarse time step Δt. However, it is not necessary to have too many coarse elements for the fine time-step to obtain a good result: one layer of coarse elements is generally sufficient.

Remark: Using the global formulation with the matrix A and the projection matrix P well describes the local time-stepping algorithm but hides the locality of the fine time step, also making it impossible to use the local time-stepping algorithm as it is in an effective computer implementation. In [22], Y. Dudouit rewrote a local version of this algorithm for discontinuous Galerkin methods for the second-order elastodynamic equation.

References

1. Hairer, E., Nørsett, S.P., Wanner, G.: Solving Ordinary Differential Equations. Series in Computational Mathematics. Springer, Heidelberg (1991)
2. Dablain, M.A.: The application of high order differencing for the scalar wave equation. Geophysics **51**(1), 54–66 (1986)
3. Cohen, G., Joly, P.: Construction and analysis of fourth-order finite difference schemes for the acoustic wave equation in inhomogeneous media. SIAM J. Numer. Anal. **33**(4), 1266–1302 (1996)
4. Shubin, G.R., Bell, J.B.: A modified equation approach to constructing fourth order methods for acoustic wave propagation. SIAM J. Sci. Comput. **8**(2), 135–151 (1987)
5. Tal-Ezer, H.: Spectral methods in time for hyperbolic equations. SIAM J. Numer. Anal. **23**(1), 11–26 (1986)
6. Cohen, G.: High Order Numerical Methods for Transient Wave Equations. Scientific Computation. Springer, Heidelberg (2001)
7. Montseny, E., Pernet, S., Ferrires, X., Cohen, G.: Dissipative terms and local time-stepping improvements in a spatial high order Discontinuous Galerkin scheme for the time-domain Maxwell's equations. J. Comput. Phys. **227**(14), 6795–6820 (2008)
8. Chevalier, M., Luebbers, R., Cable, V.: FDTD local grid with material traverse. IEEE Trans. Antennas Propag. **45**(3), 411–421 (1997)
9. Kunz, K.S., Simpson, L.: A technique for increasing the resolution of finite difference solutions of the Maxwell equation. IEEE Trans. Electromagn. Compat. **23**(4), 419–422 (1981)
10. Kim, I., Hoefer, W.J.R.: A local mesh refinement algorithm for the time domain finite difference method using Maxwell's curl equations. IEEE Trans. Microw. Theory Tech. **38**(6), 812–815 (1990). Jun
11. Prescott, D., Shuley, N.: A method for incorporating different sized cells into the finite-difference time-domain analysis technique. IEEE Microw. Guided Wave Lett. **2**(11), 434–436 (1992). Nov
12. Dumbser, M., Käser, M., Toro, E.F.: An arbitrary high-order discontinuous Galerkin method for elastic waves on unstructured meshes - V. Local time stepping and p-adaptivity. Geophys. J. Int. **171**(2), 695–717 (2007)
13. Piperno, S.: Symplectic local time-stepping in non-dissipative DGTD methods applied to wave propagation problems. ESAIM Math. Model. Numer. Anal. **40**(5), 815–841 (2006)
14. Collino, F., Fouquet, T., Joly, P.: Conservative space-time mesh refinement methods for the FDTD solution of Maxwell's equations. J. Comput. Phys. **211**(1), 9–35 (2006)

15. Bécache, E., Joly, P., Rodriguez, J.: Space-time mesh refinement for elastodynamics. Numerical results. Comput. Methods Appl. Mech. Eng. **194**, 355–366 (2005)
16. Rodriguez, J.: A spurious-free space-time mesh refinement for elastodynamics. Int. J. Multiscale Comput. Eng. **6**(3), 263–279 (2008)
17. Ezziani, A., Joly, P.: Space-time mesh refinement for discontunuous Galerkin methods for symmetric hyperbolic systems. J. Comput. Appl. Math. **234**(6), 1886–1895 (2009)
18. Rodriguez, J.: Raffinement de Maillage Spatio-Temporel pour les Equations de l'Elastodynamique, thèse de doctorat, U. Paris-Dauphine (2004)
19. Joly, P., Rodriguez, J.: An error analysis of conservative space-time mesh refinement methods for the 1D wave equation. SIAM J. Numer. Anal. **43**(2), 825–859 (2005)
20. Diaz, J., Grote, M.J.: Energy conserving explicit local time stepping for second-order wave equations. SIAM J. Sci. Comput. **31**(3), 1985–2014 (2009)
21. Golub, G.H., Van Loan, C.F.: Matrix Computations, 3rd edn. Johns Hopkins, Baltimore (1996)
22. Dudouit, Y.: Spatio-temporal refinement using a discontinuous Galerkin approach for elastodynamic in a high performance computing framework, thèse de doctorat, U. de Bordeaux (2014)

Chapter 8
Some Complex Models

Abstract This chapter presents three more complex equations with specific properties: The linearized Euler equations (LEE) which model acoustics in flow and contains a convective term, the Cauchy-Poisson problem which models gravity waves and whose evolution equation is on a boundary and two models of wave propagation in thin plates which are dispersive equations.

In this chapter, we present the approximation of some physical models which differ from the classical models treated in the previous chapter by their complexity. We first describe the linearized Euler equations (LEE) which seem to be the most realistic and accurate model for acoustic waves in flow (aeroacoustics). Their main difficulty is the presence of a convective term. In a second section, we address the Cauchy–Poisson problem which models gravity waves and whose evolution equation is written on the boundary of the domain of resolution. The third section is devoted to modeling vibration of thin plates by using Reissner–Mindlin's or Kirchhoff–Love's approach which lead to dispersive equations.

The approximation of the equations presented below is mainly focused on spectral hexahedral elements, but one can also use tetrahedral elements (see for instance [1, 2] for DGM applied to LEE and [3] for triangular FEM applied to the Kirchhoff–Love model).

8.1 The Linearized Euler Equations

Modeling acoustic waves in a fluid in motion is not an obvious problem. The simplest model proposed is the Lighthill's analogy [4] whose main drawback is that it does not separate flow and acoustic quantities. A more complex and accurate model is given by the linearized Euler equations (LEE) which are derived from a first-order

G. Cohen and S. Pernet, *Finite Element and Discontinuous Galerkin Methods for Transient Wave Equations*, Scientific Computation,
DOI 10.1007/978-94-017-7761-2_8

approximation of the Euler equations.[1] Another model, less popular, is given by the Galbrun's equation[2] [5]. A discussion on the physical models can be found in [6]. In this section, we present the LEE and their approximations, first by discontinuous Galerkin methods, as developed in [7], then by an H^1-L^2 formulation.

8.1.1 Discontinuous Galerkin Approximation

8.1.1.1 Construction of the Scheme

Let us consider the general linear hyperbolic system (which can be regarded as an extension of (4.4a)–(4.4d)):

$$\frac{\partial \underline{w}}{\partial t} - \sum_{i=1}^{d} \frac{\partial}{\partial x_i}(A_i\, \underline{w}) = \underline{F}(\underline{x}, t) \text{ in } \mathbb{R}^d, \tag{8.1a}$$

$$\underline{w}(\underline{x}, 0) = \underline{w}_0(\underline{x}) \text{ in } \mathbb{R}^d, \tag{8.1b}$$

where $\underline{w} : \mathbb{R}^d \times \mathbb{R}^+ \to \mathbb{R}^q$ and A_i are symmetric matrices of $\mathbb{R}^q \times \mathbb{R}^q\ \forall \underline{x} \in \mathbb{R}^d$.

The original LEE suffers of physical instabilities (called the Kelvin–Helmholtz instabilities). These instability are limited by the non linear terms of the Euler equations but appear in the linearized form. The model proposed in [8] has been theoretically and numerically proven to be stable and accurate for aeroacoustics. For these reasons, we shall use this model in the following. This model reads:

$$\frac{\partial p}{\partial t} + \rho_0\, c_0^2\, \nabla \cdot \underline{u} + \nabla \cdot (p\, \underline{u}_0) = f, \tag{8.2a}$$

$$\frac{\partial \underline{u}}{\partial t} + (\underline{u}_0 \cdot \nabla)\underline{u} + \frac{\nabla p}{\rho_0} = 0, \tag{8.2b}$$

where p is the acoustic pressure, $\underline{u}$ is the acoustic particle velocity, ρ_0 the density of the fluid and c_0 the velocity of the wave in the fluid at rest. $\underline{u}_0$ depends on $\underline{x}$ but is in practice either constant per element or divergence free. In these cases, $\nabla \cdot (p\, \underline{u}_0) = \underline{u}_0 \cdot \nabla p$ on each element.

By setting, in 2D:

$$A_1 = \begin{pmatrix} u_{01} & \rho_0\, c_0^2 & 0 \\ \dfrac{1}{\rho_0} & u_{01} & 0 \\ 0 & 0 & u_{01} \end{pmatrix}, \quad A_2 = \begin{pmatrix} u_{02} & 0 & \rho_0\, c_0^2 \\ 0 & u_{02} & 0 \\ \dfrac{1}{\rho_0} & 0 & u_{02} \end{pmatrix}.$$

[1] We recall that the acoustics equation is their 0th order approximation.

[2] Which was proven (in an unpublished paper) to be equivalent to LEE.

and, in 3D:

$$A_1 = \begin{pmatrix} u_{01} & \rho_0 c_0^2 & 0 & 0 \\ \frac{1}{\rho_0} & u_{01} & 0 & 0 \\ 0 & 0 & u_{01} & 0 \\ 0 & 0 & 0 & u_{02} \end{pmatrix}, \; A_2 = \begin{pmatrix} u_{02} & 0 & \rho_0 c_0^2 & 0 \\ 0 & u_{02} & 0 & 0 \\ \frac{1}{\rho_0} & 0 & u_{02} & 0 \\ 0 & 0 & 0 & u_{02} \end{pmatrix}, \; A_3 = \begin{pmatrix} u_{03} & 0 & 0 & \rho_0 c_0^2 \\ 0 & u_{03} & 0 & 0 \\ 0 & 0 & u_{03} & 0 \\ \frac{1}{\rho_0} & 0 & 0 & u_{03} \end{pmatrix}$$

and $\underline{w} = (p, \underline{u})^T$, (8.2a) and (8.2b) can be rewritten as (8.1a) and (8.1b).

We define the following domain meshed by hexahedra or quadrilaterals:

$$\Omega = \bigcup_{\ell=1}^{N} K_\ell \; (\subset \mathbb{R}^d). \tag{8.3}$$

Let us set $\underline{w}, \underline{\varphi} \in W_h^r$ with

$$W_h^r = \left\{ \underline{v} \in [L^2(\Omega)]^{d+1} \text{ such that } \forall K_\ell \subset \Omega, \; \underline{v}|_{K_\ell} \in (Q_r)^{d+1} \right\}. \tag{8.4}$$

The DGM formulation of (8.2a) and (8.2b) then reads[3]:

$$\begin{aligned} &\frac{\mathrm{d}}{\mathrm{d}t} \int_{K_\ell} \underline{w} \cdot \underline{\varphi} \, \mathrm{d}\underline{x} + \sum_{i=1}^{d} \int_{K_\ell} A_i \, \underline{w} \cdot \frac{\partial \underline{\varphi}}{\partial x_i} \, \mathrm{d}\underline{x} + \int_{\partial K_\ell} \sum_{i=1}^{d} n_i A_i \, \{\underline{w}\}_{\partial K_\ell}^{K_\ell} \cdot \underline{\varphi} \, \mathrm{d}\sigma \\ &+ \int_{\partial K_\ell} C_{\alpha,\beta} \, [\underline{w}]_{\partial K_\ell}^{K_\ell} \cdot \underline{\varphi} \, \mathrm{d}\sigma - \int_{K_\ell} \underline{F} \cdot \underline{\varphi} \, \mathrm{d}\underline{x} = 0, \end{aligned} \tag{8.5}$$

where $\{\underline{w}\}_{\partial K_\ell}^{K_\ell} = \frac{1}{2}(\underline{w}_{|K_{\ell'}} + \underline{w}_{|K_\ell})$, $[\underline{w}]_{\partial K_\ell}^{K_\ell} = (\underline{w}_{|K_{\ell'}} - \underline{w}_{|K_\ell})$, $K_\ell \bigcap K_{\ell'} = \Gamma_{\ell\ell'}$, $\underline{n}_\ell = (n_i)_{i=1}^d$ is the unit outward normal to K_ℓ and

$$C_{\alpha,\beta} = \begin{pmatrix} \alpha & 0 \\ 0 & \beta \, \mathbf{C}_n \end{pmatrix}, \quad \alpha \le 0, \; \beta \le 0,$$

with $\mathbf{C}_n = \underline{n}_\ell \, \underline{n}_\ell^T$.

8.1.1.2 Computation of the Integrals

Due to the complexity of the differential operators involved in (8.2a) and (8.2b), a mixed formulation cannot be defined.[4] So, besides the mass matrix (which is obviously diagonal), the computation of the other integrals of (8.5) is not obvious and could be rather time consuming or needing an important storage. In order to reduce the computational time, we introduce a fast matrix-vector product to compute the stiffness and jump terms.

[3] For tetrahedra or triangles, Q_r is replaced by P_r.

[4] A "hypermixed", i.e. a mixed formulation in two steps, was constructed, but it was not stable.

The basis functions support being reduced to one element K_ℓ, let us consider a basis function $\underline{\varphi}^k_{\ell,j}$ and a test function $\underline{\varphi}^{k'}_{\ell,j'}$. In these functions, the second lower index ($\in \{1..(r+1)^d\}$) indicates the interpolation point in K_ℓ and the upper index, the direction of the function. Actually, we have $\underline{\varphi}^k_{\ell,j} = \varphi_{\ell,j}\,\underline{e}_k$ and $\underline{\varphi}^{k'}_{\ell,j'} = \varphi_{\ell,j'}\,\underline{e}_{k'}$, $\underline{e}_k$ and $\underline{e}_{k'}$ being two vectors of the canonical basis of $\mathbb{R}^{d+1}$. With these notations, the second integral of (8.5) reads:

$$I = \int_{K_\ell} A_i\,\underline{w} \cdot \frac{\partial \underline{\varphi}}{\partial x_i}\,\mathrm{d}\underline{x} = \sum_{k=1}^{d+1}\sum_{j=1}^{(r+1)^d} w_j^k \int_{K_\ell} A_i\,\underline{\varphi}^k_{\ell,j} \cdot \frac{\partial \underline{\varphi}^{k'}_{\ell,j'}}{\partial x_i}\,\mathrm{d}\underline{x} = \sum_{k=1}^{d+1}\sum_{j=1}^{(r+1)^d} I_{j,k}. \quad (8.6)$$

Since $\underline{\varphi}^k_{\ell,j} = \varphi_{\ell,j}\,\underline{e}_k$ and $\underline{\varphi}^{k'}_{\ell,j'} = \varphi_{\ell,j'}\,\underline{e}_{k'}$ and by using the chain rule, we get:

$$\begin{aligned} I_{j,k} &= \int_{K_\ell} (A_i)_{k',k}\,\varphi_{\ell,j}\,\frac{\partial \varphi_{\ell,j'}}{\partial x_i}\,\mathrm{d}\underline{x} = \int_{\widehat{K}} |J_\ell|\,(A_i)_{k',k} \circ \underline{F}_\ell\,\varphi_{\ell,j} \circ \underline{F}_\ell\,\frac{\partial \varphi_{\ell,j'}}{\partial x_i} \circ \underline{F}_\ell\,\mathrm{d}\hat{\underline{x}} \\ &= \sum_{m=1}^{d} \int_{\widehat{K}} |J_\ell|\,(A_i)_{k',k} \circ \underline{F}_\ell\,\hat{\varphi}_j (DF_\ell^{-1})_{i,m}\,\frac{\partial \hat{\varphi}_{j'}}{\partial \hat{x}_m}\,\mathrm{d}\hat{\underline{x}} = \sum_{m=1}^{d} I^m_{j,k}, \end{aligned} \quad (8.7)$$

where $\widehat{K} = [0,1]^d$, $\underline{F}_\ell$ is the transform such that $\underline{F}_\ell(\widehat{K}) = K_\ell$, DF_ℓ, its Jacobian matrix and $J_\ell = \det(DF_\ell)$. On the other hand, $\forall 1 \le n \le (r+1)^d$, $\hat{\varphi}_n = \varphi_{\ell,n} \circ \underline{F}_\ell$ and, for any matrix M, $(M)_{i,j}$ is its term at the ith line and jth column.

Let us now compute $I^m_{j,k}$ by using a quadrature rule whose points $\left(\hat{\underline{\xi}}_n\right)_{n=1}^{(r+1)^d}$ coincide with the interpolation points in $\widehat{K}$ and whose weights are $(\hat{\omega}_n)_{n=1}^{(r+1)^d}$. In practice, this rule is either the Gauss or the Gauss–Lobatto rule. We have:

$$I^m_{j,k} \simeq \sum_{n=1}^{(r+1)^d} \hat{\omega}_n\,(|J_\ell|\,(A_i)_{k',k} \circ \underline{F}_\ell)(\hat{\underline{\xi}}_n)\,\hat{\varphi}_j(\hat{\underline{\xi}}_n)\,(DF_\ell^{-1})_{i,m}(\hat{\underline{\xi}}_n)\,\frac{\partial \hat{\varphi}_{j'}}{\partial \hat{x}_m}(\hat{\underline{\xi}}_n). \quad (8.8)$$

Since $\hat{\varphi}_j(\hat{\underline{\xi}}_n) = \delta_{jn}$, we finally obtain:

$$I \simeq \sum_{k=1}^{d+1}\sum_{j=1}^{(r+1)^d} w_j^k\,\hat{\omega}_j\,(A_i)_{k',k} \circ \underline{F}_\ell(\hat{\underline{\xi}}_j) \sum_{m=1}^{d} (\mathrm{adj}(DF_\ell))_{i,m})(\hat{\underline{\xi}}_j)\,\frac{\partial \hat{\varphi}_{j'}}{\partial \hat{x}_m}(\hat{\underline{\xi}}_j), \quad (8.9)$$

where $\mathrm{adj}(DF_\ell)$ is the adjugate matrix of DF_ℓ.

As defined in (3.13a) and (3.13b), $\hat{\varphi}_{j'}$ can be decomposed into a product of three one dimensional functions. Let us set $d = 3$ (same computations can be done in 2D) and $j = (j'_1, j'_2, j'_3)$ with $1 \le j'_s \le r+1$. We can write:

$$\hat{\varphi}_{j'}(\underline{x}) = \hat{\varphi}_{(j'_1,j'_2,j'_3)}(x_1, x_2, x_3) = \hat{\varphi}_{j'_1}(x_1)\,\hat{\varphi}_{j'_2}(x_2)\,\hat{\varphi}_{j'_2}(x_3) \quad (8.10)$$

By setting $\underline{\hat{\xi}}_n = (\hat{\xi}_{n_1}, \hat{\xi}_{n_2}, \hat{\xi}_{n_3})$, $\hat{\omega}_n = \hat{\omega}_{n_1}\,\hat{\omega}_{n_2}\,\hat{\omega}_{n_3}$ and taking into account (8.10), since $\hat{\varphi}_j(\underline{\hat{\xi}}_n) = \delta_{jn}$, (8.9) becomes:

$$I \simeq \sum_{k=1}^{d+1} (I_1 + I_2 + I_3), \tag{8.11}$$

where

$$I_1 = \hat{\omega}_{j_2'}\,\hat{\omega}_{j_3'} \sum_{j_1=1}^{r+1} w^k_{(j_1,j_2',j_3')}\,\hat{\omega}_{j_1}\,((A_i)_{k',k} \circ \underline{F}_\ell\,(\mathrm{adj}(DF_\ell))_{i,1})(\hat{\xi}_{j_1}, \hat{\xi}_{j_2'}, \hat{\xi}_{j_3'})\,\frac{\mathrm{d}\hat{\varphi}_{j_1'}}{\mathrm{d}\hat{x}_1}(\hat{\xi}_{j_1}),$$

$$I_2 = \hat{\omega}_{j_1'}\,\hat{\omega}_{j_3'} \sum_{j_2=1}^{r+1} w^k_{(j_1',j_2,j_3')}\,\hat{\omega}_{j_2}\,((A_i)_{k',k} \circ \underline{F}_\ell\,(\mathrm{adj}(DF_\ell))_{i,2})(\hat{\xi}_{j_1'}, \hat{\xi}_{j_2'}, \hat{\xi}_{j_3'})\,\frac{\mathrm{d}\hat{\varphi}_{j_2'}}{\mathrm{d}\hat{x}_2}(\hat{\xi}_{j_2}),$$

$$I_3 = \hat{\omega}_{j_1'}\,\hat{\omega}_{j_2'} \sum_{j_3=1}^{r+1} w^k_{(j_1',j_2',j_3)}\,\hat{\omega}_{j_3}\,((A_i)_{k',k} \circ \underline{F}_\ell\,(\mathrm{adj}(DF_\ell))_{i,3})(\hat{\xi}_{j_1'}, \hat{\xi}_{j_2'}, \hat{\xi}_{j_3})\,\frac{\mathrm{d}\hat{\varphi}_{j_3'}}{\mathrm{d}\hat{x}_3}(\hat{\xi}_{j_3}).$$

Equation (8.11) shows that the computation of (8.10) needs $d\,(d+1)\,(r+1)$ sums instead of $d\,(d+1)\,(r+1)^d$. On the other hand, the one-dimensional derivatives need to store only $r+1$ values of the derivatives of $r+1$ one-dimensional functions, i.e. the storage of $(r+1)^2$ real numbers. Finally, if we only store the values $(\mathrm{adj}(DF_\ell))_{i,m})(\underline{\hat{\xi}}_j)$, the computation of the stiffness matrix is very fast with a reasonable storage, in particular when A_i are (piecewise) constant. The jump terms can be computed as in Sect. 4.3.

8.1.1.3 Effect of the Penalty Term

Of course, as shown in Fig. 8.1 the main effect of the penalty term is to suppress the parasitic waves. If large enough, the values of α and β do not affect the accuracy of the solution. However, the presence of a penalty term decreases the CFL of the method. A systematic study seems to indicate that an optimal value would be $\alpha = \beta = -0.5$. More details are given in [7].

Remark: The penalty term can have some effect on the acoustics equation (obtained by setting $\underline{u}_0 = 0$ in (8.2a) and (8.2b)) for some extreme configurations, as shown in Fig. 8.2 but, as we saw in Sect. 4.5.2, this improvement (much less dramatical than in LEE) only comes from a better numerical dispersion induced by this term.

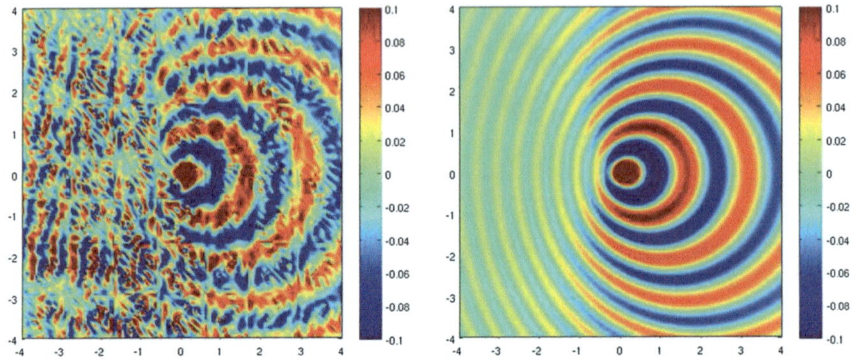

Fig. 8.1 Propagation of a spherical wave in a fluid with a constant flow. The domain is meshed by triangles split into quadrilaterals. Solution with $\alpha = \beta = 0$ (*left*) and $\alpha = \beta = -0.5$ in $C^{(2)}_{\alpha,\beta}$. Reprinted from N. Castel, G. Cohen, M. Duruflé, Application of discontinuous Galerkin spectral method on hexahedral elements for aeroacoustic, J. of Comp. Acous., vol. 17 (2), pp. 175–196, Copyright ©2009 World Scientific Publishing. Reprinted with permission

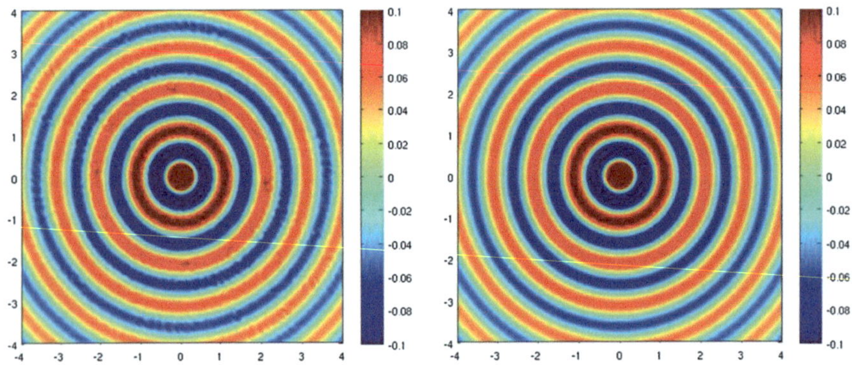

Fig. 8.2 Solution of the wave equation on a mesh composed of triangles split into three quadrilaterals without (*left*) and with (*right*) a penalty term. Reprinted from N. Castel, G. Cohen, M. Duruflé, Application of discontinuous Galerkin spectral method on hexahedral elements for aeroacoustic, J. of Comp. Acous., vol. 17 (2), pp. 175–196, Copyright ©2009 World Scientific Publishing. Reprinted with permission

8.1.2 H^1-L^2 Approximation

8.1.2.1 The Variational Formulation

In [9], an H^1 approximation of Galbrun's equation is regularized by adding a curl term. This regularization is actually an extension of the regularization of the H^1 approximation of Maxwell's equations by adding a divergence term [10]. Since, as we saw in Sect. 5.2.4, we can obtain the same effect for an H^1-L^2 approximation of Maxwell's equations by adding a dissipative term, our idea is to extend this last approach to the linearized Euler equations.[5] In this approach, the continuity of the pressure induces a gain of storage and CPU time compared to DGM.

Let Ω be an open set of $\mathbb{R}^d$. On $\partial\Omega$, we set Neumann (sound-hard wall) condition which reads:

$$\underline{u}_0 \cdot \underline{n} = \underline{u} \cdot \underline{n} = 0 \text{ on } \Gamma_N, \tag{8.12}$$

where $\underline{n}$ is the unit outward normal to Ω. Taking into account (8.12) and assuming that $\underline{u}_0$ is taken constant[6] or divergence free, we can write the following variational formulation of (8.2a) and (8.2b).

Find $p \in H^1(\Omega)$, $\underline{u} \in [L^2(\Omega)]^d$ such that

$$\frac{\mathrm{d}}{\mathrm{d}t}\int_\Omega p\,\varphi\,\mathrm{d}\underline{x} - \rho_0\,c_0^2\int_\Omega \underline{u}\cdot\nabla\varphi\,\mathrm{d}\underline{x} + \int_\Omega (\underline{u}_0\cdot\nabla p)\,\varphi\,\mathrm{d}\underline{x} = \int_\Omega f\,\varphi\,\mathrm{d}\underline{x},$$
$$\forall\varphi \in H_0^1(\Omega), \tag{8.13a}$$

$$\frac{\mathrm{d}}{\mathrm{d}t}\int_\Omega \underline{u}\cdot\underline{\psi}\,\mathrm{d}\underline{x} + \int_\Omega (\underline{u}_0\cdot\nabla)\,\underline{u}\cdot\underline{\psi}\,\mathrm{d}\underline{x} + \frac{1}{\rho_0}\int_\Omega \nabla p\cdot\underline{\psi}\,\mathrm{d}\underline{x} = 0,\ \forall\underline{\psi}\in[L^2(\Omega)]^d. \tag{8.13b}$$

8.1.2.2 The Approximate Problem

Although this formulation can be approximated by using tetrahedra (triangles) or hexahedra (quadrilaterals), its resolution in time domain is much more efficient for hexahedra which provide mass-lumping for (8.13a). So, the approximation is done on an hexahedral or quadrilateral mesh $\mathcal{T}_h$ of Ω with Gauss–Lobatto interpolation points and computation of the integrals by a Gauss–Lobatto quadrature rule. So, we have $p_h \in U_h^r \subset H^1(\Omega)$ and $\underline{u}_h \in V_h^r \subset [L^2(\Omega)]^d$, U_h^r and V_h^r defined in (3.5) and (3.33). The matrix formulation of the problem, after discretization, reads:

$$\begin{pmatrix} \mathbf{D} & \mathbf{0} \\ \mathbf{0} & \mathbf{B} \end{pmatrix}\begin{pmatrix} P \\ U \end{pmatrix} + \begin{pmatrix} \mathbf{C}_p & \mathbf{R} \\ \mathbf{R}^T & \mathbf{C}_u \end{pmatrix}\begin{pmatrix} P \\ U \end{pmatrix} = \begin{pmatrix} F \\ 0 \end{pmatrix}, \tag{8.14}$$

where $\mathbf{D}$ is a diagonal matrix and $\mathbf{B}$ a block-diagonal matrix.

Three points have to be considered in order to have a stable formulation of our approximation

1. The operator involved in the convective term of (8.13b) has no meaning since $\underline{u} \in L^2(\Omega)]^d$.
2. Suppress the parasitic modes by adding a penalty term.

---

[5]This approach is a common work of G. Cohen, A. Hüppe, S. Imperiale and M. Kaltenbacher. An article entitled "Construction and analysis of an adapted spectral finite element method to convective acoustic equations" has been accepted by the Communications in Computational Physics Journal and should be published in 2016.

[6]p being in $H^1(\Omega)$, $\underline{u}_0$ can no longer be piecewise constant.

3. Matrices $\mathbf{C}_p$ and $\mathbf{C}_u$ (which are derived from the convective terms) in (8.14) are not obviously skew symmetric, which can be troublesome for the stability of the approximate problem [11].

8.1.2.3 Additional Jump Terms

A palliative to the first point is to treat the convective term in a "DGM way", i.e. to add a jump to the volumic integral. This jump reads:

$$J_{\partial K_\ell} = \frac{1}{2} \int_{\partial K_\ell} [(\underline{n}_\ell \cdot \underline{u}_0)\, \underline{u}_h]_{\partial K_\ell}^{K_\ell} \cdot \underline{\psi}_h \, \mathrm{d}\sigma. \tag{8.15}$$

On the other hand, the penalty term reads as that applied to (8.2b), i.e.:

$$P_{\partial K_\ell} = \beta \int_{\partial K_\ell} \mathbf{C}_n [\underline{u}_h]_{\partial K_\ell}^{K_\ell} \cdot \underline{\psi}_h \, \mathrm{d}\sigma = \beta \int_{\partial K_\ell} \underline{n}_\ell \cdot [\underline{u}_h]_{\partial K_\ell}^{K_\ell} (\underline{n}_\ell \cdot \underline{\psi}_h) \, \mathrm{d}\sigma \tag{8.16}$$

where $\underline{n}_\ell$ and $[\,]_{\partial K_\ell}^{K_\ell}$ are defined as in (8.5).

So, the approximate formulation of (8.13b) is written as

$$\begin{aligned} &\frac{\mathrm{d}}{\mathrm{d}t} \int_{K_\ell} \underline{u}_h \cdot \underline{\psi}_h \, \mathrm{d}\underline{x} + \int_{K_\ell} (\underline{u}_0 \cdot \nabla)\, \underline{u}_h \cdot \underline{\psi}_h \, \mathrm{d}\underline{x} + \frac{1}{\rho_0} \int_{K_\ell} \nabla p_h \cdot \underline{\psi}_h \, \mathrm{d}\underline{x} \\ &\quad + \frac{1}{2} \int_{\partial K_\ell} [(\underline{n}_\ell \cdot \underline{u}_0)\, \underline{u}_h]_{\partial K_\ell}^{K_\ell} \cdot \underline{\psi}_h \, \mathrm{d}\sigma + \beta \int_{\partial K_\ell} \underline{n}_\ell \cdot [\underline{u}_h]_{\partial K_\ell}^{K_\ell} (\underline{n}_\ell \cdot \underline{\psi}_h) \, \mathrm{d}\sigma = 0, \\ &\forall \underline{\psi}_h \in V_h^r, \ \forall K_\ell \in \mathscr{T}_h. \end{aligned} \tag{8.17}$$

8.1.2.4 Skew Symmetry

Our second step is to check the skew symmetry of matrices $\mathbf{C}_p$ and $\mathbf{C}_u$. For this purpose, we integrate by parts the convective term of (8.17). We have:

$$\begin{aligned} \int_{K_\ell} (\underline{u}_0 \cdot \nabla)\, \underline{u}_h \cdot \underline{\psi}_h \, \mathrm{d}\underline{x} = &- \int_{K_\ell} \underline{u}_h \cdot (\underline{u}_0 \cdot \nabla)\, \underline{\psi}_h \, \mathrm{d}\underline{x} - \underbrace{\int_{K_\ell} \underline{u}_h \,(\nabla \cdot \underline{u}_0) \cdot \underline{\psi}_h \, \mathrm{d}\underline{x}}_{=0} \\ &+ \int_{\partial K_\ell} (\underline{n}_\ell \cdot \underline{u}_0)\, \underline{u}_h \cdot \underline{\psi}_h \, \mathrm{d}\sigma, \end{aligned} \tag{8.18}$$

the second volumic integral being equal to zero since, as we said above, $\underline{u}_0$ is taken constant or divergence free.

Combining (8.15) and (8.18), we get:

$$\int_{K_\ell} (\underline{u}_0 \cdot \nabla)\, \underline{u}_h \cdot \underline{\psi}_h \, \mathrm{d}\underline{x} + \frac{1}{2} \int_{\partial K_\ell} \left[(\underline{n}_\ell \cdot \underline{u}_0)\, \underline{u}_h \right]_{\partial K_\ell}^{K_\ell} \cdot \underline{\psi}_h \, \mathrm{d}\sigma$$
$$= - \int_{K_\ell} \underline{u}_h \cdot (\underline{u}_0 \cdot \nabla)\, \underline{\psi}_h \, \mathrm{d}\underline{x} + \int_{\partial K_\ell} \left\{ (\underline{n}_\ell \cdot \underline{u}_0)\, \underline{u}_h \right\}_{\partial K_\ell}^{K_\ell} \cdot \underline{\psi}_h \, \mathrm{d}\sigma, \tag{8.19}$$

where $\{\,\}_{\partial K_\ell}^{K_\ell}$ is defined as in (8.5).

Using (8.19) and (8.17) can be rewritten as

$$\frac{\mathrm{d}}{\mathrm{d}t} \int_{K_\ell} \underline{u}_h \cdot \underline{\psi}_h \, \mathrm{d}\underline{x} + \frac{1}{2} \left\{ \int_{K_\ell} (\underline{u}_0 \cdot \nabla)\, \underline{u}_h \cdot \underline{\psi}_h \, \mathrm{d}\underline{x} - \int_{K_\ell} \underline{u}_h \cdot (\underline{u}_0 \cdot \nabla)\, \underline{\psi}_h \, \mathrm{d}\underline{x} \right\}$$
$$+ \frac{1}{2} \sum_{m=1}^{2d} \int_{\Gamma_{\ell m}} (\underline{n}_\ell \cdot \underline{u}_0)\, \underline{u}_{h|K_m} \cdot \underline{\psi}_{h|K_\ell} \, \mathrm{d}\sigma + \frac{1}{\rho_0} \int_{K_\ell} \nabla p_h \cdot \underline{\psi}_h \, \mathrm{d}\underline{x} = 0,$$
$$\forall \underline{\psi}_h \in V_h^r, \ \forall K_\ell \in \mathscr{T}_h, \tag{8.20}$$

where $K_m \in \mathscr{T}_h$ is such that $K_\ell \cap K_m = \Gamma_{\ell m}$ and $\underline{u}_{h|K_m}$ and $\underline{\psi}_{h|K_\ell}$ are the values in K_m and K_ℓ of $\underline{u}_h$ and $\underline{\psi}_h$ on $\Gamma_{\ell m}$.

By summing (8.20) all over the mesh, we finally obtain:

$$\sum_{K_\ell} \left\{ \frac{\mathrm{d}}{\mathrm{d}t} \int_{K_\ell} \underline{u}_h \cdot \underline{\psi}_h \, \mathrm{d}\underline{x} + \frac{1}{\rho_0} \int_{K_\ell} \nabla p_h \cdot \underline{\psi}_h \, \mathrm{d}\underline{x} \right\} + a_h(\underline{u}_h, \underline{\psi}_h) = 0,$$
$$\forall \underline{\psi}_h \in V_h^r, \ \forall K_\ell \in \mathscr{T}_h, \tag{8.21}$$

where

$$a_h(\underline{u}_h, \underline{\psi}_h) = \frac{1}{2} \sum_{K_\ell} \left\{ \int_{K_\ell} (\underline{u}_0 \cdot \nabla)\, \underline{u}_h \cdot \underline{\psi}_h \, \mathrm{d}\underline{x} - \int_{K_\ell} \underline{u}_h \cdot (\underline{u}_0 \cdot \nabla)\, \underline{\psi}_h \, \mathrm{d}\underline{x} \right\}$$
$$+ \frac{1}{2} \sum_{\Gamma_{\ell m}} \left\{ \int_{\Gamma_{\ell m}} (\underline{n}_\ell \cdot \underline{u}_0)\, \underline{u}_{h|K_m} \cdot \underline{\psi}_{h|K_\ell} \, \mathrm{d}\sigma + \int_{\Gamma_{\ell m}} (\underline{n}_m \cdot \underline{u}_0)\, \underline{u}_{h|K_\ell} \cdot \underline{\psi}_{h|K_m} \, \mathrm{d}\sigma \right\}. \tag{8.22}$$

Let $\underline{\psi}_i$ and $\underline{\psi}_j$ be two basis functions of V_h^r. One can easily check that $(\mathbf{C}_u)_{ij} = a_h(\underline{\psi}_i, \underline{\psi}_j)$ and that (8.22) implies $(\mathbf{C}_u)_{ij} = -(\mathbf{C}_u)_{ij}$ (since $\underline{n}_\ell = -\underline{n}_m$), which proves the skew symmetry of $\mathbf{C}_u$.

Similar computations (which take into account (8.12) lead to the following approximate formulation of (8.13a):

$$\frac{\mathrm{d}}{\mathrm{d}t} \int_{\Omega} p_h \, \varphi_h \, \mathrm{d}\underline{x} - \rho_0 \, c_0^2 \int_{\Omega} \underline{u}_h \cdot \nabla \varphi_h \, \mathrm{d}\underline{x}$$
$$+ \frac{1}{2} \int_{\Omega} (\underline{u}_0 \cdot \nabla p_h)\, \varphi_h \, \mathrm{d}\underline{x} - \frac{1}{2} \int_{\Omega} \nabla \cdot (\underline{u}_0 \cdot \nabla \varphi_h)\, p_h \, \mathrm{d}\underline{x} = \int_{\Omega} f \, \varphi_h \, \mathrm{d}\underline{x}. \tag{8.23}$$

As for $\mathbf{C}_u$, one can easily deduce from (8.23) that $\mathbf{C}_p$ is skew symmetric.

Remarks:

1. We could also use the classical mixed L^2-$H(div)$ formulation to solve the LEE. However, this approach is not so attractive since the $H(div)$-elements are not easy to handle and the performance compared to DGM is not better than the H^1-L^2 approach.
2. Besides its regularizing property, the curl term introduced in [9] can have a physical meaning, i.e. modeling aeroacoustics with a rotational field. For that purpose, the Goldstein–Visser model [12] can also be used.

8.2 The Linear Cauchy–Poisson Problem

In its general form, the Cauchy–Poisson problem reads as a homogeneous Laplace problem coupled with a non-linear boundary condition depending on time. This problem, very useful for the study of tsunamis, models gravity waves propagation and its non-linear term enables to take into account crushing waves. A simpler model is a linear approximation of the boundary problem which is sufficient to model propagation of surface waves. An important issue to this model is the definition of a absorbing boundary which model crushing surface waves on the coast. For this reason, after presenting the linear problem and its approximation in this section, we discuss the modeling of absorbing boundaries. More details are given in [13].

8.2.1 *The Continuous Problem and Its Approximation*

8.2.1.1 Variational Formulation

Let Ω be an open set of $\mathbb{R}^d$ and $\partial\Omega = \Gamma_S \cup \Gamma_h \cup \Gamma_B$, $(\Gamma_S, \Gamma_h, \Gamma_B)$ being a partition of $\partial\Omega$ (Fig. 8.3). As shown in Fig. 8.3, the dth coordinate which represents the depth is downward.

On this domain, the linear Cauchy–Poisson problem reads:

$$\Delta\Phi = 0 \text{ in } \Omega, \tag{8.24a}$$

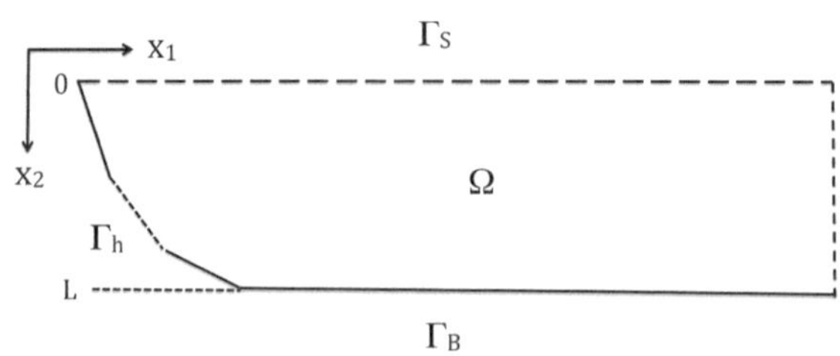

Fig. 8.3 Domain Ω of resolution of the Cauchy–Poisson problem

$$\frac{\partial^2 \Phi}{\partial t^2} + g\,\frac{\partial \Phi}{\partial n} = 0 \text{ on } \Gamma_S, \tag{8.24b}$$

$$\frac{\partial \Phi}{\partial n} = h \text{ on } \Gamma_h, \tag{8.24c}$$

$$\frac{\partial \Phi}{\partial n} = 0 \text{ on } \Gamma_B, \tag{8.24d}$$

where g is the gravitation constant, $h(\underline{x}, t)$ is a given function (which models the source of the wave), Φ is the velocity potential of the fluid and $\partial\Phi/\partial n = \underline{n} \cdot \nabla\Phi$, $\underline{n}$ being the unit outward normal to Ω. The shape of the surface wave is given by

$$\eta(\underline{x}, t) = \frac{1}{g}\,\frac{\partial \Phi}{\partial t}(\underline{x}, t), \;\; (x_d = 0). \tag{8.25}$$

Remark: The configuration of the free surface Γ_S shows that we actually have

$$\frac{\partial \Phi}{\partial n} = -\frac{\partial \Phi}{\partial x_2}.$$

Taking into account the boundary terms which appear by integrating by parts, we get the following variational formulation in $H^1(\Omega)$ of this problem:
Find $\Phi \in H^1(\Omega)$ such that

$$\int_\Omega \nabla\Phi \cdot \nabla\varphi \,\mathrm{d}\underline{x} + \frac{1}{g}\,\frac{\mathrm{d}^2}{\mathrm{d}t^2}\int_{\Gamma_S} \Phi\,\varphi \,\mathrm{d}\sigma = \int_{\Gamma_h} h\,\varphi\,\mathrm{d}\sigma, \tag{8.26}$$

In order to approximate this problem by spectral elements, we define a mixed form of (8.24a) which reads:

$$\nabla \cdot \underline{v} = 0 \text{ in } \Omega, \tag{8.27a}$$

$$\underline{v} = \nabla\Phi \text{ in } \Omega, \tag{8.27b}$$

$\partial\Phi/\partial n$ being replaced by $\underline{v} \cdot \underline{n}$ in (8.24b)–(8.24d).

Combining (8.27a) and (8.27b) with (8.24b)–(8.24d), we get the variational formulation of the mixed form of the problem, which reads:
Find $\Phi \in H^1(\Omega)$, $\underline{v} \in [L^2(\Omega)]^d$ such that

$$\int_\Omega \underline{v} \cdot \nabla\varphi \,\mathrm{d}\underline{x} + \frac{1}{g}\,\frac{\mathrm{d}^2}{\mathrm{d}t^2}\int_{\Gamma_S} \Phi\,\varphi \,\mathrm{d}\sigma = \int_{\Gamma_h} h\,\varphi\,\mathrm{d}\sigma, \tag{8.28a}$$

$$\int_\Omega \underline{v} \cdot \underline{\psi} \,\mathrm{d}\underline{x} = \int_\Omega \nabla\Phi \cdot \underline{\psi} \,\mathrm{d}\underline{x}. \tag{8.28b}$$

8.2.1.2 Approximation and Algorithm of Resolution

For the approximation, we use the classical spaces $U_h^r \subset H^1(\Omega)$ and $V_h^r \subset [L^2(\Omega)]^d$ defined in (3.5) and (3.33). After discretization in space, (8.26), (8.28a) and (8.28b) provide the two following discrete formulations of the linear Cauchy–Poisson problem:

$$K_h\, U + \frac{\mathrm{d}^2}{\mathrm{d}t^2} D_h^\gamma\, U = H. \tag{8.29}$$

$$R_h\, V + \frac{\mathrm{d}^2}{\mathrm{d}t^2} D_h^\gamma\, U = H, \tag{8.30a}$$

$$B_h\, V = R_h^T\, U. \tag{8.30b}$$

where $(D_h^\gamma)_{ij} = 1/g \int_{\Gamma_S} \varphi_j\, \varphi_i\, \mathrm{d}\sigma$ is a diagonal matrix, $(B_h)_{ij} = \int_\Omega \underline{\psi}_j \cdot \underline{\psi}_i\, \mathrm{d}\underline{x}$ is a block-diagonal matrix, $(R_h)_{ij} = \int_\Omega \underline{\psi}_j \cdot \nabla\varphi_i\, \mathrm{d}\underline{x}$ is a very sparse matrix which requires a local storage (see Chap. 3) and with the relation given in Sect. 3.3.2 by the theorem of equivalence

$$K_h = R_h\, B_h^{-1}\, R_h^T. \tag{8.31}$$

Since the term in time is only on the boundary, we can choose a θ-scheme (which is implicit) for the time approximation. With this approximation, (8.29) reads:

$$K_h\,[\theta\, U^{n+1} + (1-2\theta)\, U^n + \theta\, U^{n-1}] + D_h^\gamma\, \frac{U^{n+1} - 2\, U^n + U^{n-1}}{\Delta t^2} = H^n. \tag{8.32}$$

Now, by setting $\theta = 1/2$, (which corresponds to a Newmark scheme) we obtain the following value of U^{n+1}:

$$U^{n+1} = \left(\frac{\Delta t^2}{2}\, K_h + D_h^\gamma\right)^{-1} \left[2\, D_h^\gamma\, U^n - \left(\frac{\Delta t^2}{2}\, K_h + D_h^\gamma\right)\, U^{n-1} + H^n\right]. \tag{8.33}$$

Equation (8.33) shows that the resolution of (8.32) requires the inversion of the matrix defined by $K_h/2 + D_h^\gamma$ at each time-step. This inversion can be made by using a conjugate gradient algorithm based on a recurrent matrix-vector product which, taking into account (8.31), can be decomposed into the two following steps:

$$q_k = B_h^{-1}\, R_h^*\, p_k, \tag{8.34a}$$

$$\pi_k = \frac{\Delta t^2}{2}\, R_h\, q_k + D_h^\gamma\, p_k, \tag{8.34b}$$

with

$$p_0 = 2\,D_h^\gamma\,U^n - \left(\frac{\Delta t^2}{2}\,K_h + D_h^\gamma\right)\,U^{n-1} + H^n.$$

Applied element by element, this decomposition provides a fast algorithm of resolution because of the diagonal character of D_h^γ and B_h^{-1} and the sparsity of R_h.

Remark: This algorithm leads to a powerful method of resolution of the time-harmonic wave equations provided an adapted preconditioner, as discussed in [14, 15].

8.2.2 Unbounded Domains

For sake of simplicity, we set $d = 2$ in this section. The results can easily be extended to the 3D case.

8.2.2.1 Construction of ABC

A first approach for modeling unbounded domains is the construction of ABC whose first step is, of course, to write the transparent condition. For this purpose, we set $\Omega = \mathbb{R}\times]0, L[$ in (8.24a)–(8.24d). Then, we apply the Fourier transform in x_1 to (8.24a), which provides the following ODE:

$$\frac{\mathrm{d}^2\widehat{\Phi}}{\mathrm{d}x_2^2} - k_1^2\,\widehat{\Phi} = 0 \tag{8.35}$$

whose solution can be written as

$$\widehat{\Phi} = A\,\sinh(k_1\,x_2) + B\,\cosh(k_1\,x_2). \tag{8.36}$$

After applying the Fourier transform in t to (8.24b), we get $\omega^2\,B = -g\,k_1\,A$. On the other hand, (8.24d) implies that $k_1\,A\,\cosh(k_1\,L) + k_1\,B\,\sinh(k_1\,L) = 0$. Combining these two relations, we finally obtain the following dispersion relation of the problem:

$$\omega^2 = g\,k_1\,\tanh(k_1\,L). \tag{8.37}$$

As shown in Sect. 6.1.1, in order to get the transparent condition, we should get a relation of the form $k_1 = f(\omega)$. Unfortunately, Eq. (8.37) is implicit in k_1. Hence, in order to get an ABC, on must write an approximation of $k_1\,\tanh(k_1\,L)$. A first-order approximation of this function is $L\,k_1^2$. This approximation provides the first-order ABC:

$$\frac{\partial\Phi}{\partial t} + \sqrt{g\,L}\,\frac{\partial\Phi}{\partial x_2} = 0 \text{ on } \Gamma_S. \tag{8.38}$$

In [16], a second-order ABC is derived (by a different approach). This approach is proven to have no exponentially growing mode for the continuous formulation. However, numerical tests revealed instability in long time of resolution. Moreover, in frequency domain, one can prove that there is one frequency for which the problem is ill-posed. All these results lead us to compute PML for the linear Cauchy–Poisson problem.

8.2.2.2 Construction of PML

As in Sect. 6.2, we consider a domain $\Omega = \Omega_P \cup \Omega_{PML}$ where Ω_P is the physical domain defined in Sect. 8.2.1.1 and $\Omega_{PML} = \mathbb{R}^+ \times]0, L[$. In Ω_{PML}, we introduce the change of variables:

$$\tilde{x}_1 = \begin{cases} x_1 \text{ if } x_i < 0, \\ x_1 + \dfrac{\mathrm{i}}{\omega} \displaystyle\int_0^{x_1} \zeta(s)\, \mathrm{d}s \text{ otherwise}, \end{cases} \tag{8.39}$$

We apply this change of variables to the mixed formulation given by (8.27a) and (8.27b) and (8.24b)–(8.24d), in which we replace $\partial \Phi / \partial n$ by $\underline{v} \cdot \underline{n}$. After applying the Fourier transform in time, we get:

$$\frac{\partial \hat{v}_1}{\partial \tilde{x}_1} + \frac{\partial \hat{v}_2}{\partial x_2} = 0 \text{ in } \Omega, \tag{8.40a}$$

$$\frac{\partial \widehat{\Phi}}{\partial \tilde{x}_1} = \hat{v}_1, \quad \frac{\partial \widehat{\Phi}}{\partial x_2} = \hat{v}_2 \text{ in } \Omega, \tag{8.40b}$$

$$\omega^2 \widehat{\Phi} = g\, \hat{\underline{v}} \cdot \underline{n} \text{ on } \Gamma_S, \tag{8.40c}$$

$$\hat{\underline{v}} \cdot \underline{n} = h \text{ on } \Gamma_h, \quad \hat{\underline{v}} \cdot \underline{n} = 0 \text{ on } \Gamma_B. \tag{8.40d}$$

By using (6.72) (with $\zeta_1 = \zeta$), (8.40a) and (8.40b) becomes:

$$\mathrm{i}\,\omega \frac{\partial \hat{v}_1}{\partial x_1} + (\mathrm{i}\,\omega + \zeta(x_1)) \frac{\partial \hat{v}_2}{\partial x_2} = 0 \text{ in } \Omega, \tag{8.41a}$$

$$\mathrm{i}\,\omega \frac{\partial \widehat{\Phi}}{\partial x_1} = (\mathrm{i}\,\omega + \zeta(x_1)) \hat{v}_1, \quad \mathrm{i}\,\omega \frac{\partial \widehat{\Phi}}{\partial x_2} = \mathrm{i}\,\omega\, \hat{v}_2 \text{ in } \Omega. \tag{8.41b}$$

The inverse Fourier transform in time leads to

$$\frac{\partial}{\partial t} \nabla \cdot \underline{v} + \zeta(x_1) \frac{\partial v_2}{\partial x_2} = 0 \text{ in } \Omega, \tag{8.42a}$$

$$\frac{\partial \underline{v}}{\partial t} - \frac{\partial}{\partial t}\nabla \Phi + \zeta(x_1)\, v_1\, \underline{e}_1 = 0 \text{ in } \Omega, \tag{8.42b}$$

$$\frac{\partial^2 \Phi}{\partial t^2} + g\, \underline{v} \cdot \underline{n} = 0 \text{ on } \Gamma_S, \tag{8.42c}$$

$$\underline{v} \cdot \underline{n} = h \text{ on } \Gamma_h,\ \ \underline{v} \cdot \underline{n} = 0 \text{ on } \Gamma_B, \tag{8.42d}$$

where $\underline{e}_i$ is the unit vector in the x_i-direction.

8.2.2.3 Variational and Approximate Formulations of PML

In order to avoid a third-order derivative in time, we set $\widetilde{\Phi} = \partial \Phi / \partial t$. After multiply (8.42a)–(8.42d) by $\varphi \in H^1(\Omega)$ and $\underline{\psi} \in [L^2(\Omega)]^2$ and integrating by parts, we obtain the variational problem:

$$\frac{\mathrm{d}}{\mathrm{d}t}\int_\Omega \underline{v} \cdot \nabla\varphi \, \mathrm{d}\underline{x} + \int_\Omega \zeta(x_1)\, v_2\, \underline{e}_2 \cdot \nabla\varphi \, \mathrm{d}\underline{x} + \frac{1}{g}\frac{\mathrm{d}}{\mathrm{d}t}\int_{\Gamma_S} \zeta(x_1)\, \widetilde{\Phi}\, \varphi \, \mathrm{d}\sigma$$
$$+ \frac{1}{g}\frac{\mathrm{d}^2}{\mathrm{d}t^2}\int_{\Gamma_S} \widetilde{\Phi}\, \varphi\, \mathrm{d}\sigma = \int_{\Gamma_h} h\, \varphi\, \mathrm{d}\sigma,\ \forall \varphi \in H^1(\Omega), \tag{8.43a}$$

$$\frac{\mathrm{d}}{\mathrm{d}t}\int_\Omega \underline{v} \cdot \underline{\psi}\, \mathrm{d}\underline{x} - \int_\Omega \nabla\widetilde{\Phi} \cdot \underline{\psi}\, \mathrm{d}\underline{x} + \int_\Omega \zeta(x_1)\, v_2\, \underline{e}_2 \cdot \underline{\psi}\, \mathrm{d}\underline{x},\ \forall \underline{\psi} \in [L^2(\Omega)]^2. \tag{8.43b}$$

Moreover, we have:

$$\underline{w} = \zeta(x_1)\, v_2\, \underline{e}_2 \longrightarrow \int_\Omega \zeta(x_1)\, v_2\, \underline{e}_2 \cdot \nabla\varphi\, \mathrm{d}\underline{x} = \frac{\mathrm{d}}{\mathrm{d}t}\int_\Omega \underline{w} \cdot \nabla\varphi\, \mathrm{d}\underline{x}. \tag{8.44}$$

The discrete formulation obtained by using spectral elements reads:

$$\frac{\mathrm{d}}{\mathrm{d}t} R_h\, V + R_h\, W + \frac{\mathrm{d}}{\mathrm{d}t} D_h^{\gamma,\zeta}\, \widetilde{U} + \frac{\mathrm{d}^2}{\mathrm{d}t^2} D_h^{\gamma}\, \widetilde{U} = \frac{\mathrm{d}}{\mathrm{d}t} H, \tag{8.45a}$$

$$\frac{\mathrm{d}}{\mathrm{d}t} B_h\, V = R_h^T\, \widetilde{U} - B_h^{\zeta,1}\, V, \tag{8.45b}$$

$$B_h\, W = B_h^{\zeta,2}\, V, \tag{8.45c}$$

where $\widetilde{U}$, V, W are the discrete vectors corresponding to $\widetilde{\Phi}_h$, $\underline{v}_h$, $\underline{w}_h$. On the other hand,

$$(B_h^{\zeta,k})_{ij} = \int_\Omega \psi_{i,k}\, \underline{e}_k \cdot \underline{\psi}_j\, \mathrm{d}\underline{x},\ \ (D_h^{\gamma,\zeta})_{ij} = 1/g \int_{\Gamma_S} \zeta(x_1)\, \varphi_i\, \varphi_j\, \mathrm{d}\sigma.$$

Numerical experiments showed that explicit schemes are unstable. So, we discretize in time by a θ-scheme with $\theta = 1/2$.

$$\begin{aligned} R_h \frac{V^{n+1} - V^{n-1}}{2\,\Delta t} &+ R_h\, B_h^{-1}\, B_h^{\zeta,2} \frac{V^{n+1} + V^{n-1}}{2} + D_h^{\gamma,\zeta} \frac{\widetilde{U}^{n+1} - \widetilde{U}^{n-1}}{2\,\Delta t} \\ &+ D_h^{\gamma} \frac{\widetilde{U}^{n+1} - 2\widetilde{U}^{n} + \widetilde{U}^{n-1}}{\Delta t^2} = \frac{\mathrm{d}}{\mathrm{d}t} H, \end{aligned} \tag{8.46a}$$

$$B_h \frac{V^{n+1} - V^{n-1}}{2\,\Delta t} = R_h^T \frac{\widetilde{U}^{n+1} + \widetilde{U}^{n-1}}{2} - B_h^{\zeta,1} \frac{V^{n+1} + V^{n-1}}{2}. \tag{8.46b}$$

After transformation, (8.46b) reads:

$$\left(\frac{B_h}{\Delta t} + B_h^{\zeta,1}\right) V^{n+1} = R_h^T\, (\widetilde{U}^{n+1} + \widetilde{U}^{n-1}) + \left(\frac{B_h}{\Delta t} - B_h^{\zeta,1}\right) V^{n-1}. \tag{8.47}$$

Since ζ is positive, $1/\Delta t\, B_h + B_h^{\zeta,1}$ is invertible and we can plug the value of V^{n+1} given by (8.47) into (8.46a). We get:

$$\begin{aligned} \left(K_h^{\zeta} + \frac{1}{2\,\Delta t} D_h^{\gamma,\zeta} + \frac{1}{\Delta t^2} D_h^{\gamma}\right) \widetilde{U}^{n+1} &= \left(K_h^{\zeta} + \frac{1}{2\,\Delta t} D_h^{\gamma,\zeta} - \frac{1}{\Delta t^2} D_h^{\gamma}\right) \widetilde{U}^{n-1} \\ &+ \frac{1}{\Delta t^2} D_h^{\gamma}\, \widetilde{U}^{n} + R_H\, B_h^{\zeta}\, V^{n-1} + \frac{\mathrm{d}}{\mathrm{d}t} H, \end{aligned} \tag{8.48}$$

where

$$K_h^{\zeta} = \frac{1}{2} R_h \left(\frac{1}{\Delta t} I_h + B_h^{-1}\, B_h^{\zeta,2}\right) \left(\frac{B_h}{\Delta t} + B_h^{\zeta,1}\right)^{-1} R_h^T,$$

$$B_h^{\zeta} = \left(\frac{1}{\Delta t} I_h + B_h^{-1}\, B_h^{\zeta,2}\right) - \left(\frac{B_h}{\Delta t} + B_h^{\zeta,1}\right)^{-1} \left(\frac{B_h}{\Delta t} - B_h^{\zeta,1}\right).$$

B_h^{ζ} is a $d \times d$ block-diagonal matrix whose inverse can be stored.

8.2.2.4 Stability Analysis of the PML

This section is devoted to stability analysis by plane waves [17]. For this purpose, we suppose that ζ is constant on $\mathbb{R} \times]0, L[$ and we use the frozen coefficient technique described in [18]. In a first step, we derive the PML from the initial formulation of the Cauchy–Poisson problem, as written in (8.24a)–(8.24d) by using the change of variables defined by (8.39) after applying the Fourier transform in time to (8.24a)–(8.24d). After some algebra, we finally get, after inverse Fourier transform in time:

$$\frac{\partial^4 \Phi}{\partial t^2 \partial x_1^2} + \frac{\partial^4 \Phi}{\partial t^2 \partial x_2^2} + 2\,\zeta\,\frac{\partial^3 \Phi}{\partial t \partial x_2^2} + \zeta^2\,\frac{\partial^2 \Phi}{\partial x_2^2} = 0, \text{ in } \Omega_{PML} \tag{8.49a}$$

$$\frac{\partial^2 \Phi}{\partial t^2} + g\,\frac{\partial \Phi}{\partial x_2} = 0 \text{ on } \Gamma_S, \quad \frac{\partial \Phi}{\partial x_2} = 0 \text{ on } \Gamma_B. \tag{8.49b}$$

We are now seeking for a solution of the form $\Phi(x_1, x_2, t) = \Phi_{x_2}(x_2)\,\mathrm{e}^{\mathrm{i}\,(\omega\,t - k_1\,x_1)}$ with $\omega = \omega_R + \mathrm{i}\,\omega_I$. The acceptable solutions are such that $\omega_I > 0$ in order to obtain evanescent waves. By inserting this solution into (8.49a) and (8.49b), we get the following ODE:

$$\frac{\mathrm{d}^2 \Phi_{x_2}}{\mathrm{d}x_2^2} - \frac{(\mathrm{i}\,\omega\,k_1)^2}{(\mathrm{i}\,\omega + \zeta)^2}\,\Phi_{x_2} = 0, \tag{8.50}$$

with initial conditions

$$\frac{\mathrm{d}\Phi_{x_2}}{\mathrm{d}x_2}(0) = \frac{\omega^2}{g}\,\Phi_{x_2}(0), \quad \frac{\mathrm{d}\Phi_{x_2}}{\mathrm{d}x_2}(L) = 0. \tag{8.51}$$

As in Sect. 8.2.2.1, the solution of (8.50) and (8.51) leads to the relation:

$$\frac{\mathrm{i}\,\omega\,k_1}{\mathrm{i}\,\omega + \zeta}\,\tanh\frac{\mathrm{i}\,\omega\,k_1\,L}{\mathrm{i}\,\omega + \zeta} = \frac{\omega^2}{g}. \tag{8.52}$$

$\omega = 0$ obviously verifies relation (8.52). Now, when $\omega \neq 0$, this relation can be rewritten as

$$(a + \mathrm{i}\,b)\,\tanh(a\,L + \mathrm{i}\,b\,L) = \frac{\omega_R^2 - \omega_I^2}{g} + 2\,\mathrm{i}\,\frac{\omega_R\,\omega_I}{g}, \tag{8.53}$$

where

$$a = \frac{k_1\,(\omega_R^2 + \omega_I^2 - \omega_I\,\zeta)}{(\zeta - \omega_I)^2 + \omega_R^2} \in \mathbb{R} \text{ and } \frac{\omega_R\,k_1\,\zeta}{(\zeta - \omega_I)^2 + \omega_R^2} \in \mathbb{R}. \tag{8.54}$$

D_R denotes the set of (a, b) such that ω and k_1 verify (8.53). $x\,\tanh x$ being an even function, we can assume that $k_1 \geq 0$. Now, $\omega_R = 0 \Longrightarrow b = 0$. In this case, the real part of (8.53) reads:

$$a\,\frac{\mathrm{e}^{2\,a\,L} - 1}{\mathrm{e}^{2\,a\,L} + 1} = \frac{-\omega_I^2}{g}. \tag{8.55}$$

Equation (8.53) implies that $a\,(\mathrm{e}^{2\,a\,L} - 1) \leq 0$. On the other hand, for any sign of a, $a\,(\mathrm{e}^{2\,a\,L} - 1) \geq 0$. So $a = 0$ and then, $\omega_I = 0$.

Now, let us set $\omega_R \neq 0$. We compute the imaginary part of (8.53). We get:

$$\frac{b\,\mathrm{e}^{4\,a\,L} + 4\,a\,\mathrm{e}^{2\,a\,L}\,\sin b\,L\,\cos b\,L - b}{2\,\cos(2\,b\,L)\mathrm{e}^{2\,a\,L} + \mathrm{e}^{4\,a\,L} + 1} = 2\,\frac{\omega_R\,\omega_I}{g}. \tag{8.56}$$

One can easily check that $2\cos(2\,b\,L)\,e^{2\,a\,L} + e^{4\,a\,L} + 1$ is always positive. After some algebra and taking into account (8.54), we obtain:

$$\frac{k_1\,\zeta\,L\,c(a,b)}{(2\,\cos(2\,b\,L)\,e^{2\,a\,L} + e^{4\,a\,L} + 1)\,((\zeta-\omega_I)^2 + \omega_R^2)} = 2\,\frac{\omega_I}{g}, \tag{8.57}$$

where

$$c(a,b) = \frac{1}{L}\left(e^{4\,a\,L} + 4\,a\,e^{2\,a\,L}\,\frac{\sin b\,L\,\cos b\,L}{b} - 1\right).$$

Since the denominator of (8.53) is always positive, the sign of (8.53) depends on the sign of $c(a,b)$. Let us suppose that $\omega_I < 0$. Since

$$\frac{\sin b\,L\,\cos b\,L}{b\,L} \geq -1,$$

we have:

$$\frac{e^{4\,a\,L}}{L} + 4\,a\,e^{2\,a\,L} - \frac{1}{L} \leq c(a,b). \tag{8.58}$$

Equation (8.58) can be rewritten as

$$2\,\frac{e^{2\,a\,L}}{L}\,(\sinh 2\,a\,L - 2\,a\,L) \leq c(a,b). \tag{8.59}$$

Equation (8.54) shows that $\omega_I \leq 0 \Longrightarrow a \geq 0$. On the other hand, $a \geq 0 \Longrightarrow \sinh 2\,a\,L - 2\,a\,L \geq 0$ and then $c(a,b) \geq 0$ which contradicts (8.57). So, ω_I is positive.

In practice, PML were tested on very long times ($\simeq$1000 wavelengths) and remained stable.

8.3 Vibrating Thin Plates

Numerical modeling of vibrating thin plates is a particular case of the much more complex problem of vibrating shells [19] which is far beyond the scope of this book. Two models are used for thin plates: the Kirchhoff–Love model [20] which is second-order in time and in space and the Reissner–Mindlin model [21, 22] which is second-order in time and fourth-order in space. This section is mainly focused on the second model, as described in [23, 24]. However, guidelines for the Kirchhoff–Love model are given.

8.3.1 *The Continuous Models*

8.3.1.1 The Reissner–Mindlin Model

The Reissner–Mindlin model for thin plates leans on the two following assumptions:

1. Straight lines normal to the mid-surface remain straight after deformation.
2. The thickness of the plate does not change during a deformation.

Let us describe this model. We define an elastic extruded plane surface $\Omega \in \mathbb{R}^3$ whose thickness δ is small compared to the other dimensions, Σ its mid-surface and $\Gamma = \partial\Omega$. If we suppose that Σ is in the plane defined by $x_3 = 0$, Γ can be written as: $\Gamma = \Gamma^+ \cup \Gamma^- \cup \Gamma_0$, where

$$\Gamma^{\pm} = \Omega \cap \left\{ x_3 = \frac{\pm\delta}{2} \right\}, \quad \Gamma_0 = \partial\Omega \setminus (\Gamma^+ \cup \Gamma^-),$$

as shown in Fig. 8.4.

An elastic wave propagation in Ω can of course be described by the linear elastodynamics system (1.25a) and (1.25b). However, for small δ, it is cheaper to replace this system by an asymptotic system derived from (1.25a) and (1.25b). One of these systems is the Reissner–Mindlin model which reads:
Find $u : \Sigma \times [0, T] \to \mathbb{R}$, $\underline{\theta} : \Sigma \times [0, T] \to \mathbb{R}^2$ such that

$$\rho \frac{\delta^2}{12} \frac{\partial^2 \underline{\theta}}{\partial t^2} - \frac{\delta^2}{12} \mathbf{div}(\underline{\underline{\widetilde{C}\varepsilon}})(\underline{\theta}) + \mu (\nabla u + \underline{\theta}) = 0 \text{ in } \Sigma \times [0, T], \tag{8.60a}$$

$$\rho\,\delta \frac{\partial^2 u}{\partial t^2} - \delta\,\nabla \cdot [\mu (\nabla u + \underline{\theta})] = f \text{ in } \Sigma \times [0, T]. \tag{8.60b}$$

$$u = 0, \quad \underline{\theta} = 0 \text{ on } \partial\Sigma \times [0, T], \tag{8.60c}$$

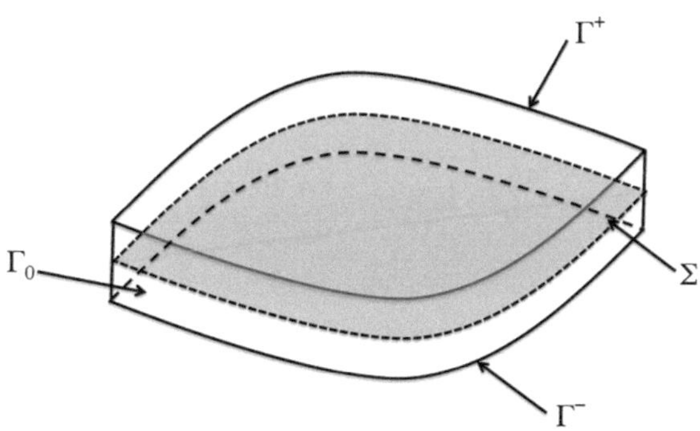

Fig. 8.4 A thin plate Ω

where $\rho(\underline{x})$ is the density, u is is the displacement of the mid-surface in the x_3-direction, θ_1 and θ_2 denote the angles made by the normal to the mid-surface and the x_3 axis. Moreover, **div** and $\underline{\underline{\varepsilon}}$ are defined as in (1.25a) and (1.25b), $\lambda = E\,\nu/(1-\nu^2)$, $\mu = E/(2\,(1+\nu))$ and $\underline{\underline{\widetilde{C}}}(\underline{x})$ is the fourth-order tensor defined by

$$\widetilde{C}_{ijk\ell} = \lambda\,\delta_{ij}\,\delta_{k\ell} + \mu\left(\delta_{ik}\,\delta_{j\ell} + \delta_{i\ell}\,\delta_{jk}\right),$$

E being the Young modulus and ν the Poisson coefficient.

8.3.1.2 The Kirchhoff–Love Model

To the assumptions of Reissner–Mindlin, the Kirchhoff–Love model adds: straight lines normal to the mid-surface remain normal to the mid-surface after deformation. So, this model is seen as a restriction of the Reissner–Mindlin model.[7] This assumption actually reads $\nabla u + \underline{\theta} = 0$, which implies $\underline{\theta} = -\nabla u$. So, after some algebra, one can write the Kirchhoff–Love equation:

$$\rho\,\delta\,\frac{\partial^2 u}{\partial t^2} - \frac{\delta^3\,\lambda}{12}\Delta(\Delta u) = f \text{ in } \Sigma \times [0,T]. \tag{8.61}$$

More details on these models can be found in [19].

Remark: The values of the coefficients are here given for an isotropic medium, but the Reissner–Mindlin model still holds for anisotropic media, with a more complex definition of $\underline{\underline{\widetilde{C}}}$. In the isotropic case, this tensor is actually the same as tensor $\underline{\underline{C}}$ of the linear elastodynamics system defined by (1.28).

8.3.2 *Plane Wave Analysis*

In this section, we assume that E and ν are real constants, $\Sigma = \mathbb{R}^2$ and $f = 0$. Following Sect. 1.1.3.2, (8.60a) can then be rewritten as

$$\rho\,\frac{\delta^2}{12}\frac{\partial^2\underline{\theta}}{\partial t^2} - \frac{\delta^2}{12}\left[\mu\Delta\underline{\theta} + (\lambda+\mu)\nabla(\nabla\cdot\underline{\theta})\right] + \mu\,(\nabla u + \underline{\theta}) = 0. \tag{8.62}$$

We now define the plane wave solution of (8.60a)–(8.60c):

$$\underline{\theta} = \underline{\theta}_0 e^{i(\omega\,t + \underline{k}\cdot\underline{x})}. \tag{8.63a}$$

[7] Although, from a chronological point of view, the Reissner–Mindlin model is a generalization of the Kirchhoff–Love model.

$$u = u_0 e^{i(\omega t + \underline{k} \cdot \underline{x})}. \tag{8.63b}$$

By plugging (8.63a) and (8.63b) into (8.60a)–(8.60c) and taking into account (8.62), we get:

$$\rho \frac{\delta^2}{12} \omega^2 \underline{\theta}_0 = \frac{\delta^2}{12} [\mu |\underline{k}|^2 \underline{\theta}_0 + (\lambda + \mu) (\underline{k} \cdot \underline{\theta}_0) \underline{k}] + \mu (u_0 \, i \, \underline{k} + \underline{\theta}_0), \tag{8.64a}$$

$$\rho \omega^2 u_0 = \mu (|\underline{k}|^2 u_0 - i \, \underline{k} \cdot \underline{\theta}_0). \tag{8.64b}$$

Equations (8.64a) and (8.64b) leads to the generalized eigenvalues problem:

$$\omega^2 M \, V = A(\underline{k}) \, V, \tag{8.65}$$

where $V = (\underline{\theta}_0, u_0)^T$ and

$$M = \rho \begin{pmatrix} \frac{\delta^2}{12} & 0 & 0 \\ 0 & \frac{\delta^2}{12} & 0 \\ 0 & 0 & 1 \end{pmatrix},$$

$$A(\underline{k}) = \begin{pmatrix} \frac{\delta^2}{12} \mu |\underline{k}|^2 + (\lambda + \mu) k_1^2] + \mu & \frac{\delta^2}{12} (\lambda + \mu) k_1 k_2 & i k_1 \mu \\ \frac{\delta^2}{12} (\lambda + \mu) k_1 k_2 & \frac{\delta^2}{12} \mu |\underline{k}|^2 + (\lambda + \mu) k_2^2] + \mu & i k_2 \mu \\ -i k_1 \mu & -i k_2 \mu & \mu |\underline{k}|^2 \end{pmatrix}.$$

Theorem 13 *All the eigenvalues of $M^{-1} A(\underline{k})$ are real positive.*

Proof $A(\underline{k})$ being Hermitian, if $A(\underline{k}) \, V \cdot V \geq 0$, its eigenvalues are positive. Actually, if we denote $\underline{R}_1$ and R_2 the right-hand sides of (8.64a) and (8.64b), we have

$$\begin{aligned} A(\underline{k}) \, V \cdot V &= \underline{R}_1 \cdot \underline{\theta}_0 + R_2 u_0 \\ &= \frac{\delta^2}{12} [\mu |\underline{k}|^2 |\underline{\theta}_0|^2 + (\lambda + \mu) (\underline{k} \cdot \underline{\theta}_0)^2] + \mu (|\underline{k}|^2 u_0^2 + |\underline{\theta}_0|^2) \geq 0. \end{aligned}$$

On the other hand, since M is diagonal positive, we obviously have:

$$M^{\frac{1}{2}} = \sqrt{\rho} \begin{pmatrix} \frac{\delta\sqrt{3}}{6} & 0 & 0 \\ 0 & \frac{\delta\sqrt{3}}{6} & 0 \\ 0 & 0 & 1 \end{pmatrix}, \quad (M^{\frac{1}{2}})^T = M^{\frac{1}{2}}. \tag{8.66}$$

Now, let us set:

$$V = M^{-\frac{1}{2}} \, U. \tag{8.67}$$

By inserting (8.67) in (8.65), we get:

$$\omega^2\, U = M^{-\frac{1}{2}}\, A(\underline{k})\, M^{-\frac{1}{2}}\, U. \tag{8.68}$$

Taking into account (8.66), one can see that $M^{-\frac{1}{2}}\, A(\underline{k})\, M^{-\frac{1}{2}}$ is Hermitian, definite, positive and then, its eigenvalues are positive. Let ω_i^2 be one of these eigenvalues, U_i the corresponding eigenvector and $V_i = M^{-\frac{1}{2}}\, U_i$. By using (8.67), we have:

$$\omega_i^2\, U_i = \omega^2\, M^{\frac{1}{2}}\, V_i, \tag{8.69a}$$

$$M^{-\frac{1}{2}}\, A(\underline{k})\, M^{-\frac{1}{2}}\, U_i = M^{-\frac{1}{2}}\, A(\underline{k})\, V_i. \tag{8.69b}$$

Combining (8.69a) and (8.69b), we finally get:

$$\omega_i^2\, M\, V_i = A(\underline{k})\, V_i, \tag{8.70}$$

which shows that ω_i^2 is an eigenvalue of $M^{-1}\, A(\underline{k})$ and all the eigenvalues of this matrix are positive. ◇

Let us now compute these eigenvalues. The resolution of the characteristic polynomial being tedious, we directly solve the problem in ω defined by (8.64a) and (8.64b). For

$$\rho\, \omega^2 - \mu\, |\underline{k}|^2 \neq 0, \tag{8.71}$$

Equation (8.64b) provides:

$$u_0 = \frac{\mathrm{i}\, \mu\, \underline{k} \cdot \underline{\theta}_0}{\mu\, |\underline{k}|^2 - \rho\, \omega^2}. \tag{8.72}$$

By inserting (8.72) into (8.64a) and multiplying by $\underline{k}$, we get, after some algebra:

$$\rho\, \omega^4 - \rho\, \omega^2 \left[(\lambda + 3\, \mu)\, |\underline{k}|^2 + \frac{12\, \mu}{\delta^2} \right] + \mu\, |\underline{k}|^4\, (\lambda + 2\, \mu). \tag{8.73}$$

The two roots of this equation in ω^2 read:

$$\omega_i^2 = \frac{1}{2\, \rho} \left[(\lambda + 3\, \mu)\, |\underline{k}|^2 + \frac{12\, \mu}{\delta^2} \pm \sqrt{D} \right], \quad i = 1, 2, \tag{8.74}$$

where

$$D = \left((\lambda + 3\, \mu)\, |\underline{k}|^2 + \frac{12\, \mu}{\delta^2} \right)^2 - 4\, \rho\, \mu\, |\underline{k}|^4\, (\lambda + 2\, \mu).$$

ω_i being positive, we finally get two dispersion relations:

$$\omega_1 = \sqrt{\frac{(\lambda + 3\,\mu)\,|\underline{k}|^2 + \dfrac{12\,\mu}{\delta^2} + \sqrt{D}}{2\,\rho}}, \tag{8.75}$$

$$\omega_2 = \sqrt{\frac{(\lambda + 3\,\mu)\,|\underline{k}|^2 + \dfrac{12\,\mu}{\delta^2} - \sqrt{D}}{2\,\rho}}. \tag{8.76}$$

A third relation is derived from (8.71):

$$\omega_3 = |\underline{k}|\sqrt{\frac{\mu}{\rho}}. \tag{8.77}$$

To these relations of course correspond three velocities given by $\omega_i/|\underline{k}|$, $i = 1..3$. Equations (8.75) and (8.76) provide dispersive velocities, i.e. velocities depending on $\underline{k}$ or, in other words, on the frequency. That means that a thin plate is a dispersive medium which means, in particular, that a polychromatic source produces several waves whose frequencies induce different velocities, as shown in Fig. 8.5.

Remark: The dispersion relation of the Kirchhoff–Love model given in (8.61) obviously reads:

$$\rho\,\omega^2 = \frac{\delta^2\,\lambda}{12}\,|\underline{k}|^4, \tag{8.78}$$

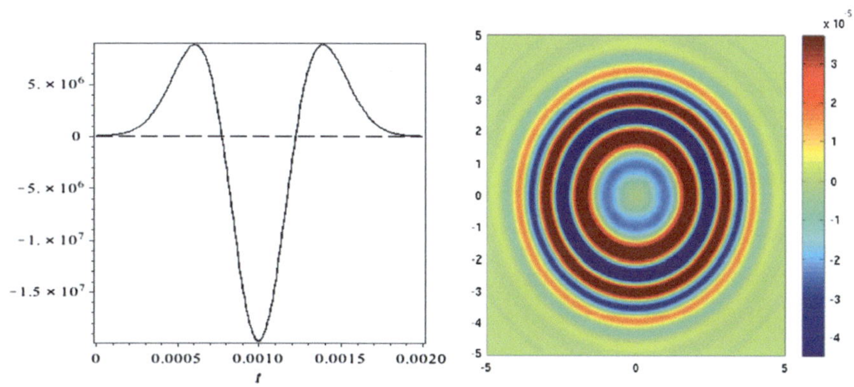

Fig. 8.5 Propagation of a second-order Ricker source (*left*) in a thin plate (*right*)

which leads to the two velocities:

$$c_i = \frac{\omega_i}{|\underline{k}|} = \pm \frac{\delta\sqrt{3\,\lambda}}{6}\,|\underline{k}|^2, \quad i = 1, 2. \tag{8.79}$$

Equation (8.79) shows that the medium is also dispersive for this model.

8.3.3 *Mixed Spectral Element Approximation*

8.3.3.1 Mixed Formulation of the Problem

By setting in (8.60a) and (8.60b) $\underline{Q} = \mu\,(\nabla u + \underline{\theta})$ and decomposing $\underline{\underline{\widetilde{C}\varepsilon}}(\underline{\theta})$ as in (3.260a)–(3.260c), we get:
Find $u : \Sigma \times [0, T] \to \mathbb{R}$, $\underline{\theta} : \Sigma \times [0, T] \to \mathbb{R}^2$, $\underline{\gamma}_i : \Sigma \times [0, T] \to \mathbb{R}^2$, $\underline{\gamma}_{ij} : \Sigma \times [0, T] \to \mathbb{R}^2$ such that

$$\rho\,\frac{\delta^2}{12}\,\frac{\partial^2\theta_i}{\partial t^2} - \frac{\delta^2}{12}\sum_{j=1}^{d}\nabla\cdot\underline{\gamma}_{ij} + \mu\,Q_i = 0, \quad \forall i = 1..d, \text{ in } \Sigma \times [0, T], \tag{8.80a}$$

$$\underline{\gamma}_i = \nabla\theta_i, \quad \forall i = 1..d, \text{ in } \Sigma \times [0, T], \tag{8.80b}$$

$$\underline{\gamma}_{ij} = A_{ij}\,\underline{\gamma}_i, \quad \forall i, j = 1..d, \text{ in } \Sigma \times [0, T], \tag{8.80c}$$

$$\rho\,\delta\,\frac{\partial^2 u}{\partial t^2} - \delta\,\nabla\cdot\underline{Q} = f \text{ in } \Sigma \times [0, T], \tag{8.80d}$$

$$\underline{Q} = \mu\,(\nabla u + \underline{\theta}) \text{ in } \Sigma \times [0, T]. \tag{8.80e}$$

The approximate problem by a mixed spectral element method of (8.80a)–(8.80e) reads:
Find $u_h(., t) \in \left[U_h^r\right]^d$, $\underline{\theta}_h(., t) \in \underline{V}_h^r$, $\underline{\gamma}_{ih}(., t) \in \underline{V}_h^r$ and $\underline{\gamma}_{ijh}(., t) \in \underline{V}_h^r$ such that

$$\rho\,\frac{\delta^2}{12}\,\frac{\mathrm{d}^2}{\mathrm{d}t^2}\int_\Omega \theta_{ih}\,\varphi_{1h}\,\mathrm{d}\underline{x} + \sum_{j=1}^{d}\int_\Omega \underline{\gamma}_{ijh}\cdot\nabla\varphi_{1h}\,\mathrm{d}\underline{x}$$
$$+ \int_\Omega Q_{ih}\,\varphi_{1h}\,\mathrm{d}\underline{x} = 0, \ \forall\varphi_{1h} \in U_h^r, \ \forall i = 1..d, \tag{8.81a}$$

$$\int_\Omega \underline{\gamma}_{ih}\cdot\underline{\psi}_{1h}\mathrm{d}\underline{x} = \int_\Omega \nabla v_{ih}\cdot\underline{\psi}_{1h}\mathrm{d}\underline{x}, \quad \forall\underline{\psi}_{1h} \in \underline{V}_h^r, \ \forall i = 1..d, \tag{8.81b}$$

Table 8.1 Comparison of the performance for different orders of approximation

Order	Nb. elements	Nb. d.o.fs	Δt	CPU (s)	L^2-error (%)
2	10000	40401	0.604e-5	556	5.2
3	1296	11881	0.882e-5	84	5.4
4	484	7921	0.908e-5	49	4.8
5	196	5041	0.961e-5	26	4.9
6	100	3721	0.971e-5	18	5.1
7	64	3249	0.931e-5	16	5

Reprinted from G. Cohen, P. Grob, Mixed higher order spectral finite elements for Reissner–Mindlin equations, SIAM J. Sci. Comp., vol. 29 (3), pp. 986–1005,

$$\int_{\Omega} \underline{\gamma}_{ijh} \cdot \underline{\psi}_{2h} \mathrm{d}\underline{x} = \int_{\Omega} A_{ijh} \underline{\gamma}_{ih} \cdot \underline{\psi}_{2h} \mathrm{d}\underline{x}, \quad \forall \underline{\psi}_{2h} \in \underline{V}_h^r, \ \forall i, j = 1..d, \tag{8.81c}$$

$$\rho\,\delta\,\frac{\mathrm{d}^2}{\mathrm{d}t^2} \int_{\Omega} u_h\,\varphi_{2h}\,\mathrm{d}\underline{x} + \delta \int_{\Omega} \underline{Q}_h \cdot \nabla \varphi_{2h}\,\mathrm{d}\underline{x} = \int_{\Omega} f\,\varphi_{2h}\,\mathrm{d}\underline{x}, \ \forall \varphi_{2h} \in U_h^r, \tag{8.81d}$$

$$\int_{\Omega} \underline{Q}_h \cdot \underline{\psi}_{3h} \mathrm{d}\underline{x} = \int_{\Omega} \mu\,(\nabla u + \underline{\theta}) \cdot \underline{\psi}_{3h} \mathrm{d}\underline{x}, \quad \forall \underline{\psi}_{3h} \in \underline{V}_h^r, \tag{8.81e}$$

where $U_h^r \subset H^1(\Omega)$ is defined in (3.5) and $\underline{V}_h^r \subset [L^2(\Omega)]^d$ is defined in (3.33).

The matrix formulation of (8.81a)–(8.81e) is obtained as in Sect. 3.2.

8.3.3.2 Numerical Results and Locking Phenomenon

In Table 8.1, we give the data of the approximation of a plate of thickness equal to 0.1 cm with an error around 5 %. One can notice that orders greater or equal to 4 provide very good performance compared to lower order. Actually, approximation of a thin plate by low-order finite elements is not easy. This is mainly due to a bad approximation of $\nabla u + \underline{\theta}$. Several solutions were proposed to avoid this problem called "locking phenomenon" [25–27] but it seems that using high-order elements naturally solves this problem.

8.3.3.3 Some Words on the Kirchhoff–Love Model

The fourth-order character of the biharmonic operator in (8.61) a priori suggests to write the variational formulation in $H^2(\Omega)$, which would imply the use of Hermite finite elements [28]. The use of such elements being not easy, an alternative is to split the biharmonic operator into two Laplace operators [29] as follows:

$$\rho(\underline{x}, t)\, \delta\, \frac{\partial^2 u}{\partial t^2} - \frac{\delta^3\, \lambda}{12} \Delta v = f \text{ in } \Sigma \times [0, T]. \tag{8.82a}$$

$$v = \Delta u, \tag{8.82b}$$

and to solve each equation of (8.82a) and (8.82b) in $H^1(\Omega)$, which enables us to use Lagrange finite elements. A third and very efficient approach is to split each Laplace operator into a gradient and a divergence, as in (3.31a) and (3.31b). The variational formulation of this approach reads:
Find $u(., t) \in H^1(\Omega)$, $u_1(., t) \in H^1(\Omega)$, $\underline{v}_1(., t) \in [L^2(\Omega)]^2$ and $\underline{v}_2(., t) \in [L^2(\Omega)]^2$ such that

$$\rho\, \delta\, \frac{\mathrm{d}^2}{\mathrm{d}t^2} \int_\Omega u\, \varphi_1\, \mathrm{d}\underline{x} + \frac{\delta^3\, \lambda}{12} \int_\Omega \underline{v}_1 \cdot \nabla \varphi_1\, \mathrm{d}\underline{x} = \int_\Omega f \varphi_1\, \mathrm{d}\underline{x},\ \ \forall \varphi_1 \in H^1(\Omega), \tag{8.83a}$$

$$\int_\Omega \underline{v}_1 \cdot \underline{\psi}_1\, \mathrm{d}\underline{x} = \int_\Omega \nabla u_1 \cdot \underline{\psi}_1\, \mathrm{d}\underline{x},\ \ \forall \underline{\psi}_1 \in [L^2(\Omega)]^2, \tag{8.83b}$$

$$\int_\Omega u_1\, \varphi_2\, \mathrm{d}\underline{x} = \int_\Omega \underline{v}_2 \cdot \nabla \varphi_2\, \mathrm{d}\underline{x},\ \ \forall \varphi_2 \in H^1(\Omega), \tag{8.83c}$$

$$\int_\Omega \underline{v}_2 \cdot \underline{\psi}_2\, \mathrm{d}\underline{x} = \int_\Omega \nabla u \cdot \underline{\psi}_2\, \mathrm{d}\underline{x},\ \ \forall \underline{\psi}_2 \in [L^2(\Omega)]^2. \tag{8.83d}$$

The finite element approximation is then made as in (3.32a) and (3.32b)

Remarks:

1. We could also use, as for the acoustics equation, a decomposition based on first-order operators in time.
2. Obviously, the Kirchhoff–Love model does not suffer from the locking phenomenon [3].
3. Equations (8.83a)–(8.83d) provide a very efficient method to solve a steady biharmonic problem.

References

1. Bernacki, M., Piperno, S.: A dissipation-free time-domain discontinuous Galerkin method applied to three dimensional linearized Euler equations around a steady-state non-uniform inviscid flow. J. Comput. Acoust. **14**(4), 445–467 (2006)
2. Delorme, P., Mazet, P.A., Peyret, C., Ventribout, Y.: Computational aeroacoustics applications based on a discontinuous Galerkin method. C. R. Acad. Sci. Paris Ser. IIb - Mechanics **333**(9), 676–682 (2005)
3. Bécache, E., Derveaux, G., Joly, P.: An efficient numerical method for the resolution of the Kirchhoff-Love dynamic plate equation. Numer. Methods Partial Differ. Equ. **21**(2), 323–348 (2005)

4. Lighthill, M.J.: On sound generated aerodynamically I. General theory. Proc. R. Soc. Lond. **211**, 564–587 (1951)
5. Galbrun, H.: Propagation d'une onde sonore dans l'atmosph terrestre et thorie des zones de silence, vol. 2, no. 2. Gauthier-Villars, Paris (1931)
6. Kaltenbacher, M.: Numerical Simulation of Mechatronic Sensors and Actuators. Springer, Heidelberg (2015)
7. Castel, N., Cohen, G., Duruflé, M.: Application of discontinuous Galerkin spectral method on hexahedral elements for aeroacoustic. J. Comput. Acoust. **17**(2), 175–196 (2009)
8. Bogey, C., Bailly, C., Juvé, D.: Computation of flow noise using source terms in linearized Euler's equations. AIAA J. **40**(2), 225–243 (2002)
9. Bonnet-BenDhia, A.S., Legendre, G., Luneville, E.: Analyse mathématique de l'équation de Galbrun en écoulement uniforme. C. R. Acad. Sci. Paris Ser. IIb - Mechanics **329**(8), 601–606 (2001)
10. Assous, F., Degond, P., Heintze, E., Raviart, P.-A., Segre, J.: On a finite-element method for solving the three-dimensional Maxwell equations. J. Comput. Phys. **109**(2), 222–237 (1993)
11. Morinishi, Y.: Skew-symmetric form of convective terms and fully conservative finite difference schemes for variable density low-Mach number flows. J. Comput. Phys. **229**(2), 276–300 (2010)
12. Goldstein, M.E.: Unsteady vortical and entropic distortions of potential flows round arbitrary obstacles. J. Fluid Mech. **89**(3), 433–468 (1978)
13. Cohen, G., Imperiale, S.: Perfectly matched layer with mixed spectral elements for the propagation of linearized water waves. Commun. Comput. Phys. **11**(2), 285–302 (2012)
14. Cohen, G., Duruflé, M.: Non spurious spectral-like element methods for Maxwell's equations. J. Comput. Math. **25**(3), 282–304 (2007)
15. Duruflé, M.: Intégration numérique et éléments finis d'ordre élevé appliqués aux équations de Maxwell en régime harmonique, thèse de doctorat, U. de Paris-Dauphine (2006)
16. Dgayguy, K., Joly, P.: Absorbing boundary conditions for linear gravity waves. SIAM J. Appl. Math. **54**(1), 93–131 (1994)
17. Bécache, E., Fauqueux, S., Joly, P.: Stability of perfectly matched layers, group velocities and anisotropic waves. J. Comput. Phys. **188**(2), 399–433 (2003)
18. Kreiss, H.-O., Lorenz, J.: Initial-boundary value problems and the Navier-Stokes equations. Commun. Pure Appl. Math. **136** (1989)
19. Chapelle, D., Bathe, K.-J.: The Finite Element Analysis of Shells-Fundamentals. Computational Fluid and Solid Mechanics, 2nd edn. Springer, Heidelberg (2011)
20. Love, A.E.H.: On the small free vibrations and deformations of elastic shells. Philos. Trans. R. Soc. (Lond.) Ser. A **179**, 491–546 (1888)
21. Mindlin, R.D.: Influence of rotatory inertia and shear on flexural motions of isotropic, elastic plates. J. Appl. Mech. **18**, 31–38 (1951)
22. Reissner, E., Stein, M.: Torsion and transverse bending of cantilever plates, Technical Note 2369. National Advisory Committee for Aeronautics, Washington (1951)
23. Cohen, G., Grob, P.: Mixed higher order spectral finite elements for Reissner-Mindlin equations. SIAM J. Sci. Comput. **29**(3), 986–1005 (2007)
24. Grob, P.: Méthodes numériques de couplage pour la vibroacoustique instationnaire: Eléments finis spectraux d'ordre élevé et potentiels retardés, thèse de doctorat, U. de Paris-Dauphine (2006)
25. Arnold, D.N., Brezzi, F.: Locking-free finite element methods for shells. Math. Comput. **66**(217), 1–14 (1997)
26. Brezzi, F., Fortin, M., Stenberg, R.: Error analysis of mixed-interpolated elements for Reissner-Mindlin plates. Math. Models Methods Appl. Sci. **1**(2), 125–151 (1991)
27. Chapelle, D., Stenberg, R.: An optimal low-order locking-free finite element method for Reissner-Mindlin plates. Math. Models Methods Appl. Sci. **8**(3), 407–430 (1998)
28. Ciarlet, P.G.: The Finite Element Method for Elliptic Problems. North-Holland, Amsterdam (2002)
29. Ciarlet, P.G., Glowinski, R.: Dual iterative techniques for solving a finite element approximation of the biharmonic equation. Comput. Methods Appl. Mech. Eng. **5**(3), 277–295 (1975)

MIX
Papier aus verantwortungsvollen Quellen
Paper from responsible sources
FSC® C105338
FSC
www.fsc.org

If you have any concerns about our products,
you can contact us on
ProductSafety@springernature.com

In case Publisher is established outside the EU,
the EU authorized representative is:
Springer Nature Customer Service Center GmbH
Europaplatz 3, 69115 Heidelberg, Germany

Printed by Libri Plureos GmbH
in Hamburg, Germany